MW01600495

AN INTRODUCTION TO
PAPER INDUSTRY
INSTRUMENTATION
REVISED EDITION

REVISED EDITION
REVISED EDITION
REVISED EDITION
REVISED EDITION
REVISED EDITION
REVISED EDITION
REVISED EDITION
REVISED EDITION

AN INTRODUCTION TO
PAPER INDUSTRY
INSTRUMENTATION

JOHN R. LAVIGNE, CONSULTANT
PULP AND PAPER INDUSTRY DIVISION
THE FOXBORO COMPANY
FOXBORO, MASSACHUSETTS

MILLER FREEMAN
PUBLICATIONS

TO MY WIFE, INEZ,
AND OUR CHILDREN,
JOHN JR., SUSAN, AND CAROL

CONTENTS

PREFACE

It has been five years since the first edition of this book was published. During this time, the book has been widely used by those associated directly and indirectly with the pulp and paper industry as a general reference source on instrumentation and controls. The response has been most gratifying. With this use of the book in mind, each chapter has been updated and expanded to include new information made available since the first edition was published. In addition, a new section in the Appendix deals with conversion factors, presenting units and conversion tables found most useful to industry personnel involved with instrumentation.

In view of the increasing interest in on-line sensors brought about by the ever expanding use of computer-coordinated control systems in the manufacturing processes of the paper industry, a chapter on quality measurement has been added. Although many of these new sensors are still considered to be in the development stage, it was felt that their operating principles should be described since they all possess the potential of becoming generally accepted measuring devices in the very near future.

A number of vocational and technical schools as well as colleges teaching pulp and paper courses are using the book as a source of information in introducing students to the subject of instrumentation. Also, many mills are using the book as a text in the training of new instrument technicians. To extend its usefulness in this area, a summary questionnaire has been prepared for readers who wish to use the text as a home-study course. Similar to the study questions at the end of each chapter, these summary questions have been produced in a separate pamphlet for distribution with the book. In this form, the reader can use the questions to test his general understanding of the information presented in the text without defacing the book. After answering the questions, he may send the pamphlet to the publisher or the author for evaluation of his work.

Except for minor changes in the selection of typefaces, the original treatment of text and illustrative material remains the same as that given in the first edition.

The author would like to repeat the expression of his appreciation to all fellow personnel of The Foxboro Company involved in the preparation of the material for this book. Special thanks go to John B. Prendergast for his help in coordinating the many activities associated with this project.

JOHN R. LAVIGNE

PREFACE TO FIRST EDITION

Automation is the technique of making a process or system automatic through the proper application and manipulation of equipment and machinery, which is primarily accomplished through instruments that measure and control the various operations. In the early days of papermaking, instruments played only a minor role. What few there were consisted mainly of single instruments directly connected to the process or mounted on posts or on walls adjacent to the process equipment, such as digesters, washers, storage chests, refiners, and paper machines.

As a result of the comparatively recent rapid emergence of the technology of process systems and automatic control in the paper industry, the contrast with today's installations is startling. Sophisticated control systems, utilizing more advanced pneumatic and electronic long-distance transmission techniques of measurement and control signals, centralized in modern efficient control centers which are pushing instrumentation budgets up into the multi-million dollar levels, are becoming quite common in the new and more progressive mills. To this must be added the ever-increasing interest in "on-line" digital process control computers and their role in the modern pulp and paper mill.

Increasing operating efficiencies, reducing production costs, and improvement of product quality have been the prime motives behind mill management's decision to use more and more instrumentation. These objectives are achieved by maintaining balanced conditions within the pulp and papermaking process in accordance with predetermined established values. In order to do this, proper means must be provided to determine the state of these conditions during the operation of the process. Basically, this consists of making on-line continuous or intermittent measurements of the process variables, such as flow, level, pressure, temperature, pH, conductivity, ORP, capacitance, basis weight, caliper, moisture, humidity, viscosity, density and speed, that affect the desired conditions and correct them to within predetermined limits, to assure that the required conditions are maintained.

This book was written in response to a need that has existed in the pulp and paper industry for a long time: that is, the lack of published elementary information which brings the uninitiated aboard instrumentation and control in the paper industry and introduces him to digital computers. Therefore, the book was organized with the intention of covering the elementary and basic consideration of instruments and controls as used in the pulp and paper processes. It has been written for the reader who has had no prior knowledge of the field and provides a good grounding on the subject.

The format is developed around the four major areas in an automatic control system, namely: the primary measurements, signal transmission, the automatic controller and the final control elements. How these areas work together as systems to control pulp and papermaking processes is shown by describing typical installations under applications. The book closes with a basic introduction to computers and their use in the paper industry.

The author would like to express his appreciation and gratitude to all fellow Foxboro personnel for their assistance and for providing the initiative to assume and complete the task of writing this book. Many thanks go to Mary Cinto, Cynthia Adamic, and Josephine Elder for typing the manuscript. To David H. Fuller for his technical editing of the text, and to all others who have contributed by proofreading, constructive criticism, and finalizing the art work, I am greatly indebted. Special credit is due to J. Robert Palmer for his aid in coordinating many of the activities in this endeavor.

JOHN R. LAVIGNE

1. Flow Measurement

About one-half of the process measurements made in industry today are flow measurements, and in the pulp and paper industry flow measurement accounts for an even greater proportion than that. Table 1-A lists typical flow measurements in a pulp and paper mill. Information obtained from flow measurements is used to control conditions of the process that are dependent on this variable. Such information also provides greater guidance and is of value in determining quantities of materials used and processed for inventory control and for accounting.

All flow measurement systems are composed of primary and secondary devices. The primary device, a sensing element, is in contact with the flowing medium and by interacting with it provides a measure of this flow. Such an interaction could be the differential pressure sensed by changes in velocity of a material flowing in conduit or by changes in exchange of head with corresponding velocity changes. The secondary device is the device that translates the interaction between flowing material and primary device in values of volume, mass, or other rates of flow so they can be used for indicating, recording, and/or control purposes.

Generally, flow measurement systems are divided into two major categories: (1) flow rate measurement systems whose initial primary measurement is based on devices responsive primarily to rate of material flow and (2) total quantity measurement systems whose initial primary measurement is based on devices responsive to measurement of more or less completely isolated flow quantities of material during intervals of short duration. Table 1-B lists common flow rate measuring systems and their primary and secondary devices.

HEAD METERING SYSTEMS

Head flowmeters, the most common of the flow rate type measurement systems, do not measure flow directly but relate flow to the differential pressure or head induced by a suitable restriction to fluid flow in a pipeline. The primary device, some form of restriction in the flow line, induces the head or differential pressure. The secondary device is connected to the dif-

TABLE 1-A

TYPICAL FLOW MEASUREMENTS, PULP AND PAPER INDUSTRY

A. Wood preparation
1. Water
2. Steam

B. Chip preparation
1. Steam

C. Digesters
1. Wood chips
2. Cooking chemicals
3. Steam
4. Pulp

D. Washers
1. Water
2. Steam
3. Wash filtrates
4. Pulp

E. Pulp mill chemical preparation
1. Water
2. Dry chemicals
3. Liquid chemicals
4. Steam

F. Bleaching
1. Pulp
2. Bleaching chemicals
3. Fresh water
4. Washer filtrates

G. Paper mill chemical preparation
1. Water
2. Dry chemicals
3. Liquid chemicals
4. Dyes
5. Additives
6. Steam

H. Stock preparation
1. Water
2. Pulp
3. Steam

I. Stock blending
1. Pulp
2. Dyes

3. Additives
4. Chemicals

J. Paper machine
1. Stock
2. Water
3. Dry chemicals
4. Liquid chemicals
5. Additives
6. Steam

K. Evaporators
1. Liquor
2. Steam

L. Recovery boiler
1. Condensate
2. Water

M. Recausticizing
1. Green liquor
2. White liquor
3. Water
4. Lime mud

N. Lime kiln
1. Gas
2. Oil
3. Air

O. Power boiler
1. Condensate
2. Water
3. Steam
4. Chemicals
5. Fuels

P. Water treatment
1. Water
2. Dry chemicals
3. Liquid chemicals

Q. Waste treatment
1. Effluent
2. Dry chemicals
3. Liquid chemicals
4. Steam
5. Chemical
6. Liquor

TABLE 1-B

FLOW RATE METERING SYSTEMS

Type	Primary Devices		
		Velocity	Propeller
			Turbine
Head	Venturi		Electromagnetic
	Orifice plate		Sonic
	Flow nozzle		
	Pitot tube		**Secondary Devices**
	Elbow taps		
		Wet	Liquid manometers
			U-tube
Area	Taper tube and		Well or reservoir
	float (rotameter)		Inclined
	Cylinder and piston		Mercury float
			Liquid seal
Head Area	Weirs		Inverted bell
	Flumes		Ledoux bell
			Ring balance
Force	Target	Dry	Bellows
	Vane		Force balance

ferential head and measures it to determine flow rate (Figure 1-1). Often the secondary device consists of two separate devices: a differential pressure transmitter and a remote receiving instrument.

Primary Devices

Theory. Primary devices in head or differential pressure metering systems operate on Bernoulli's theorem which states that the total energy at any point in a pipeline or a conduit is equal to the total energy at any other point if friction losses between the two points are neglected (Figure 1-2). The energy balance between Points 1 and 2 in a pipeline can be expressed as follows:

$$\frac{P_2}{\rho} + \frac{(Vm_2)^2}{2g} + Z_2 = \frac{P_1}{\rho} + \frac{(Vm_1)^2}{2g} + Z_1$$

where
P = static pressure absolute
Vm = fluid stream velocity
Z = elevation of center line of the pipe
ρ = fluid density
g = acceleration due to gravity

The operation of the differential flow devices can be more easily understood if the energy is considered to be in two forms: static head and dynamic or velocity head.

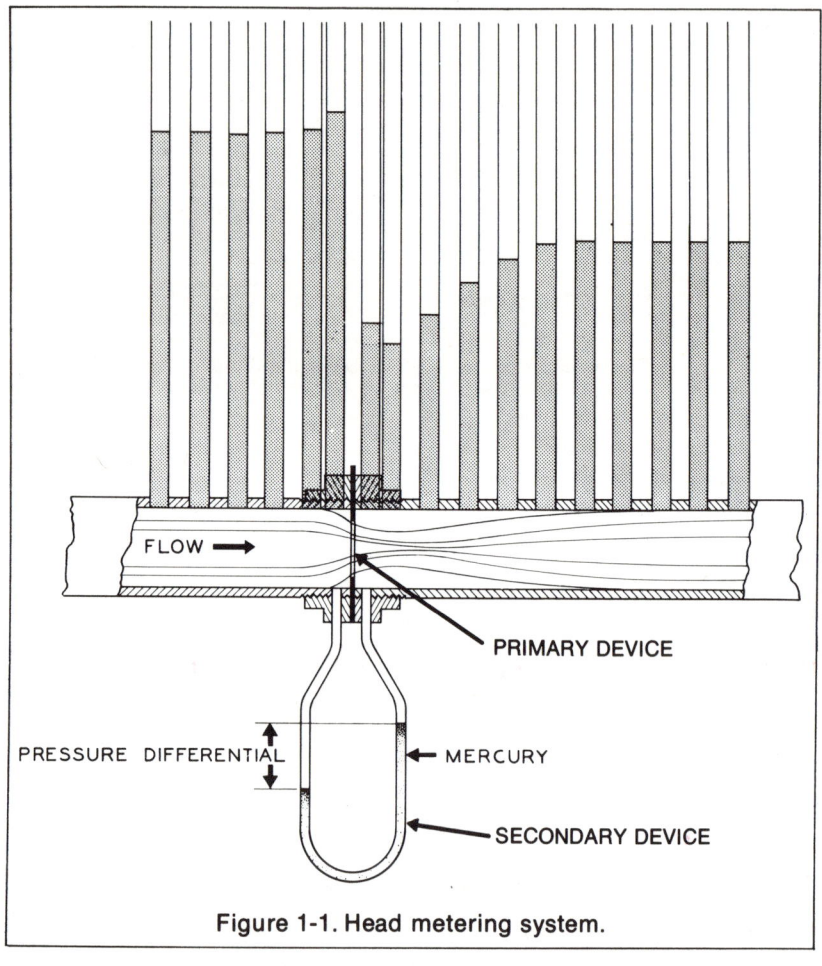

Figure 1-1. Head metering system.

With a differential pressure type primary device such as an orifice (Figure 1-2), the change in cross-sectional area between the pipe and orifice produces a change in flow velocity—flow increases through the orifice. Since total energy at the inlet to and at the throat of the orifice remains the same (neglecting losses), the velocity head at the throat must increase causing a corresponding decrease in static head. Therefore, there is a head differential between a point immediately ahead of the restriction and a point within the restriction or downstream from it. The resulting differential head or pressure is a function of velocity which can be related to flow by the secondary device (see Table 1-B).

History. Although there is no known record of the first application of differential pressure to measure flow, it is known that the Romans used an orifice to measure water to householders during the days of Caesar. Venturi

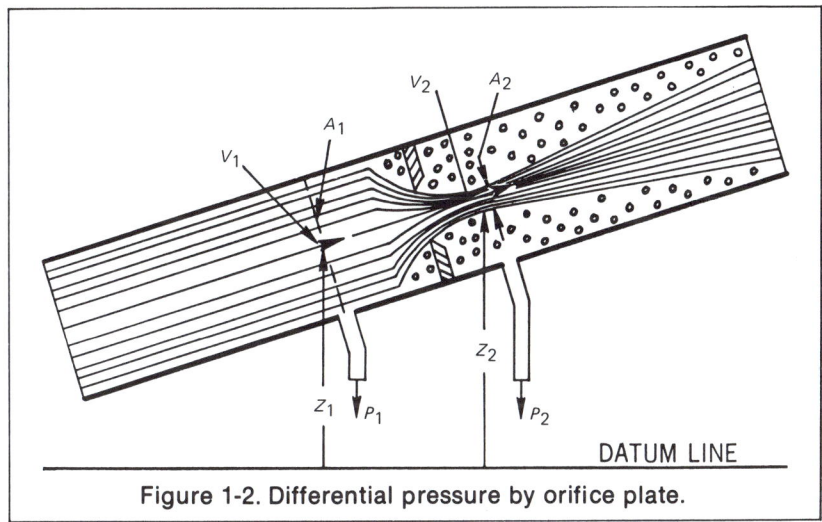

Figure 1-2. Differential pressure by orifice plate.

did his basic work on the Venturi tube in 1791. However, it was not until 1887 that Clemens Herschel, using Venturi's work, developed the commercial Venturi tube from which all primary devices used in differential measuring systems are derived.

The Venturi Tube. Herschel was primarily interested in a device which would produce a large differential with a small head loss. His effort became known as the Herschel-Venturi tube. Critical dimensions of the classic Venturi tube are shown in Figure 1-3.

This primary device is recommended if the measured fluid contains large amounts of suspended solids or if favorable pressure recovery characteristics are of prime concern. It has been used extensively in the pulp and paper industry to measure large water flows in water treatment and paper mill

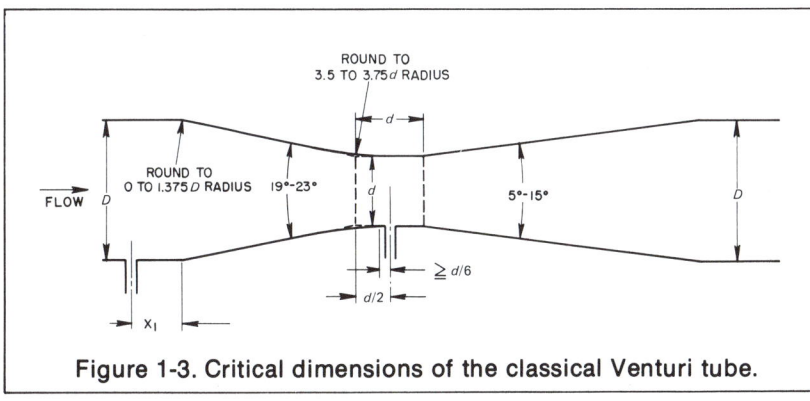

Figure 1-3. Critical dimensions of the classical Venturi tube.

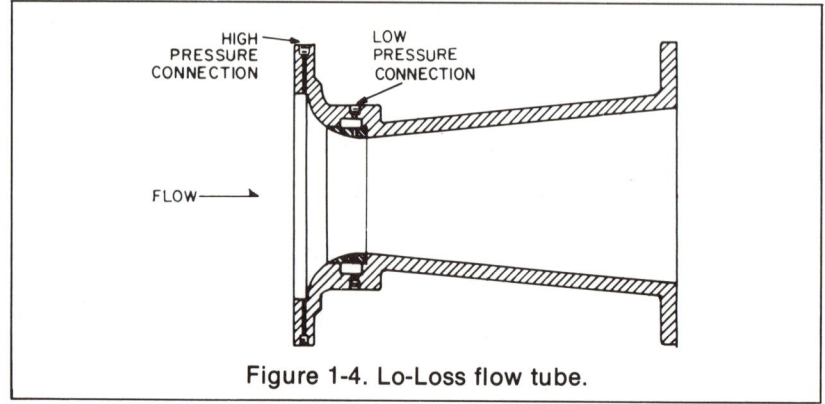
Figure 1-4. Lo-Loss flow tube.

areas. Mill effluent flows containing considerable amounts of suspended solids have also been successfully measured by Venturi tubes.

The paper industry also uses Venturi tubes which have been modified to shorten the length of the tube without too greatly increasing overall pressure loss. These are proprietary primary devices and include Dall tubes, Foster flow tubes, and Lo-Loss flow tubes. Figure 1-4 shows a typical device, the Lo-Loss flow tube.

One modification, the Nozzle-Venturi tube, has a much shorter exit section than the regular Venturi tube. The insert type Nozzle-Venturi tube which fits inside the pipeline and is installed between two flanges is pictured in Figure 1-5.

Orifice Plates. Orifice plates, originally developed for use on gas flows by Thomas Weymouth of the American Gas Association, are the most frequently used primary device in the paper industry for the measurement of fresh water, steam, clean chemicals, and uncontaminated gases. They are applicable to all clean fluids but are not applicable, except in a limited

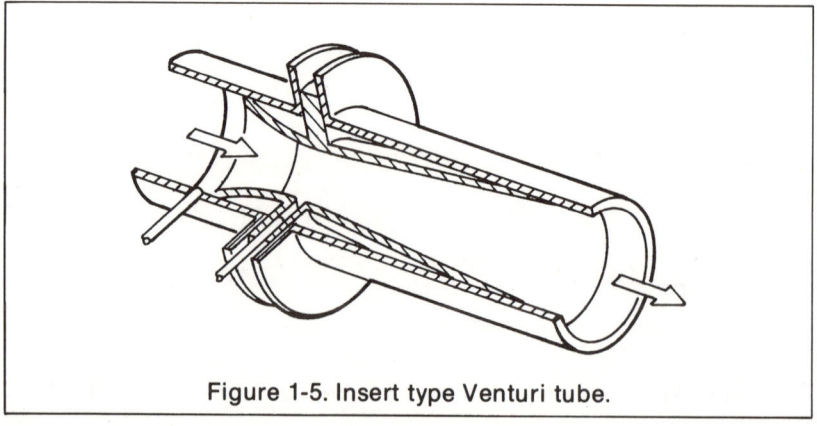
Figure 1-5. Insert type Venturi tube.

sense, to fluids containing solids in suspension and where pumping costs and line pressure losses are serious factors.

A typical thin-gauge, square-edged, concentric orifice plate is usually installed in a pipeline between flanges (Figure 1-6) to restrict the flow and cause an increase in fluid velocity at that point. The velocity change produces a decrease in line pressure downstream from the orifice creating a differential pressure across the plate. This pressure differential, which can be related accurately to flow, is sensed through precisely located connections known as *taps*, which are determined by accuracy requirements and convenience of installation and maintenance.

For specific applications in which orifice plates are used to measure flow of fluids containing limited amounts of solids or suspended matter, modifications referred to as *eccentric* and *segmental* (Figure 1-7) are used.

These modified orifice plates have orifices of a configuration different from the concentric orifice. The orifice plate is located so that the bottom of the hole is nearly flush with the bottom inside of the pipe. Accordingly, a different calibration requiring special flow coefficients is involved.

Flow Nozzles. The flow nozzle is another primary flow device which may replace an orifice plate when fluids being measured contain moderate amounts of suspended solids and sediment. A flow nozzle, compared schematically with the orifice plate in Figure 1-8, is, in a sense, an orifice with a flared or rounded approach section. Line pressure loss is less than that obtained with an orifice plate but greater than that resulting from the use of a Venturi tube. Cost is between that of an orifice plate and a Venturi tube.

Flow nozzles measure air and water in many mills and are frequently attached to the end of a pipe so as to discharge freely into the air. In the case of water the discharge may be into the air or below the surface in a tank, that is, submerged discharge.

Pitot Tube. The Pitot tube is used when fluid velocity is the prime consideration. There are two types: short single opening, and combined as

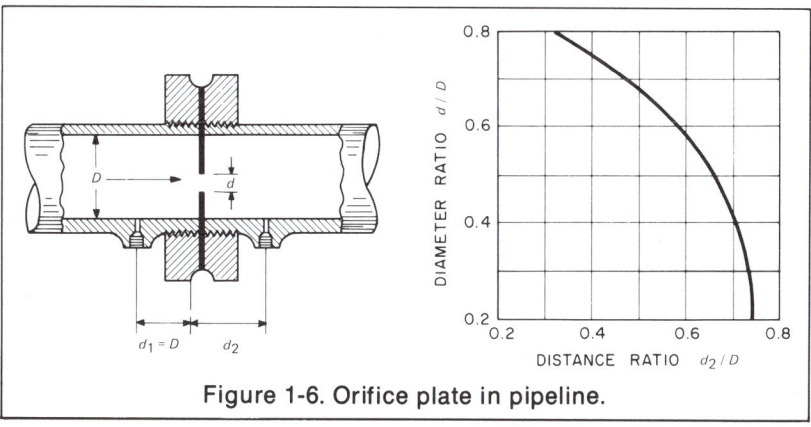

Figure 1-6. Orifice plate in pipeline.

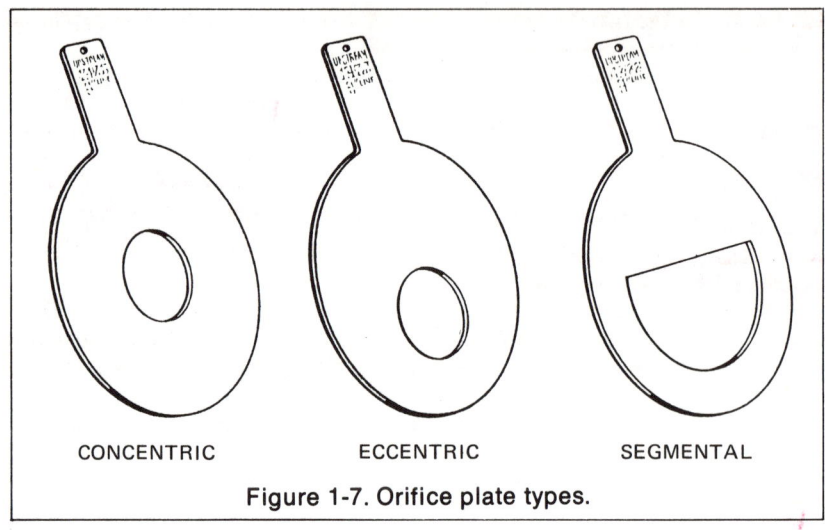

CONCENTRIC   ECCENTRIC   SEGMENTAL

Figure 1-7. Orifice plate types.

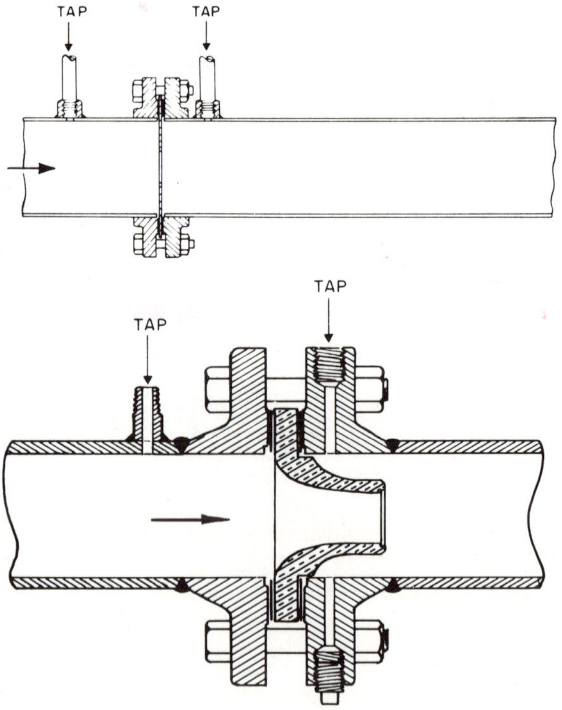

TAP   TAP

TAP

TAP

Figure 1-8. Pipeline installations.

shown in Figure 1-9. They both have two pressure connections. One type faces into the flow, sensing total or impact pressure; the other is perpendicular to flow to sense the static pressure. The difference is indicative of the velocity of the fluids at that one point in a line. Because the Pitot tube senses velocity at one point, a traverse of a cross section of the pipe must be made to obtain an accurate measurement of flow. Velocities obtained from a traverse are averaged and flow is determined by multiplying it by the cross-sectional area of the line.

The Pitot tube consumes very little energy and is relatively inexpensive. It has a tendency to foul quickly when foreign material is in the flowing fluid and, therefore, is not practical for fluids containing solids in suspension. It is suitable for measuring flows of air, gas, and water in large ducts, stacks, and water lines and sometimes is used to measure the velocity of rivers and streams.

Elbow Taps. Certain conditions require relative rather than absolute values of flow rates with good repeatability. A pipe elbow in the flow line can be used as a primary device (Figure 1-10).

The centrifugal force of a fluid flowing through an elbow creates the differential pressure for the secondary device, the flowmeter.

A major limitation of elbow flowmeters: low differential pressure is often insufficient for practical, reliable operation.

Installation. The primary device used must be inserted in the flow line. Standard equipment consisting of various types of flanges, holding rings, and fittings is available for this purpose.

To obtain the most accurate flow measurement sometimes requires that the effects of swirling and turbulence in the fluid flow be eliminated. This involves a piping arrangement referred to as a *meter run*, which is basically composed of standardized lengths of straight pipe located upstream and downstream of the ordinary device. Data have been tabulated for the various meter runs and associated primary devices. When the required straight run of pipe is not feasible in an installation, straightening vanes can reduce

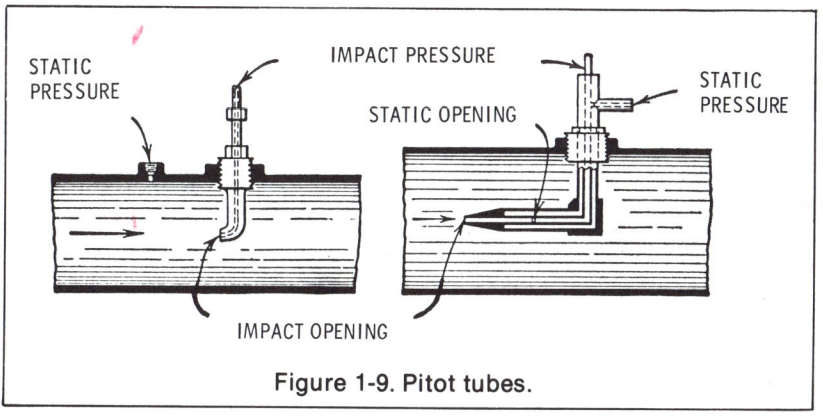

Figure 1-9. Pitot tubes.

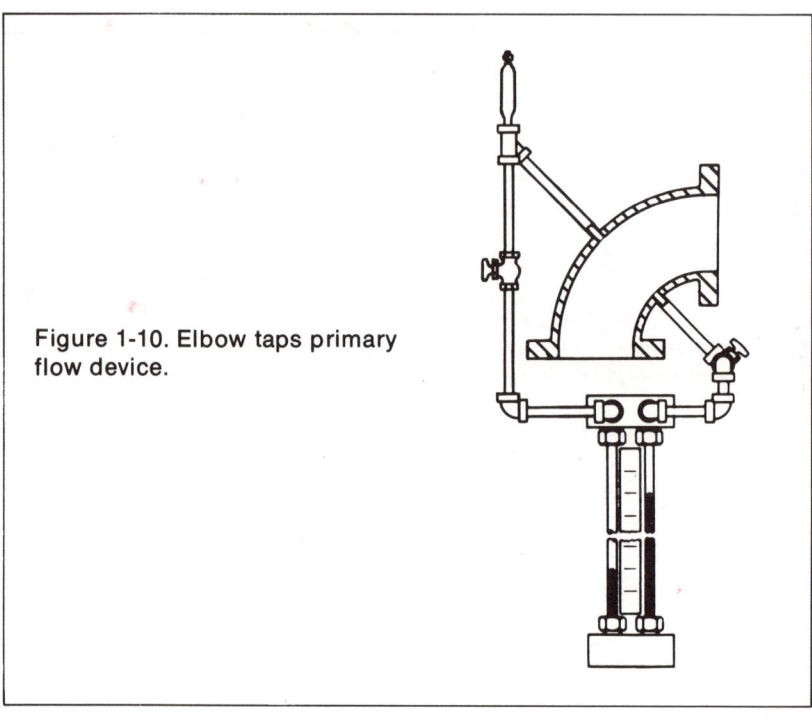

Figure 1-10. Elbow taps primary flow device.

the length of straight pipe in the approach to the primary device. Vanes, which consist of a bundle of tubes or channels, are installed in the inlet section to eliminate the swirling and turbulent flow. The two most common types are tubular and radial (Figure 1-11).

Secondary Devices

Secondary devices used in differential pressure or head type measuring systems may be divided into two general groups (Table 1-B): (1) wet flowmeters where the flowing fluid which exerts the differential pressure is in contact with mercury or another liquid in the device; (2) dry flowmeters which use no liquid for the process fluid to contact.

Wet Flowmeters

Liquid Manometer Types. There are basically two types of wet flowmeters: liquid manometer and liquid seal. Liquid manometers are the oldest, simplest, and in many respects still the most accurate and reliable of the wet flowmeters used to measure differential pressure. They are available in a variety of configurations.

Where only visual indication is needed and where static pressures are reasonably low, visual manometers with transparent tubes are used. The four most common visual manometers are the simple U-tube, well or reservoir, well with zeroing adjustment, and inclined (Figure 1-12).

Where high pressures exist and the fluids being measured are hazardous, a manometer with mercury as the liquid is used. A cylindrical steel chamber forms one leg of the manometer and contains a steel disc which floats on the surface of the mercury. The level of the liquid is determined by measuring the position of the float.

In a typical mercury float manometer (Figure 1-13), differential pressure is applied to the two sides of the manometer. With zero differential the mercury in the two legs of the manometer is at the same height. One of the legs is connected to the high pressure being measured, the other to the low. As pressure increases in the float chamber, the mercury level falls in this chamber and rises in the low-pressure or range chamber. The resulting mercury motion is changed into pen motion through the float on the surface of the mercury. Mercury manometers are provided with a check valve arrangement which prevents the mercury from leaving the float chamber upon sudden differential pressure changes or when a lead line break occurs.

Liquid Seal Types. Liquid seal type secondary devices of the wet flowmeter class are applied on services which involve low differential pressures. There are three basic types being used in the paper industry (Figure 1-14). The inverted bell flowmeter is designed to operate in a differential pressure range of 0-1 through 0-10 inches of water. The force developed by this differential pressure acting on the bell is opposed by a force developed by a spring. The resultant motion of the bell, which is governed by the stiff-

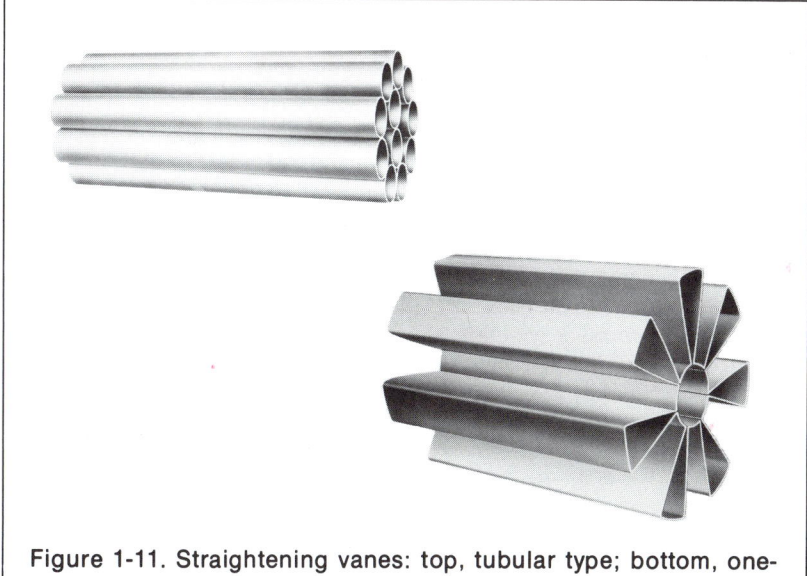

Figure 1-11. Straightening vanes: top, tubular type; bottom, one-piece radial type.

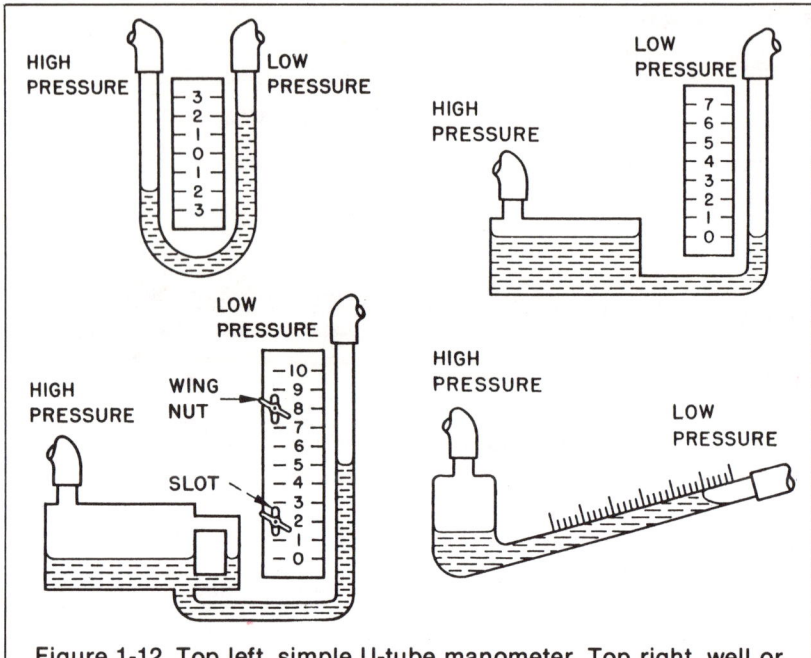

Figure 1-12. Top left, simple U-tube manometer. Top right, well or reservoir type. Bottom left, well type with zeroing adjustment. Bottom right, inclined manometer.

ness of the spring, measures the differential pressure. A sizable bell will develop considerable force from limited differential pressure.

In the Ledoux bell flowmeter the force developed by differential pressure is balanced by the gravity acting on a bell floating in mercury. The bell is shaped so that its rising and falling motion is proportional to the square root of the differential pressure and, therefore, linear to flow. The Ledoux bell is commonly found in power plants where a uniform flow scale extending over a wide range of steam flow is needed.

The ring-balance flowmeter operates on the same principle as the bell flowmeter. A ring formed from steel tubing and located in a vertical plane is pivoted at its central axis. Differential pressure exerts unequal forces on opposite closed ends of the ring causing the ring to rotate. A weight attached to the bottom of the ring develops an opposing torque as it moves with the ring. Mercury or other liquid filling the lower half of the ring isolates high- and low-pressure sections.

Dry Flowmeters

The mercury float type differential gauge was at one time generally accepted as the standard for measurement of differential pressure. However,

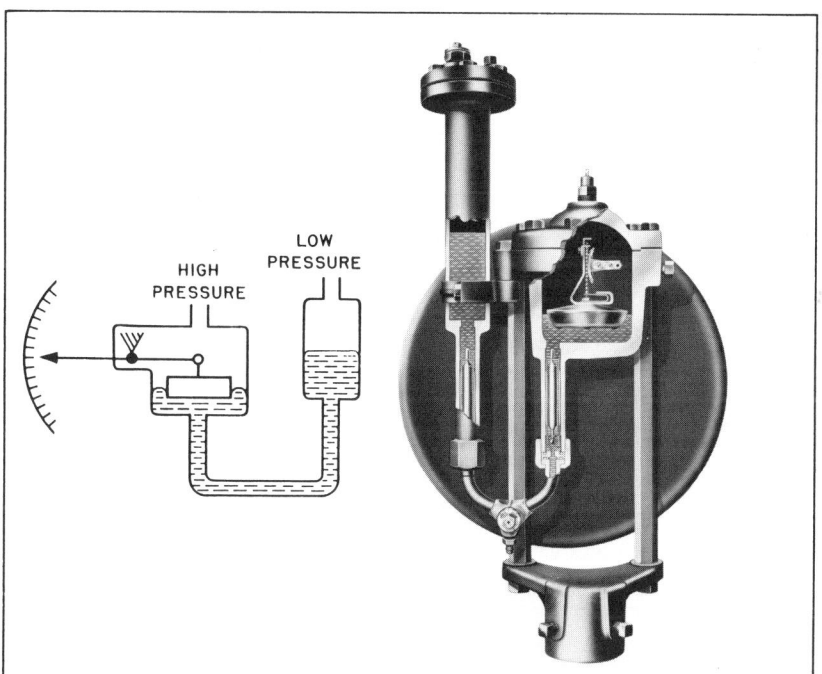

Figure 1-13. Mercury float type flowmeter. Note mercury height differential.

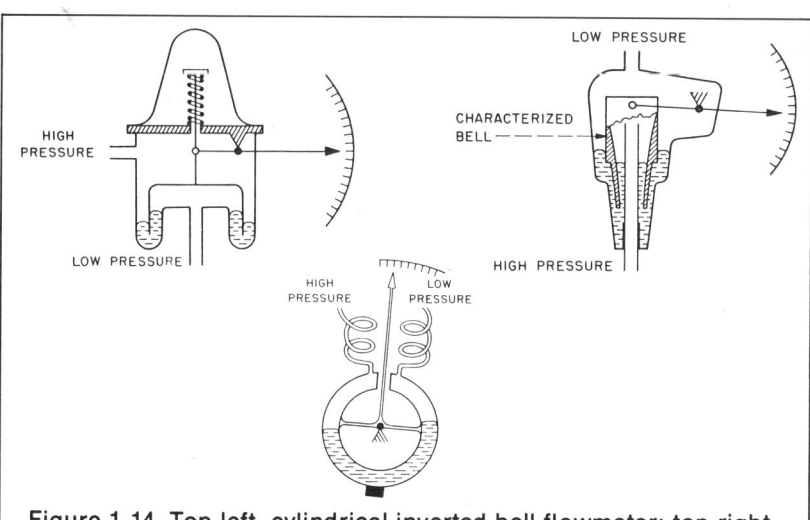

Figure 1-14. Top left, cylindrical inverted bell flowmeter; top right, inverted bell flowmeter with uniform flow scale characteristic; bottom, ring-balance flowmeter.

Flow Measurement   23

to attain high sustained accuracy these flowmeters were large and expensive. The increasing cost of mercury accentuated this cost disadvantage. Various types of mercuryless dry flowmeters were developed to overcome the disadvantages of mercury flowmeters; these devices have been developed to give excellent performance over a wide range of operating conditions in the industry.

These flowmeters are commonly divided into two types: (1) direct-connected motion type and (2) force-balance or small motion type. For both types, usually two or more diaphragms or bellows are used, with the intervening space filled with liquid. The liquid transmits the fluid pressure and, in addition, provides damping. The liquid is sealed in the structure; it never contacts the measured fluid; it is not subject to loss or contamination except in case of structure failure. Since there is no physical contact with this filling fluid in normal operation, calibration, or maintenance, these units are commonly included in the class of dry flowmeters.

Motion-Balance Bellows Type. The bellows flowmeter is generally used where a direct-operated indication or record of the differential pressure is required. Figure 1-15 shows the basic structure of the bellows flowmeter. Details of construction vary; almost all designs include double inclosed bellows filled with liquid. The bellows flowmeter, with its bellows and spring combination, depends for sustained accuracy upon repeatability of this mechanical system in contrast to the inherent repeatability of the basic mercury manometer principle. Bellows flowmeters are available which have proven by experience to be comparable to the mercury float type differential gauge in accuracy and stability.

The principle of operation is shown in Figure 1-15. The differential pressure is applied across the bellows, producing a proportional force. This force is opposed by a calibrated spring. The resultant motion is directly proportional to differential pressure. Protection of the bellows against overrange differential pressures up to the full static rating of the instrument is provided. This may be accomplished by valves operated by the motion of

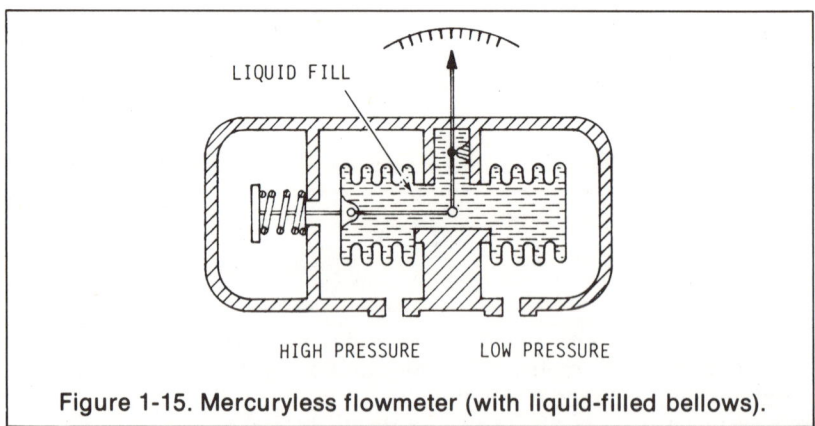

LIQUID FILL

HIGH PRESSURE   LOW PRESSURE

Figure 1-15. Mercuryless flowmeter (with liquid-filled bellows).

the bellows which closes off flow of the fill liquid when pressures exceed the working range. Another design employs a bellows structure designed to withstand overrange to full line pressure in either direction. The liquid-filled bellows is fabricated from a number of diaphragm discs and steel spacer rings. When the outside of the bellows is subjected to pressure overrange the diaphragm discs "nest" and the steel spacer rings, welded to the inner edges of the discs, butt to form a solid stop. The linear bellows motion may be converted to a rotary motion through a link and lever and then brought out through a seal, commonly a torque tube; or the linear motion may be brought out through a bellows seal. If, as shown, this motion is brought out from the liquid-filled chamber, the pressure seal must be absolutely leakproof. Damping is provided by restricting the flow of the filling liquid. Damping is customarily adjustable to make available a smooth, readable record even from noisy or pulsating flows. It is essential that an adjustment of damping also be completely leakproof. The linear motion of the bellows and range spring is transmitted through a link to the inner end of a drive bar. The outer end of the drive bar, pivoting on a bellows-sealed flexure, moves in the opposite direction to drive a pen which records differential pressure on the instrument chart.

Bellows flowmeters using a torque tube pressure seal are limited by the characteristics of the torque tube. Output motion is limited to a few degrees angular motion. This output is adequate for indication, recording, and for operation of pneumatic or electric motion transmitters. However, any considerable output load that may be imposed by secondary responsive devices tends to diminish accuracy and, particularly, responsiveness. This limitation can be overcome by use of other types of pressure seals; however, no other seal as simple as the torque tube has been developed.

Bellows flowmeters are available in stainless steel and with other corrosion-resistant structures in contact with the measured fluid. Static pressure ratings up to 10,000 psi are available as standard. A wide range of differential pressure spans is available; very low differential pressures involve basic design limitations because of the very small forces available. Similarly, there are basic problems with very high differential pressures due to limitations in the bellows. For normal differential pressures developed by the usual head metering primary devices the bellows flowmeter designs are fully adequate.

Force-Balance Types. Commonly referred to as *differential pressure transmitters,* force-balance devices are widely used in the pulp and paper industry for flow measurement where an indication and/or a record of flow is required at a location not adjacent to the primary device. Both pneumatic and electric transmissions are used.

In a typical pneumatic transmitter (Figure 1-16), the differential pressure to be measured is applied across a pair of metal diaphragms welded to opposite sides of a capsule; the space between the diaphragms and core member is filled with liquid. The force developed on the diaphragm by differential pressure is brought out of the transmitters by a rigid rod passing through a metal seal diaphragm. This force is opposed by a balancing force

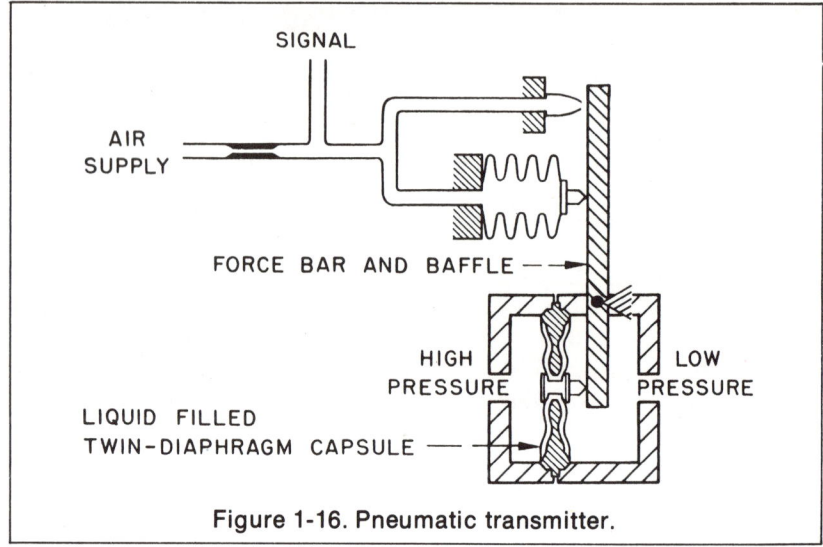

Figure 1-16. Pneumatic transmitter.

developed by pneumatic bellows. Imbalance between capsule force and pneumatic bellows force is sensed by a pneumatic nozzle-baffle. A simple pneumatic servomechanism responsive to nozzle pressure reestablishes the balance. As a result, pneumatic pressure is maintained exactly proportional to differential pressure and is used as output signal; a more or less standardized signal is 3 to 15 psi.

Electric force-balance systems operate on much the same principle. An electric current flowing in a coil which is supported in a permanent magnetic field develops a force. The difference between this force and the force developed by differential pressure produces a motion which is detected by a highly sensitive electrical unit. Output from the electrical unit operating through an electronic circuit maintains the electrical current in the coil at a value which exactly balances the force produced by the differential pressure. Current is thus a direct measure of differential pressure and is used as the transmission signal. Typical electrical transmission signals are 10 to 50 mA and 4 to 20 mA dc.

Integral Orifice Transmitter

The integral orifice transmitter is a unique measuring system which combines primary and secondary devices in one unit. A series of small orifices mounted on the high-pressure side of a force-balance differential pressure transmitter, usually sized from 0.002 to 0.350 inches in diameter with rounded inlet edges, permits measurements of very small rates of flow of fluids down to 0.00002 gpm fluid flow corresponding to water. These fluids must be clean and free of suspended materials and sediment. A typical assembly is shown in Figure 1-17.

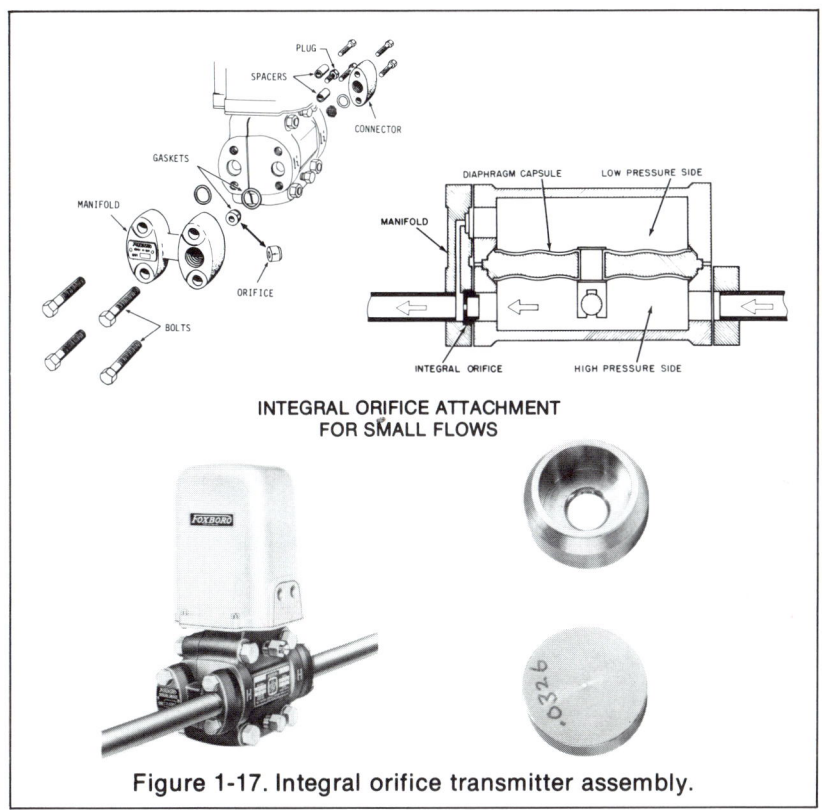

Figure 1-17. Integral orifice transmitter assembly.

THEORY OF FLUID FLOW RELATED TO HEAD

Actual flow is related to differential pressure for gas and liquid by two basic formulas, 1 and 2 below:

(1) $V = \sqrt{2gH}$

where V = velocity in ft/sec
            g = acceleration due to gravity (32.17 ft/sec^2)
            H = height, in feet, of column of fluid caused by the differential pressure across a primary device

The formula, which is for the velocity of a falling body, relates average fluid velocity to the height of a fluid column. An open tank filled with liquid and having an orifice at the bottom illustrates this relationship. The velocity of liquid flowing through the orifice is proportional to the square root of the height of the liquid in the tank.

To express (1) in terms of equivalent differential, h, in inches of fluid H,

Flow Measurement   27

is replaced by $h/12 (G_f)$ where G_f is the specific gravity of the fluid at the flowing temperature:

(1a) $$V = \sqrt{2g \frac{h}{12 \times G_f}}$$

The second basic formula states that volumetric flow rate of a fluid is the product of the cross-sectional area through which the flow passes and the average velocity of the fluid:

(2) $$Q = AV$$

where
Q = volumetric flow in cu ft/sec
A = cross-sectional area of the orifice or throat of the primary device in sq ft
V = velocity in ft/sec

Substituting the value of V in (1a), Equation 2 becomes:

(2a) $$Q = A\sqrt{2g \frac{h}{12 \times G_f}}$$

For liquids, Q is expressed in gallons per minute. The area of the orifice or throat is normally expressed in terms of its diameter (inches). Therefore, (2a) may be written:

(3) $$Q \text{ (gpm)} = 60 \times 7.4805 \times \frac{\pi d^2}{4 \times 144} \sqrt{\frac{2 \times 32.17}{12}} \times \sqrt{\frac{h}{G_f}}$$

which reduces to:

(3a) $$Q \text{ (gpm)} = 5.667 \times d^2 \times \sqrt{\frac{h}{G_f}}$$

To account for factors such as contraction of the jet, frictional losses, viscosity and the viscosity of the approach, (3a) is modified by a discharge coefficient, K. K is the actual flow rate divided by theoretical flow rate through the primary device. Basic reference books carry K factors for most primary devices in accordance with various tap locations and d/D ratios where d is the diameter of the orifice and D is the inner diameter of the pipeline in which the device is mounted. Applying K to (3a):

(4) $$Q \text{ (gpm)} = 5.667 K d^2 \sqrt{\frac{h}{G_f}}$$

Since K and d are unknown, another value, S (sizing factor), is set equal to $K(d/D)^2$. Kd^2 then equals SD^2 and:

(5) $$Q \text{ (gpm)} = 5.667 S D^2 \sqrt{\frac{h}{G_f}}$$

Similar equations can be developed for steam or vapor flow in weight units such as pounds per hour and for gas flow in volume units such as standard cubic feet per hour (scfh).

Practically all flow measurements in the pulp and paper industry are calculated at operating conditions. In other industries it is desirable to know equivalent volume flow at a stated reference temperature, usually 60°F. In the latter case, equations must be corrected by factors published in technical manuals.

AREA METERING SYSTEMS

Area flow primary devices operate on the same basic principles as other differential head flowmeters such as those employing orifices. The orifice flowmeter contains a fixed aperture and flow is indicated by a drop in differential pressure. In the area flowmeter the aperture is variable and the pressure drop is relatively constant. Thus, with the area flowmeter flow is indicated by the area of the annular opening through which the fluid must pass.

In flowmeters of this class, area variation is produced by the rise and fall of a floating element and the flowmeters must be mounted so that the element moves vertically and frictional resistance is minimized.

The outstanding advantage of an area flowmeter is that it provides direct indication of flow rate. The position of the floating element may be observed easily.

Area flowmeters are used to measure the flow of corrosive and viscous tar-like liquids. There are two general types of area primary devices: tapered tube and float, and cylinder and piston.

Tapered Tube and Float. Commonly referred to as a *rotameter*, the tapered tube and float flowmeter (Figure 1-18) is probably the most widely used type of area flowmeter. Although there may be some variation in con-

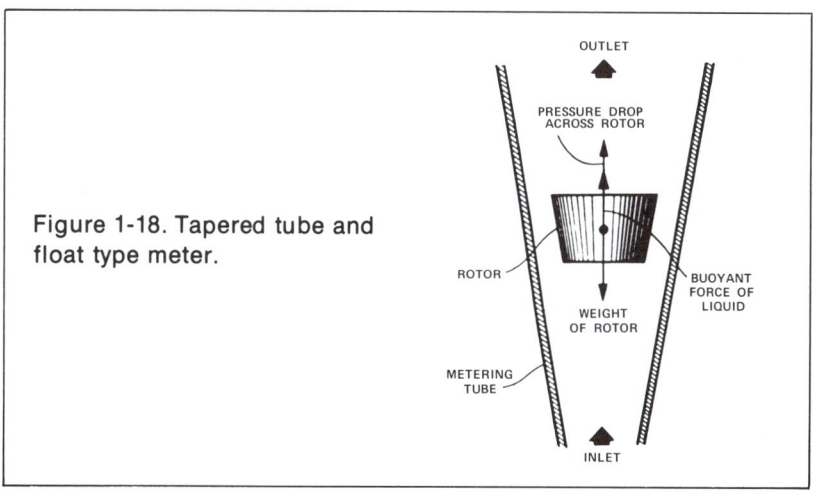

Figure 1-18. Tapered tube and float type meter.

OUTLET

PRESSURE DROP ACROSS ROTOR

ROTOR

BUOYANT FORCE OF LIQUID

WEIGHT OF ROTOR

METERING TUBE

INLET

struction, all rotameters basically consist of a submerged plummet or rotor. Fluid being measured passes through the calibrated tube from bottom to top; the rotor moves according to flow. A visual indication of flow rate is possible and rotor position, in terms of flow measurement, may be transmitted electrically or pneumatically.

The term *rotameter* stems from the earliest types of variable area flowmeters in which the float always rotated. A spinning float is more stable, easier to read, and the motion keeps the float clean. However, design improvements have made a rotating float unnecessary in some applications.

Many different rotameters are suited to different applications. The simplest rotameter (Figure 1-19) is constructed of a glass metering tube, and flow is read by noting the location of the float against a scale etched on the glass. Armored rotameters (Figure 1-20) are used for high-pressure service.

Flow measurement signals may be transmitted to a remote location by mounting a suitable pneumatic or electric transmitter on the indicating rotameter. The transmitter uses a magnetic follower arrangement to detect the position of the float or rotor. Figure 1-21 shows a typical pneumatic transmission rotameter assembly.

Electrical systems for transmitting the motion of the float vary in design. One type, shown in Figure 1-22, which uses an inductance bridge system is designed with a magnetic tube at the bottom of the rotameter. The tube allows an iron armature to be suspended from the float. A double solenoid mounted on the outside of the tube is connected to the inductance bridge of an automatic-balance instrument similar to a potentiometer. The instrument

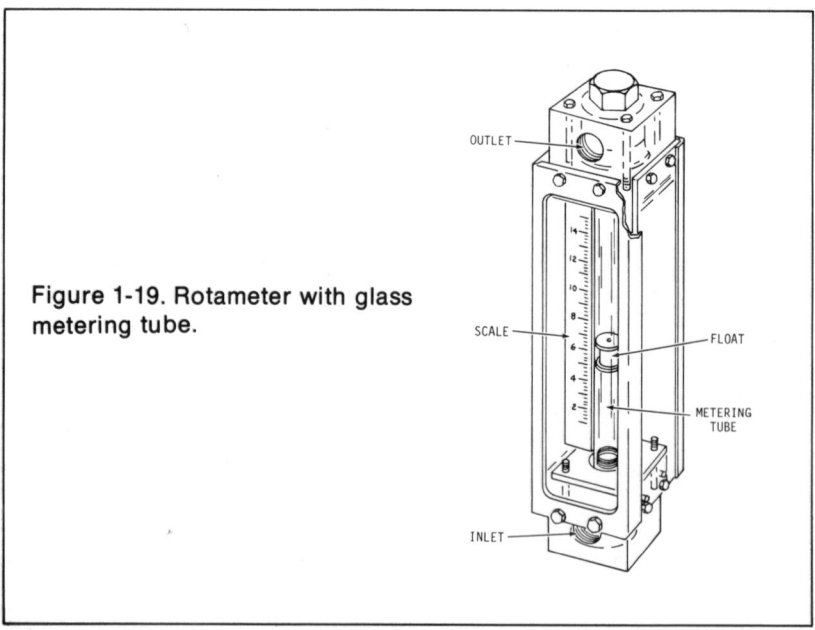

Figure 1-19. Rotameter with glass metering tube.

can be calibrated in terms of rotameter float position. Other types employ a self-balancing inductance bridge system.

The rotameter has many advantages in industrial service. The flowmeter can be built to handle specific process fluids. Proper float design and construction allow compensation for changes in fluid density and viscosity and, therefore, temperature for service on certain fluids. Pressure loss at the flowmeter is small and nearly constant. The device is quite accurate over a

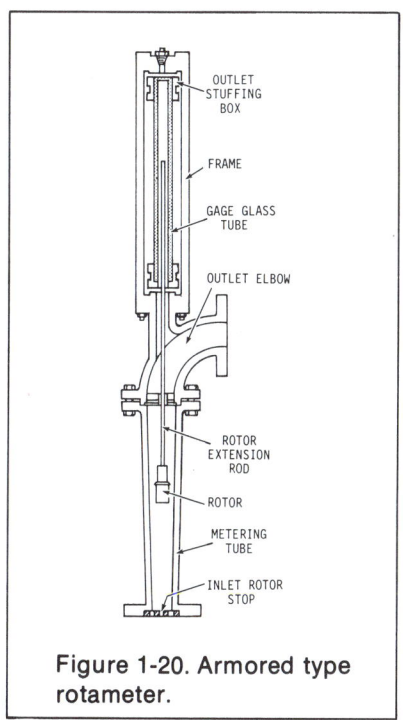

Figure 1-20. Armored type rotameter.

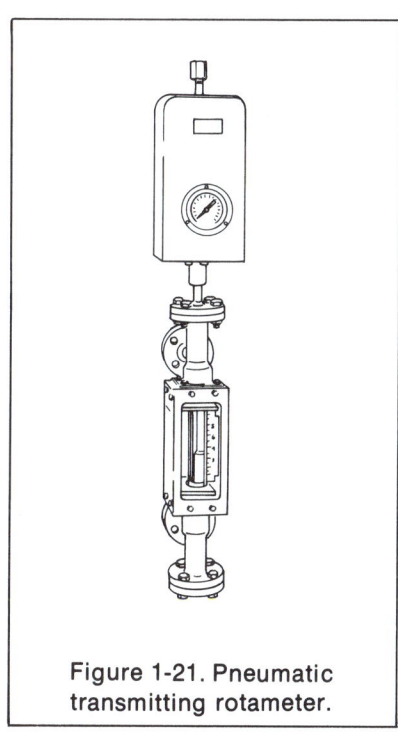

Figure 1-21. Pneumatic transmitting rotameter.

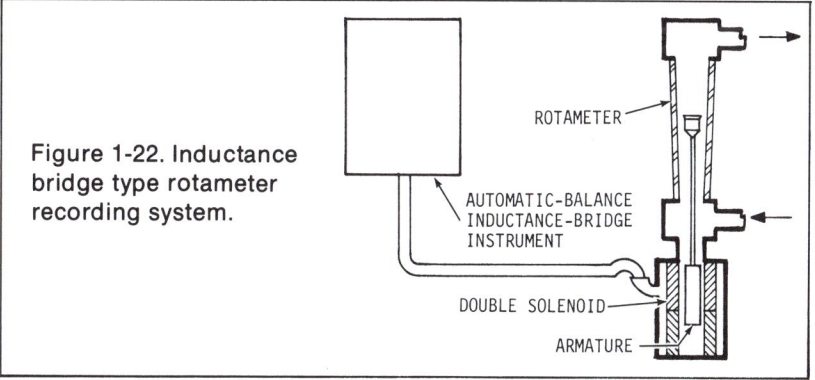

Figure 1-22. Inductance bridge type rotameter recording system.

wide range, particularly at low flows. Its main disadvantage is that it is not as rugged as head flowmeters because of the glass metering tube. Recently, metal and plastic metering tubes have been used to increase strength.

Cylinder and Piston. The difference between cylinder and piston area flowmeters and rotameters is that the area for fluid flow in the cylinder and piston type is provided by a series of openings in the wall of the cylinder. These openings are spaced helically around the cylinder in rows so that the variation in area for various heights of the piston is continuous. The pressure differential is constant because the weight of the piston is constant. Openings in the cylinder may be spaced to permit linear calibration for flow rate.

Two typical designs, shown in Figure 1-23, which differ primarily in the type of openings in the sleeve and method of loading the piston, predominate. These are the multiple-holed cylinder with weight-loaded pistons and the slotted cylinder with spring-loaded piston.

One has a cylinder with a large number of equal-sized orifices for openings. The orifices are reamed holes spaced in a uniform helical pattern. Piston movement is regulated by adjustable weights on the piston. The other has a cylinder with longitudinal slots for openings. An adjustable loading spring regulates movement of the piston.

Cylinder and piston area flowmeters are especially suited for measuring the flow of fuel oils such as Bunker C fuel oil, tar, liquid chemicals, and other such high-viscosity and corrosive fluids.

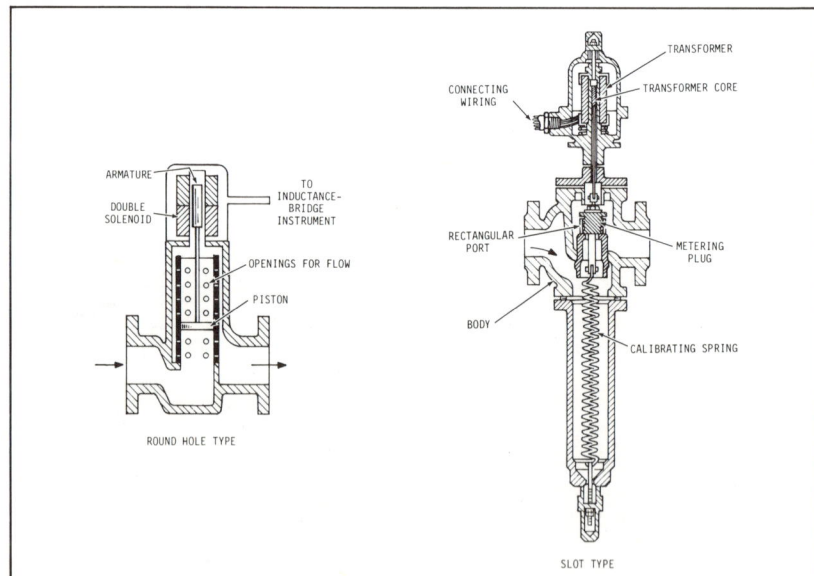

Figure 1-23. Cylinder and piston type flowmeter designs, differing in type of sleeve opening and method of loading the piston.

HEAD AREA MEASURING SYSTEMS

In head area metering systems the area of the stream and head vary; area is a function of head. These flowmeters are also distinguished from other rate flowmeters by their use in open conduits and channels, or in conduits and channels in which there is a free surface. Used almost exclusively to measure flow of water, sewage sludge, chemical wastes, and other semi-fluids, head area flowmeters are found primarily in the water and effluent treatment areas of pulp and paper mills.

As with most other metering systems, head area metering systems consist of two distinct parts: (1) a primary device on which the fluid acts to provide a source of measurement; (2) a secondary device which translates the action of the fluid on the primary device into volume, mass, or rates of flow and indicates or records the results.

Head Area Primary Devices

Most commonly used head area primary devices fall into two general categories: (1) the weir which, in effect, is a dam over which liquid flows; (2) the flume which is a formed section in a channel with a very slight slope.

The Weir. The weir is essentially a dam with a notched opening in the top through which liquid flows; as shown in Figure 1-24 it takes several shapes—rectangle, trapezoid, and V, for example. A weir box is installed

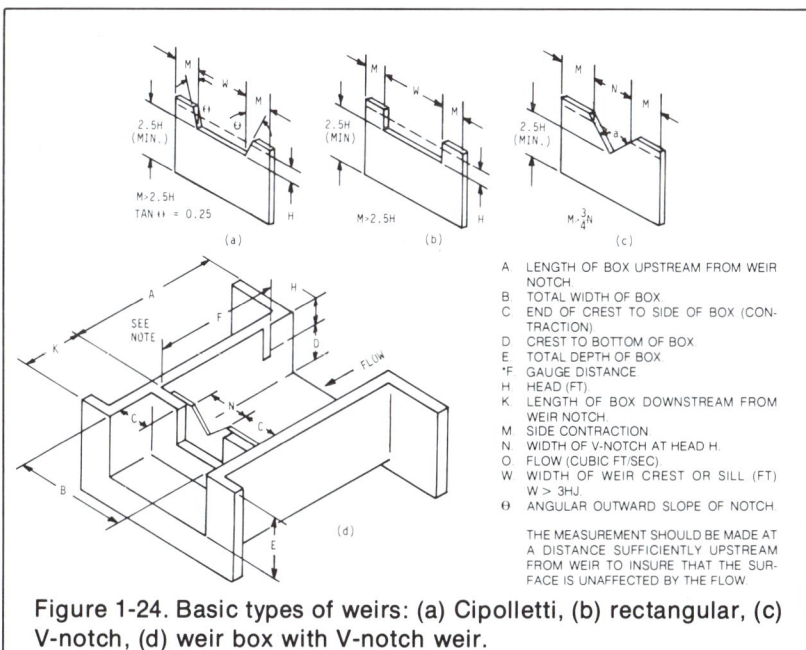

A. LENGTH OF BOX UPSTREAM FROM WEIR NOTCH.
B. TOTAL WIDTH OF BOX.
C. END OF CREST TO SIDE OF BOX (CON-TRACTION).
D. CREST TO BOTTOM OF BOX.
E. TOTAL DEPTH OF BOX.
*F. GAUGE DISTANCE.
H. HEAD (FT).
K. LENGTH OF BOX DOWNSTREAM FROM WEIR NOTCH.
M. SIDE CONTRACTION.
N. WIDTH OF V-NOTCH AT HEAD H.
O. FLOW (CUBIC FT/SEC).
W. WIDTH OF WEIR CREST OR SILL (FT)
W > 3HJ.
θ. ANGULAR OUTWARD SLOPE OF NOTCH.

THE MEASUREMENT SHOULD BE MADE AT A DISTANCE SUFFICIENTLY UPSTREAM FROM WEIR TO INSURE THAT THE SUR-FACE IS UNAFFECTED BY THE FLOW.

Figure 1-24. Basic types of weirs: (a) Cipolletti, (b) rectangular, (c) V-notch, (d) weir box with V-notch weir.

TABLE 1-C

WEIR-BOX DIMENSIONS FOR RECTANGULAR, CIPOLLETTI, AND 90° TRIANGULAR NOTCH WEIRS

(Letters refer to dimensions, Fig. 1-24)

	H	W	A	K	B	E*	C	D	F
Approximate Limits of Discharge	Maximum Head	Width of Weir Crest	Length of Box above Weir Crest	Length of Box below Weir Crest	Total Width of Box	Total Depth of Box	Distance from End of Crest to Side of Box	Distance from Crest to Bottom of Box	Gauge Distance
Second-Feet	Feet	Feet	Feet	Feet	Feet	Feet	Feet	Feet	Feet
RECTANGULAR AND CIPOLLETTI WEIRS									
1/10 to 3	1	1	6	2	4	3	1½	1½	4
1/5 to 6	1¼	1½	7	3	5	3¼	1¾	1½	4½
1/4 to 8	1¼	2	8	4	6	3½	2	1¾	5
1/3 to 17	1½	3	9	5	7	4	2	2	5½
1/2 to 23	1½	4	10	6	9	4	2½	2	6
3/4 to 35	1½	6	12	6	11½	4½	2¾	2½	6
1 to 50	1½	8	16	8	14	4¾	3	2¾	8
1 to 60	1½	10	20	8	17	5	3½	3	8
90° TRIANGULAR NOTCH WEIR									
1/10 to 2-1/2	1	–	6	2	5	3	–	1½	4
1/10 to 4-1/3	1¼	–	6½	3	6½	3¼	–	1½	5

*This distance allows for about 6 inches freeboard above highest water level in weir box.

near to and is connected with the weir. This creates a weir pond and provides somewhat of a stilling well which eliminates velocity or turbulence effects and provides a location for measuring the head from which the flow rate is determined.

In the three basic weirs and a V-notch weir with a weir box, the horizontal distances from the crest to the side walls of the weir box are referred to as *end contractions* and the vertical distance from the crest to the floor of the weir box or bed of the channel is referred to as the *bottom contraction*. When the distances from the crest to the floor are enough to cause water to pond above the weir so that it approaches the weir notch at low velocity, the weir is said to have *complete contraction*.

The rectangular weir is the simplest and the most accurate, the easiest to construct and the most popular in use today.

The triangular or V-notch weir, which comes to a point at the bottom, has no crest length; the crest is the bottom edge of the notch. This weir has a greater practical range of capacity for a given size than other types. However, it requires a greater loss of head and, therefore, it is more suitable for measuring flows of less than 4 cfs.

The trapezoidal notch, or Cipolletti weir as it is known, is a combination of the rectangular and triangular V-notch. The sides of the notch slope one horizontal to four vertical, making the end contractions set at 4:1 angle rather than being perpendicular to the edge of the weir. The value of this slope is such that the additional discharge through the added triangular portions of the notch will exactly compensate for effects of end contractions.

Weir boxes are designed and installed with certain size and position relationships. Significant ones may be generalized as:

1. The width of box: 3 times width of notch.
2. End contractions: 3 times maximum head.
3. Bottom contraction: greater than 2 times maximum head.
4. Crest width: greater than 6 inches.
5. Maximum head: greater than 0.1 foot and less than 1 foot.
6. Head measured upstream: greater than 4 times maximum head value.

Table 1-C lists the actual weir-box dimensions for rectangular, Cipolletti, and 90° triangular notch weirs for typical flow conditions.

In any weir, the sheet of liquid passing through the notch and falling over the weir crest is called *nappe*. When the liquid surface downstream from the weir plate is far enough below the crest for the air to have free access beneath the nappe the flow is said to be *free*; otherwise it is submerged. In weir measurements, the nappe, or profile of water over the weir, must be completely aerated in order to have precise flow measurements (see Figure 1-25). When maximum and minimum flows to be expected are known, the type and size of weir can be selected from published data such as those shown in Figure 1-26.

Other types of weirs of lesser importance are being used to a very limited extent today for more or less special purposes. These are characterized by

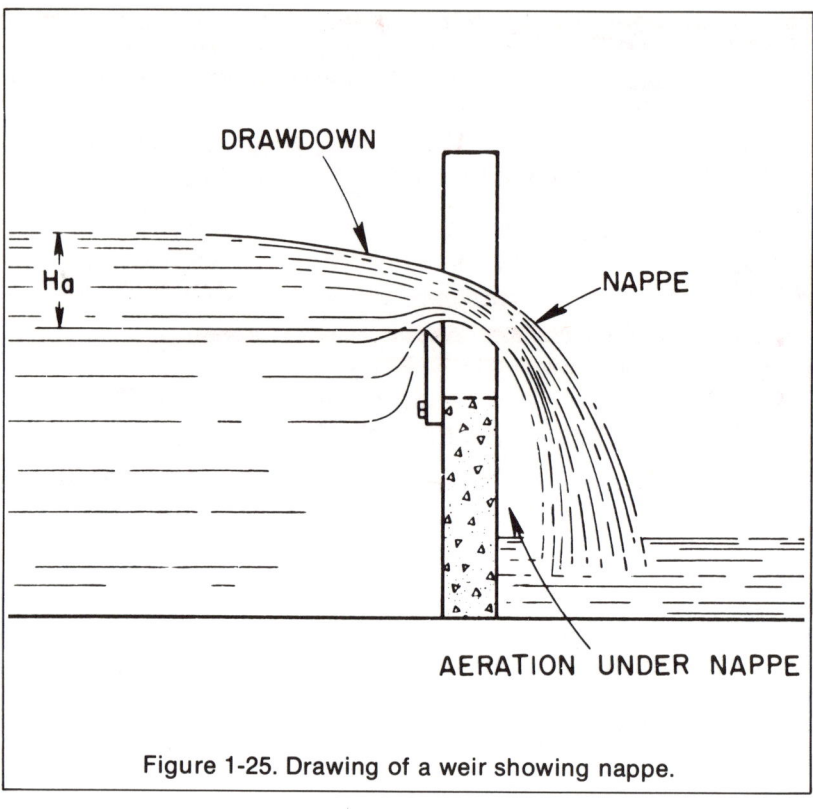

Figure 1-25. Drawing of a weir showing nappe.

the shape of the notches such as circular, proportional, exponential, parabolic, hyperbolic, and cyclodial.

The Flume. A further development of the basic weir concept, flumes are designed primarily to reduce the head loss that is experienced with the weir. The most common, the Parshall flume, is a special type of Venturi flume which possesses certain features that are believed to be superior to those of the standard weir. The Parshall weir consists of an entrance section with converging walls and level floor, a throat section with parallel walls and a downstream sloping floor, and an outlet section with diverging walls and a rising floor. Figure 1-27 shows a plan and elevation of a concrete Parshall measuring flume.

Published information and technical data are available for small and large flumes. Figure 1-28 gives flow curves for selecting and sizing Parshall flumes.

Other types of flumes or special forms used to a limited extent are more commonly known as the Palmer-Bolus flume and the parabolic discharge flume.

Open Channel Flow Nozzle. A combination of flume and weir, open channel flow nozzles can be used to measure large flows of mill effluents

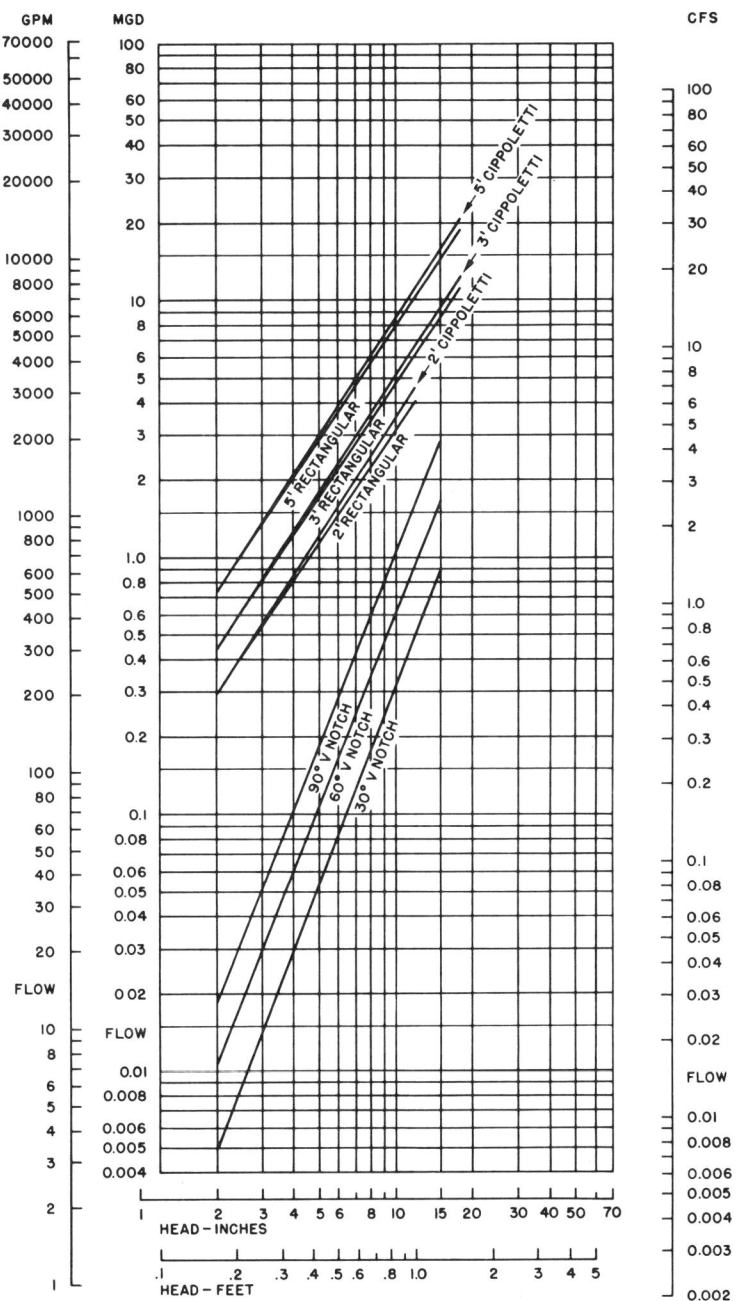

Figure 1-26. Weir capacities.

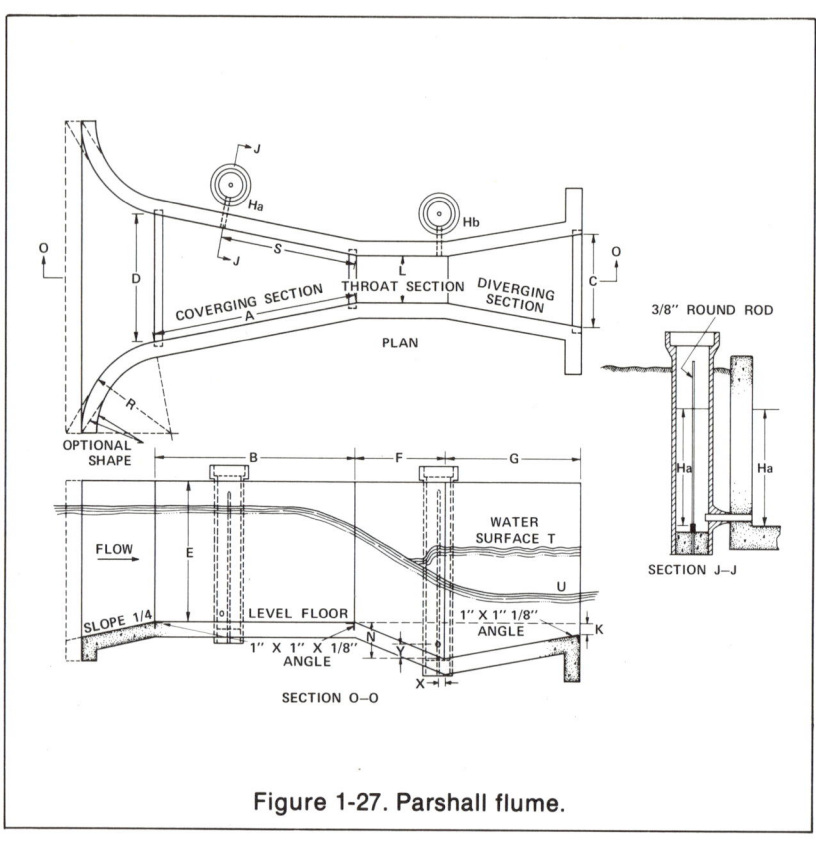

Figure 1-27. Parshall flume.

with moderate amounts of solids. The unique cross-section shape of the nozzle produces a linear relationship between head and flow. A typical unit, the Kennison nozzle, is shown in Figure 1-29.

Head Area Secondary Devices

Because level is a function of flow, the float and cable instrument is the traditional secondary device used to measure flow through head area primary devices. Float and cable instruments require a float well when used with a flume or nozzle; for a weir they require a restraint to prevent horizontal movement of the float. Figure 1-30 shows a typical installation of such a device on a V-notch weir application.

The float and cable secondary instrument is sometimes arranged with the float inside the flume.

Bubble Tube Type. Air-purged secondary devices do not require a float but measure head directly with a bubble tube. Combined with pneumatic or electronic level measuring secondary devices, they can transmit measurements to remote locations (Figure 1-31).

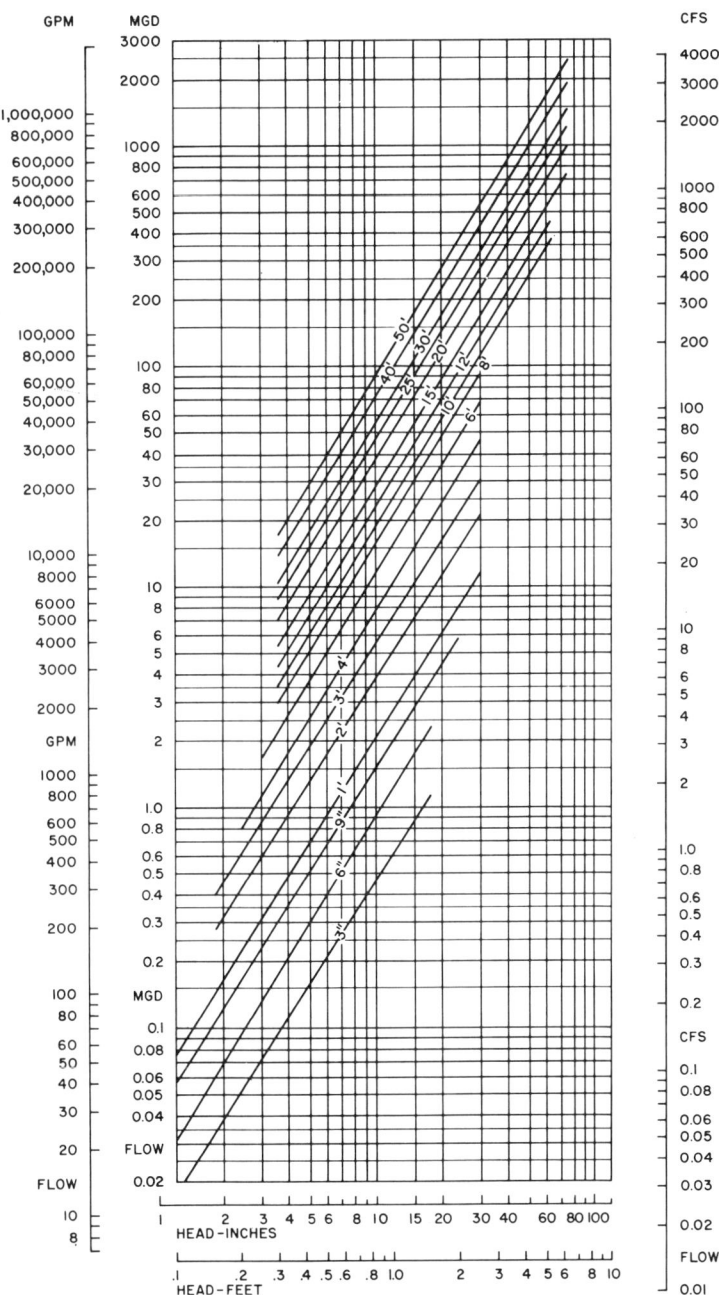

Figure 1-28. Parshall flume selection chart.

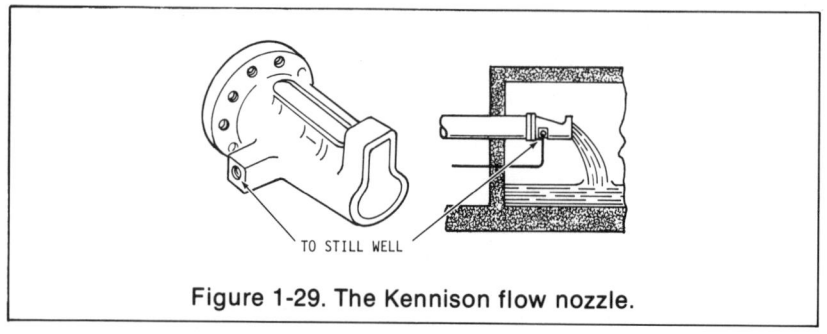

TO STILL WELL

Figure 1-29. The Kennison flow nozzle.

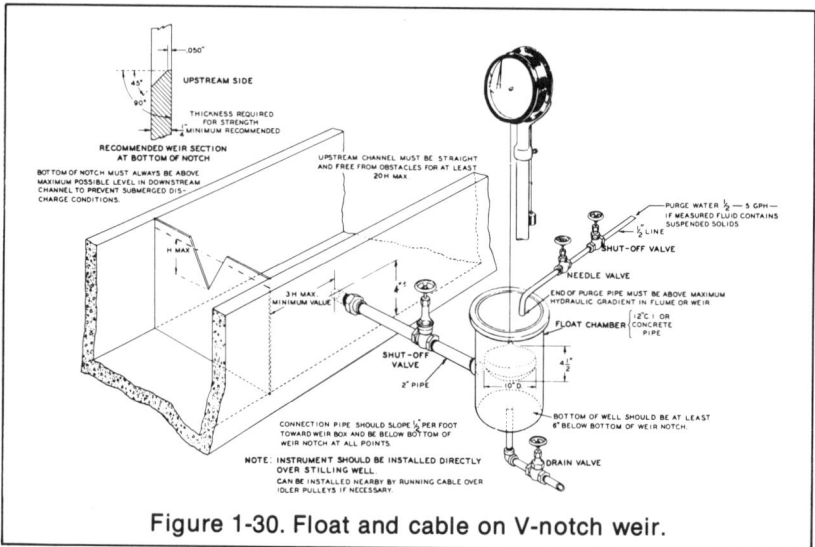

Figure 1-30. Float and cable on V-notch weir.

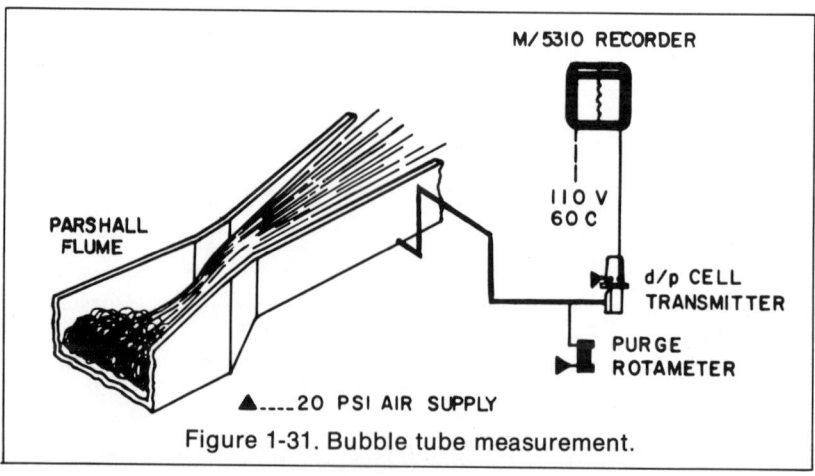

M/5310 RECORDER

110 V
60 C

PARSHALL FLUME

d/p CELL TRANSMITTER

PURGE ROTAMETER

▲----20 PSI AIR SUPPLY

Figure 1-31. Bubble tube measurement.

Flow Calculations

In a rectangular notch weir installation, the velocity of the flow is proportional to the depth at the weir and is expressed by the equation:

(1) $$V = \sqrt{2gZ}$$

where V = velocity at given lamina
Z = depth from surface to lamina
g = acceleration due to gravity

The total flow is the product of velocity times area.

(2) $$q = \int_o^h \sqrt{2gZ} \cdot b \; dz$$

and

(3) $$q = \frac{2}{3} b \sqrt{2gh^3}$$

where q = flow rate
h = head over weir crest
b = width of rectangular notch

Actual flow requires use of a flow coefficient which is $0.644 \pm 5\%$:

(4) $$q = (0.644 \pm 5\%) \, \frac{2}{3} b \sqrt{2gh^3}$$

Similarly, the equation of flow through the V-notch can be developed:

(5) $$q = (.0605 \pm 6\%) \left(\frac{8}{15} \tan\frac{\theta}{2} \right) \sqrt{2gh^5}$$

The equation for the Parshall flume can be developed in almost the same manner as for the rectangular weir; the chief difference is in the coefficient:

(6) $$q = (0.63 \pm 3\%) \frac{2}{3} b \sqrt{2gh^3}$$

FORCE METERING SYSTEMS

Operation of force flowmeters depends on the force produced by moving fluid on a swinging component of the primary device. The force tends to deflect the component from its undisturbed position.

Target Type. Target flowmeters actually combine in one unit a primary and a force-balance flow rate secondary device, the transmitter (Figure 1-32). The force on a disc or "target," which is mounted in the line of the flowing fluid, developed through the annular orifice, may be measured directly,

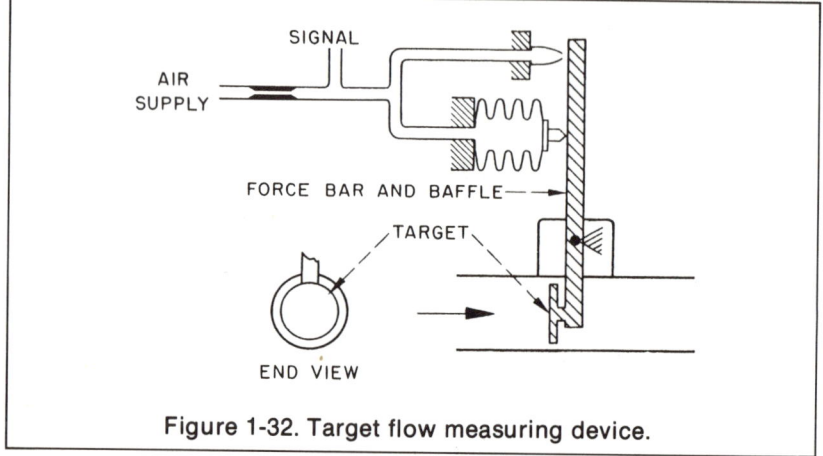

Figure 1-32. Target flow measuring device.

This arrangement eliminates external fluid connections or differential taps and the need for seals or purge systems. It also makes the target flowmeter ideal for measuring heavy viscous fluid hydrocarbons, slurries, sticky and dirty materials as well as clean fluids.

Figure 1-32 illustrates how the force on the target is converted into a pneumatic signal which is proportional to the square of the flow rate. An electronic signal is produced in somewhat the same way.

The force on a disc or "target," which is mounted in the line of the flowing fluid, developed through the annular orifice, may be measured directly. said this differential pressure is proportional to the square of the flow velocity and, hence, flow rate.

The following basic equation or relationship determines transmitter capabilities:

$$Q = K\sqrt{F}$$

where $Q =$ flow
              $K =$ target coefficient
              $F =$ force, typically 2-16 pounds

Vane Type. This type, shown in Figure 1-33, employs a swinging vane or gate which causes a material obstruction in the stream and, consequently, induces a loss of head across the primary device. In this respect, the flowmeter is somewhat similar to an area type flowmeter. The vane can be loaded by gravity or by an adjustable spring.

VELOCITY METERING SYSTEMS

Velocity metering systems are characterized by having a rotating primary device which is kept in motion by the direct movement (or velocity) of the fluid stream, or by having stationary primary devices.

Rotating Devices

Propeller Type. The heavy-duty propeller type velocity flowmeter shown in Figure 1-34 has a turbine or propeller as the primary device. The motion of the stream continuously rotates the propeller which is connected by gearing to a secondary device. Sometimes called a *register*, the secondary device is calibrated to read in total flow quantity. The flowmeter can also transmit readings by electrical means to a remote receiver for indicating, recording, or control purposes.

Propeller type flowmeters have almost no practical upper limit to capacity, cause very little pressure loss, and can be used for liquids containing abrasive materials. Flowmeters can also be arranged to subtract or reverse flows, a feature not usually found in other flowmeters.

Turbine Type. Used almost exclusively for liquid measurements, the turbine flowmeter has a bladed rotor mounted in a special passage through which the fluid stream is directed. Fluid movement against the blades rotates the rotor. There are three general designs of rotors based on the fluid flow through the rotor blades: (1) radial—fluid flows at right angle to the rotor axis; (2) helical—fluid flows axially or in parallel to the rotor axis; (3) mixed—fluid flow is partly radial and partly helical.

A typical turbine flowmeter of the helical design, basic type, consists of a section of pipe, a multibladed rotor mounted in the center of the straight-through passage, and a pick-off coil containing a permanent magnet mounted externally to the fluid passage (see Figure 1-35). A cantilever shaft held in place by fixed radial vanes supports the rotor assembly. As flow spins the rotor, movement of each rotor blade past the face of the pick-off coil changes the total flux through the coil and induces a pulse. The frequency of pulses, which is proportional to flow rate, is linear with flow and can be used directly in digital totalizing or counting. Flowmeter output can be converted to an electrical signal for remote indicating, recording, or control.

Turbine flowmeters are adaptable to a variety of fluids. However, precautions must be taken to prevent damage from entrained solids.

There are other types of velocity metering systems such as anemometers but they are not found in general use in the paper industry.

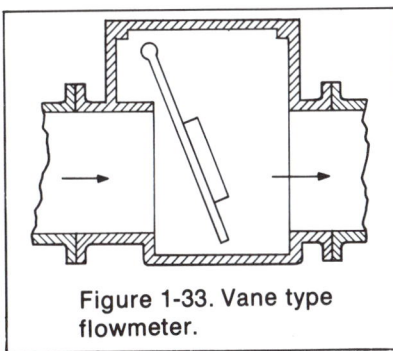

Figure 1-33. Vane type flowmeter.

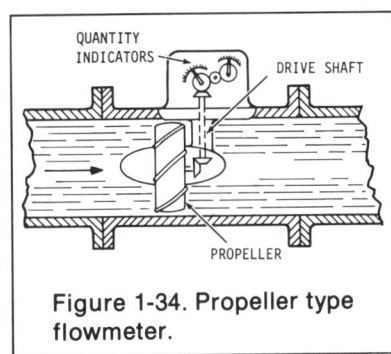

Figure 1-34. Propeller type flowmeter.

Flow Measurement   43

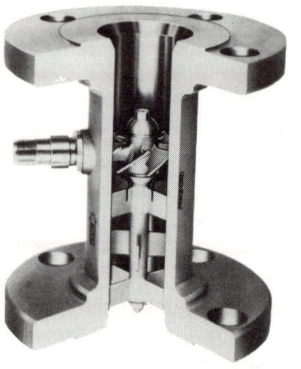

Figure 1-35. Turbine type flowmeter.

Stationary Element

Electromagnetic Type. Although originally conceived by Faraday in 1831, it was about 1956 that the practical application of the electromagnetic type primary flow device was developed in the United States.

Since then it has become widely accepted throughout industry as one of the most efficient means to measure flows of slurries of a dirty, viscous, or sludgy nature. Specifically, it has generally become the standard method of accurately measuring the flow of fibrous pulp and stock, and other fluids which are difficult to measure in the pulp and paper industry, because it does not require a restriction in the pipeline for creating a measurement signal. This permits an obstructionless flow in the pipeline. Corrosion resistance is provided by the use of special body materials and liners in the flowtube.

The magnetic flow metering system, despite its title, is a relatively simple tool. It consists of a primary device or flowtube which is connected by electric cable to a secondary instrument for indicating, recording, controlling, or converting to other compatible signals.

It is based on Faraday's law of electromagnetic induction which states that the voltage induced in a conductor of fixed length moving through a magnetic field is proportional to the velocity of the conductor. This same principle makes possible the operation of power generators, tachometer generators, and similar electrical equipment. Figure 1-36 illustrates the principle for a conductive wire and compares it with a disc or a thin section of conductive liquid across the pipe which is the conductor of the magnetic flowmeter. A voltage generated by the movement of the disc through the magnetic field in the pipe is sensed by two point type electrodes located diametrically opposite and flush with the inside of the pipe, as in Figure 1-37.

Electromotive force measured is the total voltage generated by all discs as they pass by the electrodes. The faster the discs move through the magnetic field, the greater the voltage generated. The direct, linear measurement of liquid flow provided can be found on a millivoltmeter type secondary unit.

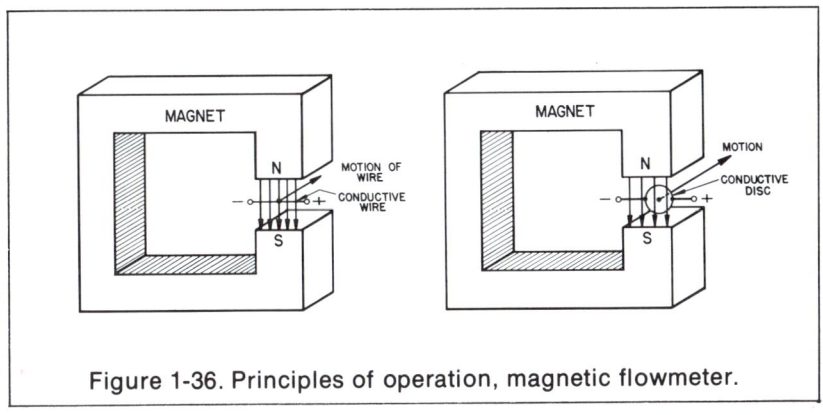

Figure 1-36. Principles of operation, magnetic flowmeter.

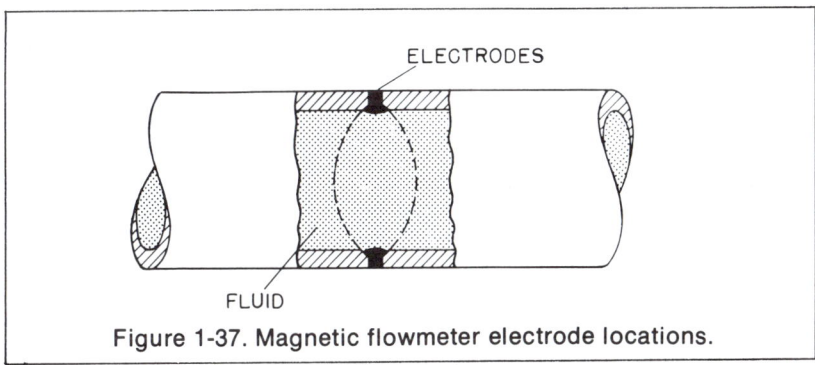

Figure 1-37. Magnetic flowmeter electrode locations.

Conductor length is equal to the diameter of the disc which is the internal diameter of the tube. With the magnetic field remaining effectively constant, actual tests and the application of Faraday's law of electromagnetic induction prove that the voltage generated is proportional to the average velocity which, in turn, is directly proportional to the volume rate of flow. Stated mathematically:

$$E = K\ VHd/A$$

where E = generated voltage measured at the electrodes
           K = a constant for units of measurement
           V = volume of liquid per unit of time
           H = field strength between electrodes
           d = distance between electrodes
           A = cross-sectional area inside pipe

For all practical considerations, K, H, d, and A remain constant and, therefore, E varies directly as V changes.

A typical secondary instrument used with an electromagnetic primary device is shown in Figure 1-38.

Flow Measurement   45

Electromagnetic primary devices have been manufactured in many sizes ranging from 0.1 inch in diameter for use on dyes, additives, and chemicals in stock preparations and stock blending in the paper industry to over six feet in diameter to measure flow of effluent sewage in the water and waste industry. There is no theoretical limit to flowmeter size; larger ones are manufactured on site.

Table 1-D lists some of the important features and advantages of this flow measuring system in such pulp and paper process areas as digesters, washing, bleaching, stock preparation, paper machines, evaporators, recovery furnaces, and recausticizing.

Ultrasonic Type. Another flow measuring device that has no moving parts is the ultrasonic flowmeter. There are several techniques in applying this principle to measure flow. However, they all depend on the time difference between upstream and downstream sound wave travel in a moving fluid. The velocity of the flowing process stream is detected by the following relationship:

$$V = \left(\frac{c^2 \tan \theta}{2D} \right) \Delta t$$

where V = velocity of flowing stream
   c = velocity of sound in process fluid
   θ = angle of sonic beam
   D = inside pipe diameter
   Δt = difference between upstream and downstream transit times

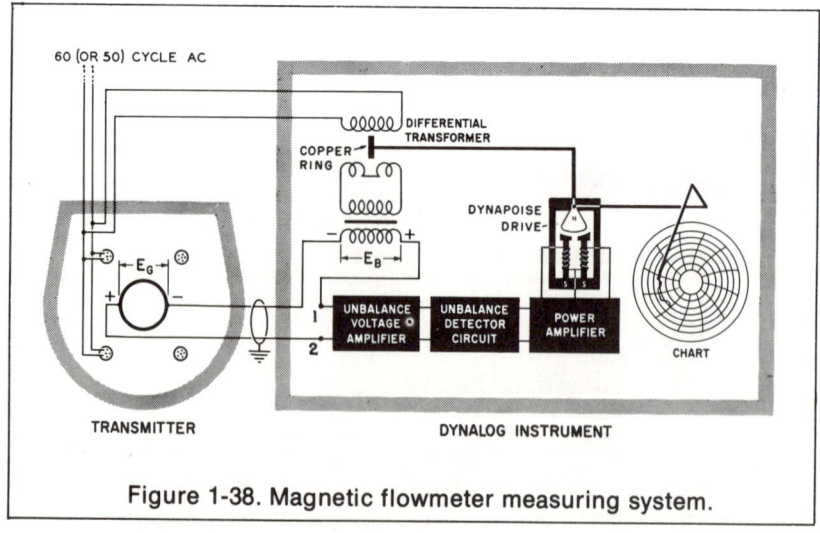

Figure 1-38. Magnetic flowmeter measuring system.

TABLE 1-D

FEATURES AND ADVANTAGES OF MAGNETIC FLOWMETERING

1. Freedom from restriction in the line—no plugging or dewatering.

2. No moving parts and adjustments in the primary device.

3. No connecting lines, therefore no contamination.

4. No straight runs of pipe required up or downstream.

5. No correction needed for viscosity. Will measure any consistency.

6. No reasonable limit to operating pressure.

7. No correction needed for conductivity changes.

8. Field calibration of primary device unnecessary.

9. Range can be changed without changing flowtube.

10. Measures flow in either direction.

11. Measures different types of liquid without recalibration.

12. Can be mounted in any position.

13. Low installation cost.

14. Corrosion-resistant, due to variety of liners that can be used.

The ultrasonic flowmeter is applicable to any liquid that gives a stable flow profile that can propagate sound waves in an undistorted manner.

In a typical unit, shown in Figure 1-39, a pair of transducers, which are devices that receive and emit sound waves, are mounted in a pipeline opposite one another at a 45° angle to the direction of the flow. Sound bursts are propagated alternately in opposite directions between the transducers. Because the upstream signal is delayed and the downstream signal is speeded up by the moving fluid, the alternate bursts yield a frequency difference, thereby canceling out the speed of sound. This difference is then electronically translated into a measurement of flow.

As in the electromagnetic type, nothing extends into the pipe to obstruct the flow of the liquid. The transducer assembly presents minimum contact area with the process liquid, and measurement is essentially independent of fluid temperatures, density, viscosity, and pressure.

Unlike the electromagnetic type flowmeter, the ultrasonic type can be used on nonconductive liquids, of which very few are found in a pulp and paper mill.

The signal output from the ultrasonic flowmeter can be fed directly to indicating, recording, controlling, and computing equipment to obtain dimensions of flow such as gallons or cubic feet per minute, and also for control purposes.

Vortex Flowmeter. These flow sensing devices are used to measure the flow of gases, liquids, and slurries, and utilize the phenomenon that a strong, regular series of vortices occurs when these materials flow past a

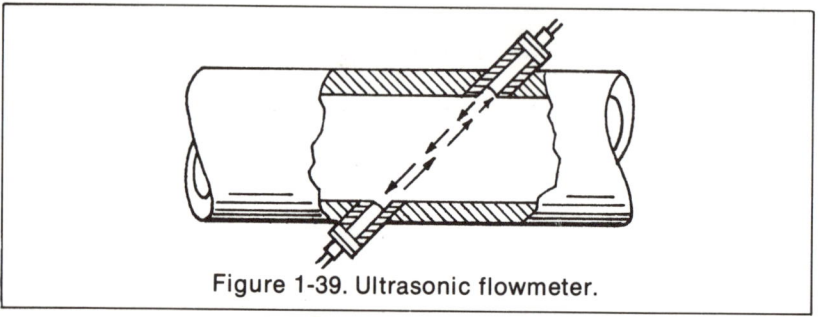

Figure 1-39. Ultrasonic flowmeter.

specially shaped object mounted across a pipeline.

Vortex shedding is a natural phenomenon which can occur when a fluid is made to flow past a bluff or non-streamlined body. If this body is cylindrical in shape, as shown in Figure 1-40, the flow does not follow the shape of the cylinder on the downstream side but separates from the cylinder surface causing eddies to form. These eddies or vortices grow in size until they become too large to remain attached to the cylinder. They then break away and are shed downstream at a frequency determined by the flow rate. The growth-shed cycle occurs alternately on either side of the cylinder in a periodic fashion so that the downstream flow pattern is a staggered arrangement or trail of vortices. This is the phenomenon that causes a flag to flutter in the breeze behind a flagpole, the "singing" of telephone wires in the wind, and the whistle of wind through tall grass.

Therefore, above a certain flow rate, the number of vortices shed depends directly on the volumetric flow rate, and a count of the vortices shed establishes the total flow. The flow measuring devices are designed to count the vortices as they are formed and create an electrical pulse output whose frequency is proportional to the flow rate. They differ primarily in the shape of the bluff body used to create the vortices and the techniques used to detect their frequency.

A typical vortex shedding flowmeter is shown in Figure 1-41. The shape of the bluff body in this example is a modified delta with its base facing upstream as shown in Figure 1-42.

The ratio of frequency to flow rate is governed by the width across the bluff body's face and the inside diameter of the process pipe. Shedding from the delta-shaped body varies the direction at which the oncoming flow

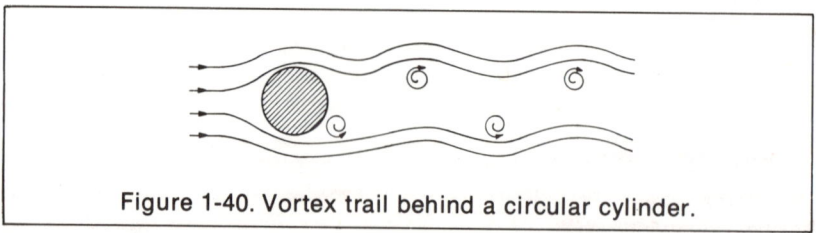

Figure 1-40. Vortex trail behind a circular cylinder.

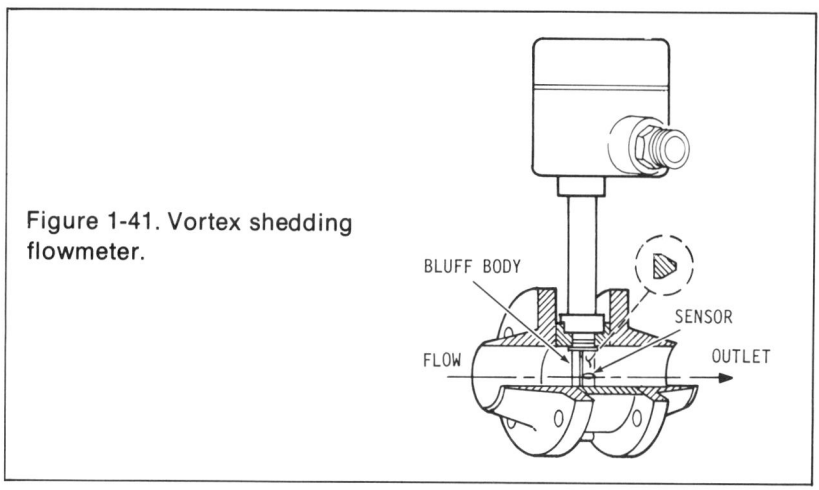

Figure 1-41. Vortex shedding flowmeter.

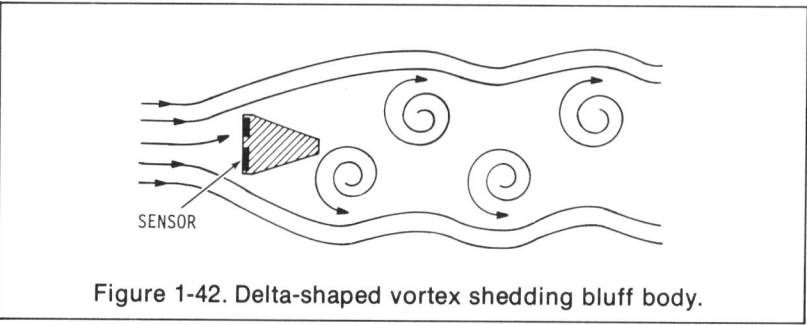

Figure 1-42. Delta-shaped vortex shedding bluff body.

stream hits the front face. Thus the sensor which detects the shedding vortices and converts them into electrical pulses is put on the front face. The sensor consists of two electronically self-heated resistance elements whose temperatures vary as a result of the velocity variations on the front face. This results in a corresponding change in the resistance of the elements. The changes in resistances are conditioned and converted to electrical pulses or into a continuous analog signal by an electronic system shown schematically in Figure 1-43. These signals are then used by indicators, recorders, and controllers for display and control purposes.

Figure 1-44 illustrates how a compound body design is used in the bluff-body shape. This is another typical vortex shedding flowmeter. The vortex shedding primary device is contoured so that the tail section is located in a region where vortices created by the first body section are forming and directs part of their energy across the gap between the body sections. The alternating fluid forces act on the vortex shedding body's tail section.

A strain gauge is bonded within the vortex shedding body to sense these alternating forces. Its signal is electronically conditioned and transmitted to suitable indicating, recording, and control devices.

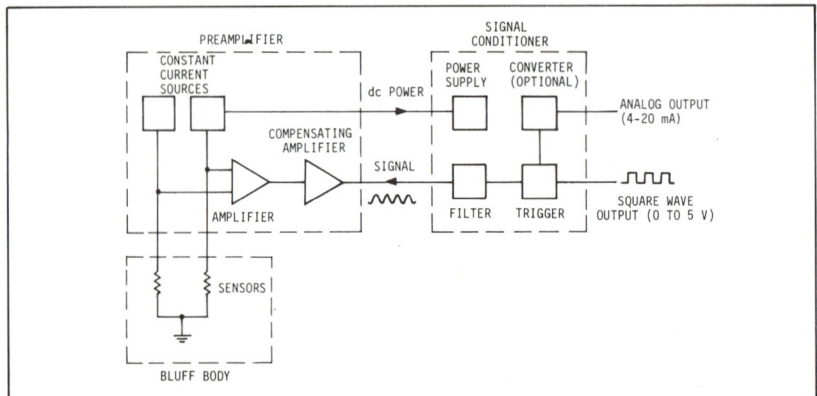

Figure 1-43. Diagram of electronic system for vortex shedding flowmeter.

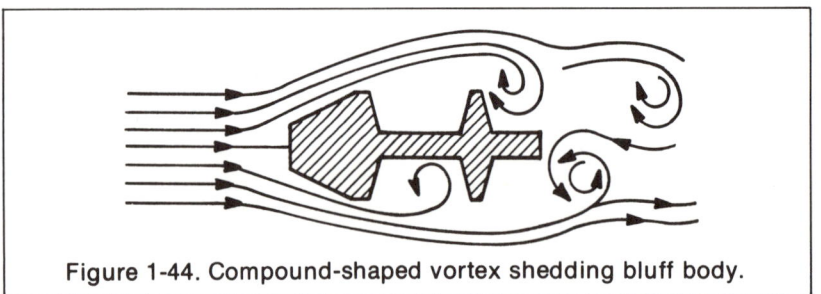

Figure 1-44. Compound-shaped vortex shedding bluff body.

A third design used in shaping the bluff body in a vortex shedding flowmeter is shown in Figure 1-45. In this configuration an alternating pressure difference between the two sides of the element is produced. Communication between the two sides is established by a diaphragm-sealed, fluid-filled passageway which encapsulates a strain detector. The piezoelectric detector reacts to the alternating pressure difference on the capsule caused by each alternating pair of shedding vortices. The rate of these voltage pulses is directly proportional and linear with respect to volumetric flow rate over a designated flow range. The signal from the strain detector is conditioned and transmitted to appropriate indicating, recording, and controlling devices for display and control purposes.

QUANTITY MEASURING SYSTEMS

Quantity flowmeters measure fluid in separate and distinct increments by filling and emptying containers of known weight or volumetric capacity. The quantity of flow is determined easily by totalizing the number of increments. The secondary device of a quantity flowmetering system includes a counter with suitably graduated dials for registering the total quantity that

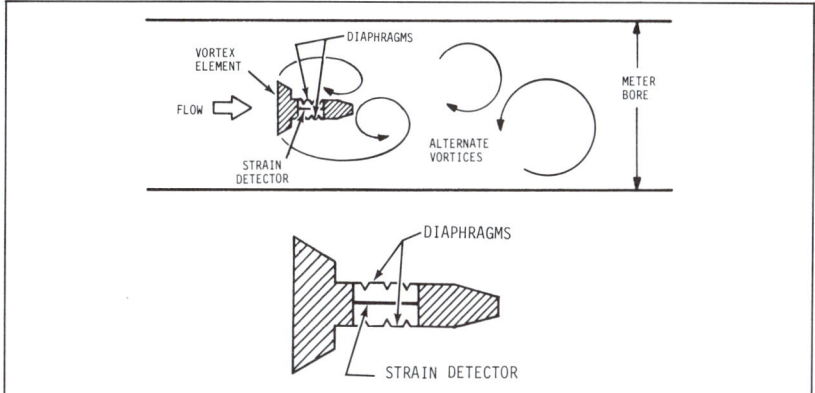

Figure 1-45. Vortex shedding bluff body design using the diaphragm/strain detector shown in detail in separate drawing.

has passed through the flowmeter. This reading can be conditioned to transmit signals to remote locations for indicating, recording, and controlling purposes.

There are two basic kinds of quantity flowmeters: (1) weighing types, those that weigh discrete increments of fluid; (2) volumetric types, those that transfer discrete increments of fluid.

Weighing Type Flowmeters

There are two general kinds of weighers, neither of which is used commonly in the pulp and paper industry. The first depends on the principle of upsetting the equilibrium of a container by a rise of the center of gravity as the container is filled. The second employs a container suspended from a counterbalanced seal beam. A trap or bucket is automatically upset when the center of gravity passes over a stable equilibrium point. Use of two tilting traps or temporary diversion of liquid into a holding chamber during the dumping operation achieve continuous flow.

Volumetric Flowmeters

Commonly referred to as positive displacement flowmeters, the volumetric group of quantity flowmeters separates fluid into specific portions. Various types are designed for metering liquids and gases.

Typical flowmeters designed for liquids may be summarized as follows:

1. Reciprocating piston type flowmeters, shown in Figure 1-46, use one or more members which have a reciprocating motion and are located in one or more fixed chambers. Adjustment of quantity per cycle can be effected by varying the magnitude of movement of one or more of the reciprocating members or by varying the relationship between the

Flow Measurement   51

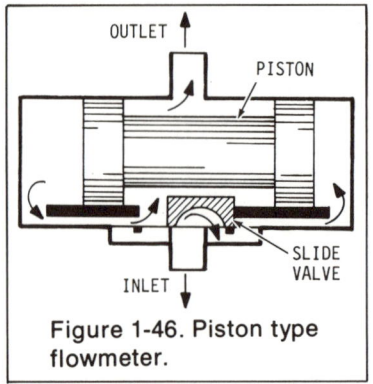

Figure 1-46. Piston type flowmeter.

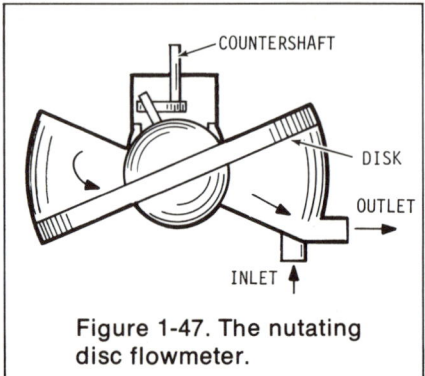

Figure 1-47. The nutating disc flowmeter.

primary and secondary readout devices. These so-called piston flowmeters are limited to use with noncorrosive and light (low viscosity) liquids in flow rates ranging from 10-1000 gallons per minute.

2. Nutating disc type flowmeters, shown in Figure 1-47, are widely used for measuring flow quantities of liquids; a common model is the domestic water meter. A circular disc, pivoted at its geometric center, nutates in a circular chamber with a conical top and bottom. The vertical shaft of the disc generates a cone with the apex downward. Liquid enters the chamber where differential pressure causes a quantity of fluid to roll around to an outlet. Inlet and outlet chambers are separated by a partition which fits into a slot in the disc. The flowmeter measures flow rates from about 15-500 gallons per minute.

3. Rotary-vane chamber type flowmeters, shown in Figure 1-48, use a drum that rotates about its own center which is eccentric to the meter body. Differential pressure across the flowmeter rotates a drum. Vanes are pushed outwards by springs to form separate sealed chambers. A register on the drum shaft indicates total flow quantity. Rotary bucket and rotary piston flowmeters operate on similar principles.

Typical volumetric flowmeters designed for metering gases may be listed as follows:

1. Sealed-drum type flowmeters, shown in Figure 1-49, consist of a motor with spiral-like vanes contained in a horizontal cylindrical drum. The meter is filled with water to a height slightly over the center hub. The gas enters through the rotor hub and passes through the water to the chamber above which is formed by adjacent spiral vanes. When a vane emerges from the water, the gas is released to the outlet.

2. Lobed-impeller types as shown in Figure 1-50 have two rotors, each having lobes arranged to mesh very much like gears. The impellers rotate in close-fitting chambers. Fluid trapped in the space between the lobes is

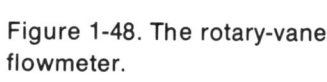

Figure 1-48. The rotary-vane flowmeter.

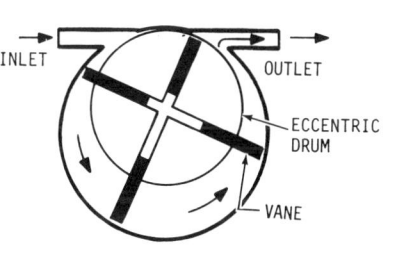

Figure 1-49. Sealed-drum flowmeter for gas.

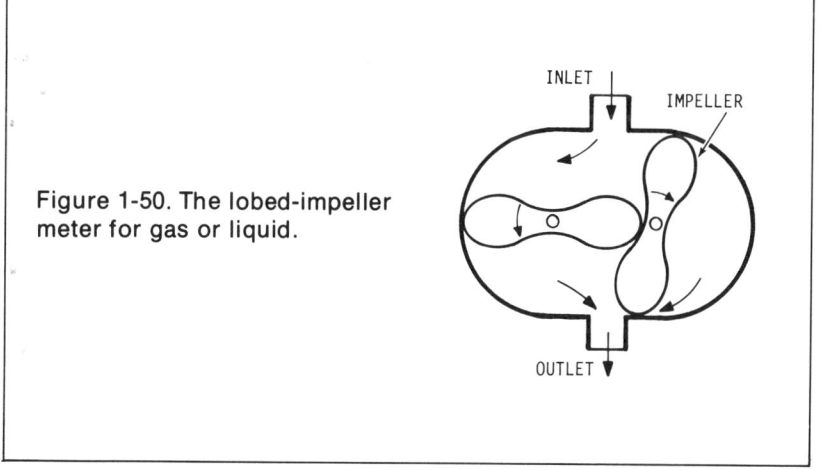

Figure 1-50. The lobed-impeller meter for gas or liquid.

passed from inlet to outlet. These meters can also be used on liquids.

Bellows flowmeters and divided bell flowmeters are two other types especially designed for gas flow measurements.

SOLIDS METERING SYSTEMS

Continuous determination of the flow rate of solid materials such as wood chips in wood preparation; bark and coal in the power boiler area; dry broke, pulp, and powdered chemicals in the stock preparation area are necessary primarily for inventory and control processes in the pulp and paper industry.

Measurement of solid materials is somewhat more direct than measurement of gas and liquid flows. However, they fall under the same general classifications: volume and weight. Screw bucket conveyors operating at constant speed furnish flow in terms of volume. Conveyors measure weight.

Weighing Type Flowmeters

Flowmeters for weighing flow of dry materials determine weight of a material passing over a given point. Weighing flowmeters are designed for particular kinds of material flowing in pipes or on open conveyors. Flowmeters similar to the propeller and tilting flowmeters for liquids are sometimes used to measure flow of dry powdered or granular material. However they are not commonly found in the paper industry.

Conveyor Flow

Conveyor flow devices actually weigh a section of a conveyor belt. Essential components of these flow measurement systems are: (1) suspension or load receiving element, which is considered the primary device; (2) sensor or measuring and transmitting apparatus to couple the primary device to (3) the secondary device, which is an indicator, recorder, or controller.

In a typical system shown in Figure 1-51, flow is measured in mass units or tons per hour by a spring balance. Integrators determine total flow; suitable compensation is made for conveyor belt speed.

Other conveyor systems use automatic electric or pneumatic scales instead of the spring balance.

Another conveyor system introduced in recent years utilizes the absorption of radiation to measure belt loading. A detector on one side of the belt converts the radioactive energy, such as gamma rays which pass through the belt and material, into an electric current. The amount of radioactivity detected is inversely proportional to the density of the material being measured plus a section of the belt. By linking the speed of the belt to the measuring system, mass per unit length of belt can be measured. Figure 1-52 shows a diagram of the arrangement used on such a measuring system.

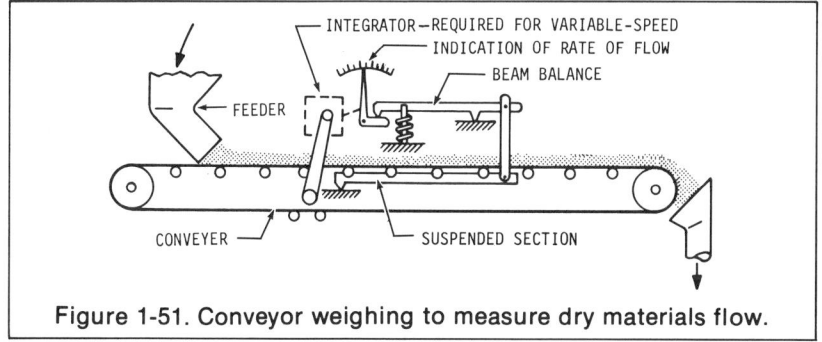

Figure 1-51. Conveyor weighing to measure dry materials flow.

MISCELLANEOUS METERING SYSTEMS

Many other devices using various design concepts and physical phenomena have been developed for making flow measurements. These include linear resistance or friction types of instruments using capillary tubes and porous plugs, thermal devices, sound velocity measurement devices, and magnetic resonance. These are primarily highly specialized devices and have not achieved generally wide acceptance in industry.

SELECTING FLOWMETERS

Accurate flow measurement depends greatly on the careful matching of flowmeters to applications and operating conditions. Relative capabilities, dependability, and reproducibility of the measurement signal needed in different phases of the operations must be balanced against cost.

Table 1-E contains guidelines which help select a primary device and/or secondary transmitter for flow measurement. Ratings in most cases are relative. The listing, though not complete, covers general flow transmitter applications commonly used in the pulp and paper industry.

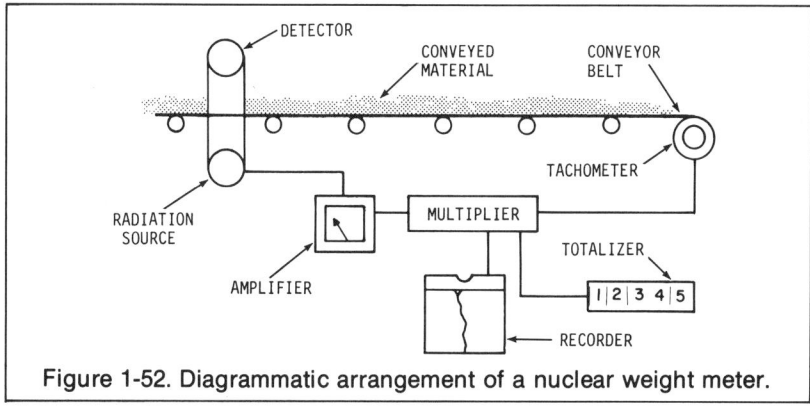

Figure 1-52. Diagrammatic arrangement of a nuclear weight meter.

TABLE 1-E

PRIMARY FLOW MEASURING INSTRUMENTS - SELECTOR GUIDE

	ORIFICE PLATES	INTEGRAL ORIFICE	VENTURI TUBES	FLOW NOZZLES	PITOT TUBES	ELBOW TAPS	TARGET	ELECTROMAGNETIC[1]	TURBINE	POSITIVE DISPLACEMENT	VARIABLE AREA	WEIRS AND FLUMES[2]
Minimum Permanent Pressure Loss	G	A	Be	G	--	--	G	Bt	A	A	G	--
For Pulp Stock Flow	N	N	N	A	N	N	N	Bt	N	A	A	A
For Slurry Flow	N	N	N	G	N	G	Be	Bt	N	A	A	A
For Liquids Containing Suspended Materials & Sedmts.	N	N	N	G	N	G	Be	Bt	N	A	A	A
Dirty Service	N	N	Be	G	N	Be	Bt	Bt	A	A	A	A
Liquids Containing Entrained Vapors	G[3]	U	Bt	G	A	A	Bt	Bt	G	N	G	G
Vapors Containing Liquids	G[4]	U	Bt	G	N	A	Bt	--	U	N	G	--
Viscous Flows	A	G	A	A	G[5]	U	Be	Bt	A	Bt	G	G
Flows with Changing Viscosities	A	A	A	A	N	A	A	Bt	A	A	Be	A
Low Flow Rates	N	Bt	N	N	N	N	N	Bt	Bt	Be	Be	N
Ease of Changing Capacity	Bt	Bt	Be	Be	A	A	Be	Be[6]	Be[6]	Be[6]	A[6]	A

TABLE 1-E (Continued)

	ORIFICE PLATES	INTEGRAL ORIFICE	VENTURI TUBES	FLOW NOZZLES	PITOT TUBES	ELBOW TAPS	TARGET	ELECTROMAGNETIC[1]	TURBINE	POSITIVE DISPLACEMENT	VARIABLE AREA	WEIRS AND FLUMES[2]
Ease of Installation	Be	Bt	Be	Be	Be	Bt	Bt[7]	Bt[8]	Bt[7]	Be[8]	Bt	G
Ease of Maintenance (Cleanability)	A	A	Be	Be	A	Be	Be	Bt	G	G	Be	Be

RATINGS CODE:

A	= Adequate		U	= Unknown
G	= Good		L	= Lower Initial Cost
Be	= Better		M	= Medium Initial Cost
Bt	= Best		H	= Higher Initial Cost
N	= Not Recommended			

NOTES:

1. For use with conductive fluids.
2. For measurement in open channels.
3. In vertical line if flow is upward.
4. In vertical line if flow is downward.
5. For measuring velocity at one point. For measuring total flow accuracy depends on velocity traverse of pipe.
6. Larger size required if new maximum flow exceeds maximum rating of installed meter.
7. Reducers and a section of process pipe the same size as the meter required.
8. Reducers required if meter size is different from pipe size.

BIBLIOGRAPHY

ASME. *Fluid Meters—Their Theory and Application*. 5th ed. 1959.

Behar, M. F. *The Handbook of Measurement and Control*. Pittsburgh: The Instruments Publishing Company, 1951.

Benedict, R. *Fundamentals of Temperature, Pressure, and Flow Measurements*. New York: John Wiley & Sons, 1975.

Cusick, C. J. *Flow Meter Engineering Handbook*, 4th ed. Ft. Washington, PA: Honeywell, Inc., Industrial Division, 1968.

Eckman, D. P. *Industrial Instrumentation*. New York: John Wiley & Sons, 1950.

Foxboro Magnetic Flowmeter. Foxboro, MA: The Foxboro Company, TI 27-71g.

Soisson, H.E. *"Instrumentation in Industry*. New York: John Wiley & Sons, 1975.

Spink, L. K. *Principles and Practices of Flow Meter Engineering*. 9th ed. Foxboro, MA: The Foxboro Company, 1967.

Zoos, A., and Delahooke, C. *Industrial Process Control*. Albany, NY: Delmar Publishers, Inc., 1961.

Select answer or answers which correctly apply to the following statements:

1. The proportion of process measurements represented by flow in the pulp and paper industry is
   a) 1/4.
   b) 1/2.
   c) over 1/2.

2. All flow measurement systems consist of a primary element and
   a) recording device.
   b) secondary device.
   c) transmitting device.

3. Flow measurements are divided into two major categories, flow rate measurement systems and
   a) head metering systems.
   b) area metering systems.
   c) total quantity measurement systems.

4. Head flowmeters
   a) measure flow directly.
   b) relate flow to differential pressure.
   c) measure total pressure.

5. The most frequently used primary element in the paper industry for the measurement of fresh water, steam, clean chemicals, and uncontaminated gases is
   a) the Venturi tube.
   b) the orifice plate.
   c) the flow nozzle.

6. Secondary elements used in differential pressure or head type measuring systems may be divided into two general groups: wet flowmeters and
   a) dry flowmeters.
   b) mercury type flowmeters.
   c) bellows type flowmeters.

7. Liquid seal type secondary elements of the wet flowmeter class are applied on services which involve
   a) high differential pressures.
   b) low differential pressures.
   c) high static pressures.

8. Bellows or diaphragm flowmeters are generally used where
   a) indirect-operated indication of differential pressure is required.
   b) mercury or other sealing fluids are harmful to the process.
   c) fluid to be measured contains suspended material.

9. The most common methods of transmission for force balance type flow measurement systems are pneumatic and
   a) electric.

b) mechanical.
c) hydraulic.

10. Integral orifice transmitters are used to measure very small flow of fluids which

a) contain suspended materials.
b) contain entrained air.
c) are reasonably clean.

Indicate whether the following statements are True or False by inserting T or F in parentheses:

11. In area type meters the aperture is constant and the pressure drop is variable. (   )

12. Area flowmeters are used to measure flow of corrosive and viscous tar-like liquids. (   )

13. In head area metering systems, the area of the stream and head remain constant. (   )

14. The weir is essentially a dam with a notched opening in the top which takes several shapes. (   )

15. The five basic weir shapes are the V-notch, rectangle, trapezoid, circular and square. (   )

16. The height of the sheet of liquid upstream of the weir is called the nappe. (   )

17. The traditional secondary device used with a head area metering system is the float and cable. (   )

18. Air purged secondary elements do not require a float and measure head directly. (   )

19. Velocity metering systems are characterized by spinning floats. (   )

20. Turbine flowmeters are designed especially for use on fluids with entrained solids. (   )

Select answer or answers which correctly apply to the following statements:

21. The operation of a magnetic flowmetering system is based on Faraday's law of electromagnetic induction, which states that voltage induced in a conductor moving through a magnetic field is primarily proportional to

a) material of the conductor.
b) diameter of the conductor.
c) velocity of the conductor.

22. This same principle makes possible the operation of
a) power generators.
b) steam engines.
c) centrifugal pumps.

23. The direct, linear measurement of liquid flow provided by magnetic flow-meters can be measured on
a) differential pressure type secondary instruments.
b) current type secondary instruments.
c) millivoltmeter type secondary instruments.

24. In the formula $E = KVHd/A$, E is the
a) generated voltage measured at the electrodes.
b) field strength between electrodes.
c) electromagnetic force.

25. Magnetic flowmeters are commonly used in the paper industry for meas-uring such flows as
a) steam.
b) gases.
c) pulp slurries, dyes, additives, and chemicals.

26. There are two basic kinds of quantity metering systems, the weighing type and the
a) gravimetric type.
b) volumetric type.
c) hydraulic type.

27. Volumetric type flowmeters are commonly referred to as
a) positive displacement flowmeters.
b) rotameters.
c) turbine flowmeters.

28. Nutating disc meters are widely used for measuring flow quantities of
a) gas.
b) solids.
c) liquids.

29. Lobed-impeller type flowmeters are used to measure
a) gas.
b) solids.
c) liquids.

30. Conveyor type flow measurements are made in units of
a) volume.
b) weight.
c) pressure.

(See Appendix for answers)

2. Pressure Measurement

The measurement of pressure is nearly as important as the measurement of flow in the manufacture of pulp and paper. It might even be considered the basic process variable in that it is utilized for measurement of flow (difference of two pressures), level (liquid pressure or back pressure from a bubble tube), and temperature (fluid pressure in a filled thermal system).

Pressure measurement can be found in practically every area of pulp and paper manufacture, from wood preparation to paper making and finishing (Table 2-A).

BASIC THEORY OF PRESSURE

Pressure is the action of one force against another force and is used here to mean thrust either distributed over a surface or acting against a given surface within a closed container. Usually it is measured as force per unit area. In the English system force is measured in pounds and a common unit of pressure is pounds per square inch (abbreviated psi). Pressure, particularly lower pressure, is frequently expressed as height of liquid supported in a column. Therefore, a pressure may be stated in inches or feet of water, or centimeters or inches of mercury. These are not actual pressure units but represent the pressure necessary to support the liquid column at that height. Since one inch of water represents 0.036 pounds per square inch, it is more convenient to use inches of water for low pressures.

Other common units are atmospheres, bars, and millibars. The pressure of the atmosphere at sea level under standard conditions is 14.696 psi absolute, often rounded to 14.7 psi absolute. In the older metric system the unit of force was the dyne and pressures were sometimes expressed in dynes per square centimeter. Since this was a very small unit, the bar was created, equaling one million dynes per square centimeter. Another common unit in the older system is the kilogram-force per square centimeter which equals 14.223 psi.

In modern practice in the International System of Units (Systems Internationale, commonly SI) now used throughout most of the world, the unit of force is the newton, and the unit of pressure is the newton per square meter. It has been given the special name of pascal (Pa). A major advantage of SI units is that all values, small or large, can be expressed with the same unit by giving it standardized prefixes. Thus, the kilopascal, 1000 pascals, covers most values

TABLE 2-A

TYPICAL PRESSURE MEASUREMENTS
PULP AND PAPER INDUSTRY

A. Wood preparation
  1. Water
  2. Steam

B. Chip preparation
  1. Steam

C. Digesters
  1. Steam
  2. Cooking chemicals
  3. Digester cooking
  4. Digester relief

D. Washers
  1. Water
  2. Steam

E. Pulp mill chemical
    preparation
  1. Water
  2. Liquid chemical
     lines
  3. Steam

F. Bleaching
  1. Water
  2. Steam
  3. Bleaching
     chemicals

G. Paper mill chemical
    preparation
  1. Water
  2. Liquid chemical
     lines

  3. Dyes and additive
     lines
  4. Steam

H. Stock preparation
  1. Water
  2. Steam
  3. Refiners
  4. Stock lines and
     headers

I. Stock blending
  1. Stock lines and
     headers

J. Paper machine
  1. Stock lines and
     headers
  2. Water
  3. Steam
  4. Liquid chemical
     lines
  5. Dyer and additive
     lines
  6. Dryers

K. Evaporators
  1. Liquor lines
  2. Steam
  3. Evaporator
     effects
  4. Condensate line

L. Recovery boiler
  1. Condensate line
  2. Water
  3. Steam
  4. Liquor

M. Power boiler
  1. Condensate
  2. Water
  3. Steam
  4. Fuel

N. Recausticizing
  1. Water
  2. Steam
  3. Green liquor line
  4. White liquor line

O. Lime kiln
  1. Oil
  2. Gas
  3. Steam

P. Water treatment
  1. Water
  2. Liquid chemical
     lines

Q. Waste treatment
  1. Effluent lines
  2. Liquid chemical
     lines

of pressure used in industry, those now expressed in pounds per square inch and in inches of water as well as the many other units used in various applications. One pound per square inch equals 6.895 kilopascals (kPa). One inch of water equals 0.25 kilopascal. High pressure such as pressure ratings of valves, vessels, etc., are commonly expressed in megapascals (MPa), one million pascals or one thousand kilopascals. This avoids the use of many zeros and makes the precision of the value more evident. For instance, 5000 psi becomes 34 MPa, and the question as to whether the rating is ±1000 psi or ±1 psi is avoided since ±1 MPa equals ±150 psi unambiguously.

Conversion can be made by actual calculations or by referring to a conver-

TABLE 2-B
MEASURABLE PRESSURES

Absolute Pressure: Measured above perfect vacuum or zero absolute. Zero absolute represents total lack of pressure. (*A-A*)

Atmospheric Pressure: Exerted by earth's atmosphere. Also referred to as barometric pressure. Atmospheric pressure at sea level is 14.7 psia, or 29.9 inches of mercury absolute. Value decreases with increasing altitude.

Gauge Pressure: Pressure above atmospheric. Represents positive difference between measured pressure and existing atmospheric pressure. Convert to absolute by adding actual atmospheric pressure value. (*B-B'*)

Differential Pressure: Difference between two pressures can be measured with an instrument with a zero difference point at midrange to permit measurement in either direction. (*C-C'*)

Vacuum: Pressure below atmospheric. (*D-D'*)

Static Pressure: Force per unit area exerted on a wall by a fluid at rest or flowing parallel to a pipe wall. Sometimes called line pressure.

Total Pressure: All pressures, including static, acting in all directions.

Velocity Pressure: Pressure exerted by the speed of flow. Also called velocity head or impact pressure.

Hydrostatic Pressure: Pressure below a liquid surface exerted by the liquid above.

Total Pounds or Tons: Expression used with hydraulic machinery to indicate force over a given area.

---

sion table found in practically all technical reference books. (See Appendix D.) Inexpensive slide rules are also becoming more generally available, particularly for conversions between SI and common units. The different types of measureable pressures are listed in Table 2-B and shown in Figure 2-1.

PRESSURE MEASUREMENT SYSTEMS

Like flow measurement systems, all pressure measurement systems consist of two basic parts. A primary device is in direct or indirect contact with the pressure medium and interacts with changes in pressure. The secondary device translates the interaction into appropriate pressure values for use in indicating, recording, or controlling.

Pressure systems or instruments can be divided into two major categories: (1) those that employ mechanical means to detect and communicate pressure information from the process and secondary device and (2) those that rely on electrical phenomenon or relationship to carry out this function.

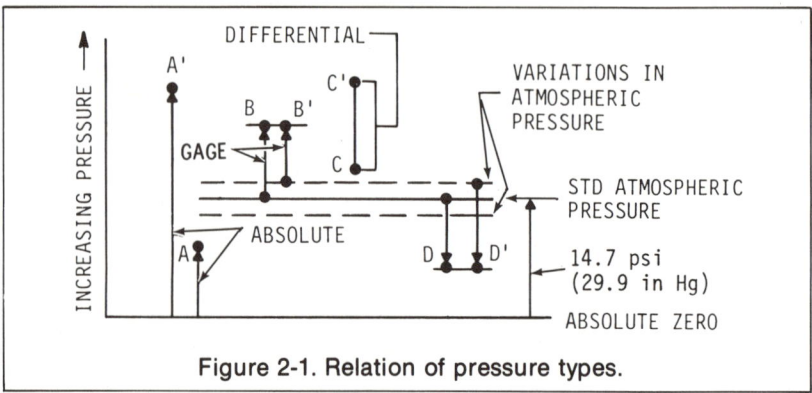

Figure 2-1. Relation of pressure types.

MECHANICAL PRESSURE MEASUREMENT SYSTEMS

Pressure is determined in mechanical systems by balancing a sensor against the unknown force. This can be done by another pressure or force. The two most common pressure-balance sensors are the manometer and the weight-loaded piston or deadweight tester.

The most common force-balance sensors are:

1. Those that require elastic deformation. They are bellows, diaphragm, and Bourdon metallic devices.
2. Those that do not depend on elastic deformation for operation. This group includes the bell and limp diaphragm devices. Table 2-C shows a classification of mechanical pressure measurement systems.

Pressure-Balance Systems

Generally, pressure balances are used for the measurement of low-pressure ranges in liquids where the pressure range may be from a few inches of water to several pounds per square inch.

TABLE 2-C

MECHANICAL PRESSURE
MEASURING SYSTEMS

I. Pressure balanced	1. Bourdon spring	3. Bellows
A. Manometer	a) C-type	a) Pressure
B. Weight-loaded	b) Spiral	b) Absolute pressure
	c) Helical	B. Nonelastic
II. Force balanced		1. Bell
A. Elastic	2. Diaphragm	2. Limp diaphragm

Manometer. The simplest pressure-balance system is the manometer or liquid column pressure gauge in which the pressure created by a column of liquid is used to balance the pressure to be measured. Pressure exerted at the base of a column of liquid is equal to the density of the liquid multiplied by the height of the liquid column. The pressure reading is the difference in height from the top of the pressure column to the top of the vented column. Mercury is normally used in manometers. The barometric pressure is commonly expressed in inches of mercury. The column reading in height of liquid may be converted into pounds per square inch by multiplying inches of mercury by 0.491. Sketches of some typical liquid column gauges are shown in Figure 2-2. Figure 2-2 *A* is a schematic drawing of an absolute pressure gauge consisting of a U-tube with one end closed and evacuated. The open end admits the absolute pressure being measured. The liquid normally used is mercury. The absolute pressure measurement is the product of the density and liquid level difference of the legs. Figure 2-2 *B* is a fixed-cistern type barometer with one end closed and evacuated. The other end is open and inverted in a pool of mercury. It is used to measure atmospheric pressure which is the difference between the top of the mercury column and the level of the mercury in the container holding the pool. Figure 2-2 *C* shows a typical U-tube manometer. This gauge has both ends open. The tube has mercury in the body. It is used to measure differential

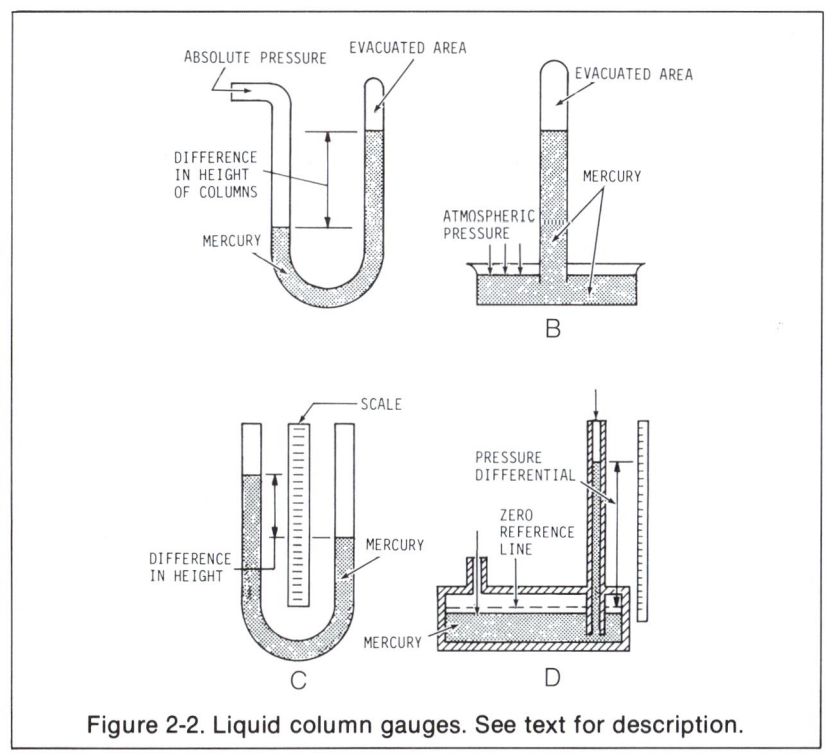

Figure 2-2. Liquid column gauges. See text for description.

pressures as determined by differences in height of mercury in the two legs. Figure 2-2*D* shows a well manometer consisting of a single tube and well. The well replaces one leg of the U-tube manometer. Its use is to measure differential pressure by determining differences in mercury height in leg and well.

There are other types of manometers also, such as the inclined-tube manometer which is similar to the well manometer in theory but consists of a single, tilted tube to provide greater reading accuracy through the use of a longer tube. Another is the ring-balance manometer which consists of a circular tube, split at the top, pivoted in the center, and balanced at the bottom. The measurement in the latter case is differential pressure with the pressures admitted to each side of the ring-tube creating an unbalance.

Weight-Loaded Devices. The most common weight-loaded device is the deadweight tester which works on the basis of Pascal's law: any pressure communicated to the surface of a confined liquid is transmitted unchanged to every part of the liquid. The deadweight tester employs a low-friction piston open to the surface of a contained liquid. A pump displaces liquid, forcing the piston upward. By the addition of weights of known value, the pressure within the liquid then becomes a function of the weight being supported divided by the area of the piston. This pressure is usually pounds per square inch.

Force-Balance Systems

Elastic Types. This group of pressure devices is based on the principle that any material will be deformed or distorted when any force is applied to it. The amount of deformation, movement, or distortion can be used as a measure of force and consequently of the pressure which created the force.

Bourdon Springs. The C-type spring (Figure 2-3) is the earliest and simplest pressure measuring device. Generally referred to as the Bourdon tube, it was invented by the French engineer Eugene Bourdon. This device is found in one of the most common instruments in the pulp and paper industry, the pressure

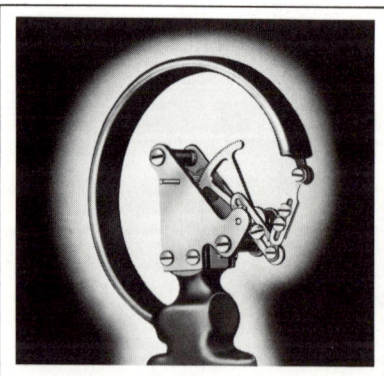

Figure 2-3. C-type Bourdon pressure element.

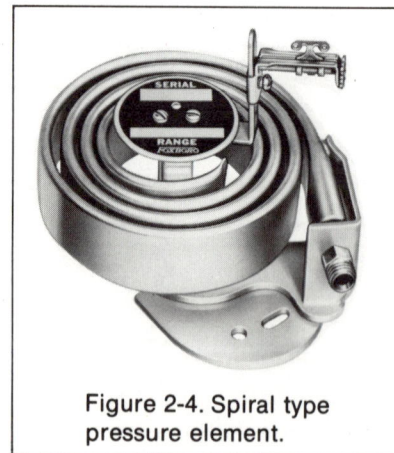

Figure 2-4. Spiral type pressure element.

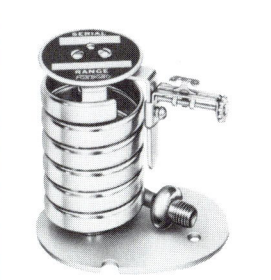

Figure 2-5. Helical type pressure element.

Figure 2-6. Diaphragm type pressure element.

gauge or indicator. It is semicircular in shape with an elliptical cross section. One end of the tube is sealed, the other is connected by piping (or directly) to the process. As the pressure in the process increases, it is applied to the open end and the tube tends to straighten out. The resultant motion is transferred through a linkage or rack-and-pinion mechanism to an indicating pointer. The resulting change in motion is proportional to the amount of pressure applied.

Spiral and helical devices were later developments of the Bourdon tube of which there are two simple variations. The spiral element (Figure 2-4) is actually a long C-type tube resembling a flat coil. The metal tubing is coiled like a fire hose. One end is rigid, and the movement of the free end is linked to a pointer or pen arm of an indicator, recorder, or controller.

The helical element (Figure 2-5) is essentially another long C-type spring, wound like a vertical spring. The added length of the helical coil makes it more sensitive to small pressure changes. The cumulative movement of the long spring produces sufficient movement of the spring tip to operate a pen arm with multiplication through linkages and gears.

Bourdon tube devices are not considered the best for very low pressures. In these cases diaphragm and bellows type elements are used. The diaphragm element (Figure 2-6) is made up of several discs which have been pressed out of sheet metal and fastened together in pairs. The bellows element (Figure 2-7) is

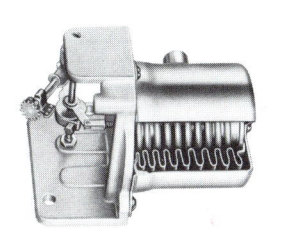

Figure 2-7. Bellows type pressure element.

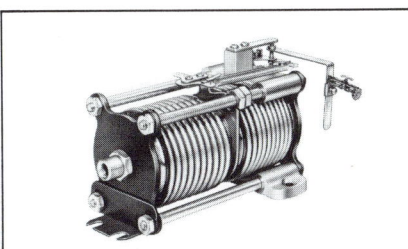

Figure 2-8. Opposed bellows type absolute pressure element.

formed from a homogenous piece of seamless tubing. When pressure is applied to one end, the circular sections expand axially and this motion is used as a measure of the pressure. If more convolutions are added the axial movement or stroke will increase. Increasing the diameter decreases the force required to move the bellows making it possible to measure smaller pressures. By designing the bellows assembly so that it is opposed by an identical evacuated and sealed unit, the measurements can be referenced to absolute zero, automatically compensating for altitude and barometric fluctuations (Figure 2-8), thereby measuring absolute pressure.

The type of elastic force-balance pressure measurement system used depends on the measurement range desired and element materials. Table 2-D can be used to select the proper element for representative measurement ranges.

Nonelastic Types. Nonelastic type meters are used extensively to measure low pressure in the 1 inch to 10 inches of water range encountered primarily around recovery and power boilers in pulp and paper mills. The bell type pressure gauge is typical of this group. The theory of the bell type gauge is extremely simple (Figure 2-9). When no pressure above atmosphere or other reference pressure is applied, the bell will sink. As pressure is applied to the inside of the bell, the difference between the external atmospheric pressure and the inside pressure becomes a lifting force. The bell will rise until the forces are balanced again. The gravitational pull (weight) plus the exterior force must equal the internal lifting force plus the buoyant force. Actually, the bell gauge measures the differences in pressure that are applied to both sides of the bell.

Another type of nonelastic pressure measuring device is the limp or slack diaphragm (Figure 2-10). This type of instrument is found throughout the mill measuring low pressures or vacuums; indicating, recording, and controlling furnace drafts and air duct pressures which seldom exceed 20 inches of water. It employs a flexible, nonmetallic diaphragm usually made of high-quality leather, goldbeater's skin, or a thin neoprene-like material. The pressure acts against the effective area of the diaphragm causing it to deflect against the force of a flat spring. The resulting displacement is multiplied by a suitable linkage to a pointer or pen arm. Differential pressure can be measured by applying the second pressure to the other side of the diaphragm and using a sealed linkage to detect motion of the diaphragm.

PRESSURE ELEMENT SEALS

It is often desirable to prevent the process fluid from coming in contact with the pressure measuring element. Such conditions as high temperatures, corrosiveness, and sludgy, semisolid, viscous, and solidifying materials could ruin the instrument. Therefore protection is provided by isolating the pressure element from the process. Connecting piping or capillary tubing are frequently used. There are several ways to accomplish this seal. The universal method for protecting elements from the high-temperature steam is to use a single coil siphon (Figure 2-11 A). The coil prevents steam and high temperature from reaching the pressure element.

TABLE 2-D

HOW TO SELECT THE MEASURING ELEMENT

A. Determine the operating pressure range of your process and decide what maximum pressure you expect to measure.

The chart shows the scope of *maximum* pressures for each type of element. (The bottom scale pressure will be zero unless an expanded or compound range is selected.) Your selection is determined, always, by the maximum pressure to be measured.

B. From the chart below select the proper type of element to cover this range.

C. If there is a choice of materials of construction for the type of element you select, use the table below to help you determine what material is best suited to your process conditions.

BRASS, BRONZE OR BERYLLIUM COPPER
Most commonly used materials for pressure elements. Ideal for general purpose measurements where corrosive vapors are not likely to contact the elements. Beryllium copper offers highest accuracy.

TYPE 316 STAINLESS STEEL
Strong, durable material for applications where corrosive vapors are expected to contact the pressure element. Good elastic properties for high-, medium-, or low-pressure measurements.

NI-SPAN C ALLOY
Noncopper-bearing alloy especially suited to applications where highest measurement accuracy must be maintained in spite of wide variations in the surrounding temperature. Ni-Span has practically zero thermoelastic coefficient. Low hysteresis and corrosion-resistant.

CU-NI-MN ALLOY
An excellent spring alloy especially adapted to pressure measurement in the low ranges. Provides extra protection for "overrange" beyond normal operating pressures.

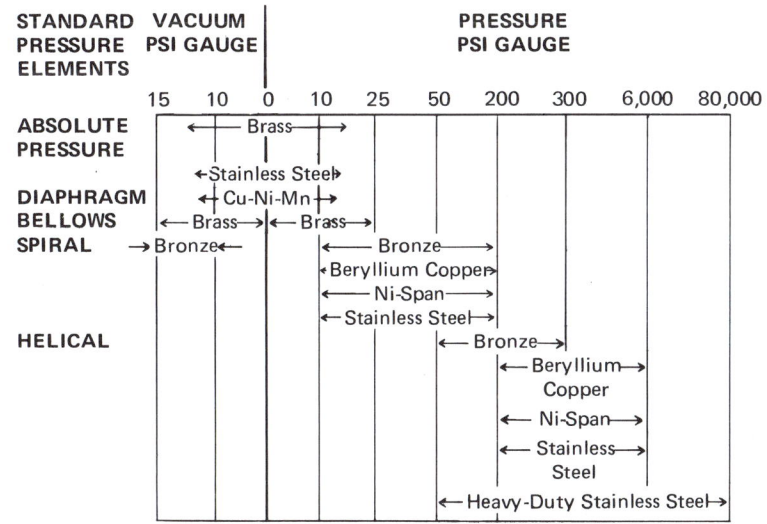

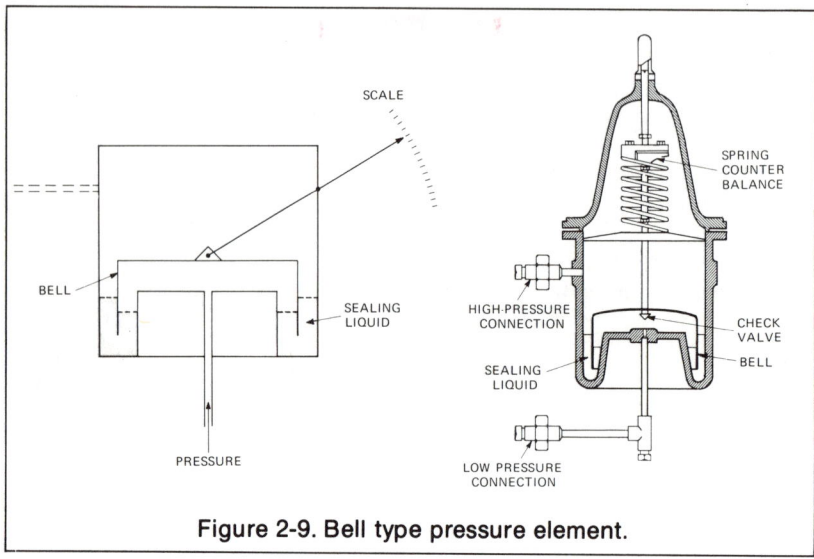

Figure 2-9. Bell type pressure element.

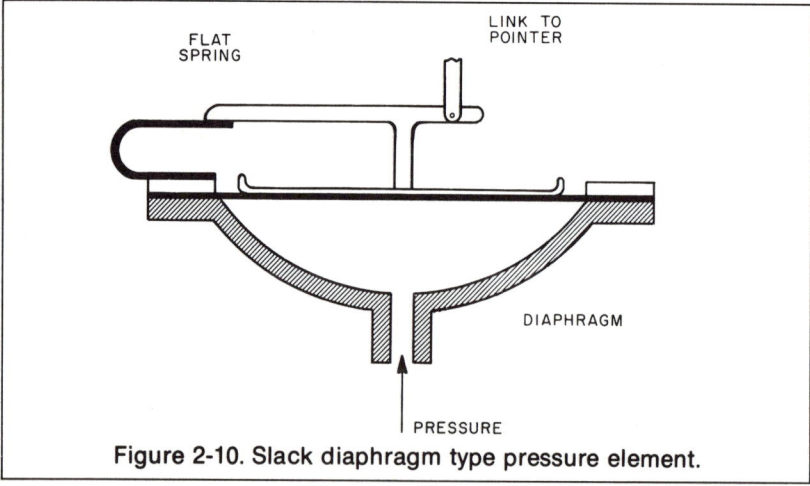

Figure 2-10. Slack diaphragm type pressure element.

Liquid Seals

Under certain conditions, pressure elements can be protected from corrosive process fluids by filling the connection line with an inert fluid. A seal pot forms a reservoir for the fluid (Figure 2-11 *B*). The sealing fluid in a direct-connected seal pot must have a density greater than that of the process fluid when the element is mounted below the point of measurement and less if it is mounted above. The type of sealing fluids used depends on the chemical and physical characteristics of the process fluid. Sealing liquids for various process fluids can be found in standard technical reference books.

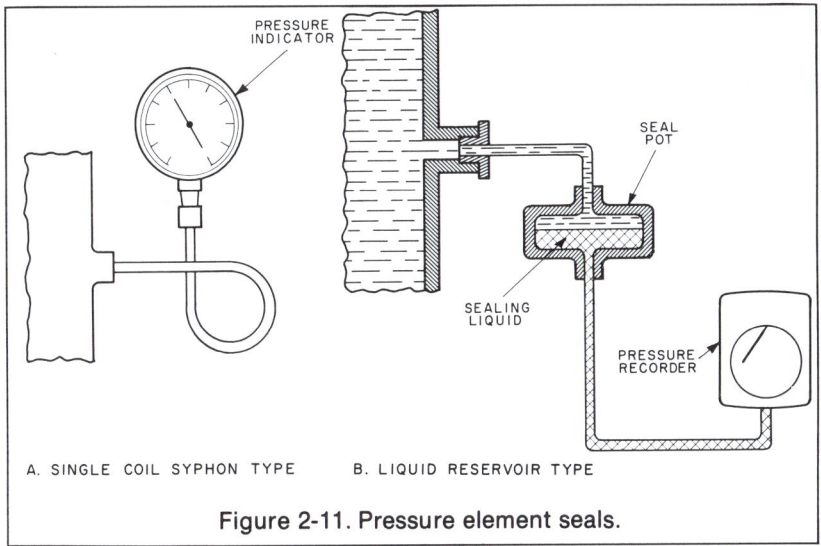

A. SINGLE COIL SYPHON TYPE   B. LIQUID RESERVOIR TYPE

Figure 2-11. Pressure element seals.

Volumetric Seals

On applications involving not only corrosive but also viscous and solidifying fluids, a volumetric sealing system is used. It consists of the pressure element connected by a capillary tubing to a flexible sealed chamber which comes in contact with the process fluid. The entire system is completely filled with a suitable noncompressible liquid. Pressure is applied to the flexible seal member of the chamber causing the liquid fill to be forced into the capillary and pressure element. This expands the element and activates the pen or pointer. Figure 2-12 includes sketches of three representative groups of the more common volumetric seals used today in the paper industry.

Purge Systems

There are times when a small and continuous flow of air, inert gas, or water may be sufficient to maintain the line free from sludge. Under these circumstances a purge system can be used (Figure 2-13). The air or water is supplied at the necessary pressure through a constant flow regulator. With the flow of air through the line constant, the pressure measuring device is calibrated to compensate for the constant pressure drop in the connection line.

ELECTRICAL PRESSURE MEASUREMENT SYSTEMS

Systems that employ electrical elements to produce the desired pressure measurement can be generally classified on the basis of electrical parameters employed: strain, resistance, magnetic, capacitance, piezoelectric, oscillometric, photoelectric, thermoelectric, and ionization-conductance.

Pressure Measurement   71

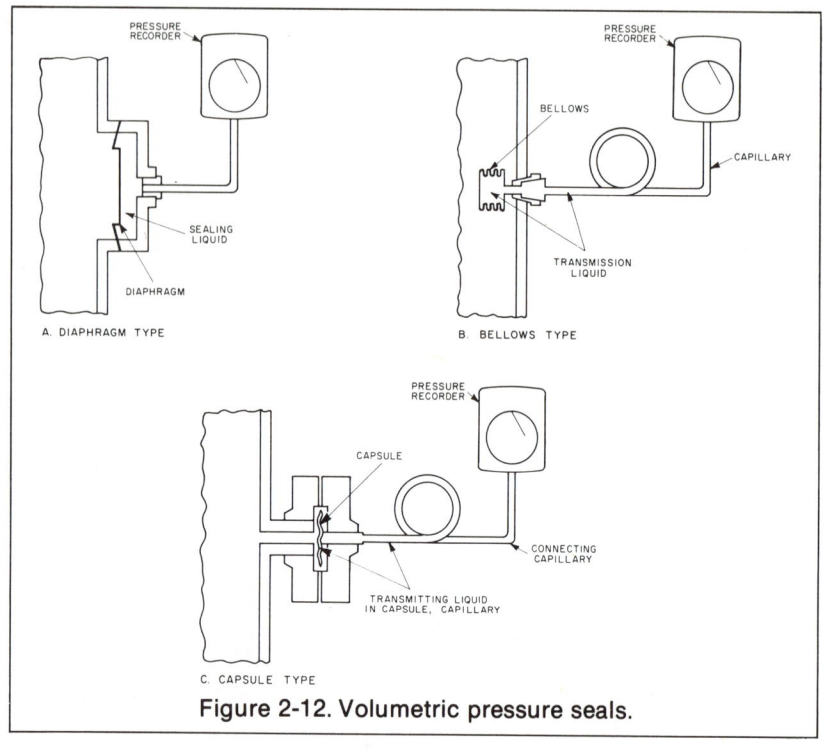

Figure 2-12. Volumetric pressure seals.

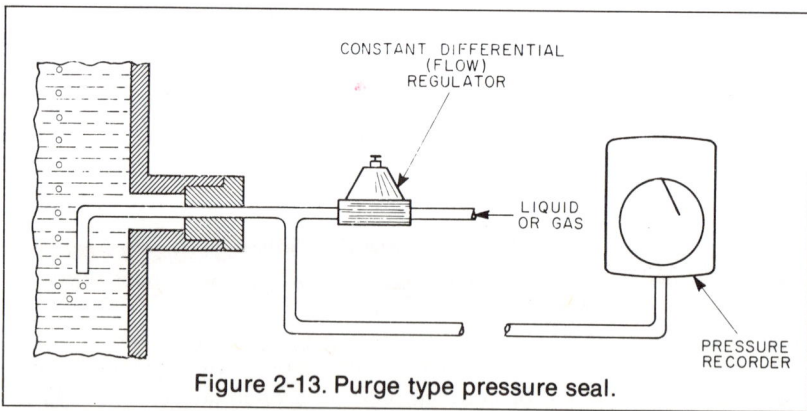

Figure 2-13. Purge type pressure seal.

Strain Gauges. The only electrical strain gauge of pulp and paper significance is used to measure loading pressures on such equipment as calender stocks and presses. Strain gauges operate on the phenomenon that when a wire is stretched, its length and diameter are altered resulting in a change in its electrical resistance. This electrical resistance can then be related to the force or pressure causing the distortion.

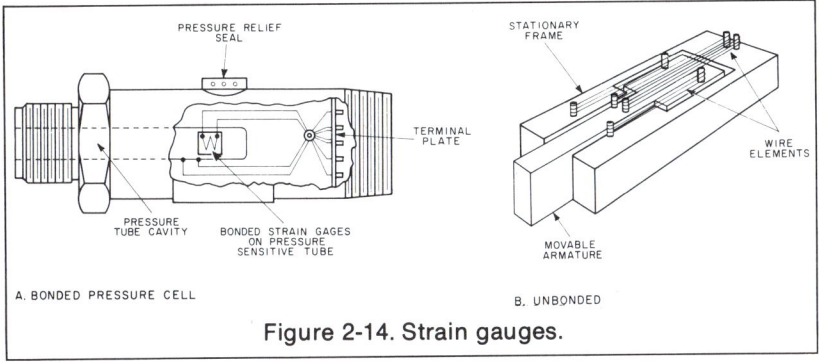

Figure 2-14. Strain gauges.

There are two basic types of wire strain gauges in general use: bonded and unbonded. The bonded type usually takes the form of a flat grid. It is normally cemented directly to the flexible member under load. For ease of handling, the wire is usually embedded in a thin carrier cloth or paper patch. Figure 2-14 *A* shows how it would be mounted on a pressure-sensitive tube to form a pressure cell.

The unbonded strain gauge consists of a frame and an armature that moves with respect to the frame. Four sets of wires under tension are normally used (Figure 2-14 *B*).

BIBLIOGRAPHY

Benedict, R. *Fundamentals of Temperature, Pressure, and Flow Measurements*. New York: John Wiley & Sons, 1975.

Considine, D. M. *Process Instruments and Controls Handbook*. New York: McGraw-Hill Book Company, 1957.

Considine, D. M., and Ross, S. D. *Handbook of Applied Instrumentation*. New York: McGraw-Hill Book Company, 1964.

Forman, D. P. *Industrial Instrumentation*. New York: John Wiley & Sons, 1950.

Fribance, A. E. *Industrial Instrumentation Fundamentals*. New York: McGraw-Hill

Book Company, 1962.

Pressure Seals. Foxboro, MA: The Foxboro Company, TI 13-91a.

Soisson, H. E. *Instrumentation in Industry*. New York: John Wiley & Sons, 1975.

Stephenson, J. N. *Pulp and Paper Manufacture*, vol. 4. New York: McGraw-Hill Book Company, 1955.

Tyson, F. C. *Industrial Instrumentation*. Englewood Cliffs, NJ: Prentice-Hall, 1961.

Zoos and Delahooke, *Industrial Process Control*. Albany, NY: Delmar Publishers, Inc., 1961.

STUDY QUESTIONS

Indicate whether following statements are True or False by inserting T or F in parentheses:

1. Pressure is the action of one force against another force. ( )

2. Inches or feet of water and centimeters or inches of mercury are actual units of pressure. ( )

3. The pressure of the atmosphere at sea level is 14.696 pounds per square inch absolute. ( )

4. Pressure is determined in mechanical systems by balancing a sensor against the unknown force. ( )

5. Generally, pressure balances are used for the measurement of high-pressure ranges in liquids. ( )

6. The most common weight-loaded device is the Bourdon spring. ( )

7. Spiral and helical elements actually are variations of the C-type element. ( )

8. The bell type pressure gauge is used primarily in the higher pressure ranges. ( )

9. Liquid seals, volumetric seals, and purge systems are used to protect the pressure element from the process fluid. ( )

10. The strain gauge is an electrical pressure measurement system. ( )

(See Appendix for answers)

3. Level Measurement

Level is an important variable in the pulp and paper industry, not only for proper process operation but for cost accounting and inventory purposes. Level measuring devices vary in complexity from simple visual gauges to local or remote reading instruments, depending on whether indication, recording, or automatic control is required.

In the pulp and paper industry, level measurements are made on both liquids and solids.

LIQUID LEVEL

Typical liquid level measurements are made in open tanks or storage chests and closed tanks or process equipment throughout the pulp and paper mill (Table 3-A). *Open tank* refers to process vessels and equipment open to the atmosphere; and *closed tank* refers to process vessels and equipment not open to the atmosphere and, in a sense, airtight.

Levels in Open Tanks

Open tank level measuring instruments used today fall into several general categories: visual, pressure or hydrostatic head, direct-contact or float, and other types.

Visual. This group is one of the earliest and simplest methods for the continuous measurement of liquid level within a tank or vessel. It is used only when direct-mounted local indication is required and when the liquid is comparatively clean. Sight and gauge glasses essentially consist of a simple transparent glass or plastic tube attached to the vessel in such a way that the liquid height in the tube is equal to the level of the liquid in the vessel. A calibrated scale marked on the tube or mounted beside it provides a convenient means to read the level in inches or feet, or volume in gallons, cubic feet, etc. Figure 3-1 shows a typical sight and gauge glass installation.

Hydrostatic Pressure. A liquid column creates a hydrostatic pressure directly proportional to the height of the liquid above a reference point. An appropriate pressure measuring element (see Chapter 2), properly connected to the process, measures liquid level in appropriate units for which the element can be calibrated. The two hydrostatic pressure type level instruments gener-

TABLE 3-A

TYPICAL LIQUID LEVEL MEASUREMENTS
PULP AND PAPER INDUSTRY

Area	Type	
	Open Tank	Closed Tank
Wood preparation	Log ponds	Cooking vessel
Digester	Black and white liquor measuring tanks	
	Blow tank	
Washers	Unwashed and washed pulp storage	
	Washer vat	
	Filtrate tank	
Pulp mill chemical preparation	Sodium hydroxide storage	Chlorine tank cars
	Sulphuric acid storage	Sodium hydroxide tank cars
	Sodium chlorate tanks	Sodium chlorate tank cars
	Sodium hypochlorite storage	ClO_2 generator
Bleaching	Unbleached and bleached pulp storage	
	Washer vat	
	Seal boxes	
	Bleach towers	
Paper mill chemical preparation	Chemical storages	
	Liquid dye and additive storages	
	Pulpers	
Stock blending	Pulp storages	
	Blend chest	
Paper machine	Machine chest	Vacuum receiver tank
	Seal boxes	Headbox
	Save-all vat	Dryer drum condensate
	White water pit	
	Fan pump pit	
	Couch pit	
Evaporators	Thin black liquor storage	Evaporator effects
	Thick black liquor storage	Steam chest condensate
Recovery boiler	Thick liquor storage	Cascade evaporator
	Precipitator	
	Dissolving tank	
	Boiler drum	
Recausticizing	Green liquor storages	
	Slaker	
	Clarifier	
	White liquor storages	
	Lime mud storage	
Lime kiln	Lime mud filter vat	
	Slurry tank	

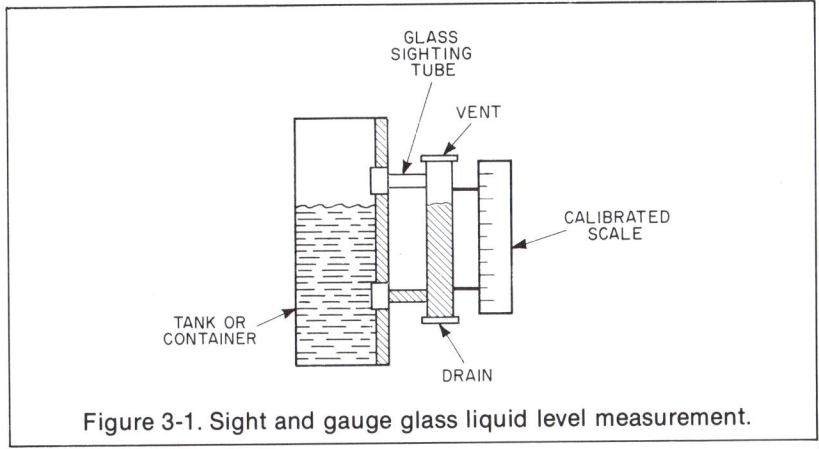

Figure 3-1. Sight and gauge glass liquid level measurement.

ally found in pulp and paper mills employ a continuous purge or bubble tube, or a diaphragm box.

Continuous purge level instruments are used where liquids are corrosive or have suspended solids. Where air or another suitable fluid is available as purge and remote reading is required, this is the simplest, least expensive, and most dependable and widely used type of liquid level instrument. Figure 3-2 shows a

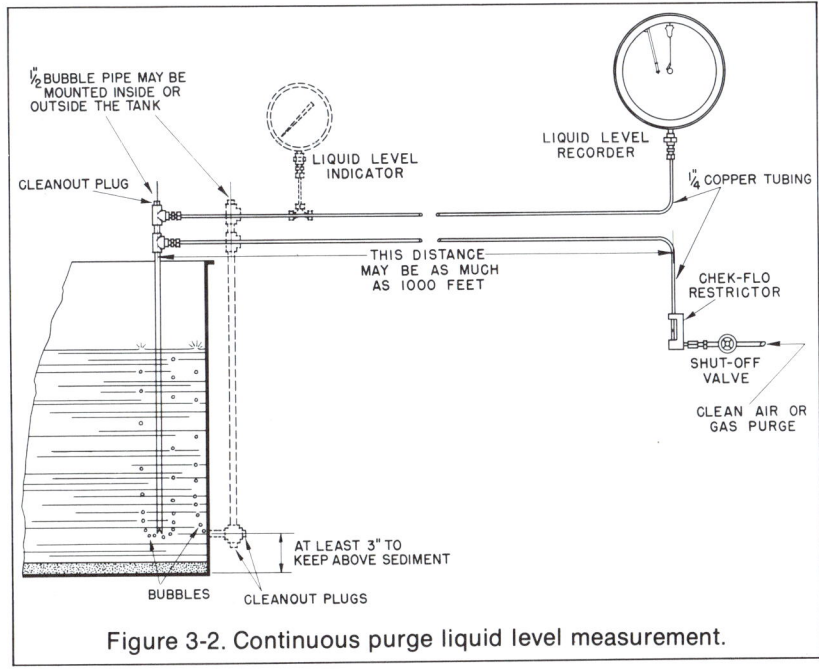

Figure 3-2. Continuous purge liquid level measurement.

typical installation of a continuous purge level measuring system. A length of open pipe is lowered into the vessel to a point about 3 inches above the bottom or a connection is made into the side of the tank at the same height. This height is not critical, but the point at which it is made becomes the reference point used in calibration of the hydrostatic pressure element in level or volume and it should be above any sediment which might collect. Compressed air or any suitable fluid is supplied through a valve and restrictor, purge rotameter, or sight feed bubbler so that the purge fluid escapes from the open end of the pipe. The air pressure in the pipe exactly corresponds to the depth of the liquid. As liquid level changes, the air pressure in the bubble pipe changes to correspond.

A pressure instrument connected to the bubble pipe registers the air pressure on the chart or indicating dial in units of liquid level (inches or feet) or units of volume (cubic feet, gallons, etc.) as required.

Diaphragm box systems are used to measure liquid levels in open tanks where air or another suitable fluid is not available or where a purge is undesirable (Figure 3-3). The diaphragm box is essentially a cup covered with a flexible diaphragm which is protected by a guard ring. The box is made up of two sections, with the diaphragm inserted between them and clamped in place airtight. A capillary connecting tube enters the top section of the diaphragm box which is located at some selected reference level in the tank. The pressure caused by the column of liquid above the reference level acts on the diaphragm to compress the air in the sealed pressure system to an amount equivalent to the actual liquid head. Level variations cause proportional pressure changes in the air system. These pressure changes actuate a pressure spring in the instru-

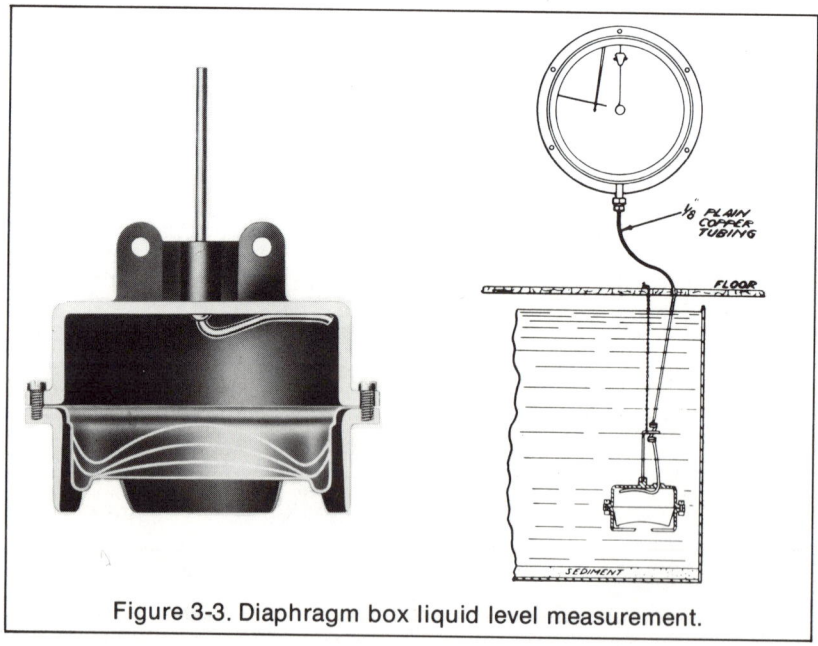

Figure 3-3. Diaphragm box liquid level measurement.

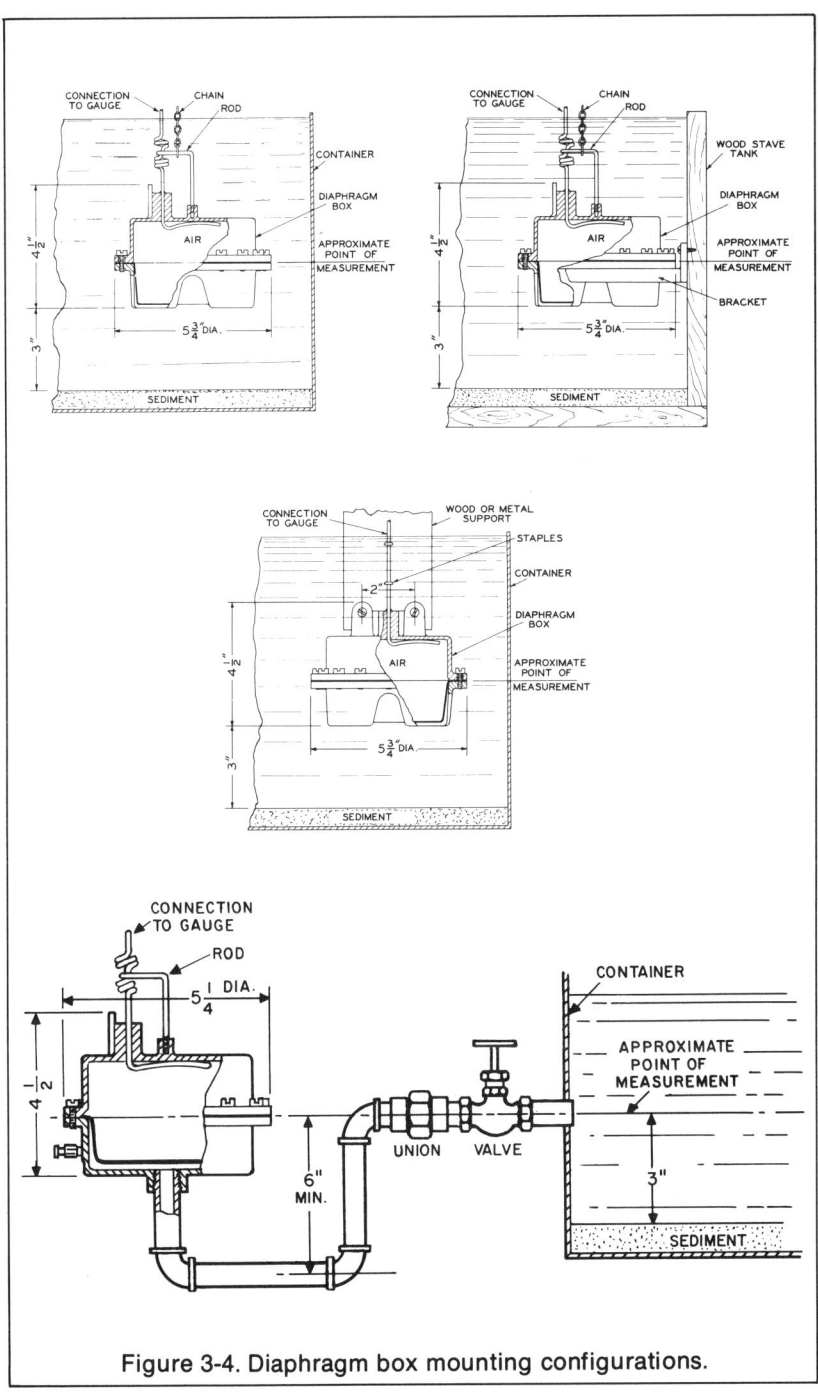

Figure 3-4. Diaphragm box mounting configurations.

ment which is connected to the recording pen or indicating pointer.

Diaphragm boxes may be mounted in various configurations (Figure 3-4).

A widely accepted diaphragm type measuring device for pulp stock and chemical slurry is the flanged differential pressure transmitter. Two versions are shown in Figure 3-5: the flush diaphragm and the extended diaphragm. The flush type mounts directly on a pad on the wall of the tank as shown in Figure 3-6. The extended version is used on liquids and slurries that would solidify or compact in pockets formed between the flush diaphragm and inner surface of the wall. The extended diaphragm capsule is mounted so that it is flush with the inside wall of the tank or vessel. The liquid head exerted on the diaphragm is simultaneously balanced by a pneumatic or electrical feedback force in the top works. This feedback force represents the liquid level and is used for remote indicating, recording, or control.

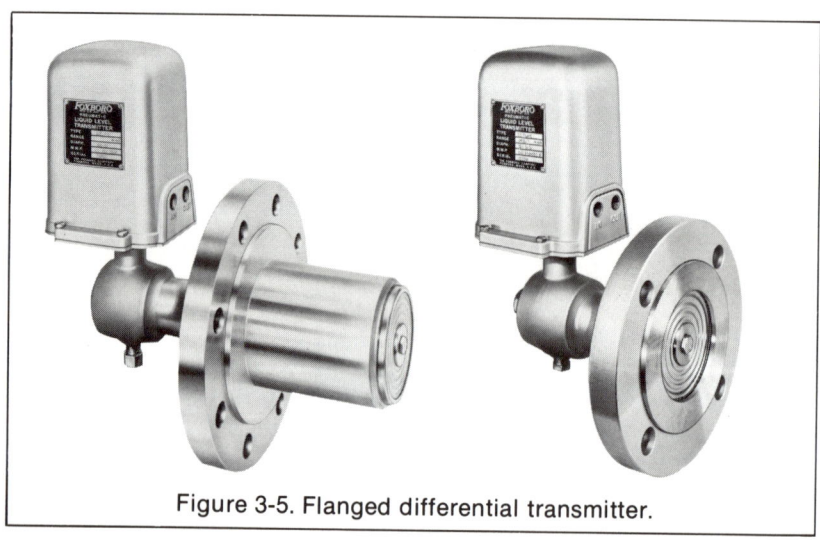

Figure 3-5. Flanged differential transmitter.

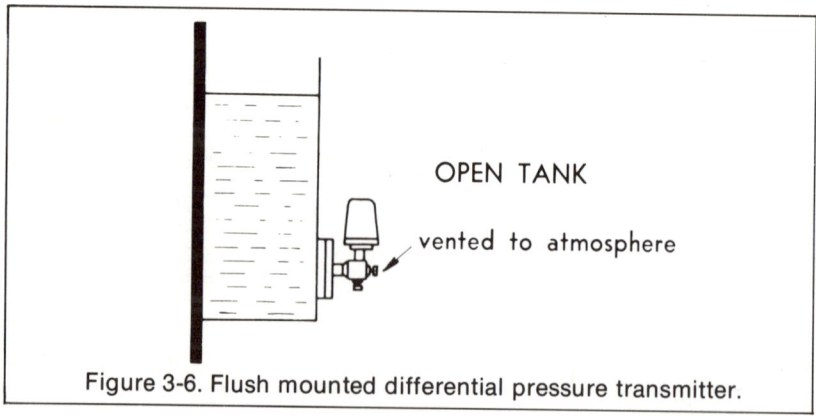

Figure 3-6. Flush mounted differential pressure transmitter.

Float and Cable. Direct measurement of liquid level in open vessels is often made by the float and cable method. Use of this method, however, is somewhat limited in the pulp and paper industry because many of the process fluids are either slurries of suspended fibers, viscous, corrosive, or sticky. Float and cable level measurements are most commonly found in the water and waste treatment areas of the mill. They do not depend on hydrostatic pressure to measure level. Instead, they are self-powered instruments which operate directly from the movement of a float on the surface of a liquid (Figure 3-7).

There are many versions of float and cable instruments. Fundamentally they consist of a float and counterweight connected by a cable operating over a drum. Motion of the float causes the drum to rotate. In this way the vertical motion of the float is translated to a uniform measurement by a reduction mechanism, which can be used for indication, recording, or control purposes.

Miscellaneous. Other types of open tank liquid level devices are sometimes used on a somewhat specialized basis. More elaborate and expensive equipment is involved in these cases. Some use radioactive sources, capacitance probes, sonic waves, strain gauges, and load cells.

Levels in Closed Tanks

When it is necessary to measure liquid level in closed tanks such as digesters, paper machine headboxes, Deculators, condensate receiver tanks and evaporators, a simple hydrostatic pressure measurement cannot be used.

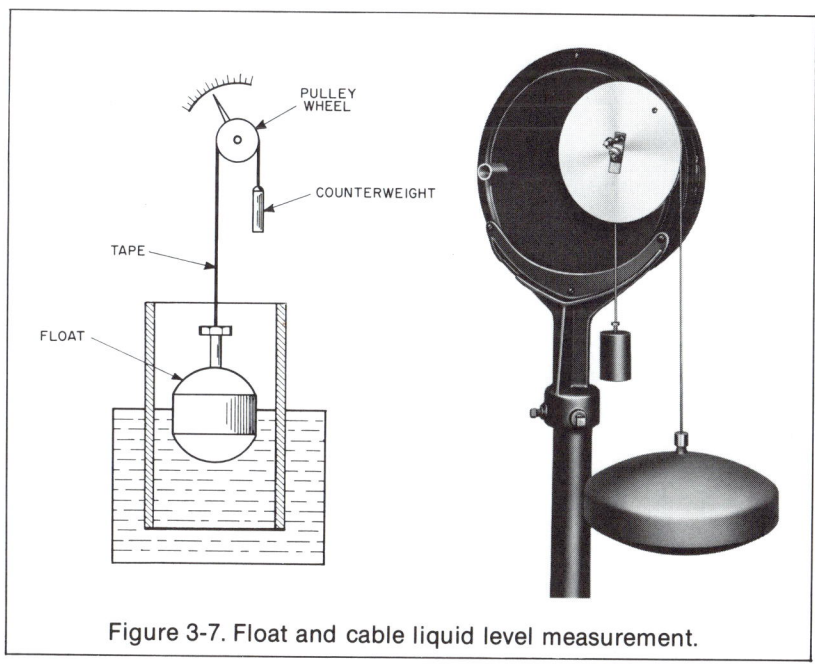

Figure 3-7. Float and cable liquid level measurement.

Enclosed tank pressure influences include both weight or pressure of liquid and pressure or vacuum of the atmosphere above the liquid. Under these conditions, liquid level measurements more commonly involve differential pressure, float position and displacement, nuclear radioactive devices, sonic devices, capacitance, and other electrical devices.

Differential Pressure. The liquid level in this case is inferred by measuring total pressure and correcting it for pressure above the liquid which gives the difference between the two pressures. Here, differential pressure measuring secondary elements are used, such as manometers, bellows, and force-balance differential pressure cells. Figure 3-8 shows how the mercury U-tube manometer is used for measuring level of clean liquids in closed tanks. Liquid seals filled with appropriate inert sealing fluid are added to protect the meter if the liquid is corrosive, contains suspended solids, or is highly volatile. Purge systems are sometimes used in place of seals.

Figure 3-9 shows the use of force-balance differential pressure transmitters for liquid level measurements in closed vessels. Installation is usually much simpler than for the mercury manometer. Because of its corrosion-resistant construction and the negligible displacement of liquid in the lines, it can be used to measure the level of many corrosive liquids without a seal or purge system.

Displacement Float. The displacement float can be used on open or closed tanks. The operation is based on Archimedes' principle which states that the resultant force of a fluid on a body immersed in it acts vertically through the center of gravity of the displaced fluid and is equal to the weight of the fluid displaced. The resultant upward force on the body is called buoyancy and can be counterbalanced by another force so that the movement of the body can be

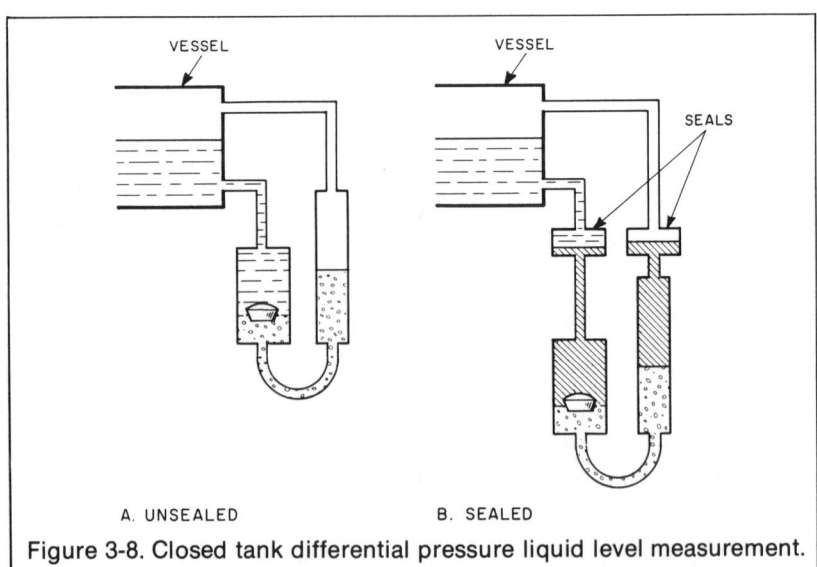

Figure 3-8. Closed tank differential pressure liquid level measurement.

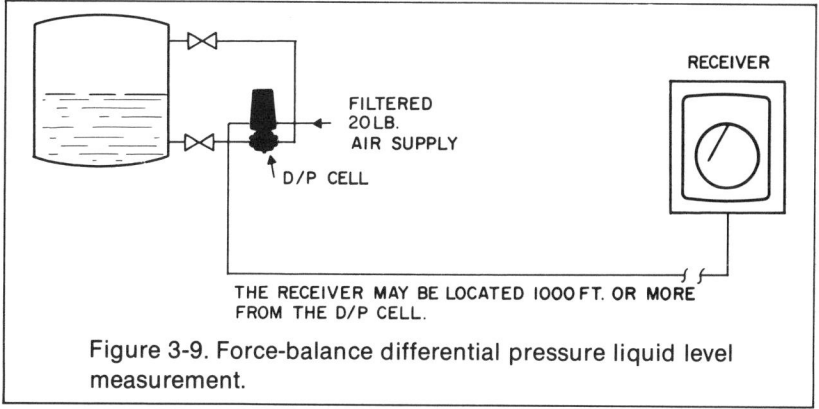

Figure 3-9. Force-balance differential pressure liquid level measurement.

FILTERED 20LB. AIR SUPPLY

D/P CELL

RECEIVER

THE RECEIVER MAY BE LOCATED 1000 FT. OR MORE FROM THE D/P CELL.

used as a measurement of level. Figure 3-10 is a schematic arrangement of a displacement type liquid level measurement with the buoyancy force counterbalanced by a spring. The buoyancy force can also be counterbalanced by a pneumatic or electrical force-balance system which produces a corresponding liquid level signal which can be transmitted for remote indication, recording, or control. A typical unit is shown in Figure 3-11.

Nuclear Radioactive. Radioactive systems are used for measuring liquid level in closed tanks. They consist of a measuring assembly and an amplifier-indicator. The measuring assembly contains a radioactive source such as radium, cesium, or cobalt and the radiation detection cell which can be a form of Geiger counter or a specially designed gas ionization cell. The detection cell produces an electrical signal proportional to the intensity of the radioactive rays reaching it, which is amplified to produce a measurement in appropriate units of level.

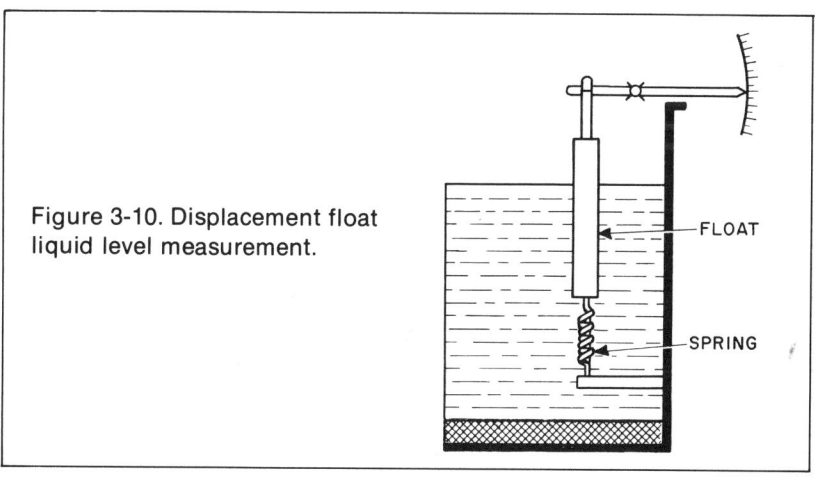

Figure 3-10. Displacement float liquid level measurement.

FLOAT

SPRING

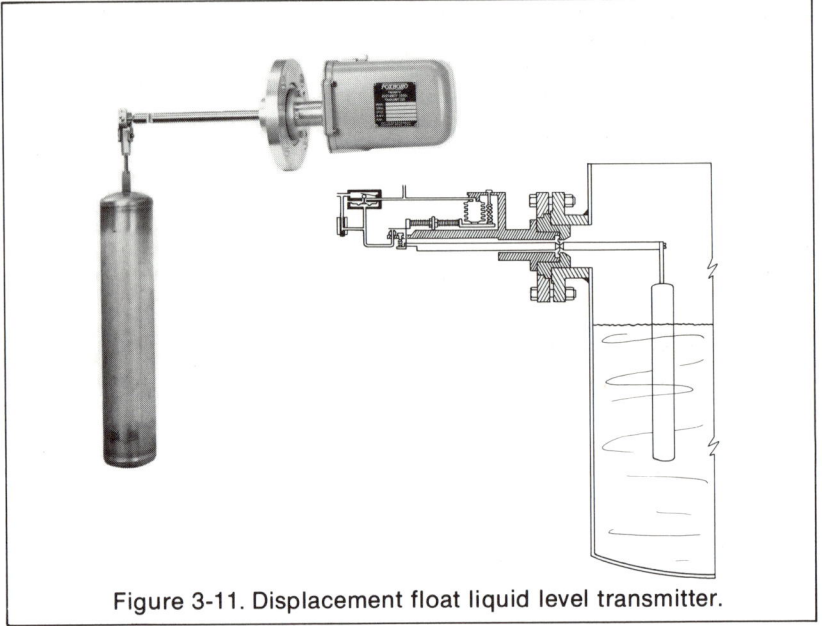

Figure 3-11. Displacement float liquid level transmitter.

Special arrangements are used in the installation of these systems as shown by Figure 3-12.

A. The radioactive source can be fixed within or on the side of the tank at a low level, with the detection cell fixed to the outside at a higher level. As the level rises within the vessel, the increase of absorbing material between the radiation source and detector decreases the radiation intensity received by the detector. The detector then converts radiation received into an electrical signal which is amplified and converted to a signal suitable for use as a level measurement.

B. A similar system employs a source mounted in a float. The intensity of the radiation is a function of the distance between the source and detector and is directly proportional to level change.

C. Another type involves the use of a strip type source placed vertically in a column and corresponding measuring cells located on the tank or vessel so that the liquid whose level is being measured interposes itself between the two columns as the level rises. The output current from the detection cells, which varies with the level, is inversely proportional to level change.

Sonic. This is another method of measuring liquid level in closed tanks which can also be used on open tank applications. Sonic level measurement is based on the emission of sound wave pulses from a source, transmission of these energy waves through the liquid phase or vapor phase in the vessel, and the reflection of the waves back from the surface to the receiver. The transit time

84   Paper Industry Instrumentation

of this impulse is used as a measure of liquid level. Sonic type level measurements are shown in Figure 3-13.

With the liquid phase arrangement, pulses of ultrasonic energy are directed upward to the surface of the liquid being measured. The pulses rebound from the surface to the receiver. The transit time from the source to the receiver is a measure of distance between the surface and the pulse emission source and directly proportional to liquid level. This transit time is electronically converted to a measurement of liquid level in standard units.

The vapor space arrangement is based on the rebound or echo principle through the vapor or gas phase above the liquid. In operation, it is similar to the liquid phase except the elapsed time of sound pulse transmission is inversely proportional to the liquid level being measured.

Capacitance. Another electrical method to measure liquid level in an enclosed tank is capacitance. Basically, a capacitance level measuring system consists of a primary measuring element or probe and a secondary instrument which transforms a capacitance variation into display of the liquid level in the tanks (Figure 3-14).

In its simplest form, the capacitance level probe is a metal rod mounted in the vessel. The rod must be electrically insulated from the vessel wall. The capacitance of the system is formed by the rod which can be considered as the

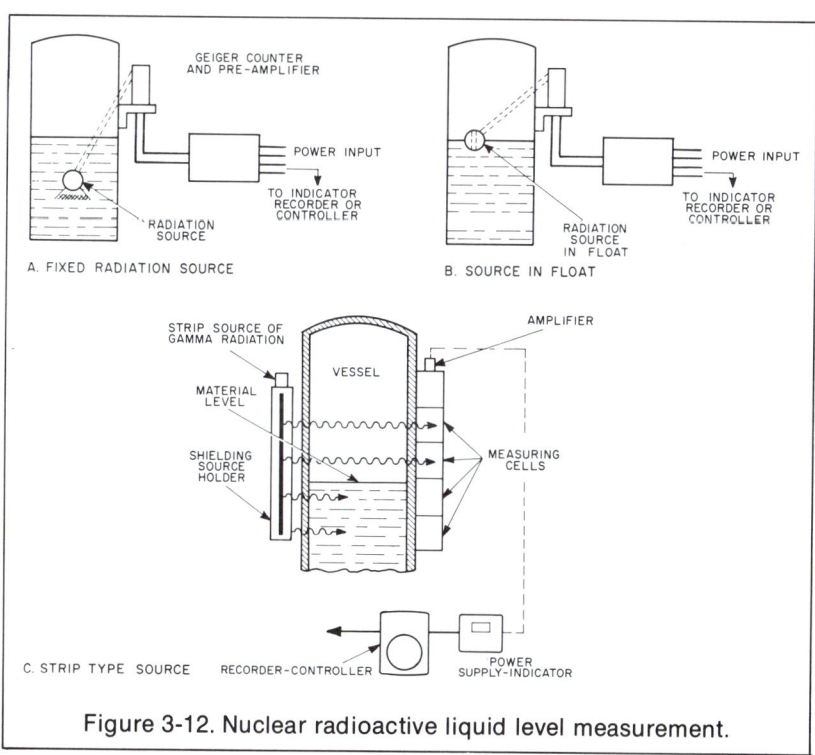

Figure 3-12. Nuclear radioactive liquid level measurement.

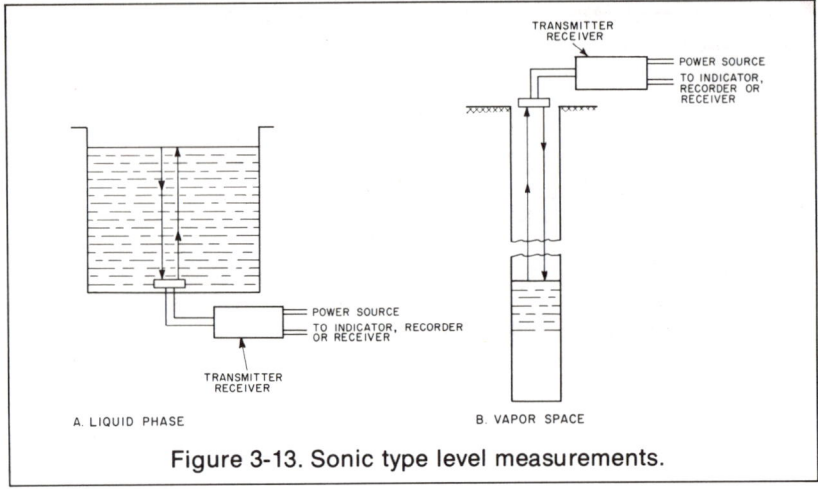

TRANSMITTER
RECEIVER

POWER SOURCE
TO INDICATOR,
RECORDER OR
RECEIVER

POWER SOURCE
TO INDICATOR, RECORDER
OR RECEIVER

TRANSMITTER
RECEIVER

A. LIQUID PHASE

B. VAPOR SPACE

Figure 3-13. Sonic type level measurements.

"live" electrode and the vessel wall which serves as the "grounded" electrode. As the probe, liquid, and wall form a capacitive network, a change in capacitance caused by a change in liquid level is sensed by an indicating or recording instrument which expresses it in terms of liquid level.

Miscellaneous. Other methods find limited use in measuring liquid level in closed vessels. They include the temperature probe for one point measurement, expansion tube type, oscillator type, conductivity, and combination of photoelectric and gauge glass.

SOLIDS LEVEL

It is often necessary to measure the level of bulk or dry solid material in certain areas of the pulp and paper process. These include such level measurement applications as wood chip bins, high-density stock storage chests, dry chemical tanks, salt cake silos, soap storage tanks, and lime storage tanks.

Figure 3-14. Capacitance level measurement.

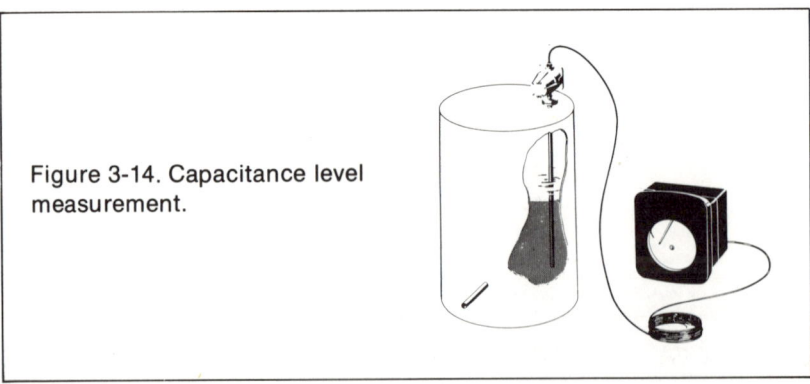

Primary devices used for this purpose are considered as either *fixed point* type or *continuous* type.

Fixed Point Devices

These solid level detecting devices provide measurement at one or several specific levels. They are used largely for the actuation of alarms or conveyors. They include diaphragm, paddle wheel, pendant cone, and probe level measurement devices.

Diaphragm Type. This device employs a flexible diaphragm which is exposed to the bulk material in the storage bin. As the solids level rises, pressure caused by the weight of the dry material forces the diaphragm against a counterweighted lever mechanism which mechanically actuates a switch (Figure 3-15). The switch can energize alarm circuits or machinery such as conveyors or slide valves.

Paddle Wheel Type. A typical paddle wheel level measurement is illustrated in Figure 3-16. The paddle shaft of the unit is driven by a synchronous motor. When rotation of the paddle is resisted by a solid material, it causes the motor support and gear housing to rotate in a horizontal plane which actuates two miniature switches in consecutive order. The first switch actuates equipment such as an alarm circuit; the second cuts off the power to the paddle which then stays in a locked position. When the level drops and the dry material falls away, a spring pushes the assembly back to its original position and the two miniature switches are released.

Pendant Cone Type. As shown in Figure 3-17, this device consists of a switch enclosed in a dust-tight case which incorporates a universal pivot collar

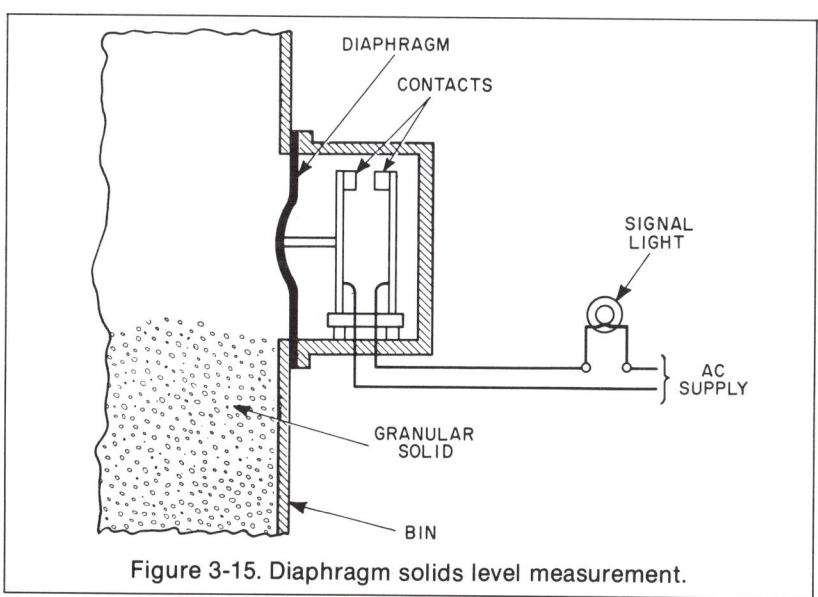

Figure 3-15. Diaphragm solids level measurement.

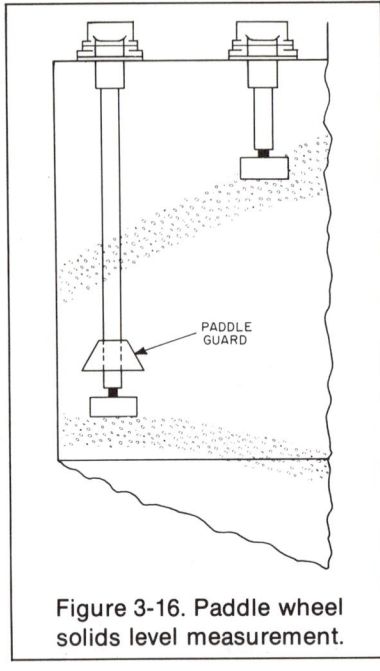

Figure 3-16. Paddle wheel solids level measurement.

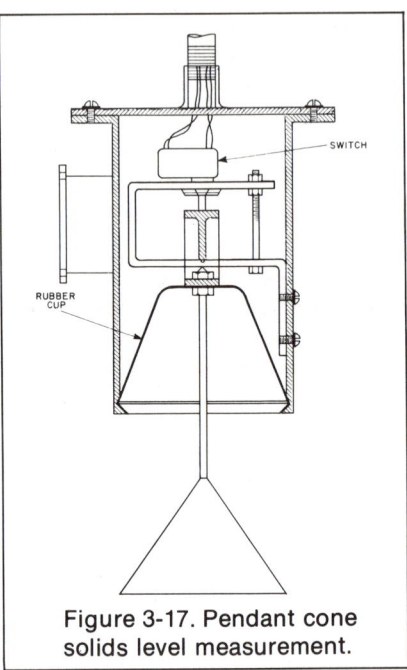

Figure 3-17. Pendant cone solids level measurement.

from which a pendant cone is suspended. The pendant rod terminates in either a plastic float ball or metallic cone. When the level of solid material rises and comes in contact with the cone, a switch is actuated that energizes an alarm circuit or other equipment.

Probe Type. When the solid material to be measured has a high electrical conductivity an electrical probe is often used (Figure 3-18). When the material makes contact with the probe it completes an electrical circuit which, with proper amplification, can be used to actuate an alarm circuit or other equipment. Another type of probe depends on the deflection of a tapered rod to actuate the switch mechanism.

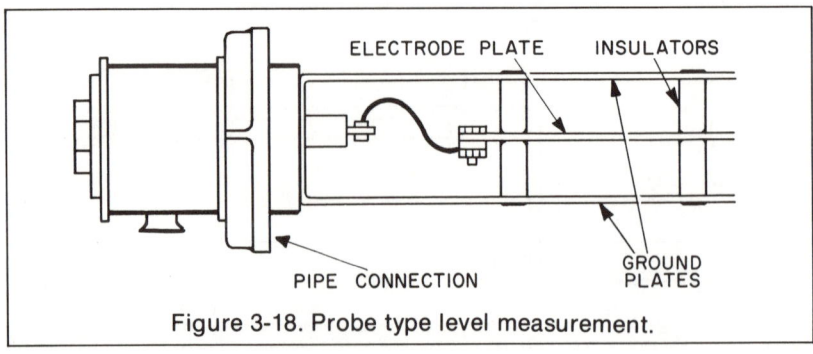

Figure 3-18. Probe type level measurement.

Other types of fixed spring solid level detecting devices such as electric capacitance, vibrating reed, etc., are not commonly found in the pulp and paper industry.

Continuous Devices

Continuous level measuring instruments provide a continuity of measurement throughout the range of level changes in a storage vessel. They are particularly important when close control of level is of primary importance (where batch storage is part of a continuous process).

Resistance Probe Type. This method has primarily been used in the paper industry to measure levels of wood chips and pulp. However, applications on other dry materials such as clays, starch, and titanium dioxide are also possible, as well as measurements of liquids including polyvinyl alcohols, general chemicals, and water levels in tanks, wells, and streams. As shown in Figure 3-19, the sensing device is usually gravity suspended through a small access hole in the storage compartment roof and consists of a tape-like transducer, sensitive along its entire length. Typically ranging anywhere from 2 to 200 feet in length, it extends from the top to the bottom of the tank, silo, or well. It is acted upon by the pressure of the surrounding material causing the sensor resistance to vary as material moves up and down its length. Having a very

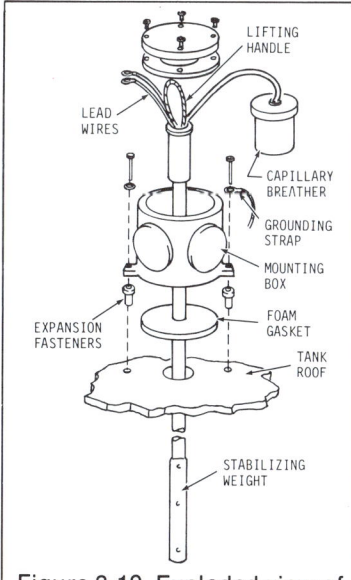

Figure 3-19. Exploded view of a typical resistance level sensor shown for mounting through an access hole in the tank roof.

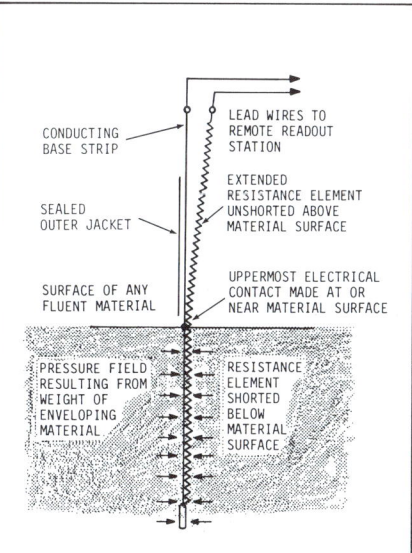

Figure 3-20. Simplified schematic of a resistance type level sensor showing the action of fluent material to partially short out the extended resistance element.

small cross section, the sensor moves to follow the line of material flow without interference. A typical sensing device consists of a precision-wound resistance helix having either 24 or 48 contacts for every foot of length. The jacket envelope consists of multiple layers of selected functional materials serving to enclose and protect the electrical system and to act as a pressure receiving diaphragm forming a tightly sealed envelope. A breather at the top of the sensor contains chemicals for drying and corrosion protection, and completes the isolation of the system from the external environment.

Figure 3-20 illustrates the principle of operation of this device. The gravity pressure of the surrounding fluent material acts upon the jacket diaphragm and causes progressive contact of the extended resistance device against the conducting base strip at substantially all points below the material surface. The resistance device remains unshorted above the material surface and it is this upper resistance arm that is metered to determine location of the material surface. The element is connected to remotely mounted readout devices with lead wires to be used for indicating, recording, or control purposes.

Other Methods. Nuclear radioactive sources, capacitance, and sonic devices as previously described for closed tank liquid level measurements are equally applicable and differ only in details of installation.

Continuous solid level measuring is also done by grid response units and actual weighing; however, these are not generally used in pulp and paper applications.

BIBLIOGRAPHY

Anderson, N. A. *Instrumentation for Process Measurement and Control.* Philadelphia: Chilton Book Company, 1972.

Behar, M. F. *The Handbook of Measurement and Control.* Pittsburgh: The Instruments Publishing Company, 1951.

Considine, D. M. *Process Instruments and Controls Handbook.* New York: McGraw-Hill Book Company, 1957.

Diaphragm Box Type Liquid Level Instruments. Foxboro, MA: The Foxboro Company, TI 13-40a, 1973.

Eckman, D. P. *Industrial Instrumentation.* New York: John Wiley & Sons,1950.

Lipták, B. G., ed. *Instrument Engineer's Handbook,* vol. 1, "Process Measurement." Philadelphia: Chilton Book Company, 1969.

Liquid Level Transmitters, Foxboro, MA: The Foxboro Company, Bulletin H-10, 1967.

Rhodes, T. J., and Carroll, C. G. *Industrial Instruments for Measurements and Control.* New York: McGraw-Hill Book Company, 1972.

STUDY QUESTIONS

Select answer or answers which correctly apply to the following statements:

1. There are two general types of liquid level measurements made in the paper industry
   a) hydraulic and mechanical.
   b) direct and indirect.
   c) open tank and closed tank.

2. The earliest and simplest method for continuous measurement of liquid level within a tank is
   a) sight and gauge glass.
   b) hydrostatic pressure type.
   c) float and cable.

3. The two most common hydrostatic pressure type level instruments generally found in pulp and paper mills are the bubble tube and
   a) mechanical.
   b) electrical.
   c) diaphragm box.

4. Diaphragm box systems are used primarily to measure liquid level in
   a) open tanks.
   b) pressurized tanks.
   c) liquids with suspended solids.

5. The flanged type differential pressure cell was developed for liquid level measurements of
   a) clean fluids.
   b) turbulent fluids.
   c) slurries.

6. Float and cable type liquid level measurements are used in
   a) pulp slurries.
   b) water and waste treatment areas.
   c) sticky fluids.

7. Nuclear radioactive level measurement systems are primarily
   a) electrical.
   b) pneumatic.
   c) mechanical.

8. Sonic level measurement systems depend on the transit time of sound waves
   a) to penetrate liquid.
   b) to reflect back from surface of the liquid.
   c) to bounce back from surface of the vessel.

9. Capacitance is another liquid level measurement method of the
   a) mechanical type.
   b) pneumatic type.
   c) electrical type.

10. Solids measurement systems are used to measure level of
    a) bulk materials.
    b) slurries.
    c) suspensions.

(See Appendix for answers)

4. Temperature Measurement

Another important process variable in the pulp and paper industry is temperature. Proper temperatures throughout the entire process from wood preparation through pulp manufacturing, chemical recovery, and final papermaking insure efficient, economical, and safe operation. Like flow, pressure, and level, temperature measurements are found in general use throughout the entire pulpmaking and papermaking process (Table 4-A).

BASIC THEORY OF TEMPERATURE

Temperature is most simply defined as a measure of the level of heat energy in a body. The term *temperature* is generally used to denote the relative heat of a body determined by its ability to lose this heat energy to its surroundings. The measurement of temperature is accomplished by observing the effect of heat on a known substance. The entire system is commonly known as a thermometer. The scale is a quantitative representation of the temperature of a body in arbitrary units.

Temperature scales are usually determined by choosing a particular physical property of a body such as the thermal expansion of a gas at constant volume or the change of electrical resistance of wire with changing temperature. Two thermal equilibrium points are assigned numbers and the interval is subdivided into a number of equal parts. Since such scales depend on different physical properties of the body substance, the temperature intervals need not be identical even though the numbers on the scales agree.

A scale that is independent of the substance used to define it is called a *thermodynamic scale*. A scale was first devised by Lord Kelvin and became known as the Kelvin scale, abbreviated K. It assigns a value of 273.16 K to the equilibrium condition of ice and water at standard pressure which is called the ice or freezing point; 373.16 K is assigned to the equilibrium condition of water and steam at standard pressure and is referred to as the *steam* or *boiling* point.

There are a number of other temperature scales based on the ideal thermodynamic scale but they use different units.

1. The Fahrenheit scale, abbreviated F, designates 0°F as the lowest point of a specific salt and ice mixture, 32°F to the ice or freezing point, and 212°F to the steam or boiling point.

TABLE 4-A

TYPICAL TEMPERATURE MEASUREMENTS, PULP AND PAPER INDUSTRY

A. Wood preparation
  1. Long pond water

B. Pulping
  1. Pulp grinder pit
  2. Digester cooking
  3. Blow heat recovery water

C. Washing
  1. Wash water
  2. Filtrate

D. Chemical makeup
  1. Chlorine vaporizer
  2. Chemical storage tanks
  3. ClO_2 generator
  4. Water

E. Bleaching
  1. Chemical mixers
  2. Bleach towers
  3. Water

F. Evaporators
  1. Black liquor

  2. Steam chest
  3. Steam
  4. Condensate

G. Recovery boiler
  1. Evaporator
  2. Black liquor heaters
  3. Steam

H. Recausticizing
  1. Green liquor heater
  2. Slaker
  3. Water heater

I. Lime kiln
  1. Kiln
  2. Molten Lime

J. Papermaking
  1. Paper machine headbox
  2. Felt conditioners
  3. Wire pit
  4. Felt dryers
  5. Vacuum condensers
  6. Calender surface

2. The Celsius scale is also called the Centigrade scale. Both are abbreviated C and assign 0°C to the ice point; 100°C to the steam point.
3. The Rankine Scale, abbreviated R', is also referred to as the Fahrenheit absolute scale; it designates 491.69°R' to the ice point and 671.69°R' to the steam point.
4. The Réaumur scale, abbreviated R, assigns the ice point to 0°R and the steam point to 80°R.

Figure 4-1 shows the relationship of the various temperature scales.

The Fahrenheit scale is the most frequently used in the United States, but the Celsius scale is rapidly being adopted. The other scales are practically never used. The equation to convert Celsius measurement to Fahrenheit is: °C = 5/9 (°F-32).

It can also be used in reverse. This conversion can also be accomplished by the use of a common conversion table (Table 4-B).

Other equations and tables can be used to convert other temperature scales.

TEMPERATURE MEASUREMENT SYSTEMS

All temperature measurements in the pulp and paper industry are of two basic groups: (1) the nonelectric types, which include the bimetallic, liquid-

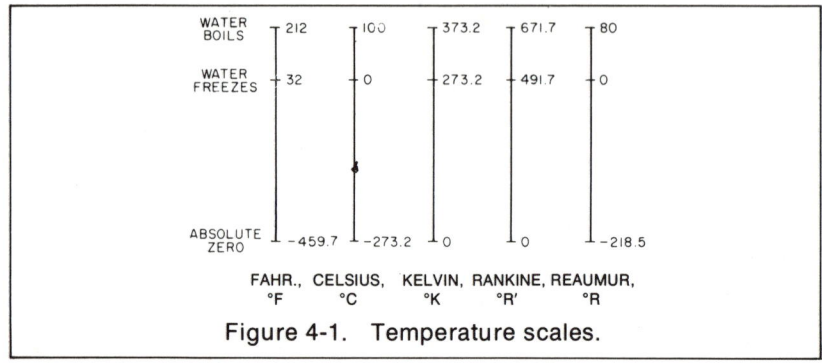

	WATER BOILS	212	100	373.2	671.7	80
	WATER FREEZES	32	0	273.2	491.7	0
	ABSOLUTE ZERO	−459.7	−273.2	0	0	−218.5

FAHR., CELSIUS, KELVIN, RANKINE, REAUMUR,
°F   °C   °K   °R′   °R

Figure 4-1. Temperature scales.

in-glass, and mechanical or filled thermal systems; (2) the electrical types, which depend on voltage or resistance changes caused by either a thermocouple, resistance element, or thermistor element. On several specific applications, radiation and optical pyrometers are also used.

TABLE 4-B
FAHRENHEIT/CELSIUS

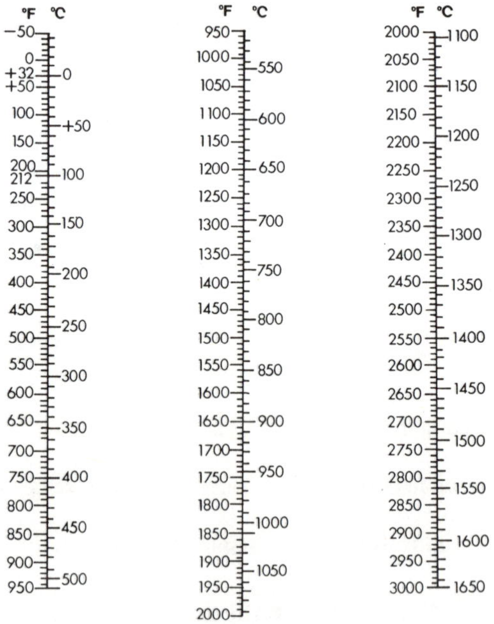

°F	°C		°F	°C		°F	°C
−50			950			2000	1100
0			1000	550		2050	
+32	0		1050			2100	1150
+50			1100	600		2150	
100	+50		1150			2200	1200
150			1200	650		2250	
200	100		1250			2300	1250
212			1300	700		2350	
250			1350			2400	1300
300	150		1400	750		2450	
350			1450			2500	1350
400	200		1500	800		2550	1400
450			1550			2600	
500	250		1600	850		2650	1450
550			1650			2700	
600	300		1700	900		2750	1500
650			1750			2800	
700	350		1800	950		2850	1550
750	400		1850	1000		2900	1600
800			1900			2950	
850	450		1950	1050		3000	1650
900			2000				
950	500						

Degrees $C = 5/9$ (Degrees $F − 32$)
Degrees $F = 9/5$ (Degrees $C + 32$)

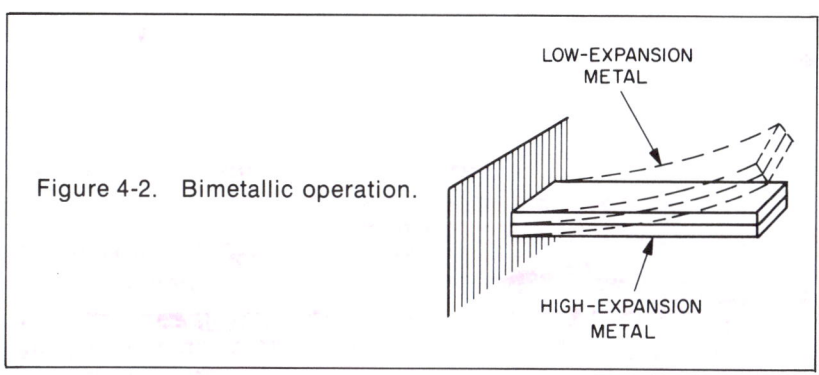

Figure 4-2. Bimetallic operation.

LOW-EXPANSION
METAL

HIGH-EXPANSION
METAL

Bimetallic Thermometer

Many temperature measurements need not be monitored continually or used for remote transmission to continuous recorders or controllers. These serve only as guides to satisfactory operation. One method used in making these on-site local temperature determinations is the bimetallic thermometer.

The operation of the bimetallic thermometer depends on the difference in expansion ratios of two unlike metals. A welded unit of different metals will change its curvature when subjected to a change in temperature (Figure 4-2). Temperature change causes the free end of a straight bimetallic cantilever beam to deflect. This deflection can be related quantitatively to the temperature change. The deflection is nearly linear with temperature, depending mainly on the coefficients of linear thermal expansion.

When this principle is employed in the industrial bimetallic thermometer, the bimetal is wound in the form of a helix with one end permanently fastened and the other end connected to a pointer which sweeps over a circular dial, indicating the temperature in accordance with the desired scale values. Figure 4-3 shows a bimetallic thermometer.

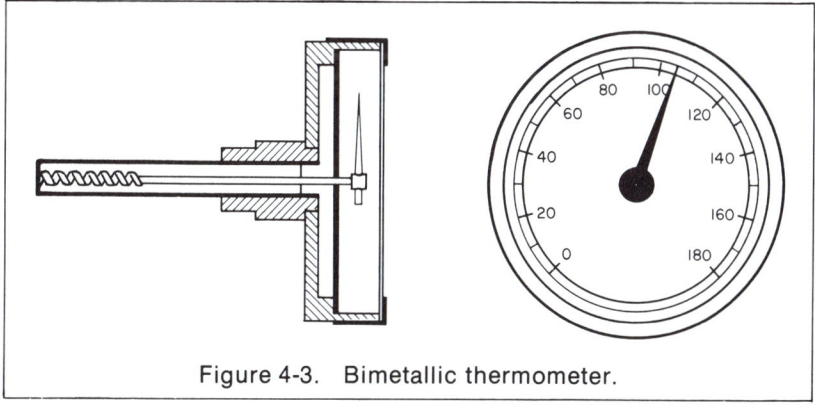

Figure 4-3. Bimetallic thermometer.

Temperature Measurement   95

The three basic forms of bimetallic thermometers are the flat spiral, the single helix, and the multiple helix (Figure 4-4).

Liquid-in-Glass

Another thermometer for on-site local temperature measurements in the general range of −50 to +1000°F is the liquid-in-glass type. Its operation depends on the characteristic of a liquid's volume to change as temperature changes. Therefore, a liquid-filled glass tube can be suitably calibrated to express temperature accurately. Many liquids could be used, but only a few—colored alcohol, hydrocarbons, and usually mercury—are common.

When the liquid is allowed to expand into a calibrated glass tube the temperature can be determined by personal observation. The upper limit cannot greatly exceed the boiling point of the liquid in a sealed tube because of the pressure that would be created. Thus, this method is limited to on-the-spot reading and lower temperature ranges. Typical industrial mercury-in-glass thermometers are shown in Figure 4-5. Their bulbs are glass envelopes containing mercury enclosed in a metal well. As heat is transferred through the well and metal stem to the mercury, the mercury expands into the capillary above.

Industrial mercury-in-glass thermometers are found in such pulp and paper mill applications as open tanks containing liquid, starch, and size cooking kettles, steam lines, pipelines for fluid flow, air ducts, or wherever a simple, inexpensive temperature indicating device with reasonable accuracy, precision, and speed is required.

Mechanical (Filled) Thermal Systems

The operation of filled thermal systems depends on the response of an expandable liquid, vapor, or gas contained in a completely closed system. A mechanical or filled thermal system consists of a temperature-sensitive bulb connected by capillary tubing to an expansible Bourdon, spiral, or helical element like those described in the chapter on pressure measurement. As the temperature of the fill material in the bulb rises, the increased volume and/or

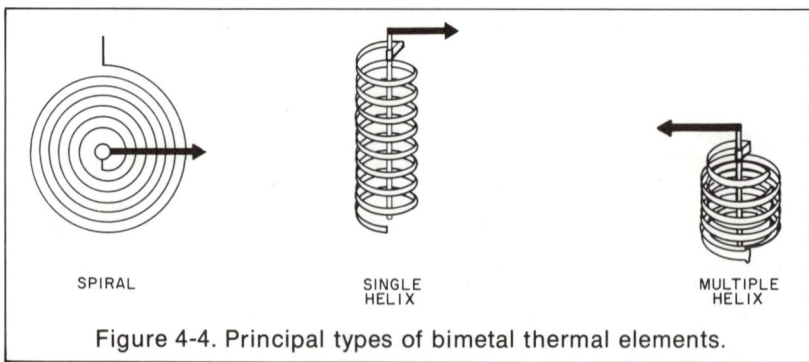

SPIRAL   SINGLE   MULTIPLE
                HELIX   HELIX

Figure 4-4. Principal types of bimetal thermal elements.

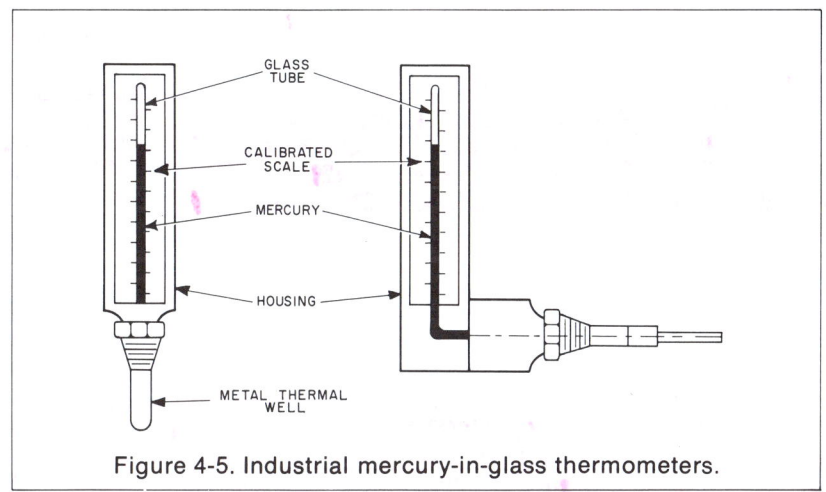

Figure 4-5. Industrial mercury-in-glass thermometers.

pressure is transmitted through the capillary tube to the element which re-sponds to the change in volume or pressure, thereby positioning a recording pen or indicating pointer that can be used for display or control. The basic components of an indicating temperature system employing a helical type element are illustrated in Figure 4-6. Filled thermal systems are ideal for

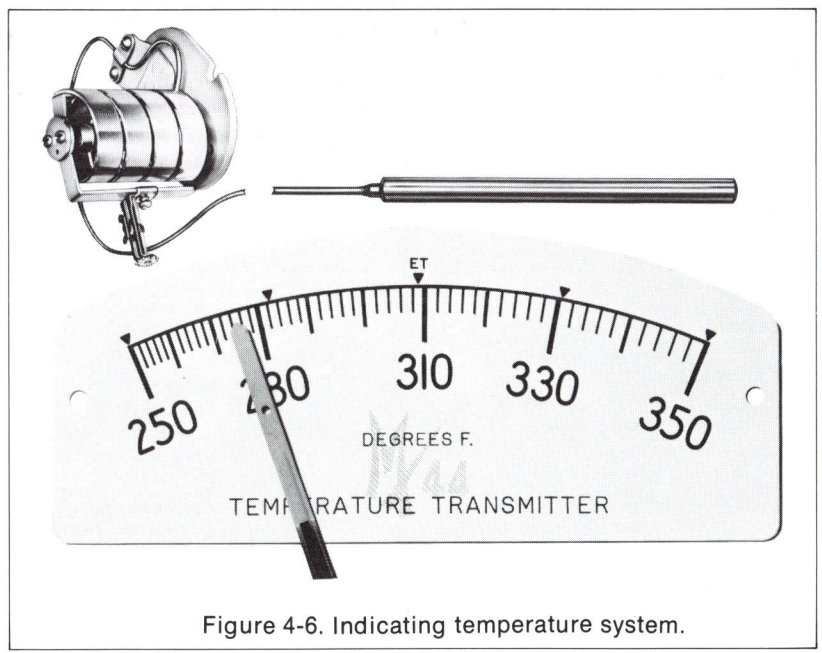

Figure 4-6. Indicating temperature system.

measuring large temperature changes. They are by far the most commonly used in pulp and paper mills today.

All filled thermal systems may be divided into two fundamental groups: (1) those that respond to volume changes and (2) those that respond to pressure changes. Those that respond to volume changes are completely filled with a liquid such as alcohol, hydrocarbon, or mercury. Pressure-responsive systems are filled with a gas or partially filled with a volatile liquid. In order to avoid confusion when specifying thermometers, American instrument manufacturers use the Scientific Apparatus Makers Association (SAMA) classification as standard (Table 4-C).

Class I Thermal Systems

General. Class I thermal systems consist of a closed unit completely filled with a liquid under pressure. (Liquid metals such as mercury are not used.) They include a bulb connected by capillary to an element wound in the shape of a C-spring, spiral, or helix, located in a remote indicating, recording, or control instrument case. An increase in temperature causes expansion of the liquid. To allow for the increasing volume, the element uncoils and the instrument pen or pointer is moved accordingly. These systems are available in narrower spans and smaller bulb size, which varies with span, than other filled systems.

Temperature changes at the element and along the tubing can affect the temperature measurement. Some liquid-filled thermal systems incorporate compensation for these variations by modifications of the basic Class I system. Several SAMA subclasses are the result.

Class IA (Fully Compensated). These systems make temperature measurements where the connecting tubing and responsive element are exposed to different ambient (surrounding) temperatures (Figure 4-7). They automatically compensate for changes in ambient temperature by using two capillaries and two elements which are matched in volume. Temperature variations at the element and along the capillary tubing produce equal and opposite motions from the element. These motions nullify each other so that only the motion

---

TABLE 4-C

A. Volumetric Expansion			IIA	Bulb temperature always above rest of system
	Class I	Liquid filled	IIB	Bulb temperature always below rest of system
	IA	Fully compensated		
	IB	Case compensated	IIC	Bulb temperature alternately above or below rest of system
	Class V	Mercury filled		
	VA	Fully compensated	IID	Bulb temperature at or close to rest of system
	VB	Case compensated		
			Class III	Gas filled
B. Pressure			IIIA	Fully compensated
	Class II	Vapor filled	IIIB	Case compensated

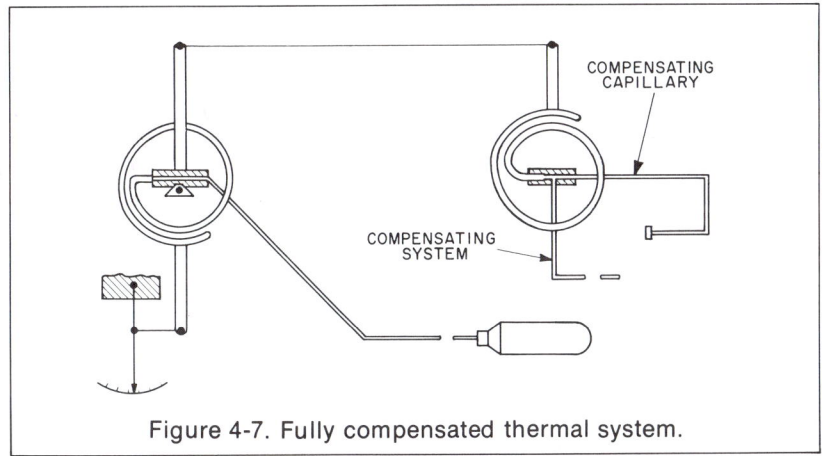

Figure 4-7. Fully compensated thermal system.

produced by varying the bulb temperature actuates the recorder pen. One hundred feet is considered the maximum practical length of capillary for this class of thermal system and the temperature ranges fall between −300°F and +600°F. Figure 4-8 shows a helical type Class IA element so constructed that the measuring helical floats on a movable base, the position of which is governed by the compensating helical.

Class IB (Case Compensated). This system is used when the capillary tubing and responsive element are at the same ambient temperature. It is referred to as case compensation because compensation is provided at the element location only, or within the instrument case. Compensation is accomplished by means of a bimetallic element which responds to variations in

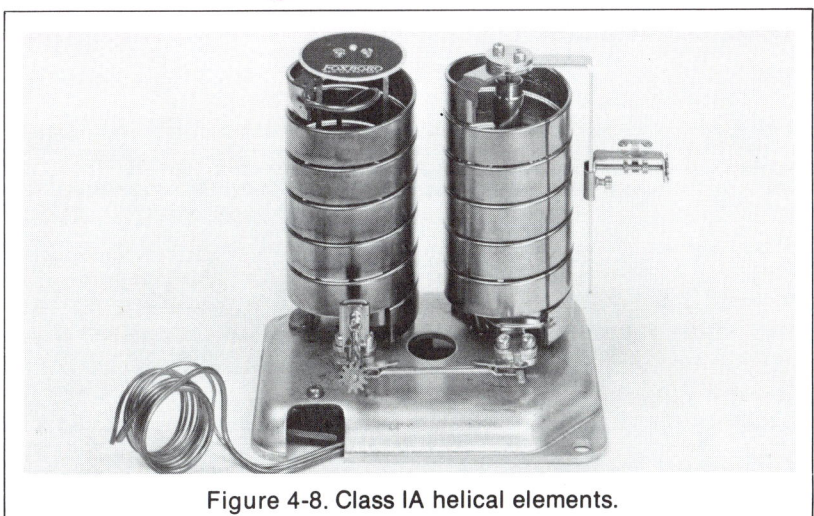

Figure 4-8. Class IA helical elements.

case temperature and nullifies measuring system motion caused by changing ambient temperature (Figure 4-9). Because the element and tubing remain at the same ambient temperature, the bimetallic compensator can be calibrated to correct for both tubing and case temperature variations. In order to maintain a reasonable accuracy, tubing lengths on Class IB systems are limited to 20 feet and temperature ranges are kept between −200°F and +600°F. With longer lengths the probability of temperature differences between case and tubing and the error from such a difference increases. Figure 4-10 shows a helical type Class IB element with the bimetallic compensator shaped as a helix accomplishing its compensating function in the same manner as the Class IA system with a floating measuring element.

Class II Thermal Systems

General. These systems operate on the principle of vapor pressure. The system is evacuated and then partially filled with a volatile liquid, the most common being methyl chloride, ethyl ether, ethyl alcohol, sulfur dioxide, and toluene. The pressure in any vapor pressure system is a function of the temperature at the vapor-liquid boundary. The system volumes are chosen so that under working conditions the vapor-liquid boundary is always in the bulb. The pressure, therefore, is a measure of the temperature of the bulb. Vapor pressure type thermal systems, therefore, require no compensation for changes in ambient temperature at the element or along the tubing. The vapor space in the system allows for expansion of the liquid and the pressure in the system remains unaffected, being governed only by the temperature at the vapor-liquid boundary which is maintained in the bulb. Temperature indication is a nonlinear function of bulb temperature because the vapor pressure grows at an increasing rate with rise in absolute temperature. This results in an expanding display scale which provides easier-to-read and more accurate temperature readings in the upper portion of the range. It is the simplest, most rugged, and least expensive of the filled thermal systems and is recommended

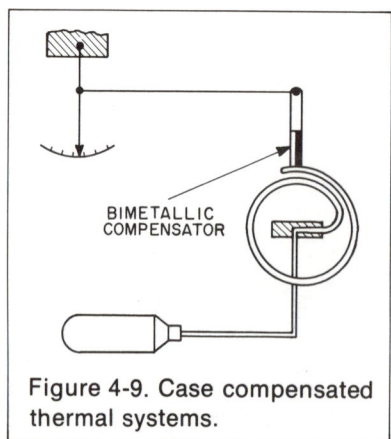

Figure 4-9. Case compensated thermal systems.

Figure 4-10. Class IB helical elements.

for general use on ranges between approximately −300°F and +600°F.

Class II vapor pressure thermal systems are made in four subclassifications: IIA, IIB, IIC, and IID (Figure 4-11) to meet the specific application conditions found in pulp and paper mills.

Class IIA. This is the most common construction and is used under conditions where the bulb temperature is always above the temperature of the rest of the system. It has liquid in the capillary tubing and element, and vapor and liquid in the bulb.

Class IIB. This system is used when the bulb temperature is always below the temperature of the rest of the system. It has vapor in the element and tubing, and liquid and vapor in the bulb.

Class IIC. This construction is used on applications where the bulb temperature may be alternately above or below the temperature of the rest of the thermal system. In operation it has either liquid or vapor in the tubing and element depending on relative bulb temperatures as described under Class IIA and Class IIB. This system will not give reliable temperature indication during transfer of the liquid into or out of the element, occurring when bulb temperature crosses the ambient temperature of element and capillary tubing.

Class IID. When a temperature measurement is usually at or close to ambient a Class IID thermal system is generally used. This is essentially a double filled system with a volatile and nonvolatile liquid. The actuating volatile liquid is confined to the bulb by a transmitting nonvolatile liquid of low vapor pressure which fills the capillary tubing and element. The bulb capacity is large so that the actuating liquid will never enter the capillary tubing.

Class III Thermal Systems

General. These systems are filled with a gas and operate on the principle of pressure change with temperature change. Class III systems operate on the basis of Charles' law, which states that the absolute pressure of a confined gas is proportional to the absolute temperature. An appreciable amount of gas

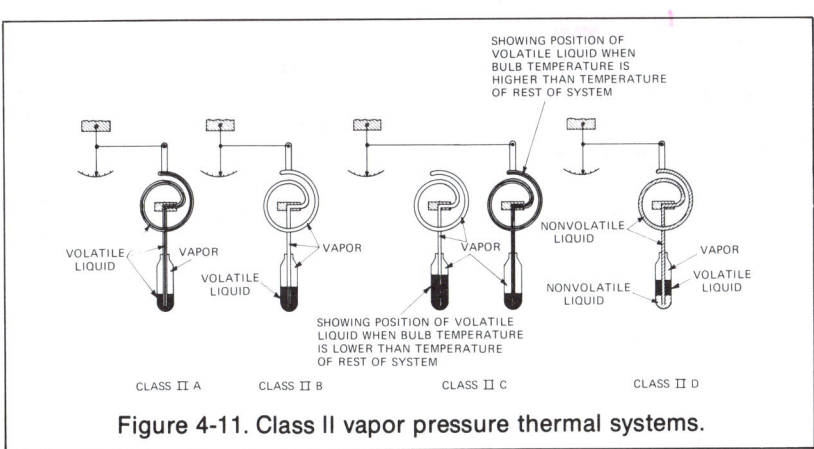

Figure 4-11. Class II vapor pressure thermal systems.

mass within the bulb is prevented from flowing out as the bulb temperature is increased across its range by making the capillary and element small compared with bulb volume. Gas-filled systems can measure the widest range of temperatures of all filled systems and perform best as wide-range devices with 300°F to 1000°F spans between the limits of approximately −450°F and +1400°F. With relatively large bulb volumes, they may be used to measure average temperatures of gaseous products flowing through large ducts, hoods, dryers, ovens, and other similar applications.

The construction of the gas thermal system is similar to that of the liquid-filled system, the major difference being the use of a confined, nontoxic, inert gas instead of a liquid to generate a pressure which is proportional to the temperature. Nitrogen and helium are the most common gases used but carbon dioxide, hydrogen, oxygen, and air could also be used.

Gas thermal systems are the most difficult to compensate accurately for ambient temperature changes due, in part, to the fact that gas density in the element increases with bulb temperature. Thus, a nonuniform rate of compensation is required. However, there is usually no need for compensation because the *volume* of the sensing bulb is many times greater than the *volume* of its connecting tubing (a ratio exceeding 40:1). Only a negligible pressure change results in the system from a change in ambient temperature.

In an effort to apply smaller bulbs, other methods of compensation have been attempted by modification. These have resulted in systems designated as Classes IIIA and IIIB.

Class IIIA (Fully Compensated). This version accomplishes compensation by means of a second thermal system without the bulb in the same manner as the Class IA thermal system shown in Figure 4-7.

Class IIIB (Case Compensated). In this case compensation is accomplished by constructing the system with a bimetallic element similar to the Class IB thermal system shown in Figure 4-9.

Class V Thermal Systems

General. The Class V thermal systems are completely filled with mercury or mercury-thallium amalgam and operate on the principle of liquid expansion. Basically, Class V is the same system as the Class I liquid-filled system except that it uses mercury as the filling fluid. Temperature ranges are between −38°F (freezing point of mercury) and +1200°F. As in Class III, compensation is by construction modifications—referred to as Classes VA and VB.

Class VA (Fully Compensated). Compensation is accomplished with a second system without the bulb in the same manner as Class IA, Figure 4-7.

Class VB (Case Compensated). With this system, compensation is provided by a bimetallic element like the Class IB system in Figure 4-9.

APPLICATION COMPARISON OF THERMAL SYSTEMS

A comparison of various filled thermal systems is summarized in Table 4-D. This information is an approximate guide only. Dimensional, functional, and physical data vary depending on the manufacturer.

TABLE 4-D

APPLICATION DATA FOR FILLED MEASUREMENT SYSTEMS

Type	Vapor Pressure	Gas Pressure	Mercury Expansion	Expansion Liquid
S.A.M.A. Class	IIA, B, C, D	IIIB	VA, VB	IA, IB
Temperature limits	−425 to +600 F[1]	−450 to +1400 F	−38 to +1200 F	−300 to +600 F
Minimum span	70 F[1]	200 F[6]	100 F[6]	40 F
Maximum span	400 F	1000 F	100 F	600 F
Limits of Overrange	IIA, C, D 50 F-above top scale temp. IIB-to 120 or 212 F	1400 F[7]	200% of span	100% of span
Tubing, Length, max.[2]	150 ft	100 ft	VA-100 ft VB-50 ft	IA-100 ft IB-20 ft
Bulb size, maximum	6 x 5/8	10 x 7/8	6 x 5/8	6 x 3/8
minimum	2 x 3/8	6 x 5/8[4]	3 x 1/24	3 x 1/4
63% Time constant[3]	2 sec	2-8 sec[4]	2-6 sec[4]	6 sec
Relative cost	low	medium low	medium high	high
Scale	non-uniform[5]	uniform	uniform	uniform

Notes:
1. Spans as narrow as 20 F are possible under certain application conditions, particularly very low temperatures. Minimum temperature of −425 F possible with special construction.
2. Longer lengths possible, but unwieldy bulb sizes or poor ambient temperature compensation usually result.
3. Time for temperature to reach 63 percent recovery constant of a step change for bulbs immersed in well-agitated liquid baths. Short tubing lengths and minimum bulb diameters are required to obtain these minimum figures.
4. Lowest value generally attainable only with force balance pneumatic transmitters. These instruments have bulbs as small as 6 x 3/8 (gas systems) and 3 x 3/8 (mercury systems).
5. Uniform motion or output with temperature may be accomplished for certain ranges by mechanical means.
6. Minimum gas and mercury system span for force balance pneumatic transmitters is 50 F.
7. Reduce to 250 F for narrowest spans.

Temperature Measurement   103

THERMAL BULBS

Temperature bulb designs vary from one instrument manufacturer to another. Basic components that make up a typical filled thermal bulb assembly are shown in Figure 4-12.

In an effort to simplify identification of bulbs, SAMA recommends that X be used to represent the sensitive portion of the bulb; U, the insertion length; J, the extension required to place the bulb in the best possible position; and Y, the diameter of the bulb. Figure 4-13 shows the basic bulb constructions.

The bulbs are used in the following conditions with reference to the illustrations shown in Figure 4-13:

A. Plain bulbs used for general application on shallow pots, open kettles, and tanks where no threaded support is required. Available in any thermal system except Class III.

B. Plain bulb with bendable extension for open vessel application where bending of extension positions the sensitive portion of bulb for best results.

C. Union bulb with bendable extension for use in pressure vessels or pipelines as bare bulb up to 1000 psi.

D. Union bulb with rigid extension for use as bare bulb only where strong lateral forces from high velocity or agitation are present.

E. Preformed coiled capillary bulb provides high speed of response for measuring average temperatures of ducts, ovens, dryers, and hoods.

F. Preformed loop type capillary bulb also used to measure average temperature of ducts, ovens, dryers, and hoods.

G. Lead-sheathed or plastic-coated bulb for use for corrosion resistance and electrical insulation in acid.

H. Flush mounting bulb for tanks or vats using agitators. Prevents damage by mounting flush with inside wall of tank.

CONNECTING TUBING

Capillary tubing and protective armor are available to meet the requirements of specific applications. They are manufactured in a variety of config-

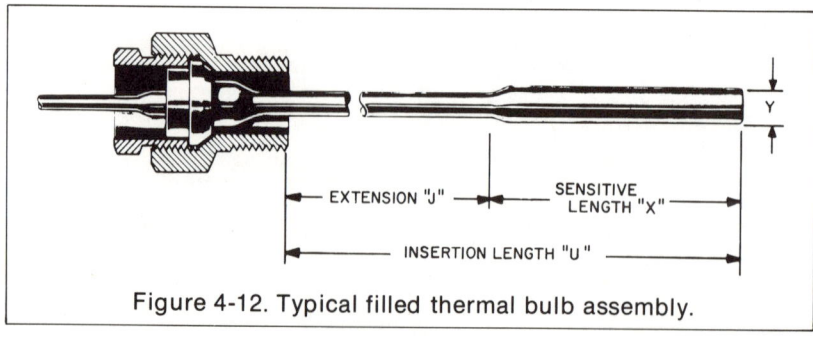

Figure 4-12. Typical filled thermal bulb assembly.

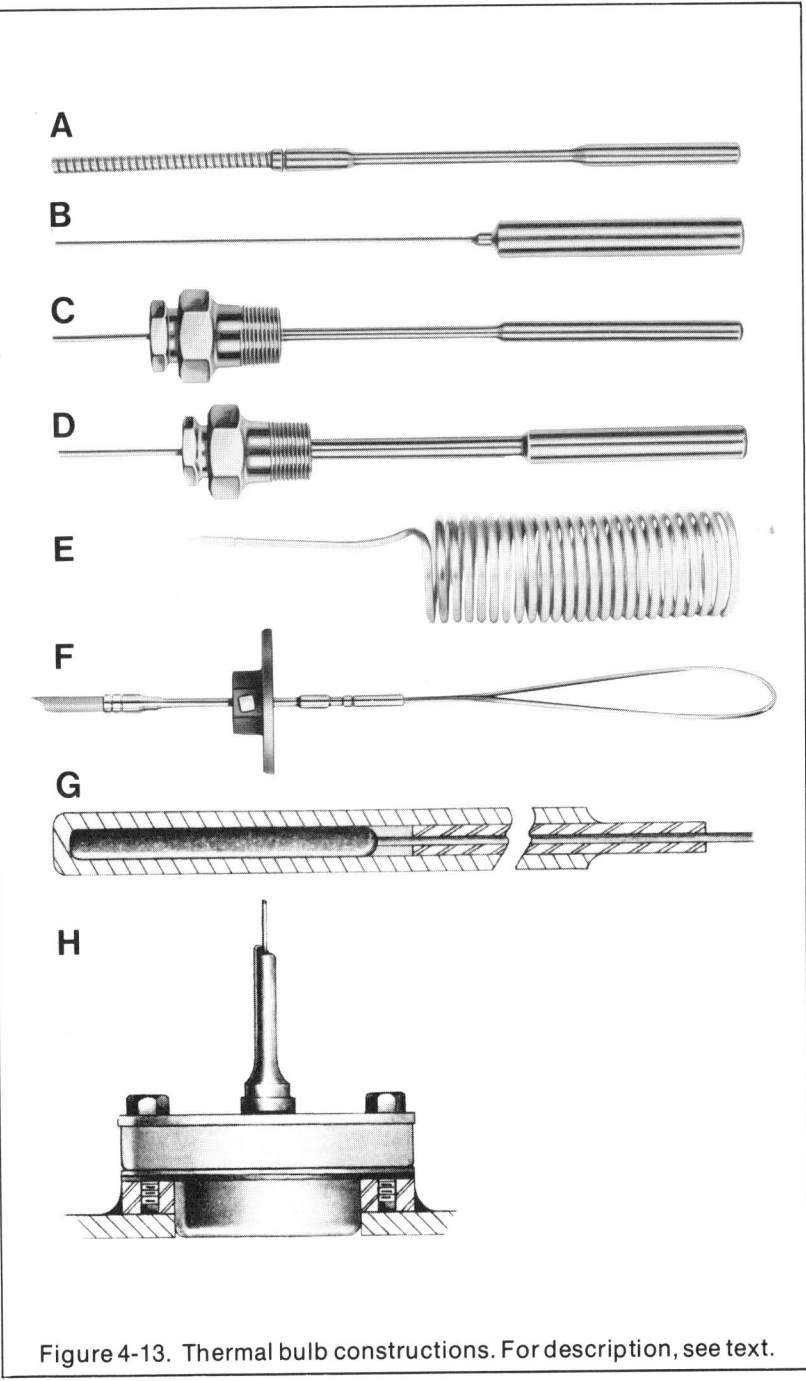

Figure 4-13. Thermal bulb constructions. For description, see text.

urations (Figure 4-14). These configurations are:

A. Heavy-duty ⅛-inch OD, Type 316 stainless steel capillary tubing for all filled thermal systems except Class IA. This is the strongest capillary tubing available.
B. ⅛-inch OD, Type 316 stainless steel sheathed dual for Class IA systems.
C. Copper capillary with flexible bronze armor for ranges to 600°F max.
D. Copper capillary with extruded vinyl plastic-covered flexible bronze armor. Maximum temperature rating of vinyl covering is 220°F.
E. Type 316 stainless steel tubing with Type 304 stainless steel flexible armor. More flexible than ⅛-inch OD tubing.
F. Type 316 stainless steel capillary with lead armor for sheathed bulb.

A stainless steel capillary with extruded vinyl plastic-covered flexible stainless steel armor ⅜-inch OD is also available. Maximum rating of vinyl covering is 220°F.

THERMAL WELLS

Pressure-tight protective thermal wells or sockets are used to protect bulbs from corrosion, abrasion, erosion, impact, and high pressure. They also allow the removal of the bulb without interrupting the process. The air space between the bulb and the well is kept to a minimum in order to assure ultimate speed of response by minimizing the thermal lag. Some typical thermal wells (Figure 4-15) are listed as:

A. Standard well, suitable for most installations. Lengths from 6 inches up; external threads ½ to 1 NPT.

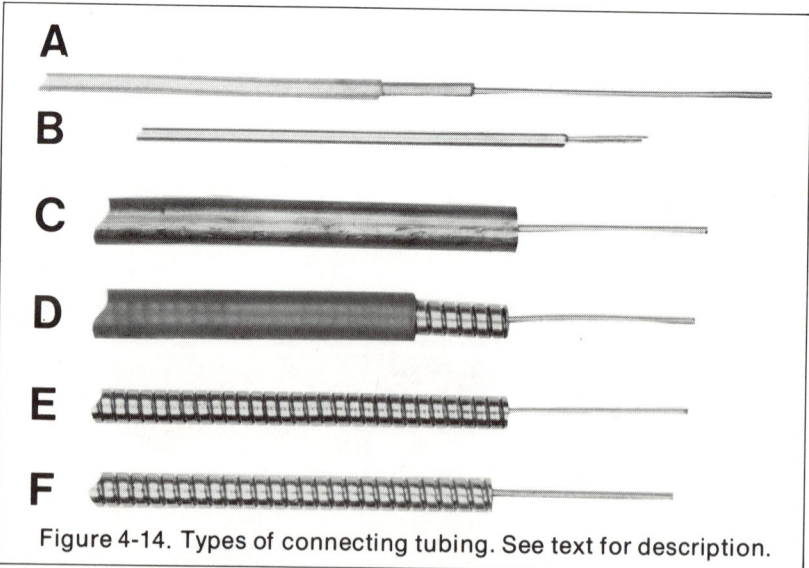

Figure 4-14. Types of connecting tubing. See text for description.

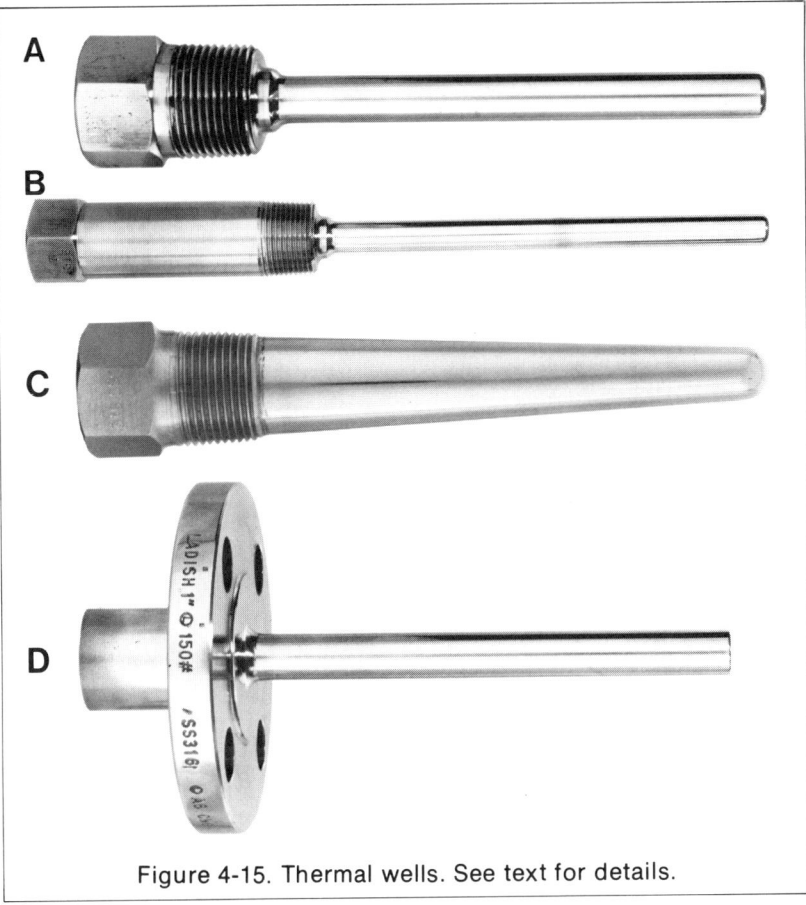

Figure 4-15. Thermal wells. See text for details.

B. Standard well with lagging extension to allow well to be used on lagged pipe installations.

C. Taper well, used in high-velocity pipelines, abrasive service, or any installation that requires high lateral or root strength such as steam lines and stock towers.

D. Flanged wells, used under conditions which favor flange mounting of the well rather than screw mounting.

Bulb flanges are used in thermal bulb assemblies to provide mounting support in dryers, ducts, and hoods where a pressure-tight connection is not needed. Two types are shown in Figure 4-16:

A. Clamp type for plain bulbs with bendable extensions.

B. Threaded iron flange for union-connected bulbs where the hole in the vessel cannot be threaded.

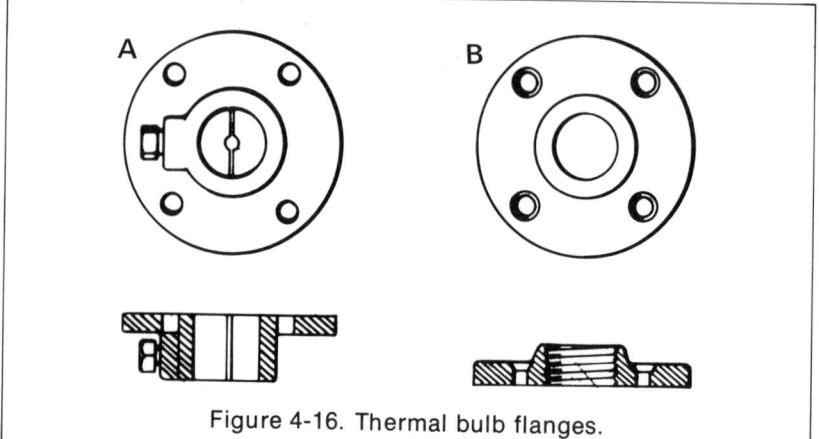

Figure 4-16. Thermal bulb flanges.

Typical bulb installation methods are illustrated in Figure 4-17:

A. Union bendable bulbs, where the well is screwed into a half coupling welded to the pipe wall.
B. Bulb union is installed in a tee through a reducing bushing.
C. Well with lagging extensions. Screwed or welded into half coupling welded to the pipe wall.
D. Union bendable bulb, bent and threaded into the wall.
E. Plain bendable bulb hanging on the edge of the vessel.
F. Union bendable bulb. Threaded flange bolted to the vessel cover.
G. Union bendable bulb in well, threaded into the sidewall.
H. Union bendable bulb installed through threaded hole in the sidewall.

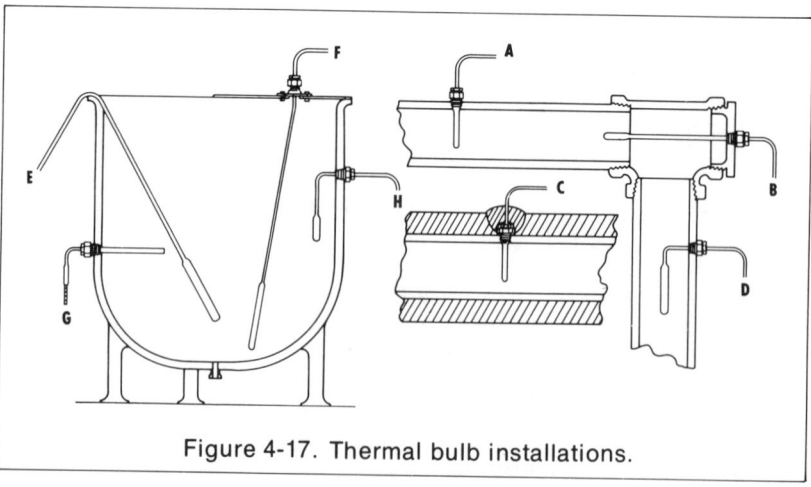

Figure 4-17. Thermal bulb installations.

ELECTRICAL THERMAL SYSTEMS

Although not used as extensively in the pulp and paper industry as mechanical (filled) thermal systems, electrical thermal systems provide another important means of measuring temperature. Electrical measurement of temperature is commonly used when the temperature-sensitive bulb is over 100 feet from the recording or control instrument. Neither the speed of response nor the sensitivity of the electrical system decreases with distance. Also, the cost of connecting cable is much less than that of capillary tubing. Electrical thermal systems that operate on a voltage change utilize a primary sensing element commonly referred to as a *thermocouple*. The sensing element for resistance type systems is a resistance temperature detector (RTD). Radiation and optical measuring systems involve other special means for detecting temperature.

Thermocouples

Thermocouples are useful in the paper industry to measure hot air or gases such as SO_2 from rotary sulfur burners, lime kilns, and in recovery and power boiler stacks. They operate on the principle of thermoelectricity which was discovered by Thomas J. Seebeck in 1821. Seebeck observed that a voltage is generated causing an electric current to flow continuously in a closed circuit consisting of two dissimilar metals when their junctions are maintained at different temperatures.

Basically a thermocouple temperature measuring system consists of two dissimilar metals (such as iron and constantan) with two junctions. The measuring or hot junction is the end inserted into the medium where the temperature is to be measured; the reference or cold junction is the open end that is normally connected to instrument terminals (Figure 4-18). The electromotive force (voltage) of the thermocouple increases as the difference in temperature increases; therefore, the receiver instrument which is capable of measuring emf can be calibrated to read temperature directly. The thermal emf developed is also dependent on the metals of the thermocouple. Although any two dissimilar metals could be used for a thermocouple only a few materials have proved practical. The most commonly used types of thermocouples made of these materials are listed as follows:

Copper-Constantan. This thermocouple has a pure copper element for the positive conductor and a constantan element for the negative conductor. Constantan is a general name that covers a group of alloys which contain approximately 55 percent copper and 45 percent nickel. Copper-constantan thermocouples are used in either oxidizing or reducing atmospheres within the temperature range of $-300°F$ to $+600°F$.

Iron-Constantan. Iron is used for the positive element and constantan for the negative element in these thermocouples. They can also be used in either oxidizing or reducing atmospheres but are limited to temperatures below $1000°F$ in oxidizing atmospheres because iron is subject to increasing oxidation as temperatures increase beyond this point. They can be used to $1500°F$ in neutral or reducing atmospheres.

Temperature Measurement   109

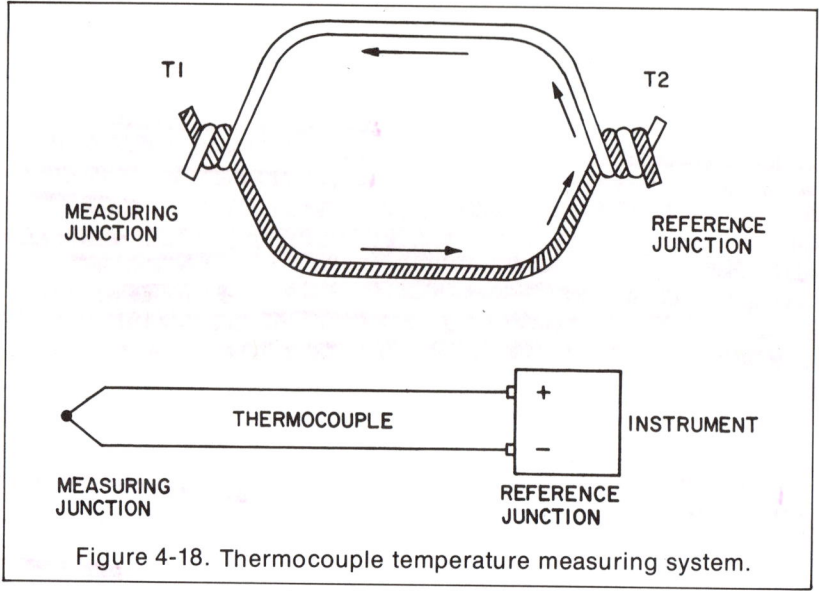

Figure 4-18. Thermocouple temperature measuring system.

Chromel-Alumel. This thermocouple consists of a chromium alloy positive element and an alumel negative element. Suitable for oxidizing atmospheres, it has an operating range of $-300°F$ to $+1600°F$.

Platinum Rhodium-Platinum. This thermocouple uses a platinum and a rhodium-platinum alloy for its elements. Temperature limits range as high as 2800°F, but their use is limited to oxidizing atmospheres because of the rapid deterioration effect of reducing atmospheres. Platinum rhodium-platinum thermocouples are highly accurate over a wide temperature range and are usually protected by a porcelain tube at high temperatures.

Generally, the thermocouple is encased in a well or some type of protective tube. Typical assemblies showing various protecting tubes are illustrated in Figure 4-19.

Proper location of the thermocouple is important for accurate temperature measurements. As shown in Figure 4-20, the couple must have sufficient immersion in the medium being measured to avoid errors due to conduction of heat to or away from the measuring junction. Although the couple should be subjected to the actual temperature being measured, it must be kept out of any direct flame for useful and reliable measurement and long life. Connection heads should not be subjected to temperatures in excess of about 400°F.

Thermocouples are subjected to various forms of contamination in service and may deteriorate in time. The accuracy of some couples becomes progressively less under such severe operating conditions as SO_2 gas from a sulfur burner. Therefore, periodic tests are advisable. Although many couples appear to be in good condition by observation, testing may show them to be sufficiently contaminated to be below minimum accuracy requirements. The most satisfactory method of checking thermocouples is while they are installed

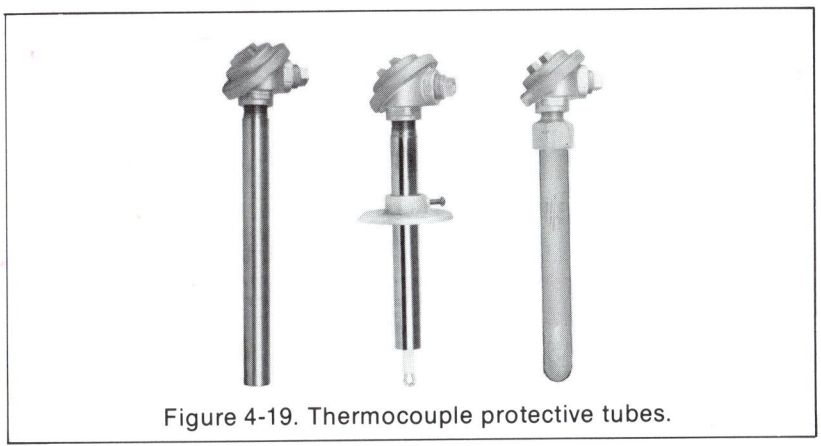

Figure 4-19. Thermocouple protective tubes.

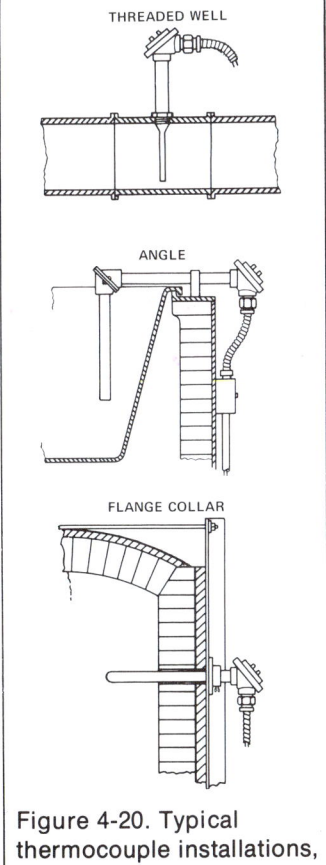

THREADED WELL

ANGLE

FLANGE COLLAR

Figure 4-20. Typical thermocouple installations, showing proper location.

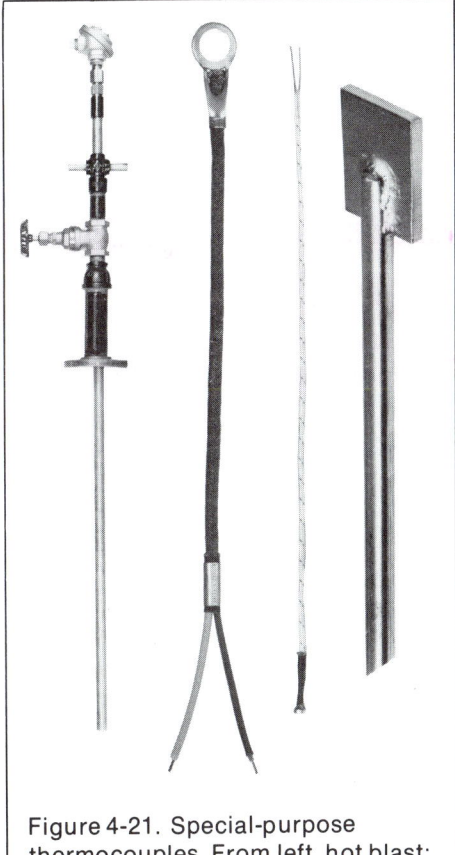

Figure 4-21. Special-purpose thermocouples. From left, hot blast; gasket; button; tube wall.

under actual operating conditions. Where space permits, plant tests are made by inserting a new check couple connected to an accurate portable instrument. The reading is then compared with the service couple measurement.

Special-purpose Thermocouples

Thermocouples are readily adaptable to many configurations and can be designed to comply with specific requirements. Figure 4-21 illustrates some typical special-purpose thermocouples. The hot blast thermocouple is especially designed for fast response and is used to measure the temperature of preheated air to a furnace. The gasket thermocouple is used for mounting on studs or bolts to measure skin temperatures of process lines, shell vessels, or other process machinery. The gasket couple is sometimes used under spark plugs on internal combustion engines. The button attaches to pipes, kettles, and process machinery for measuring surface temperatures. The tube wall thermocouple is designed for measuring furnace tube or retort wall temperatures. It is installed by welding the pad to the tube or some other surface.

Resistance Temperature Detectors

Another electrical device used in the pulp and paper industry to measure temperatures in digesters, on dryer rolls, calender rolls, etc., is the resistance temperature detector. Its operation depends on the inherent characteristic of metals to change their electrical resistances to current flow when they undergo a change in temperature. As the electrical resistance of wire varies with temperature, a coil of wire can be used as a temperature sensor and a direct relationship can be established depending on the material of construction. The sensitive portion consists of platinum, nickel, or copper wire wound on a silver or copper core (Figure 4-22) and encased in a protective shell. Curves relating resistance to temperature are known as *RTD curves* and are available for

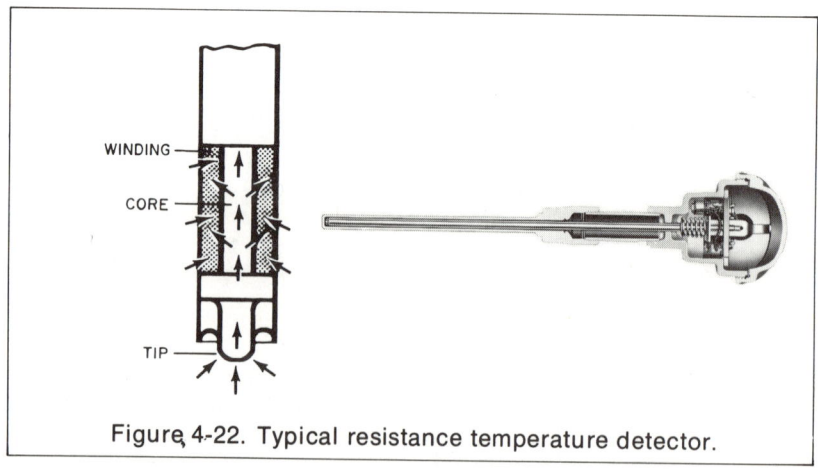

WINDING

CORE

TIP

Figure 4-22. Typical resistance temperature detector.

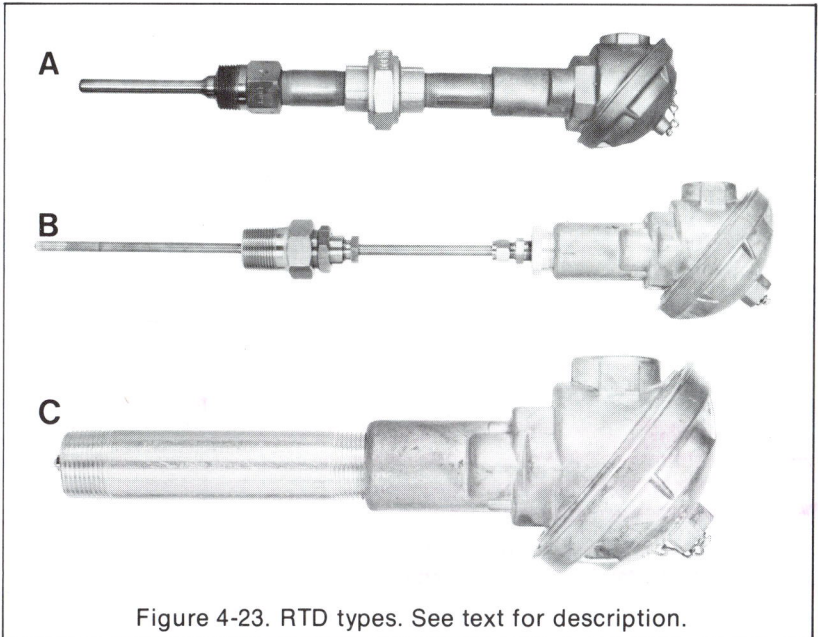

Figure 4-23. RTD types. See text for description.

resistance temperature detectors constructed of different metals.

RTDs are made in a wide variety of standard and special configurations in accordance with their intended use. Figure 4-23 shows three typical standard assemblies:

A. Well type assembly for general-purpose applications requiring accessibility and interchangeability of RTDs or electrical connectors.
B. Bare RTD assembly which threads into the sidewall of a vessel or pipeline, contacting process fluid directly for fastest response.
C. Surface temperature assembly provides terminal head convenience with the tip-sensitive RTD pressed against the pipe or vessel wall for surface temperature measurement.

Some typical special-purpose RTDs are shown in Figure 4-24. They are:

A. Surface RTD with a flat sensing element which may be taped to a flat or curved surface. This RTD is ideally suited for sheet metal structures and for installations requiring a very compact surface sensing element.
B. In the roll surface RTD, the resistance wire measuring head is suspended in a free-floating eel-slip faceplate shaped to the curvature of the surface moving at speeds up to 3000 feet per minute, such as dryer and calender rolls on a paper machine.
C. Room temperature RTD has a fully exposed resistance coil protected by a wall panel and mounted either flush or on the surface of a standard outlet box.

Temperature Measurement   113

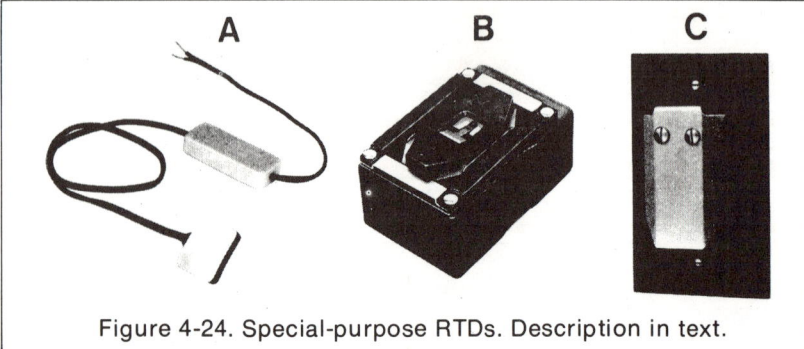

Figure 4-24. Special-purpose RTDs. Description in text.

Resistance type measurement systems are generally considered the most accurate of all for ranges between −100°F and +1000F°. A basic system consists of a temperature-sensitive primary resistance element or RTD and a receiving secondary element or instrument which converts the change to a temperature reading. The Wheatstone bridge, with various modifications required for use with an RTD, is the most common detection method used as the receiving instrument to detect and convert resistance changes to temperature. A simple bridge is shown schematically in Figure 4-25. A battery applies a potential across two of the junction points and a galvanometer connects the other two points. *A* and *B* are fixed resistors, while *R* is the resistance thermometer, and *S* is an adjustable slidewire. Adjustment of slide *C* on slidewire *S* balances the bridge which can be calibrated for temperature with slider *C* reading on a calibrated scale.

An ac capacitance bridge (Figure 4-26) is also frequently used. An ac supply of 1000 Hz is used and the balance of the bridge is detected by electronic means. The variable capacitor (C3) is used for balancing the bridge by impeding its branch. Balance of the bridge is, therefore, accomplished by adjustment of capacitor C3 against variation in resistance R1 of the RTD.

APPLICATION COMPARISON OF ELECTRICAL THERMAL SYSTEMS

A comparison of thermocouple and resistance thermal systems is summarized in Table 4-E. This information should be considered an approximate guide only, because makers' functional data and specifications vary.

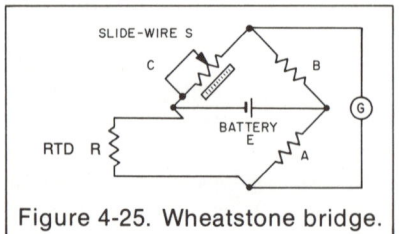

Figure 4-25. Wheatstone bridge.

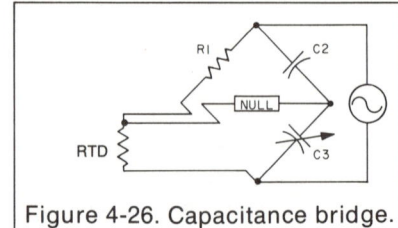

Figure 4-26. Capacitance bridge.

TABLE 4-E

APPLICATION COMPARISON OF ELECTRICAL THERMAL SYSTEMS

Type Thermocouples	IDA Type	Temperature Limits[1] Min.	Max.	Approximate Minimum Span[2] Minus 100 F to Plus 200 F	200 F to 500 F	500 F to 1000 F	1000 F to 2000 F	2000 F to 3500 F	Max. Span[3]	Maximum Overrange Temperature
Copper-Constantan	T	−300 F	650 F	250 F	200 F	160 F	–	–	A	1100 F
Iron-Constantan	J	−300 F	1400 F	200 F	175 F	160 F	150 F	–	A	1800 F
Chromel-Alumel	K	−300 F	2000 F	250 F	225 F	220 F	220 F	–	A	2200 F
Platinum, Rhodium-Platinum	R, S	+32 F	2650 F	–	900 F	800 F	700 F	650 F	A	3100 F
Resistance Bulbs										
Nickel	–	−320 F	600 F	5 F	5 F	5 F	–	–	B	660 F
Platinum	–	−420 F	1650 F	120 F	120 F	120 F	120 F	–	B	1800 F
Copper	–	−320 F	250 F	200 F	200 F	–	–	–	B	300 F

LIMITS OF ERROR OF THERMOCOUPLES, THERMOCOUPLE WIRES AND EXTENSION WIRES

Type of Thermocouple	Temperature Range Min.	Max.	Standard Limits of Error	Extension Wire	Temperature Range Min.	Max.	Standard Limits of Error
Copper-Constantan	−150 F	−75 F	±2% of reading	Copper-Constantan	−75 F	+200 F	±1½ F
	−75 F	200 F	±1½ F				
	200 F	600 F	±¾% of reading				
Iron-Constantan	0 F	530 F	±4 F	Iron-Constantan	0 F	+400 F	±4 F
	430 F	1400 F	±¾% of reading				
Chromel-Alumel	0 F	530 F	±4 F	Iron-Cupronel	+75 F	+400 F	±6 F
	530 F	2300 F	±¾% of reading	Chromel-Alumel	+75 F	+400 F	
Platinum, Rhodium-Platinum	0 F	1000 F	±5 F	Copper-Alloy	+75 F	+400 F	±12 F
	1000 F	2700 F	±½% of reading				

1. Upper temperature limits vary with construction and wire size. Smaller gauge thermocouples are more sensitive but deteriorate faster at higher temperatures.
2. Temperature change is necessary to develop approximately 5 millivolts in the specified range. Substantially narrower spans with less common minimum null balance instruments with a 1 millivolt span.
3. A is maximum span governed only by applicable temperature limits; B is maximum span generally governed by measuring instrument within temperature limits listed. The use of resistance thermometers for wide span applications (greater than 200 F) generally defeats the superior accuracy characteristics for which resistance measurement is most often selected.

OTHER TEMPERATURE MEASUREMENT SYSTEMS

Temperature measuring systems which operate on the radiation effect of a heated body are also used to a limited extent in the pulp and paper industry. Known as *radiation* and *optical pyrometers*, they are used to measure temperatures of objects, without requiring physical contact, which are inaccessible by any means other than visual, such as molten lime in lime-burning kilns, recovery and power boiler interiors, and moving paper sheet surfaces on the paper machines. Generally, there is no upper temperature limit for radiation and optical pyrometers.

Radiation Pyrometer

The ability to measure temperatures with a radiation pyrometer is based on the fact that every object emits radiant energy, and the intensity of this radiation is a function of the object's temperature. A sufficiently hot object emits a certain amount of light or *visible* radiation and the hotter the object, the brighter and whiter its color. These objects also emit a tremendous amount of *invisible* infrared radiation. The intensity of this infrared radiation is also a function of the object's temperature. The smaller fraction of the total energy radiated is visible. Below certain temperatures (1000°F for a perfect emitter), the intensity of visible radiation for most objects is so small that it cannot be seen. However, there is still a large quantity of infrared emissions below these specified temperatures. The radiant energy at every wavelength increases with increasing temperature and the determination of the radiant intensity at any wavelength establishes the object's temperature. The difference between infrared radiation and visible radiation is the wavelength of the electromagnetic wave. Red light has a longer wavelength than blue light and infrared radiations have longer wavelengths than either.

All radiation can be considered to be composed of elementary packets of energy called photons. All photons travel in straight lines at the speed of light. They can all be reflected by appropriate mirrors. Their paths can be bent and focused by the proper refractive elements or lenses, and can be interpreted as representative of total energy being radiated from a heated body. All photons will dissipate their energy as heat on being absorbed by an appropriate absorber. A measure of this heat can then be correlated to the temperature of the object under observation.

Radiation detectors take many forms but all serve the same purpose of converting an incident photon flux into an electrical signal. The two main types are the thermal detector and the quantum detector.

A typical thermal detector type radiation pyrometer consists of an optical system combined with a thermopile (a number of thermocouples connected in series). It measures the total energy radiating from a heated body. The optical system focuses the energy radiated on the thermopile, which produces an output millivoltage directly related to the temperature of the source under measurement. The basic components of a typical radiation pyrometer are shown in Figure 4-27. The radiated energy is converted into an electromotive force (voltage) by the thermopile. This voltage can then be measured and

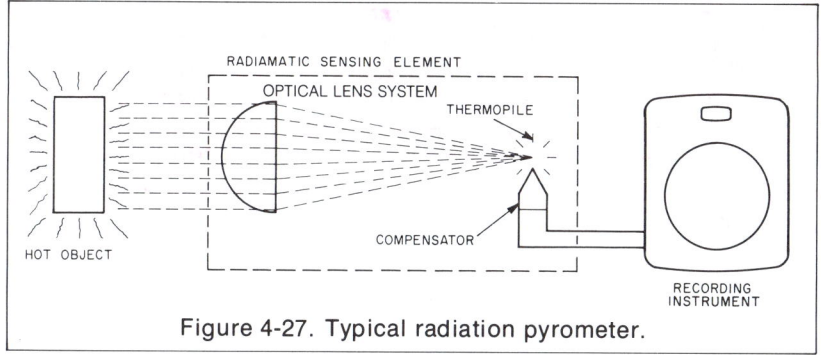

Figure 4-27. Typical radiation pyrometer.

converted to temperature in a number of ways.

The typical quantum detector type radiation pyrometer senses radiation by employing a semiconductor detecting element. The incident photon interacts with a bound electron within the crystal lattice. The photon's energy is transferred to the electron to free it from its immobile state, permitting it to move through the crystal. During the time it is free it can produce a signal voltage in the detector. This voltage is used for indicating, recording, or controlling of the temperature being measured.

Optical Pyrometer

The operation of optical pyrometers is based on the principle that a measure of the spectral radiant intensity of the radiated energy from a heated body at a given wavelength can be related to the temperature of the body. In other words, the color of an object is an indication of its temperature and the brightness of a hot object is also a measure of its temperature. In the basic circuit used with optical pyrometers (Figure 4-28), a reference filament in the system is made to match the brightness created by the radiated energy from the unknown temperature source. The two temperatures are assumed equal when the filament is indistinguishable in brightness from the hot object in the background. Commercially available radiation pyrometers vary primarily in the method used to make this match. When viewed through a suitable

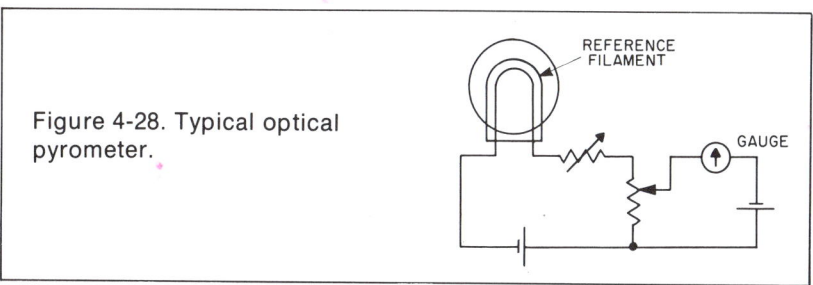

Figure 4-28. Typical optical pyrometer.

eyepiece, if the filament appears darker (colder) than the object being measured, its brightness is increased by passing more current through it. If the filament appears brighter (hotter) than the object being measured, the current through the filament must be reduced. When the current has been adjusted to the correct value the filament becomes indistinguishable from the object being measured. The voltage produced by the flow of current necessary to achieve this match is then measured by a potentiometer and converted to temperature units by a suitable scale.

BIBLIOGRAPHY

Anderson, N. A. *Instrumentation for Process Measurement and Control*. Pittsburgh: Rimbach Publications, Chilton Company, 1964.

Benedict, R. *Fundamentals of Temperature, Pressure, and Flow Measurements*. New York: John Wiley & Sons, 1975.

Class II Vapor Pressure Thermal System. Foxboro, MA: The Foxboro Company, TI 11-11a, 1963.

Considine, D. M. *Process Instruments and Controls Handbook*. New York: McGraw-Hill Book Company, 1957.

Eckman, D. P. *Industrial Instrumentation*. New York: John Wiley & Sons, 1950.

Lipták, B. G., ed. *Instrument Engineer's Handbook*, vol. 1, "Process Measurements." Philadelphia: Chilton Book Company, 1969.

Nerzfeld, C. M., ed. Temperature: Its Measurement and Control in Science and Industry, vol. 3. Huntington, NY: Robert E. Krieger Publishing Company, Inc., 1963.

Payne, H. G. *Temperature Controls—Selection and Application*. Fourth Conference on Manufacturing Automation, Purdue University, April 1920.

Temperature Conversion. Foxboro, MA: The Foxboro Company, TI 5-1a, 1961.

STUDY QUESTIONS

Indicate whether the following statements are True or False by inserting T or F in parentheses:

1. A temperature scale is a qualitative representation of the level of heat energy in a body expressed in specific and absolute values. (F)

2. In the Kelvin scale, 273.16 K is assigned to the equilibrium condition of water and steam at atmospheric pressure and is referred to as the "steam" or "boiling" point. (F)

3. 5 degrees Celsius is equivalent to $41°F$. (T)

4. Temperature measurements in the paper industry are either mechanical or electrical in nature. (F)

5. The operation of a bimetallic thermometer depends on the difference in contraction and expansion ratios of two similar metals. (F)

6. The operation of filled thermal systems depends on the response of an expandable liquid, vapor, or gas contained in a completely closed system. (T)

7. Class I thermal systems are filled with a liquid under pressure. ( )

8. Class IB thermal systems are used where connecting tubing and responsive elements are exposed to different ambient temperatures. ( )

9. Class II thermal systems operate on the principle of vapor pressure. ( )

10. Class III thermal systems are completely filled with mercury and operate on the principle of liquid expansion. ( )

Select answer or answers which correctly apply to the following statements:

11. In the identification of thermal bulbs, "X" is used to represent the.
    a) insertion length.
    b) sensitive portion.
    c) diameter of the bulb.

12. The air space between the thermal bulb and well is kept to a minimum in order to
    a) prevent entrapment of dirt.
    b) provide close tolerance fittings.
    c) eliminate thermal lag.

13. Electrical measurement of temperature would be preferred if the distance between temperature-sensitive bulb and the instrument is
    a) 10 ft.
    b) 25 ft.
    c) 500 ft.

14. Basically a thermocouple temperature measuring system consists of
    a) two dissimilar metals with one junction.
    b) two dissimilar metals with two junctions.
    c) three dissimilar metals with three junctions.

15. A common type of thermocouple is made of
    a) iron-copper.
    b) copper-constantan.
    c) platinum gold.

16. Common materials used in the sensitive portion of a resistance temperature detector (RTD) consist of
    a) platinum wire wound on silver core.
    b) nickel wire wound on copper core.
    c) gold wire on a plastic core.

17. Bare RTDs are used primarily to
    a) save the expense of a thermal well.
    b) simplify installation.
    c) achieve faster response.

18. Resistance type temperature measuring systems are generally considered most accurate between
    a) −100 F and +675 F.
    b) 750 F and 1600 F.
    c) 1600 F and 2650 F.

19. The usual radiation pyrometer consists of
    a) an optical system.
    b) RTDs.
    c) thermocouples.

20. With an optical pyrometer, temperature is measured as a function of
    a) the color of an object.
    b) brightness of the object.
    c) density of the object.

(See Appendix for answers)

5. Analytical Measurement

The purpose of analytical measurement is to provide information on the composition of the contents of a process stream—information that can be used to maintain conditions necessary to meet predetermined requirements. Analytical measurements are usually made on raw materials going to the process, materials within the process, and final products of the process. Originally, analytical measurements were made by a chemist who quantitatively and qualitatively tested representative "grab" samples in a control laboratory. But information gathered in this manner, being of an intermittent rather than a continuous nature, proved to be of limited value for control use on continuous processes. This prompted chemical and petroleum industry technologists to pioneer the development of methods to make many of these measurements continuously and on-line. Subsequently, several of these methods were adapted for use in pulp and paper processes.

Analytical measurements used to determine composition are either of a chemical or physical nature. Although there are many types of chemical analytical measurements, those most common to the pulp and paper industry are most commonly classified as electrochemical. Included in this classification are conductivity, hydrogen-ion concentration (pH), ion-selective potential, oxidation-reduction potential (ORP), amperometry, and capacitance. Of the many analytical measurements of a physical nature, those that have been used in the pulp and paper mills are turbidity and humidity.

Many of these measurements are still considered in the development stage, but practically all of them have been or are being used for process measurement in the paper industry in some form.

CONDUCTIVITY MEASUREMENT

Conductivity measurement is the determination of the solution's ability to conduct electric current; its conductance, (the reciprocal of its resistance). It is generally expressed as the conductance of a unit volume of liquid and called *specific conductance* or more usually *conductivity*. This measurement is made in the paper industry to detect electrolytic contaminants around water and waste treatment areas and in such normally low conductive liquids as condensate of digester liquor heaters and black liquor evaporator condensate. It is also used as a guide to equipment operation. An example of this would be brown stock washer filtrate conductivity, whereby information is provided for use in regulating wash water to control it at its optimum washing effectiveness. Conductivity measurement is also used in the paper industry to determine concentration of a chemical in which a relationship between conductivity and concentration has been established as, for instance, in the dilution of sodium hydroxide and alum solution from a high storage concentration to a lower desired concentration.

Basic Theory of Conductivity

Aqueous solutions of acids, bases, or salts are known as *electrolytes* and are conductors of electricity. The degree of electrical conductivity of such solutions is affected by three factors: the nature of the electrolyte, the concentration, and the temperature. A measurement of the conductivity at a fixed temperature can be a measurement of the solution concentration and can be expressed in percent by weight, parts per million, or other applicable units.

Conductivity Cells

If the conductivity value of various concentrations of an electrolyte is known, it is then possible to determine the concentration by passing current through a solution of known dimensions and measuring its electrical resistivity or conductivity. Figure 5-1 illustrates this relationship for some common electrolytes. The primary element in an electrical conductivity measurement

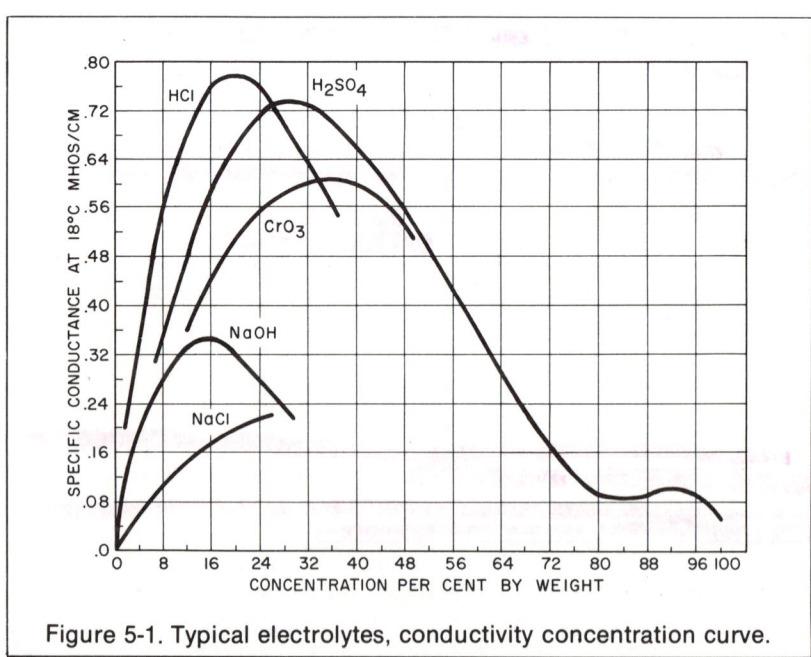

Figure 5-1. Typical electrolytes, conductivity concentration curve.

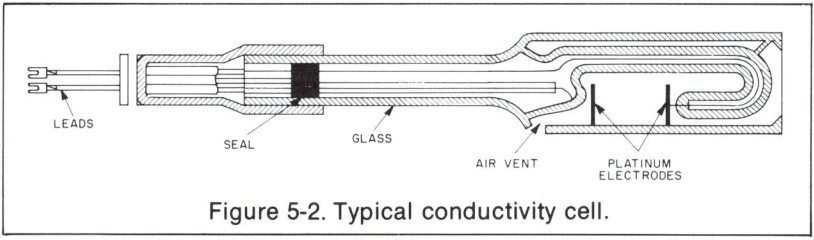

Figure 5-2. Typical conductivity cell.

system is the conductivity cell. These cells (Figure 5-2) consist of a pair of electrodes whose area and spacing are precisely fixed, a suitable insulating member to define the conducting paths of a fixed volume of liquid, and suitable fittings for supporting and protecting the cell. The conductivity measured by the cell varies as follows:

$$(1) \quad k = C\frac{l}{A}$$

where k = conductivity, mho/cm
        C = conductance, mhos
        A = area of electrodes, cm^2
        l = distance between electrodes, cm

To establish a common basis for comparing conductivity of different solutions, an idealized conductivity cell is considered with the conducting paths all confined within a volume of one cubic centimeter between electrodes, one square centimeter in area, and one centimeter apart. In this idealized cell, the conductivity is shown by equation (1) to be numerically equal to the conductance. For example, this cell would have a conductance of 0.03 mho if contained in a liquid with a conductivity of 0.03 mho/cm. (In the past it has been common practice to drop the "per centimeter" part of the conductivity unit. However, this practice leads to confusion when working with real cells and with other units and is no longer considered acceptable for technical work.)

As the physical dimensions of the cell are changed the conductivity relationship is as follows:

$$(2) \quad C = \frac{k}{F}$$

where F = cell factor, cm^{-1} (or cell constant)

It is also comon practice to omit the dimension unit when referring to cell factor, so that a cell factor of 10 cm^{-1} is said to have a cell factor of 10. Referring back to equation (1) it can be seen that F is a function of $\frac{l}{A}$ or the physical dimensions of the cell. These relationships are shown in Figure 5-3. The cell factor then serves as a multiplying factor to increase or decrease conductance or resistance output of the cell to a desirable value by selecting low constant cells

Figure 5-3. Drawing showing physical dimension relationships of basic conductivity electrodes.

for low conductivity (high resistivity) solutions and high constant cells for high conductivity (low resistivity) solutions. Table 5-A may be used as a guide in selecting the proper cell constant for a particular conductivity range.

Conductivity cells are constructed in a variety of configurations depending on application requirements. Mounting arrangements, for instance, can vary to meet specific installation needs. Configurations can also be affected by the desired range of cell constants. Some typical cells are illustrated in Figures 5-4 and 5-5.

The insertion type cell is shown in Figure 5-4. In this illustration the cell is fitted for installation in a pipeline or in the side of a tank. It is available in cell constants of 10 and 100. The sensitive portion of the cell consists of two platinum electrodes mounted in an H-shaped structure of Pyrex glass tubing. The electrodes, located in separate sections of tubing, are platinum rings centrically mounted in the tubing and flush with the inside surface of the tubing. The cell will withstand pressures up to 125 psi.

The insertion type cell illustrated in Figure 5-5 is also designed for mounting in a pipeline or in the side of a tank. But this assembly includes a valve and fittings to provide for removal of the cell without interrupting the process. In addition, this type of cell, as illustrated, employs carbon rather than metallic electrodes. They consist of two small carbon rods mounted in fluorocarbon plastic to insulate them from each other and from the cell fittings. One of the electrodes can be a carbon rod while the other can be a concentric carbon ring.

Other cell configurations for immersion type intermittent testing and so-called sample flow-through measurements are available, but they are not commonly used for industrial pulp and paper process applications.

TABLE 5-A

RANGE OF CONDUCTIVITY (MICROMHOS)

Bottom Value Desired	Top Value Desired		Cell Constant
	Minimum	Maximum	
0	1	200	0.01
0	100	2,000	0.10
0	1,000	5,000	1.00
0	5,000	200,000	10.00
0	100,000	2,000,000	100.00

Secondary Element

An indicating, recording, and/or control instrument serves as the secondary element of a conductivity measurement system. Conductivity instruments use a Wheatstone bridge circuit to measure resistance between the conductivity electrodes. The instrument may be calibrated for either resistance or conductance because:

$$(3) \quad C = \frac{1}{R}$$

where C = conductance, mhos
R = resistance, ohms

By incorporating a conductivity cell in the circuit of the secondary instrument, a measurement of conductance or resistance can be made to move a pen or pointer that has been properly calibrated in desired values in accordance with conductivity changes. These secondary instruments can be calibrated in

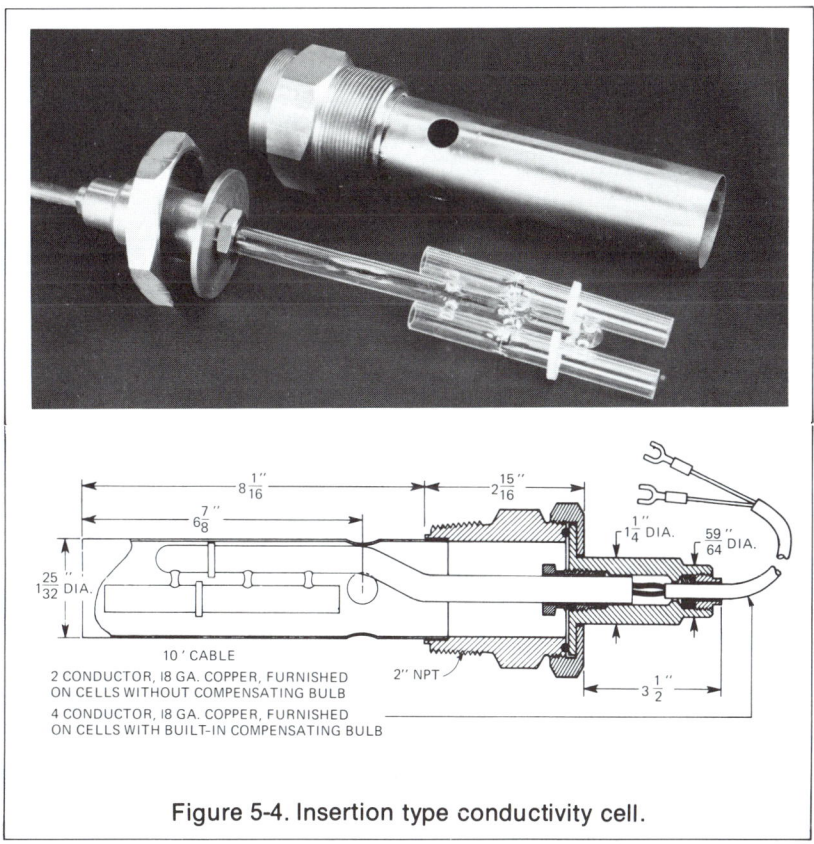

Figure 5-4. Insertion type conductivity cell.

either electrical or conductivity units or in terms of concentration of solution such as percent concentration, grams per liter, and parts per million. The conductivity of most solutions increases as the temperature increases. Therefore, if it is desired to read the measurement in concentration units and the temperature of the solution cannot be kept constant, the effect of temperature on the solution's conductivity can be compensated for manually or automatically.

Manual compensation is accomplished by a manually operated adjustment in the circuit, calibrated in the actual temperature of the solution at point of measurement.

Automatic temperature compensation consists of a resistance temperature detector (RTD) which is inserted in the vicinity of the measuring cell. As the temperature of the solution changes, the resistance of the RTD changes. This resistance is incorporated into the measuring bridge circuit and it compensates for temperature changes by adjusting the output span rather than adjusting to the zero. This is because the effect of temperature on conductivity is a percentage change per degree temperature change. The simplified schematic (Figure 5-6) illustrates how the cell and compensating RTD are incorporated in the conductivity measurement circuit.

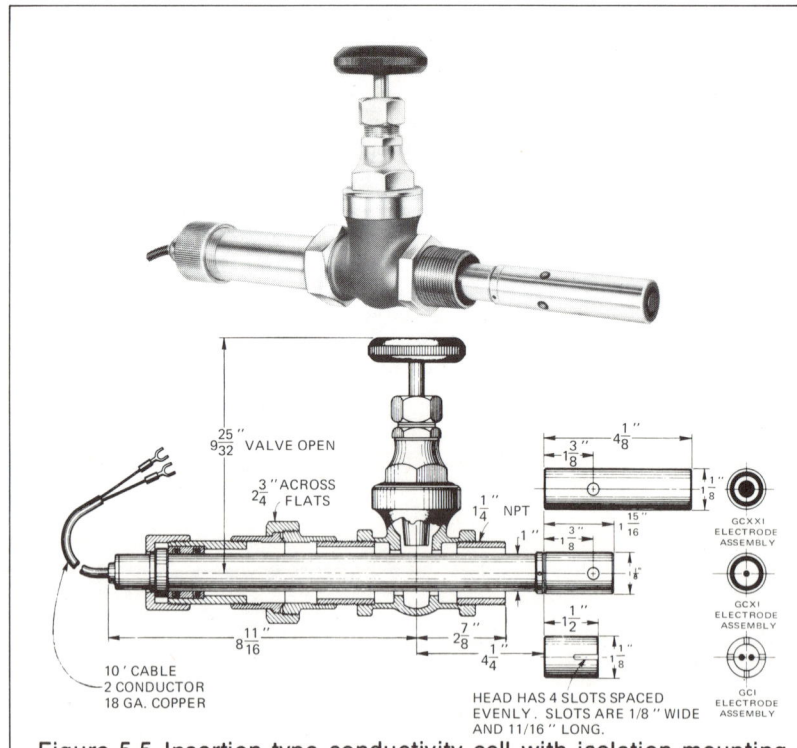

Figure 5-5. Insertion type conductivity cell with isolation mounting valve.

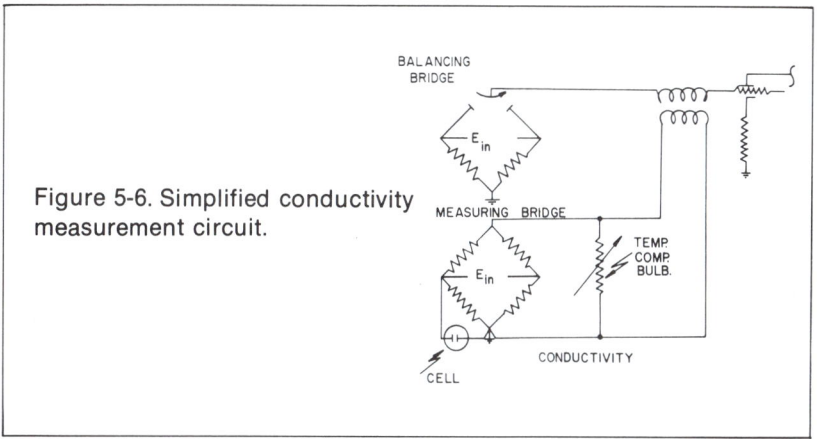

Figure 5-6. Simplified conductivity measurement circuit.

An alternating current is fed to the measuring circuit to minimize polarization effects on the conductivity of cells caused by an electrochemical action when an electric current is passed through the solution. If not minimized, these effects will result in inaccurate measurement. One of the polarization effects is electrolysis which generally produces a gaseous layer on the electrode surface setting up a back voltage and increasing the apparent resistance of the solution. Because of this, direct current voltage is not practical in conductivity measurements. As alternating current is applied in increasing frequency the polarization effect decreases. This is caused by the tendency of the gases produced to go back into the solution on the alternate half-cycle when applied voltage reverses polarity. A limit is reached where a further increase in frequency produces no further benefit. It has been found that 1000 cycles per second alternating current is approximately half the optimum operating frequency. Cycles per second are referred to as hertz (Hz).

The polarization effect is more noticeable on very smooth electrode surfaces. To minimize these effects, either porous carbon is used or the surfaces of metallic electrodes are coated with platinum black to provide a porous surface. This action increases the effective area of the electrodes, decreasing the current density and producing less gas per unit area. Also, the gases produced at the surface are spread over the porous coating, dissolve in the liquid, and diffuse back into the bulk of the solution. In use, this platinum black is gradually worn off and must be replaced periodically.

Electrodeless

In higher conductivity ranges (1 mmho/cm to 1 mho/cm), electrode polarization, fouling, and electrochemical action tend to affect the accuracy and reliability of the more conventional electrolytic conductivity measurements employing cells with electrodes in direct contact with the solution to be measured. Under these conditions, an electrodeless conductivity measurement technique is available.

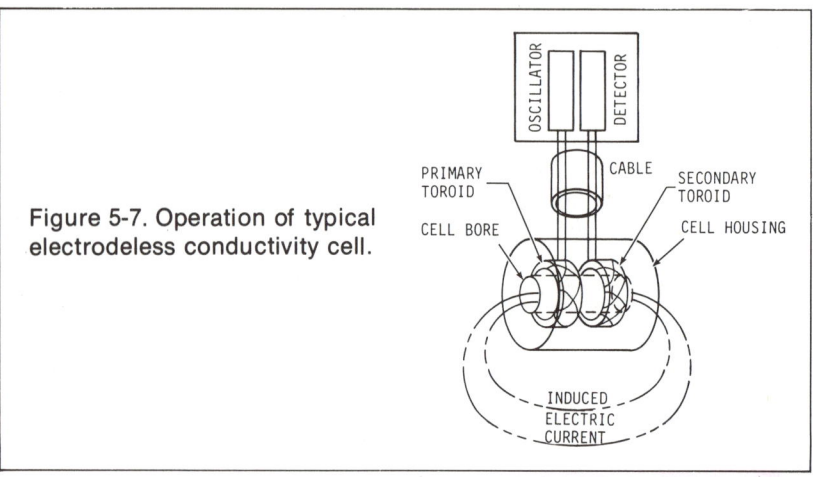

Figure 5-7. Operation of typical electrodeless conductivity cell.

In the electrodeless system (Figure 5-7), two toroidally wound coils are encapsulated in close proximity within the sensor which is immersed in the solution.

The coil winding method employed minimizes capacity effects between the coils. These coils surround the bore of the probe so that a loop of solution couples the two toroids. An ac signal produced by an oscillator is applied to one toroid. A current is generated in the loop solution which varies directly with the conductance of the solution. This current induces a secondary current in the second toroid which is fed to a current detector. An output signal is produced directly proportional to the conductance of the solution. This signal, available in the form of a standard output voltage of 0 to 10 V dc, can be used for display on an indicator or recorder, or by a control system. Typical electrodeless sensing cell configurations are shown in Figure 5-8.

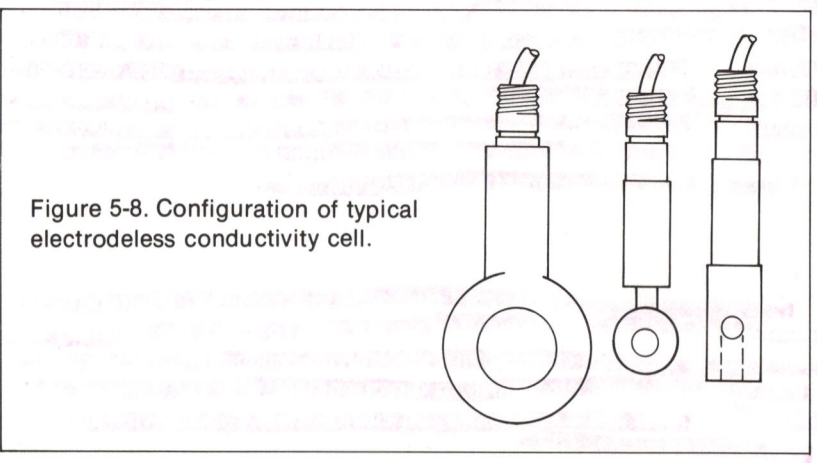

Figure 5-8. Configuration of typical electrodeless conductivity cell.

HYDROGEN-ION CONCENTRATION
(pH) MEASUREMENT

The symbol pH represents the acidity or alkalinity of a solution. It is a measure of a key ingredient of aqueous solutions of all acids and bases—the hydrogen-ion concentration. In the pulpmaking and papermaking process, pH measurements are made for a variety of reasons:

1. Checking the proper operation of process equipment by monitoring effluents from cooling towers in SO_2 plants and cooking liquors in sulfite accumulator systems.
2. Preventing corrosion by assuring that the process is in the proper pH range, such as keeping neutral conditions in ClO_2 bleach washers.
3. Maintaining those proper chemical reactions which occur under optimum pH conditions, such as in the pulp bleaching towers.
4. Helping operators maintain conditions within pH limits under which a process operates most efficiently, such as in a paper machine's wet end.
5. Aiding in the adjustment of alkaline or acid conditions in biological treatment in water and waste treatment operations associated with the plant.

Earlier techniques for measuring the hydrogen ion involved the use of a liquid indicator. When added to the sample, this indicator would produce a certain color change in relation to the value of the pH. The result could then be compared with a standard for an evaluation of the concentration.

Such a method does not lend itself well to automatic determination, nor can it be used with liquids which are normally colored or turbid. For these reasons a method of measurement was developed which was based on the potential created by a set of special electrodes in the solution. This method of measurement has actually become the standard pH measurement for indicating, recording, and/or control. However, its understanding presupposes knowledge of what pH is and familiarity with the fundamentals of the solutions' properties.

Basic Theory of pH

Stable compounds are electrically neutral. When they are mixed with water, many of them break up into two or more charged particles. The charged particles formed in the dissociation are called *ions*. The amount of dissociation of a compound varies from one compound to another and with the temperature of the solution. At a specified temperature, a fixed relationship exists between the concentration of the charged particles and neutral undissociation compound. This relationship is called the *dissociation* or *ionization constant:*

$$(1) \quad K = \frac{[M^+] [A^-]}{[MA]}$$

where K = dissociation constant
M^+ = concentration of positive ions
A^- = concentration of negative ions
MA = concentration of undissociated ions

For an acid like hydrochloric, which breaks up completely into positively charged hydrogen ions and negatively charged chlorine ions ($HCl \rightarrow H^+ + Cl^-$), the dissociation constant is practically infinity. For this reason it is called a *strong* acid. On the other hand, when an acid like acetic breaks up, very few hydrogen ions show up in the solution ($HAC \rightarrow H^+ + AC^-$). It, therefore, has a low dissociation constant and is called a *weak* acid. It can then be seen that the strength of an acid depends on the number of hydrogen ions available, which, in turn, depends on the weight of the compound and, in water, the dissociation constant of the particular compound.

Pure water dissociates into H^+ and OH^- ions ($HOH \rightarrow H^+ + OH^-$), but is considered extremely weak because very little of the HOH breaks up into H^+ and OH^- ions. The number of water molecules dissociated is so small in comparison to those undissociated that the value of (HOH) can be considered 1. The ionization constant of water has been determined to have a value of $\frac{1}{10^{14}}$. The product $(H^+)(OH^-)$ is then $\frac{1}{10^{14}}$.

If the concentration of hydrogen ions and hydroxyl ions is the same, they must be $\frac{1}{10^7}$ and $\frac{1}{10^7}$ respectively. No matter what other compounds are dissolved in the water, the product of the concentrations of H^+ and OH^- ions is always $\frac{1}{10^{14}}$.

Therefore, if a strong acid is added to water, many hydrogen ions are added and will reduce the hydroxyl ions accordingly. For HCl, the H^+ concentration becomes $\frac{1}{10^2}$ and the OH^- concentration correspondingly becomes $\frac{1}{10^{12}}$.

Because it is awkward to work in terms of small fractional concentrations like $\frac{1}{10^7}$, $\frac{1}{10^{12}}$, $\frac{1}{10^2}$, Sorenson, in 1909, proposed for convenience that the expression pH be adopted for hydrogen concentration to represent degree of acidity. He defined pH as the negative logarithm of the hydrogen-ion concentration ($pH = -\log [H^+]$) or as the log of the reciprocal of the hydrogen-ion concentration ($pH = \log \frac{1}{[H+]}$).

If the hydrogen-ion concentration is $\frac{1}{10^x}$, the pH is said to be X. In pure water where the concentration of the hydrogen ion is $\frac{1}{10^7}$, the pH is therefore 7. Accordingly, if an acid solution always has more hydrogen ions than hydroxyl ions, the pH and the concentration of the hydrogen ions will be greater than $\frac{1}{10^7}$ (viz, $\frac{1}{10^6}$, $\frac{1}{10^5}$, $\frac{1}{10^4}$) and the pH will be lower than 7 (viz, 6, 5, 4).

Conversely, if the OH^- ions exceed H ions the hydrogen-ion concentration must always be less than $\frac{1}{10^7}$ (viz, $\frac{1}{10^8}$, $\frac{1}{10^9}$, $\frac{1}{10^{10}}$) and the pH will be higher than 7 (viz, 8, 9, 10).

When OH^- ions predominate in a solution due to the complete dissociation of an alkaline compound like sodium hydroxide ($N_aOH \rightarrow NA^+ + OH^-$), it is referred to as a *strong* base. If there is very little dissociation as with ammonium hydroxide ($NH_4OH \rightarrow NH^+ + OH^-$), it is called a *weak* base.

As with the acid, the strength of a base depends on the number of OH^- ions available, which, in turn, depends on both the weight and dissociation constant of the particular compound.

The pH number itself is an exponential number and a change of a pH unit means a tenfold change in acid strength (see Table 5-B).

pH Electrode Systems

Industrial electrode systems for pH determination consist of two separate types of electrodes: (1) the active or measuring electrode which produces a voltage proportional to the hydrogen-ion concentration and (2) a reference electrode which serves as a source of constant voltage against which the output of the measuring electrode is compared.

A number of measuring electrodes have been developed for pH applications. However, the glass electrode has evolved as the one universally used for industrial process purposes. Typically, it consists of an envelope of special glass designed to be sensitive only to hydrogen ions. It contains a neutral solution of constant pH (called a *buffer solution*) and a conductor immersed in the internal solution which makes contact with the electrode lead. The buffer solution is usually a saturated solution of potassium chloride (KCl) and the present-day conductor is essentially a subelectrode of silver/silver chloride. A typical glass electrode is shown in Figure 5-9.

TABLE 5-B

pH VALUE vs. HYDROGEN ION CONCENTRATION CHART

Hydrogen Ions Grams per liter	pH Value	
1.0	0	ACID
0.1	1	
0.01	2	
0.001	3	
0.0001	4	
0.00001	5	
0.000001	6	
0.0000001	7	
0.00000001	8	
0.000000001	9	ALKALINE
0.0000000001	10	
0.00000000001	11	
0.000000000001	12	
0.0000000000001	13	
0.00000000000001	14	

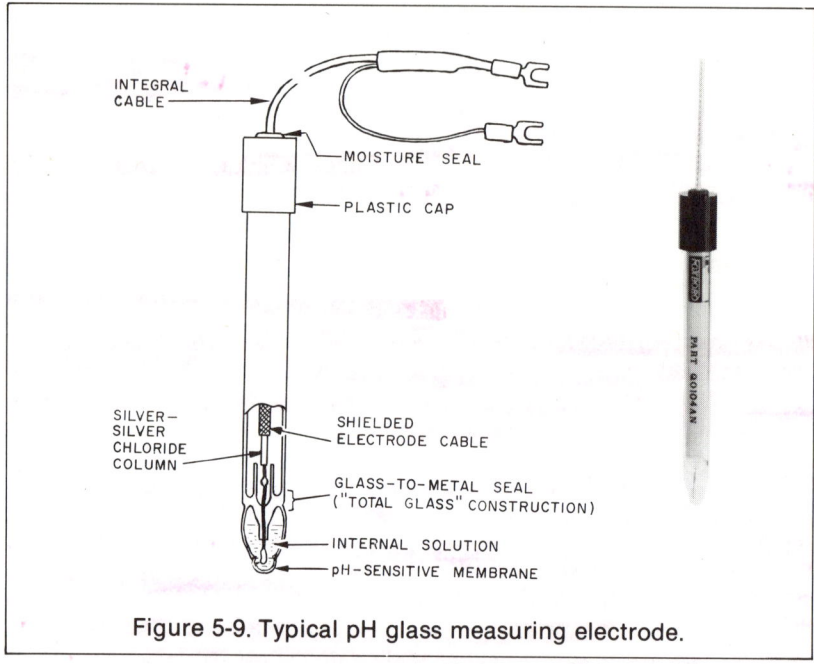

Figure 5-9. Typical pH glass measuring electrode.

The electrode operates on the principle that a potential is observed between two solutions of different hydrogen-ion concentration when they are separated by a thin glass wall. The solution within the electrode has a constant concentration of hydrogen ions. Therefore, whenever the hydrogen-ion concentration of the solution being measured is different from that of the neutral solution within the electrode, a potential difference (or voltage) is developed across the electrode. If the solution being measured has a pH of 7.0, the potential difference is 0. When the pH of the measured solution is greater than 7.0, a positive potential exists across the glass tip; when the pH is less than 7.0, a negative potential exists.

While the exact mechanism by which this happens is not known, the relationship between the potential and the hydrogen concentration follows the Nernst equation:

$$E = \frac{RT}{F} \ln \frac{[H^+] \text{ outside}}{[H^+] \text{ inside}}$$

where E = potential
R = the gas low constant
F = Faraday's number, a constant
T = absolute temperature
H^+ = hydrogen-ion concentration

A variety of glass electrodes is available and the selection of a particular electrode is usually based on the temperature range and physical characteristics of the process.

The reference electrode is used to complete the circuit so that the potential across the glass electrode can be measured. The general configuration of typical reference electrodes being currently manufactured has the connecting wire in contact with pure silver which is in intimate contact with silver chloride. The silver chloride, in turn, is in contact with a potassium chloride solution (KCl). This solution, sometimes referred to as the *salt bridge*, is in contact with the process solution. The potential from the wire KCl is fixed because the concentration from the connecting wire to the KCl is fixed, and the potential between the KCl and the process solution varies so insignificantly that overall potential of the reference is essentially constant. A typical reference electrode is shown in Figure 5-10. The body of the electrode is made of glass and the KCl liquid junction is provided by an orifice in the bottom of the electrode. Various materials can be placed across the liquid junction in order to regulate the KCl flow. Some of the more common materials used are asbestos, palladium, and ceramic. In operation, the KCl must be replenished periodically, and this is done by filling a reservoir provided for the purpose of maintaining a constant supply to the electrode.

Due to the temperature coefficient of the glass electrode, the system must be compensated for temperature if it is to continue to read pH correctly.

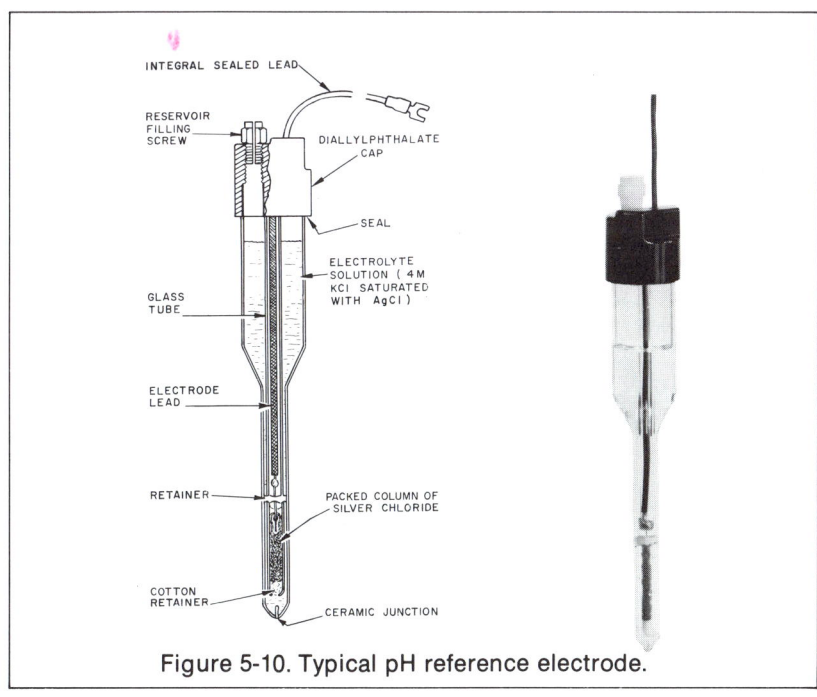

Figure 5-10. Typical pH reference electrode.

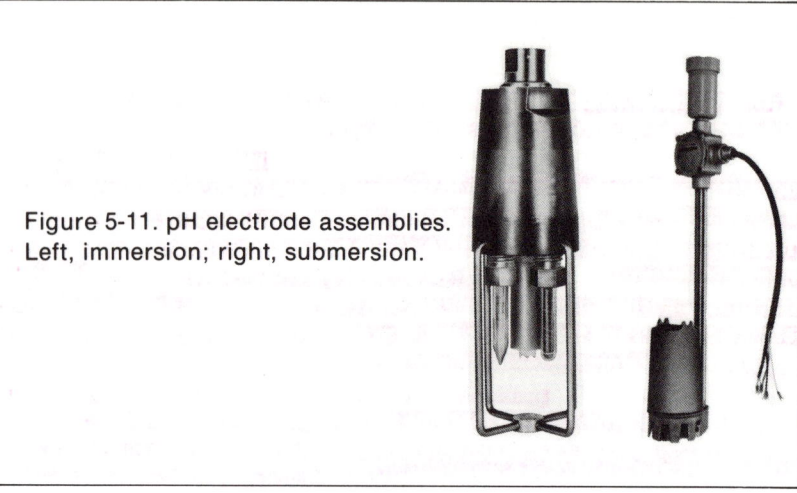

Figure 5-11. pH electrode assemblies.
Left, immersion; right, submersion.

This is done manually where temperatures do not vary widely. Otherwise, it is done automatically by means of a resistance temperature detector located in the vicinity of the electrodes and connected into the circuit. Therefore, the industrial pH assembly can consist of as many as three units: the glass measurement electrode, the reference electrode, and a resistance-thermometer device mounted in various holders of different designs. The choice of the particular design of the assembly depends on the application: immersion type electrode assemblies are used for open tanks at liquid surface level; submersion types are used for suspension beneath the liquid surface level; flow-through types are used when the process fluid or a sample can be piped directly to the assembly; and an insertion type is used for direct mounting in pipes or vessels. Examples of these assembly types are shown in Figure 5-11.

Accuracy of measurement is affected by accumulated material around the electrodes. This material should be removed in accordance with the rate of accumulation. Ultrasonic cleaning devices are now available to do this on a continuous basis.

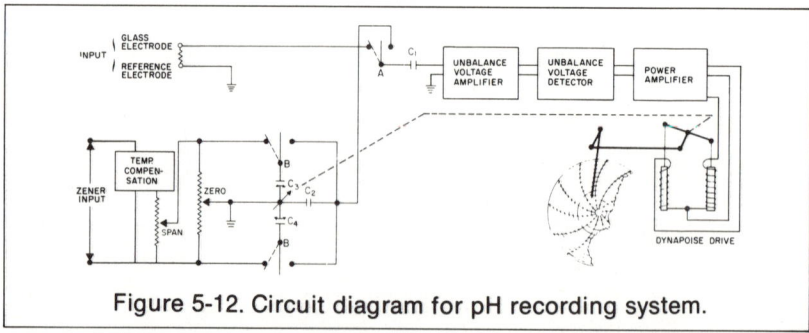

Figure 5-12. Circuit diagram for pH recording system.

Secondary Element

The cell potential is measured by means of a potentiometer type secondary instrument operated from an amplifier connected to both electrodes. The amplifier is required because essentially no current can be drawn through the cell. Older systems used a preamplifier before a conventional potentiometer but current practice uses potentiometers with specially designed high-impedance circuits. A simplified typical circuit for such an instrument is shown in Figure 5-12.

ION-SELECTIVE MEASUREMENT

An ion-selective measurement can be defined simply as one which has a relatively high degree of specificity for a single ion or a class of ions in solution. Actually, the measurement of pH was the earliest ion-selective measurement to attain widespread use. The theory and application of pH measurement utilizing the glass electrode as a hydrogen ion-selective primary measuring element has been previously discussed.

Starting in 1959, a number of additional electrodes began to be developed that were highly sensitive to other ions such as sodium, potassium, silver, fluoride, chloride, sulfide, copper, cyanide, and divalent cations such as calcium and magnesium. Many more are still in the development stage. Although all these measurements are not of particular interest to the pulp and paper industry some have been applied in a number of mill areas. Their more extensive use for making on-stream analytical measurements of selected variables appears inevitable. Applications which have been attempted with favorable results are:

1. Sulfide ion-selective measurements to control black liquor oxidation operation and to analyze cooking liquor as well as furnace stack gases.
2. Sodium ion-selective measurement to monitor millstream's soda losses.
3. Divalent cation (calcium and magnesium) measurements to provide a guide to the hardness of water used in boilers and process and to control the operation of water softening, ion exchange equipment, and water treatment plants.

The Electrode System

As in pH measurements, the electrode system consists of a measuring and reference electrode. Figure 5-13 compares a pH measurement electrode with a typical solid-state membrane electrode used for measurement of other ions. It shows how the reference electrode is common to both systems. The basic difference between the measurement electrodes is in the membrane tips.

The pH electrodes utilize a special kind of glass membrane that is particularly sensitive to hydrogen ions. The composition of this glass is responsible for the preferential selectivity of the hydrogen ion. The solid-state membrane type ion-selective electrode is constructed with a crystalline membrane sealed into the tip, whose composition, i.e., physical and chemical properties, deter-

Analytical Measurement   135

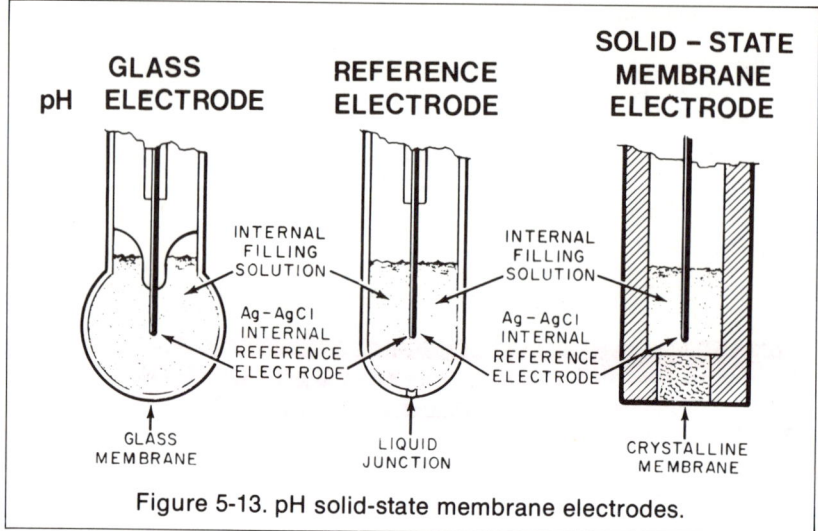

Figure 5-13. pH solid-state membrane electrodes.

mines degree of ion selectivity. Consequently, the sulfide ion-selective measuring electrode membrane is made of silver sulfide. Other types of electrode tips are of special glass that is sensitive to sodium and potassium ions.

Figure 5-14 shows the construction of still another tip used on a liquid ion exchange electrode that would be employed to measure calcium or magnesium ions. A fourth type of membrane electrode—the silicone rubber membrane electrode—is shown in Figure 5-15 in two configurations: one for a simple electrode and one for a combination which includes the reference electrode.

Basic Theory of Ion-Selective Measurements

Just as the conventional pH glass electrode develops an electrical potential in response to the concentration (which is a function of degree of dissociation or *activity*) of hydrogen ion in solution, the ion-selective electrode develops an electrical potential in response to the activity of the ion for which the electrode is selective. The relationship between the ionic activity and electrode potential is logarithmic and its performance can be expressed by a modified form of the Nernst equation:

$$E = E^\circ + 2.3 \frac{RT}{nF} \log A$$

where E = measured potential of the system

E° = a constant characteristic of the particular measuring and reference electrodes employed

RT/nF = Nernst factor = 59.16 mv at 25 C when n = 1 (monovalent ion) or 29.58 mv when n = 2 (divalent ion)

R = the gas constant

T = absolute temperature in K°

F = Faraday's constant

n = valence of the ion

A = ion activity in the solution

When the ionic activity increases, the electrode potential becomes more positive if the electrode is sensing a cation, and more negative if the electrode is sensing an anion. Therefore, it can be seen that for a tenfold change in ionic activity the electrode potential at 25°C changes by 59.16 mv if the measured ion is monovalent and 29.58 mv if the measured ion is divalent. The activity of an ion—the rate of dissociation or the rate at which it takes part in a chemical reaction—approaches the concentration in *dilute* solutions.

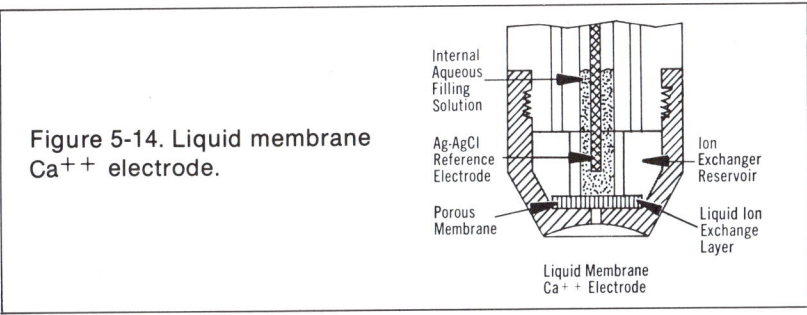

Figure 5-14. Liquid membrane Ca⁺⁺ electrode.

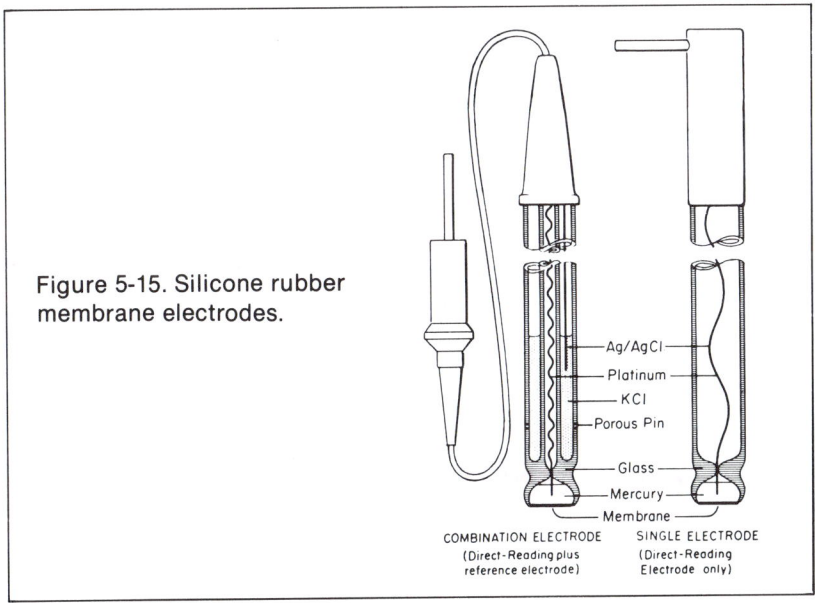

Figure 5-15. Silicone rubber membrane electrodes.

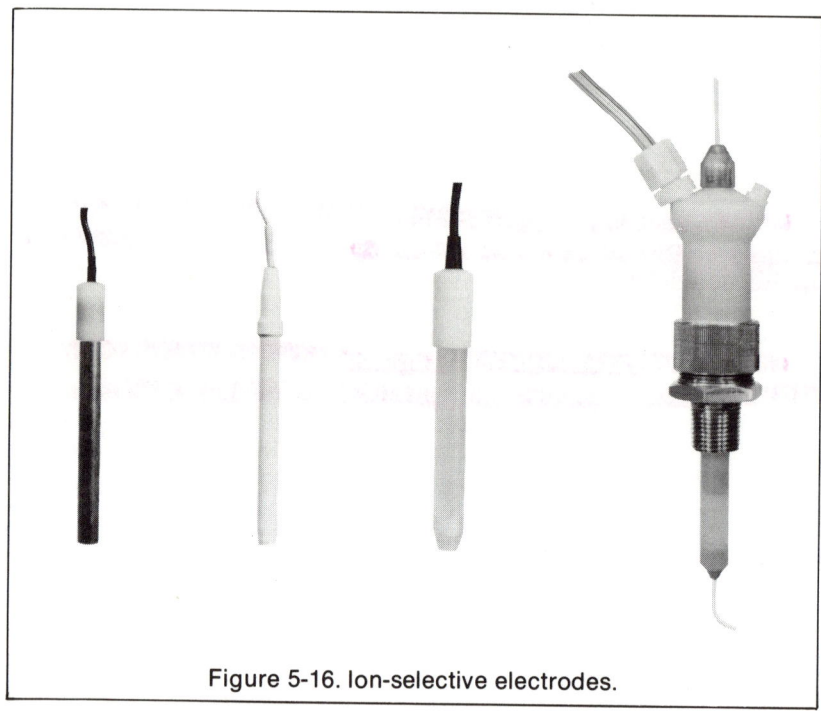

Figure 5-16. Ion-selective electrodes.

In many cases, the activity is proportional to the concentration allowing the electrode to be calibrated in terms of concentration. The relationship between activity and concentration is shown by the following equation:

$$A = \gamma C$$

where A = ion activity in the solution
  γ = activity coefficient
  C = ion concentration

The activity coefficient depends on the total ionic strength of the process solution. When the composition is known, the total ionic strength can be computed and the activity coefficient can be obtained. The total ionic strength can be computed from:

$$TIS = \tfrac{1}{2} \Sigma Z_i^2 C_i$$

where TIS = total ionic strength
  Z_i = the ionic charge
  C_i = concentration of the ion in solution

Figure 5-16 is a photograph of three typical measuring electrodes and a reference electrode. The entire assembly consists of an ion-selective measuring electrode, a reference electrode and a flow-through or immersion/submersion chamber which serves as an electrode holder and protective housing. Ion-selective electrode assembly chambers are shown in Figure 5-17.

Secondary Element

Because it is a potentiometric type measurement, the secondary indicating or recording instrument is of the potentiometer type also, and is sometimes operated from the output of an intermediate amplifier. Figure 5-18 is a diagram of a typical complete system showing the potentiometric display of ion concentration on the indicating dial and means of converting this signal into a secondary signal compatible for control and recording purposes.

OXIDATION-REDUCTION POTENTIAL (ORP) MEASUREMENT

Many chemical reactions that occur in industrial processes involve a loss or gain of electrons in the participating chemicals. Such reactions are termed *oxidation-reduction reactions*.

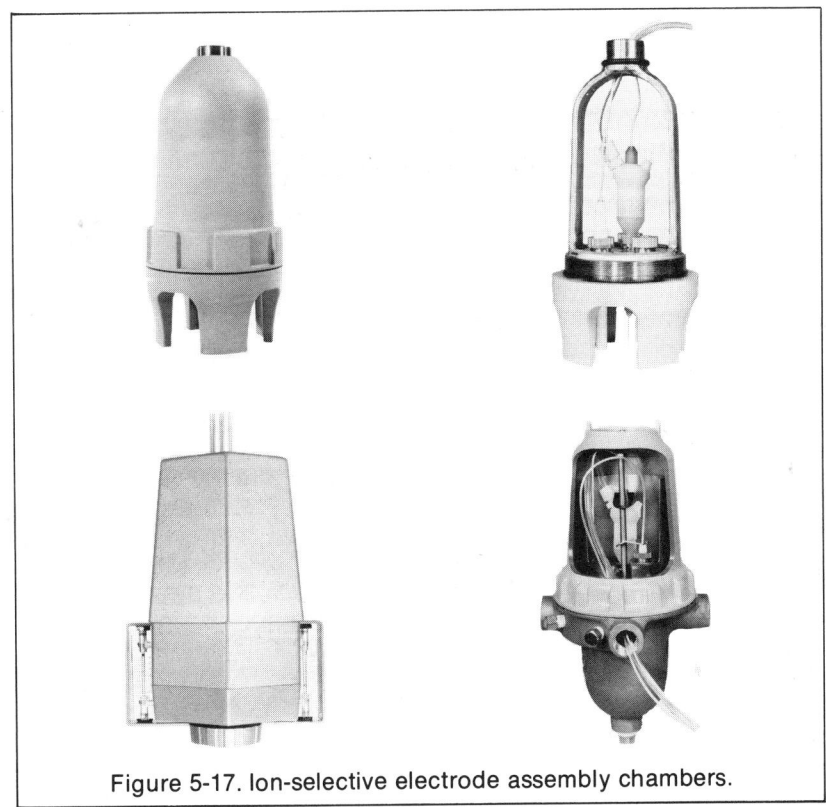

Figure 5-17. Ion-selective electrode assembly chambers.

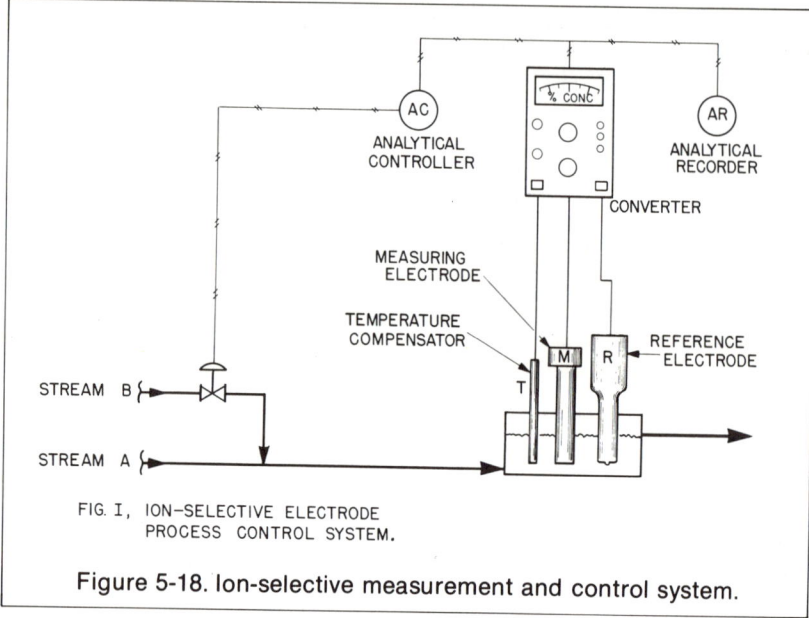

FIG. I, ION-SELECTIVE ELECTRODE
PROCESS CONTROL SYSTEM.

Figure 5-18. Ion-selective measurement and control system.

Oxidation refers to the loss of electrons by a material when an oxidant is added. Typical oxidants include oxygen, chlorine, peroxides, premanganates, dioxides, chlorates, nitric acid, sulfuric acid, and hypochlorites. Reduction refers to the gain of electrons by a material when a reductant is added. Some of the more important industrial reductants are hydrogen sulfide, ferrous oxide, carbon, aluminum, sodium, and magnesium. Reactions usually take place in aqueous solutions. When an atom, ion, or molecule is oxidized, the electrons removed must be taken up by other atoms, ions, or molecules at the same time; and these atoms, ions, or molecules are thus simultaneously reduced. The overall reaction is usually referred to as an *oxidation* or a *reduction* simply to define the primary objective. However, in every oxidation reaction there is a simultaneous reduction occurring, and vice versa.

The substance giving up the electrons is the *reductant* or *reducing agent* while the substance to which the electrons are being added is the *oxidant* or *oxidizing agent*. This chemical reaction produces a potential in the solution which is referred to as *oxidation-reduction potential* (ORP). The measurement of this potential provides a means of monitoring the degree of completion of the reaction by giving the ratio of products to reactants as the reaction proceeds. There is a definite value of measurable oxidation potential for every ratio of products to reactants throughout the reaction. Therefore, the measurement of oxidation-reduction potential can be used to control the reaction at a value considered optimum for the process.

This measurement is extensively used in the pulp and paper process to control the manufacture of sodium hypochlorite bleach solutions by mixing chlorine with sodium hydroxide; to bleach pulp slurries by chlorine; and to control addition of an oxidizing agent for neutralization of chlorine dioxide with sulfur dioxide in the bleaching process.

Basic Theory of ORP

As discussed under pH measurement, many chemicals when dissolved in water dissociate into positively charged particles which are called ions. This breaking-up phenomenon is known as dissociation or ionization. It is in the ionic form that subsequent chemical reactions take place. In the case of oxidation-reduction reactions, there is a definite measurable ratio between end product ions and reactant ions. The value of millivoltage representative of this ratio can be described by another modification of the Nernst equation.

$$(1) \quad E = E_o - \frac{RT}{nF} \log_e \frac{\text{conc. oxidation product}}{\text{conc. reduction product}}$$

where E = ORP voltage of the process

E_o = a voltage value peculiar to system under study

R = the universal gas constant

n = number of plus or minus charges lost or gained by ionic change of state under consideration

F = the Faraday constant = 96,500 Coulombs

$\log_\epsilon$ = Napierian logarithm (to the base 2.7183 instead of to the base 10), also called a natural logarithm

In a reaction between ions A and B in which A is reduced by B to form ion C, and B is oxidized by A to form ion D, these two simultaneous reactions each follow Equation (1) to produce two potentials. The observed potential is the difference between these so-called half-cell potentials and can be expressed by the following equation:

$$(2) \quad E = K_1 + K_2 T \log \frac{[C][D]}{[A][B]}$$

where E = ORP

K_1, K_2 = constants

T = absolute temperature

C = concentration of end product ion C

D = concentration of end product ion D

A = concentration of reactant ion A

B = concentration of reactant ion B

It is readily observed that the only variables are the temperature and the concentration of the ions. The end product ions are in the numerator and the reactant ions in the denominator. This ratio determines how the reaction has progressed towards completion and is represented by the ORP in millivolts.

ORP Electrode Systems

Industrial electrode systems for ORP measurements consist of two types of electrodes similar to those used in pH measurements: (1) the active or measurement electrode which detects the voltage of the solution; (2) the reference electrode which completes the electrical circuit by contacting the process solution in such a manner that a constant reproducible voltage, relatively independent of the ORP, is generated and against which the unknown voltage of the solution can be compared.

Measurement Electrode. The measurement electrode is platinum in most installations but other noble metals such as gold can be used. A typical measuring electrode, shown in Figure 5-19, is constructed of a platinum conductor incorporated with, or extending from, the bottom of a glass body. One face of the electrode is exposed to the process solution. Electrical contact is made by a noble metal lead wire running upward through the body. Carbon

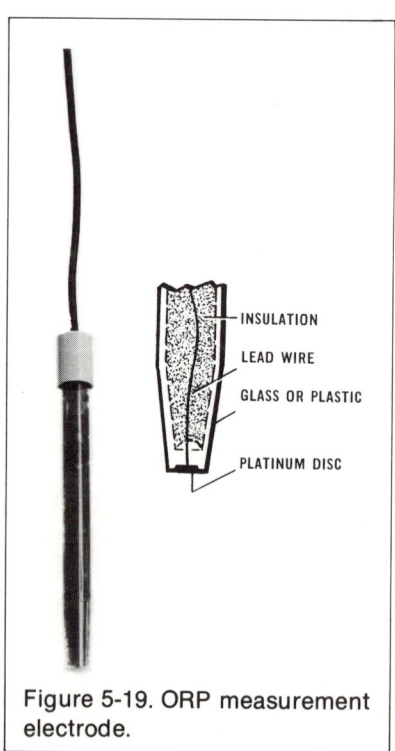

INSULATION
LEAD WIRE
GLASS OR PLASTIC
PLATINUM DISC

Figure 5-19. ORP measurement electrode.

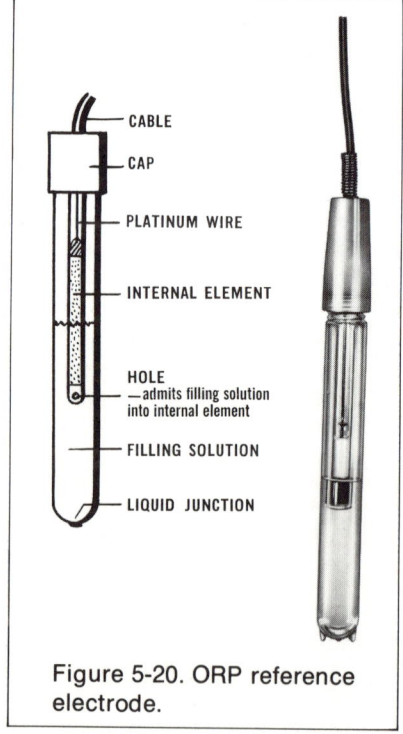

CABLE
CAP
PLATINUM WIRE
INTERNAL ELEMENT
HOLE
—admits filling solution into internal element
FILLING SOLUTION
LIQUID JUNCTION

Figure 5-20. ORP reference electrode.

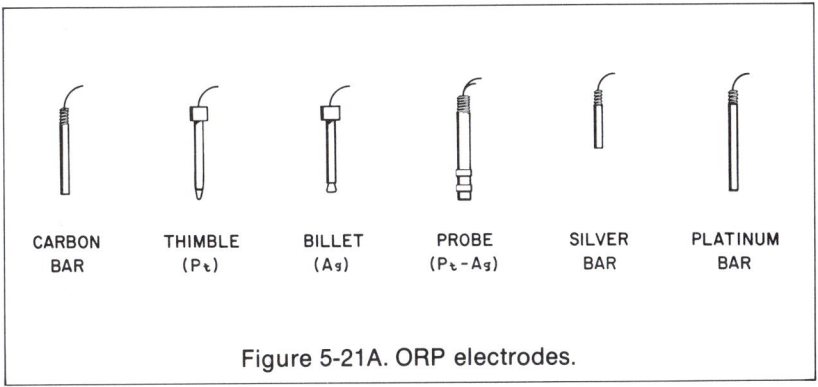

Figure 5-21A. ORP electrodes.

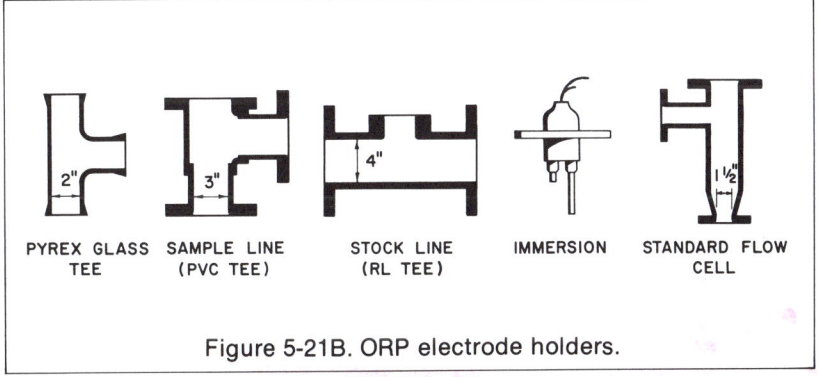

Figure 5-21B. ORP electrode holders.

measuring electrodes have also been used with considerable success in the paper industry.

Reference Electrode. The chemically insensitive, nonpolarizable reference electrode of calomel or silver/silver chloride is encased within a glass sleeve as shown in Figure 5-20. Continuous flow seepage of a filling liquid through a liquid junction orifice into the process provides electrical contact. As in the case of applications found in the paper industry, a silver electrode can be used as a reference electrode where chlorine is present in the process solution.

The ORP electrodes are supplied separately or as an assembly for installation in the process. Figure 5-21A shows a variety of individual electrodes and Figure 5-21B shows electrode holders that have been and are being used in the pulp and paper industry today.

Secondary Element

In order to complete the measuring system, a potentiometric type of secondary measuring instrument is used, which affords a simple means of providing a measurement of oxidation-reduction potentials that can be used for indicat-

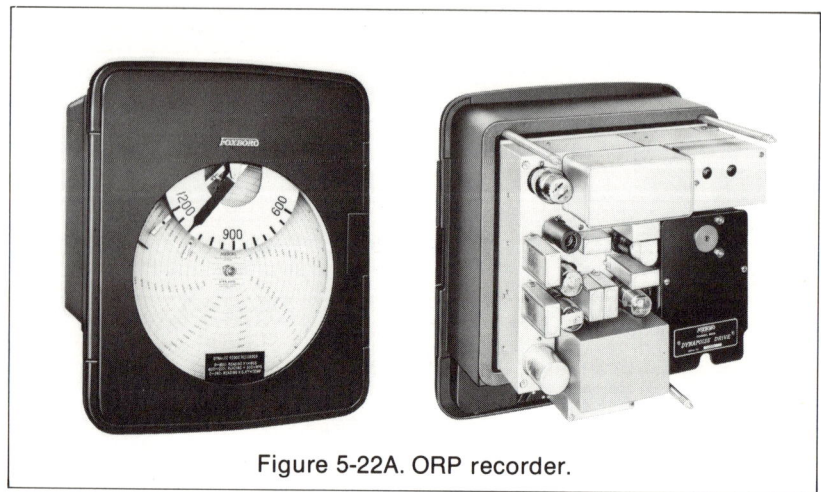

Figure 5-22A. ORP recorder.

ing, recording, and control purposes. Such an instrument is shown in Figure 5-22A, with a simplified circuit diagram shown in Figure 5-22B. A typical installation of an ORP cell probe type electrode assembly through a tank wall, such as in a bleach plant, is shown in Figure 5-23.

Electrolytic Polarization

Another electrochemical method for continuous measurement of the concentrations of reducing or oxidizing chemicals in the paper industry utilizes the effects of electrolytic polarization as the basis of its operation.

The sensor consists of a combination of three electrodes—a sensing and a

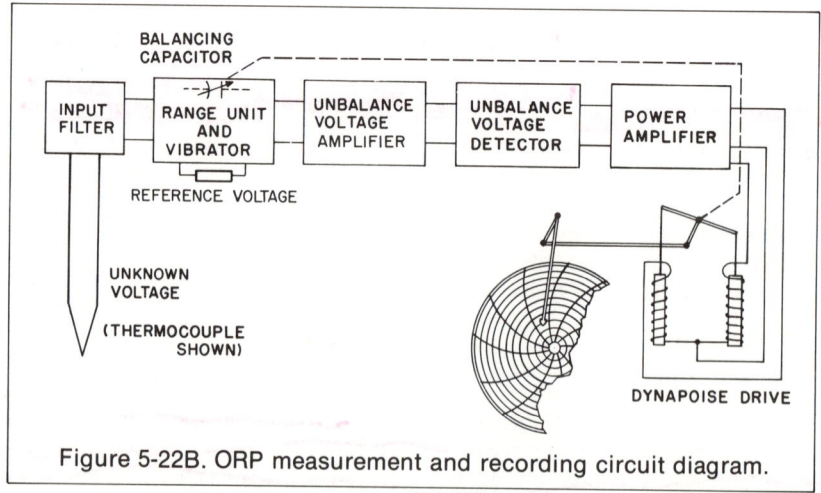

Figure 5-22B. ORP measurement and recording circuit diagram.

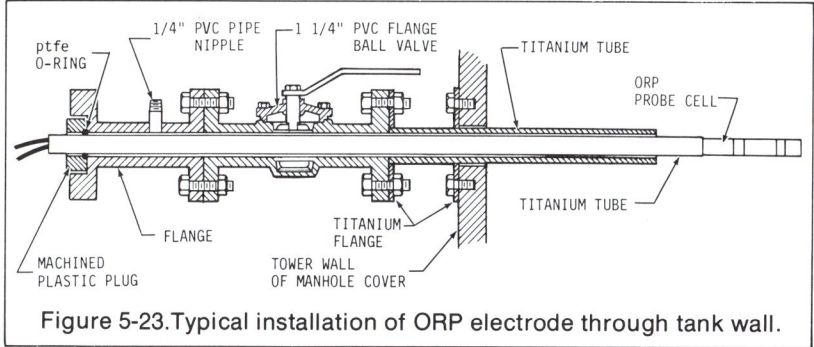

Figure 5-23. Typical installation of ORP electrode through tank wall.

counter metal electrode, plus a reference electrode—placed in a flow-through cell through which a sample, withdrawn from the process, is passed.

The principle of operation is shown in Figure 5-24. The electrodes, referred to as the working and current electrodes, are made of platinum and the reference electrode is a normal calomel electrode. Using the current from an outside source, a constant potential difference is maintained between the reference electrode and the working electrode. This is achieved by a rheostat or by a potentiostat. The diffusion of reaction chemical from the solution to the surface of the sensing electrode determines the electrolyzing current. Therefore, the current required to maintain constant potential is then proportional to the concentration of the reacting chemical to be measured. Scaling and calibration circuits in the electronic instrument modify the electrode current to produce a standard electrical output signal.

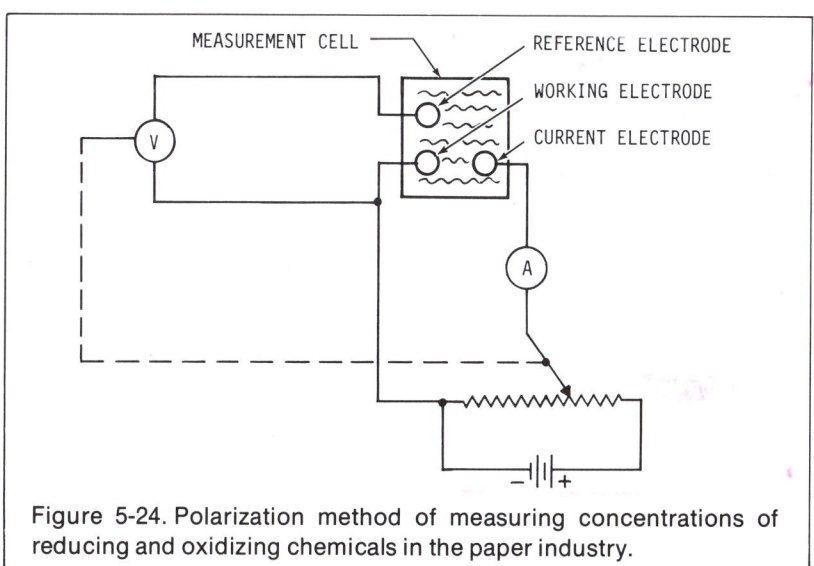

Figure 5-24. Polarization method of measuring concentrations of reducing and oxidizing chemicals in the paper industry.

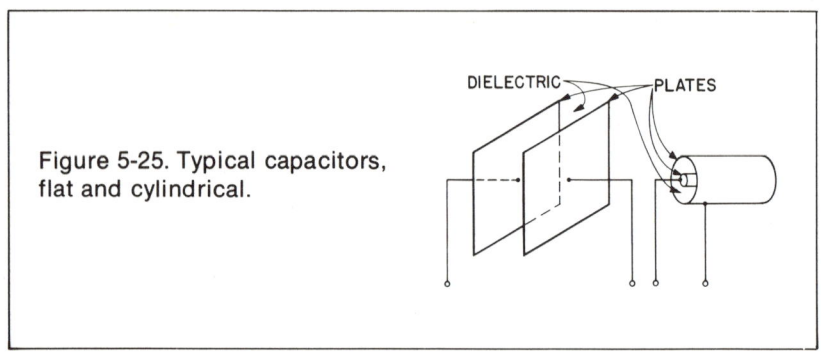

Figure 5-25. Typical capacitors, flat and cylindrical.

CAPACITANCE MEASUREMENT

Capacitance measurement, although not extensively used in the pulp and paper industry, is nonetheless important. It is used in the measurement of moisture content in paper and of such granular materials as wood chips, wood pulp, and coal. It is also used to measure level. (See Chapter 3.)

Basic Theory of Capacitance

Since capacitance is a measure of the electrical size of a capacitor, the primary measuring element of any capacitance measurement system is some form of capacitor. A capacitor consists of two conductors separated by an insulator. The conductors are its *plates*, the insulator its *dielectric*. Figure 5-25 shows two common types of capacitors, but there is a variety of other shapes. As with resistance primary measuring elements, the capacitance primary measuring element is designed so that a desired process variable changes the value of the capacitor. A characteristic property of a capacitor is its ability to accept and store an electric charge. It is this property which determines its behavior in a circuit.

Under normal conditions (see Figure 5-26) the capacitor plates are uncharged. That is, an equal number of positive and negative charges which neutralize each other are distributed on their surfaces as indicated by the voltmeter reading of zero. Two particles charged with the same polarity—for instance, both positive—physically repel each other while unlike charges attract. When the switch (Figure 5-26, diagram *b*) is closed, negative charges—electrons—are attracted to the positive side of the power source. As this electric current flows, the plate assumes an increasing positive polarity approaching that of the source. Conversely, negative charges are repelled by the negative side of the source and flow as current to plate 1, leaving it with an excess of negative charges.

Charging current continues to flow at a decreasing rate until a sufficient number of electrons have been displaced to produce the source potential across the capacitor plates. The capacitor is then fully charged and current flow stops, a condition shown in Figure 5-26, diagram *c*. Each charging opera-

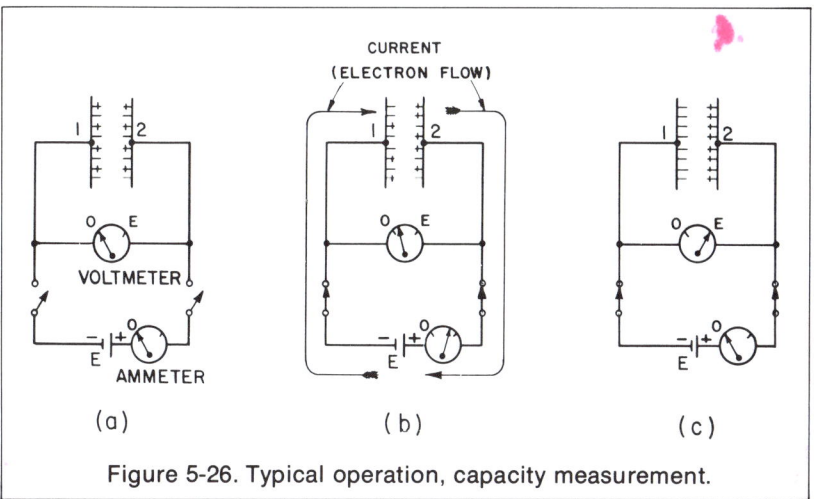

CURRENT
(ELECTRON FLOW)

VOLTMETER

AMMETER

(a)   (b)   (c)

Figure 5-26. Typical operation, capacity measurement.

tion causes a specific number of electrons to flow in the circuit. As the capacitor is made larger, thereby accepting more charge, the number of electrons increases.

The measure of the electrical size of a capacitor is its capacitance which is equal to the ratio of the amount of charge produced to the emf applied and can be expressed as:

(1) $C = Q/E$

where C = capacitance in farads
Q = quantity of charge in Coulombs
E = emf applied in volts

Therefore, if one capacitor charges to twice the number of unit charges as another when the same voltage is impressed on each, the capacitance of the first is double that of the second. The farad is a very large unit. As a matter of convenience, the microfarad (μF), millionth of a farad, and the picofarad (pF), which used to be called the micromicrofarad and is 10^{-12} farad, are the much more commonly used units.

The two basic factors affecting the electrical size of a capacitor (its capacitance) are its physical dimensions and its dielectric, which is the material between the plates. The physical dimensions involved are specifically the area of the plates and their spacing. As in Figure 5-27, for a particular spacing—designated by dimension t—each unit plate area is capable of supporting a specific number of unit charges when a particular voltage is applied. The total charge is, therefore, directly proportional to plate area A. If the spacing between the plates is halved, the intensity of the electric field between the plates is increased. Double the number of unit charges can now be sup-

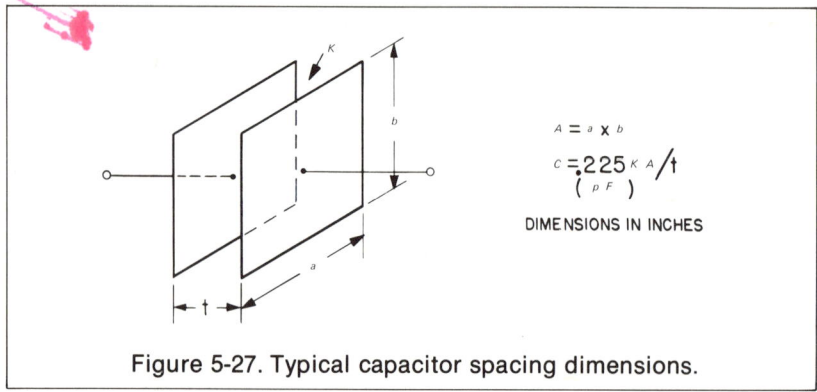

Figure 5-27. Typical capacitor spacing dimensions.

ported per unit area. Capacitance is, therefore, inversely proportional to plate spacing as shown by the standard formula for capacitance of a parallel plate capacitor:

(2) $C = .225\ KA/t$

> where C = capacitance in picofarads
> K = dielectric constant
> A = area in square inches
> t = spacing in inches

The dieletric constant (K) represents the effect of a particular dielectric used in the capacitor. When the dielectric is vacuum the constant is unity. Replacing vacuum with an insulating material will increase the capacitance by a multiplying factor—or a new value of K. For example, substituting water for vacuum produces 80 times the capacitance. Accordingly, the K value of water is 80. Every pure element and chemical compound except vacuum has a characteristic dielectric constant which is greater than unity. For most practical applications the K of air can also be considered unity and air is usually used as a reference.

The dielectric constant of material, then, is determined by the capacitance produced by the material when used as the dielectric of a capacitor. It is numerically equal to the ratio of the capacitance produced by the material to the capacitance produced by vacuum in the same capacitor. To measure the dielectric constant of a material it is only necessary to measure the capacitance produced when the material is placed between the plates of a capacitor. The capacitor is so designed that an adequate change of capacitance is produced with the desired range of dielectric constant of the material to be measured.

Primary Measuring Element

The primary measuring element converts dielectric constant variations to capacitance change. The ratio of capacitance change in pF to dielectric constant change which produces it defines the cell factor of a given element, that is:

(3) $F = \Delta C / \Delta K$

where F = cell factor

C = capacitance in pF

K = dielectric constant

For example, if a capacitance change of 2 pF is produced by a dielectric constant change of 0.2 units (e.g., 2.0-2.2), the cell factor is 10. Conversely, the

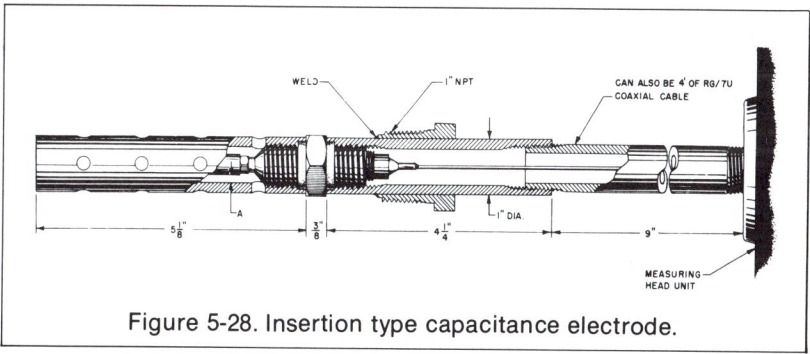

Figure 5-28. Insertion type capacitance electrode.

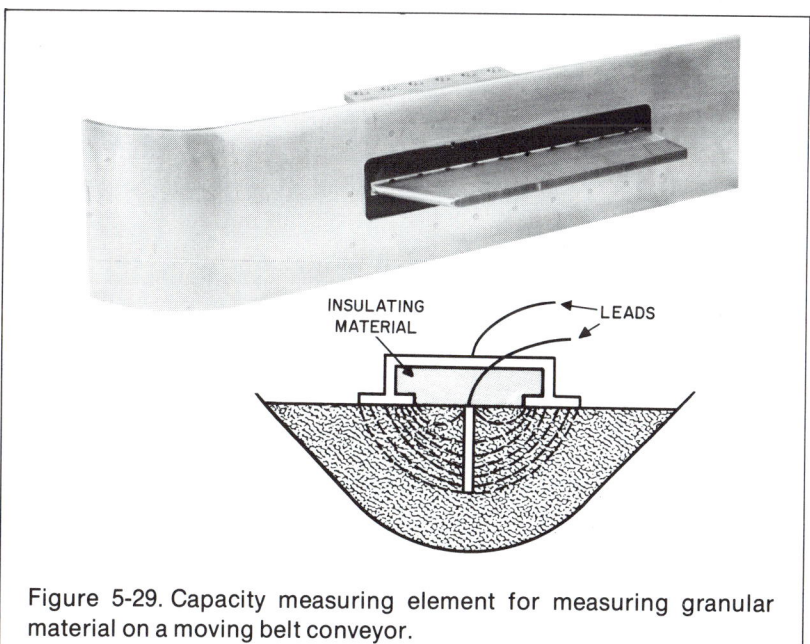

Figure 5-29. Capacity measuring element for measuring granular material on a moving belt conveyor.

product of dielectric change and cell factor yields capacitance change. Using Equation (3), the cell factor of an element designed for fluids can be determined by measuring its capacitance first in one fluid with an accurately known dielectric constant such as air (1.00) and then in another such as carbon tetrachloride (2.24 at 20°C).

A common insertion type electrode is shown in Figure 5-28. A specially designed measuring element to measure moisture in granular material is shown in Figure 5-29 and one designed for measuring moisture in paper is shown in Figure 5-30.

Secondary Measuring Instrument

In order to complete the measuring system a secondary instrument is used, one which is electrically designed to provide a measurement of capacitance in

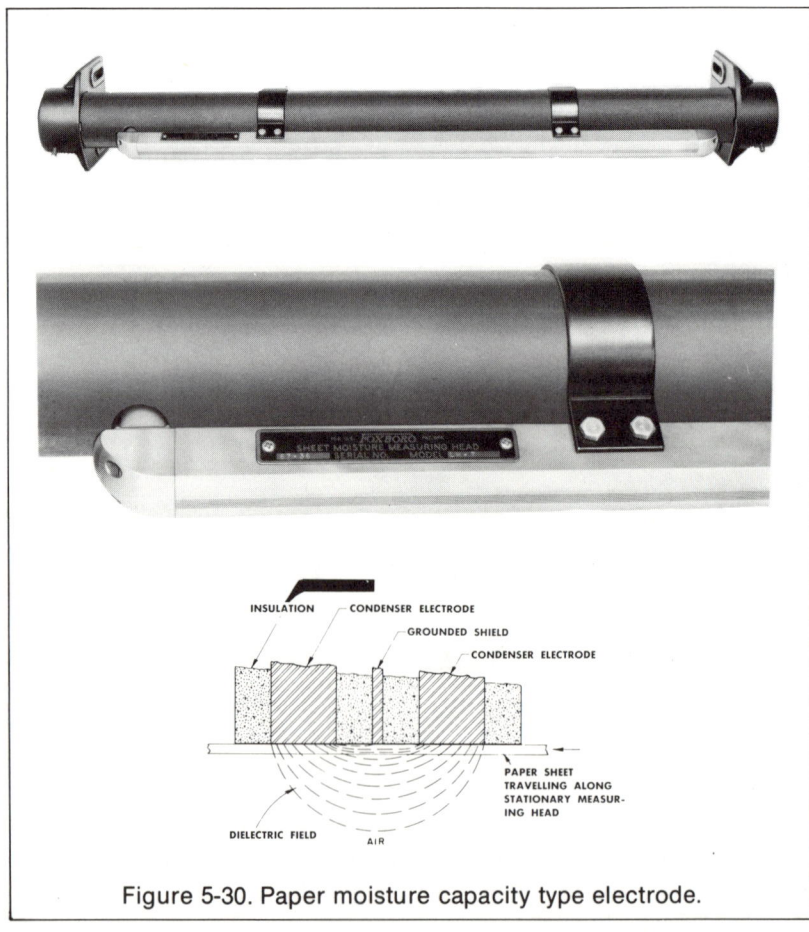

Figure 5-30. Paper moisture capacity type electrode.

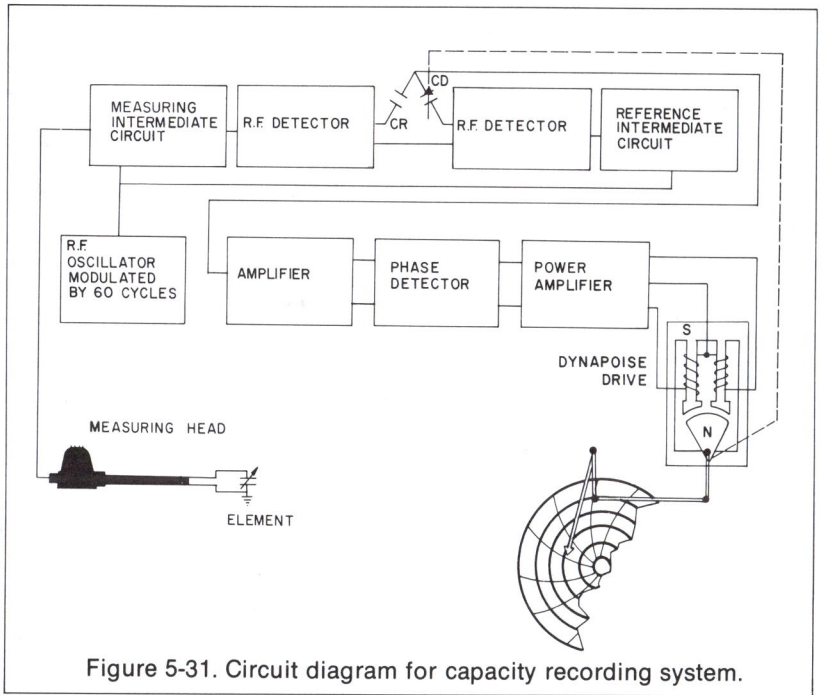

Figure 5-31. Circuit diagram for capacity recording system.

the required form to use for indicating, recording, or control purposes. A typical simplified circuit block diagram of such an instrument is shown in Figure 5-31.

OPTICAL MEASUREMENTS

Of the many optical measurements performed in process industries, only turbidity and color are used in the pulp and paper industry and then only to a limited degree. Turbidity is commonly used in the water and waste treatment facilities in the mill. It has also been tried in the recausticizing phase in the kraft pulping process to determine the calcium carbonate content of white liquor. The on-machine color measurement of paper is relatively new but it is of growing interest with an increasing number of successful applications being reported.

Turbidity Measurement

Turbidity is a physical characteristic of a solution which is caused by the presence of suspended and colloidal particles. It describes an optical property of a fluid which causes light to be scattered and absorbed rather than transmitted in a straight line through the fluid.

Analytical Measurement   151

The simplest device for measuring turbidity is the Jackson candle turbidometer (see Figure 5-32). The turbidometer measures the length of the light path through the liquid solution containing suspended or colloidal particles to determine the point at which the image of the flame of a standard candle becomes indistinguishable against the general background illumination when the flame is viewed through the suspension. The greater the turbidity the shorter the light path required for the flame to "disappear." Standardization is accomplished by comparison with natural water and suspension of a known amount of Fuller's earth. The scattering of light from the particles, rather than transmission straight through the medium, is described as the Tyndall effect after its discoverer. As indicated in Figure 5-32, the Tyndall ratio is equal to the scattered light divided by the transmitted light. This scattered light is sometimes referred to as Tyndall *light* or as a Tyndall *beam*. The Tyndall ratio increases as the turbidity increases though it is more important in low turbidity ranges.

Although the theoretically true turbidimeter (Figure 5-32) measures both Tyndall light and transmitted light, other turbidimetric techniques have been successful and are commonly used: the photometer which measures only the transmitted light and the nephelometer which measures the intensity of scattered light.

Turbidity Primary Element. Figure 5-33 shows the type of turbidimeter that measures scattered light. It consists of a light source, sample chamber, photo-

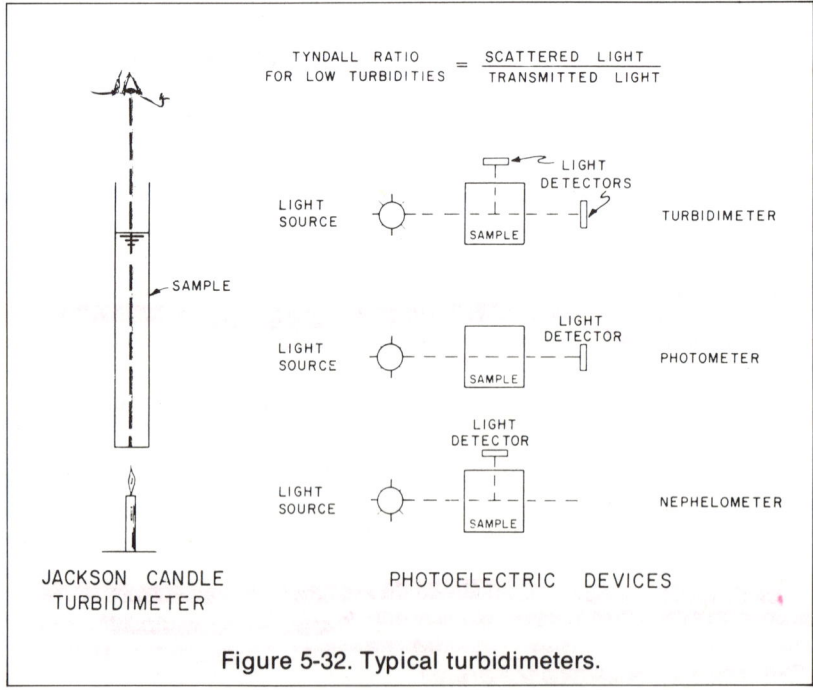

Figure 5-32. Typical turbidimeters.

multiplier, and calibrator. In operation, a sample stream is piped from the process and passed through the sample chamber, overflowing into a drain to maintain a constant flat surface. Part of a beam of collimated light is directed via the light tube to the calibration slide. The rest of the beam strikes the mirror and is directed at the surface of the sample. The light striking the sample refracts into it. As this beam strikes particles in suspension, light is scattered by them

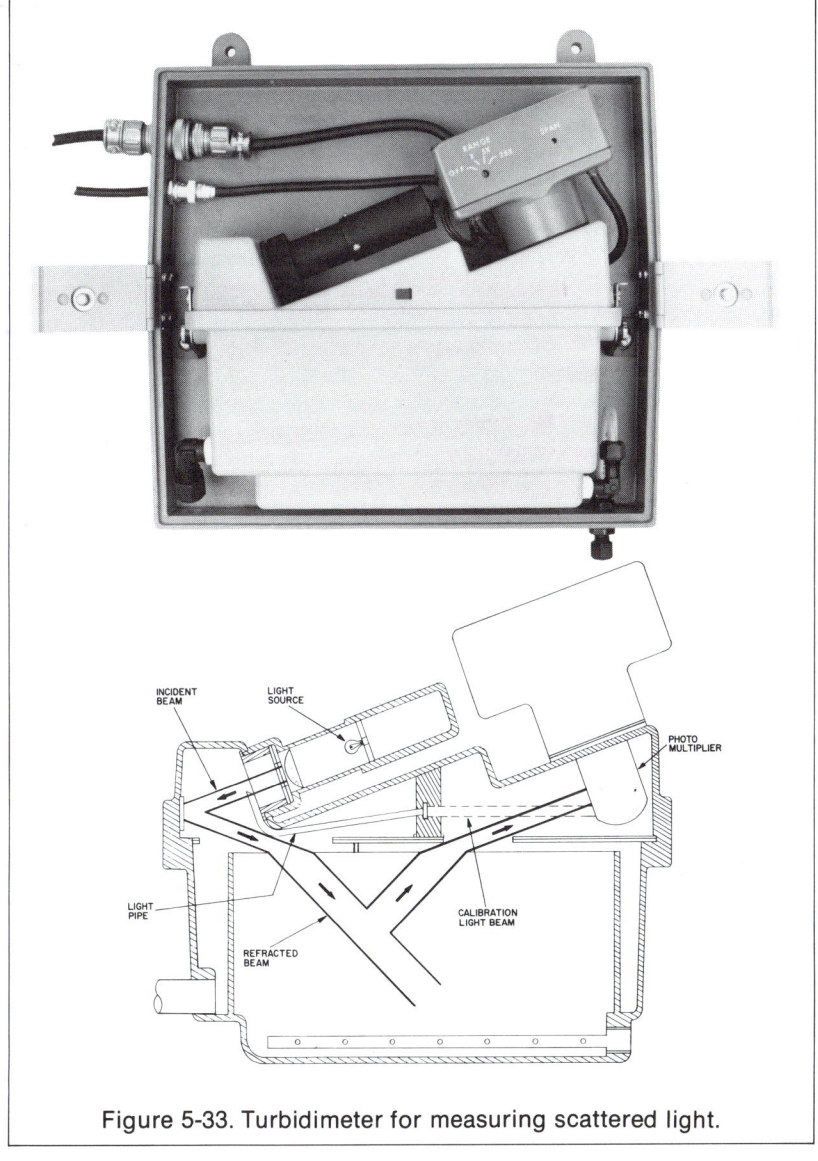

Figure 5-33. Turbidimeter for measuring scattered light.

(Tyndall light) and intercepted by the photomultiplier. Because the scattered light is proportional to the amount and nature of particles in the path of the incident beam and the output of the photomultiplier is directly proportional to the amount of light falling upon it, the voltage signal sent from it is proportional to the turbidity of the liquid.

A turbidometer that operates on the basis of a transmitted beam of light passing through the solution is shown in Figure 5-34.

Secondary Element. Since the signal from the primary measuring element is a voltage, an emf secondary measurement instrument with an electrical circuit similar to the one used with oxidation-reduction potential measurements can also be used here.

Photometric

When it is desired to measure such variables as the color of mill effluent streams and gases such as sulfur dioxide (SO_2) and chlorine dioxide (ClO_2) in the gas stream from a generator, other techniques have been used. One of these techniques is to use the principle of split beam photometric analysis. With a selected combination of light sources, optical filters, phototubes, and cell path length, a basic analyzer can measure a specified narrow or wide range of a color grading scale.

Figure 5-35 is a functional diagram of such a photometric analyzer. Radiation from the selected light source passes through the sample and then into the photometric unit, The radiation is split by a semitransparent mirror into two beams. One beam passes through an optical filter to remove all wavelengths except the *measuring* wavelengths which change in intensity as the product color changes. The other beam passes through an optical filter which transmits only the *reference* wavelengths, relatively unaffected by color changes. Logarithmic amplifiers convert phototube currents to voltage signals. These signals are proportional to the optical absorbance; and electrical subtraction of

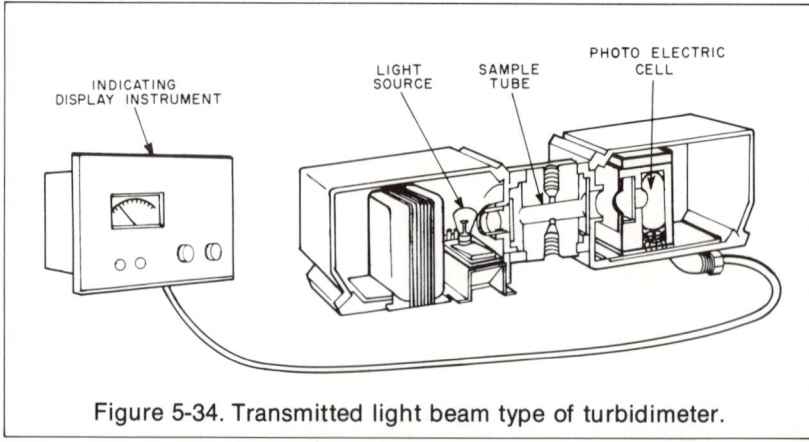

Figure 5-34. Transmitted light beam type of turbidimeter.

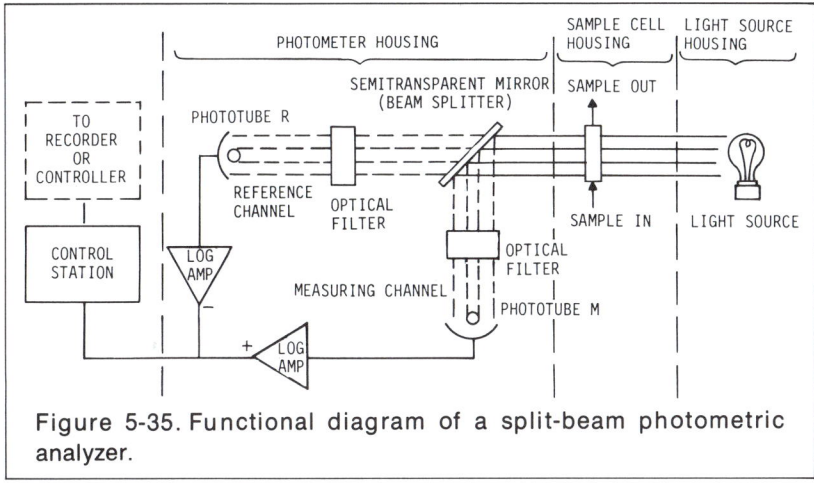

Figure 5-35. Functional diagram of a split-beam photometric analyzer.

the signal produces an output equal to the difference in optical absorbance, which is related to the color of the sample. The analyzer is designed to compensate for bubbles or foreign matter in the sample and for light intensity fluctuations because these occurrences affect both light beams equally.

ANALYTICAL GAS MEASUREMENTS

Continuous measurements for determining the concentration of one or more components in a certain gas stream are widely made in industry in general. They are of interest to the papermaking industry primarily to monitor the amount of CO_2, O_2, and/or combustibles in stack gases from lime kilns, recovery boilers, and power boilers. They are also used to monitor the amount of SO_2 in sulfur burned product gases used in the manufacture of cooking liquors for the sulfite pulping processes as well as other process areas where SO_2 gas is produced. Such measurements must be related to some response which is associated with the desired component to be measured. There are five general types of gas analyzers used in industry today: chemical, electrical conductivity, thermal conductivity, paramagnetic, and mechanical. At one time or another, all types have been used in paper mills. However, today the most common gas analyzers used are of the chemical, electrical conductivity, and paramagnetic types.

Chemical Gas Analysis

The Orsat apparatus is considered to be the standard instrument for quantitatively analyzing components of gas on an intermittent basis. Its principle of operation is based on the drawing of a known amount of gas from the stream, then passing it through a solution which absorbs only one component. The amount of that component is found by measuring the volume of sample left.

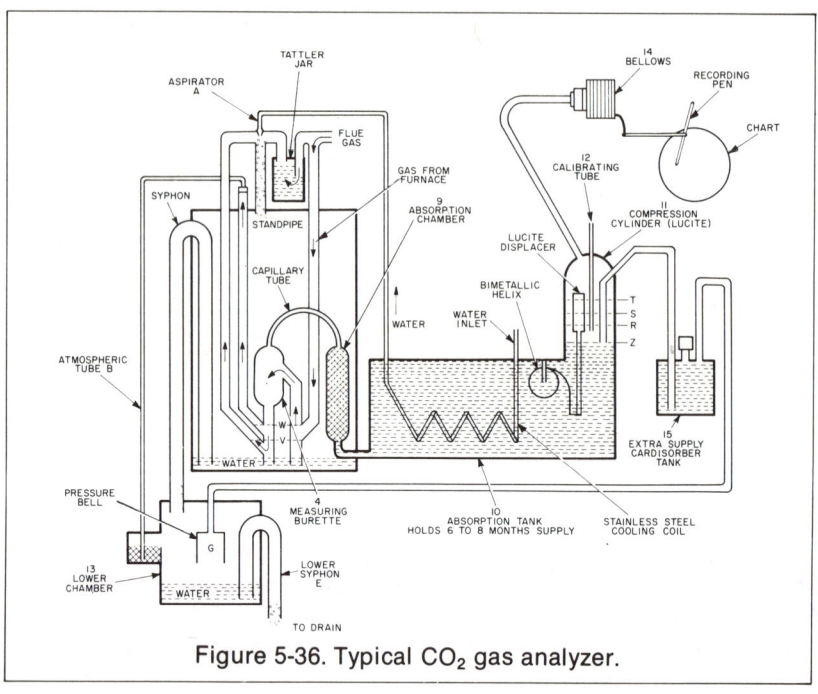

Figure 5-36. Typical CO_2 gas analyzer.

Oxygen is absorbed by a potassium pyrogallate solution. After measuring the oxygen removed, the CO_2 is absorbed into alkaline cuprous chloride. Other appropriate solutions can be used to selectively absorb other gases for analysis. The Orsat is extensively used throughout the boiler house and mill to check the accuracy of other types of continuous measurement instruments.

A typical on-line continuous chemical gas analytical measurement system used for measuring CO_2 in flue gases is schematically shown in Figure 5-36. It is a motor-driven instrument and operates on the same principle as the Orsat apparatus. The system is actuated entirely by water passing through an aspirator (A) which draws the flue gas into a measuring chamber (4). As the water accumulates in the standpipe and reaches level V, a quantity of gas is trapped and forced up into tube B through chamber C. When water reaches level W, an accurately measured sample of gas at atmospheric pressure remains in measuring chamber 4. As the water rises, the gas sample is forced from the measuring chamber into the absorption chamber (9) which is packed with both steel wool and a liquid which absorbs the CO_2, forcing some of the liquid to go into main tank 10 and causing level Z to rise in compression chamber 11; at level R calibrating tube 12 becomes sealed. Any further rise forces trapped air into bellows 14, causing a pen or pointer to move across a suitable scale or chart calibrated on the basis of the rise of liquid absorbent level due to the amount of gas remaining in the absorption chamber after the CO_2 has been absorbed.

In the time the gas sample has had its CO_2 absorbed, the water fills the standpipe and starts the siphon that empties it, allowing the unabsorbed part of the gas sample to be pushed back into the measuring chamber and preparing for a repeat cycle.

Heat of Combustion

Heat of combustion measurement makes use of the heat of combustion of combustibles in the sample to measure them and the heat of the combustion of hydrogen mixed with the sample to measure the oxygen present. It is typically used in the paper industry to measure oxygen content of power and recovery boiler stack gases to indicate excess air admitted to the combustion. It is also used to measure combustible gases as a guide to fuel air mixing performance.

Figure 5-37 shows a typical scheme for measuring percent oxygen and percent combustibles in a gaseous mixture. The gas to be analyzed is continuously drawn into the analyzer at a constant rate of flow controlled by pressure regulating valves. Regulated amounts of air and hydrogen are added to the combustibles and oxygen units, and are maintained at a temperature of 160°F by a heater in the analyzer block. Two identical noble metal catalyst filaments are mounted in each analyzing cell on a common base. The measuring filament is exposed to the gas mixture entering the cell, while the compensating filament chamber is closed on all sides except for a small hole which allows a small amount of sample to enter.

As the air-gas sample mixture passes over the filament of the red hot combustibles it burns off whatever combustible components are present. As

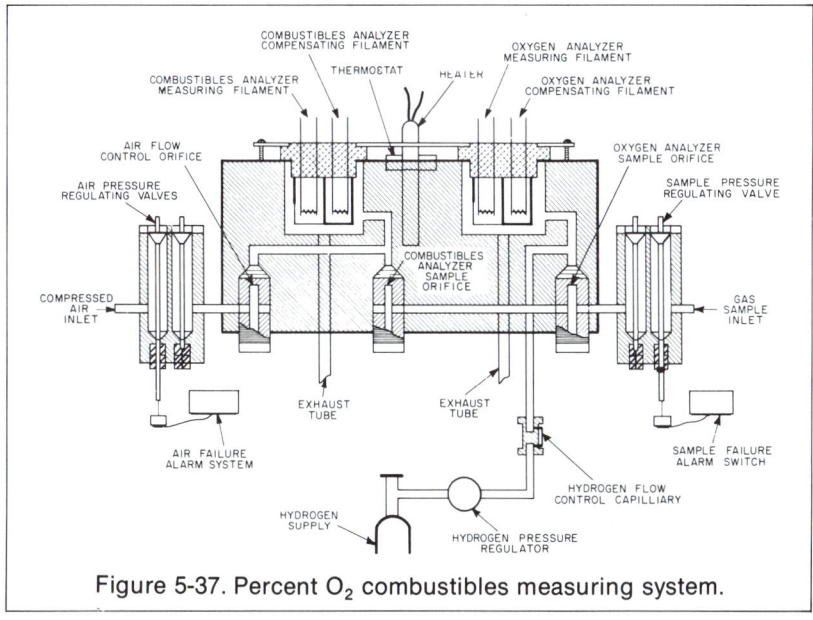

Figure 5-37. Percent O$_2$ combustibles measuring system.

the hydrogen gas sample passes over the red hot oxygen filament, the hydrogen will burn if there is no oxygen present to support combustion. The heat of combustion and the resulting temperature of the filament are proportional to the amount of oxygen or combustibles present. The rise in filament temperature increases its electrical resistance accordingly. The filaments form two legs of a Wheatstone bridge in the instrument circuit. Imbalance of the bridge circuit caused by a change in resistance of the filament produces a signal voltage which controls the operation of a balancing motor in the receiving instrument. This motor drives a linkage system to position a pointer or recording pen and/or controller.

Oxygen analyzers and combustion analyzers are available separately as well as in combination units.

Ceramic Element

Another method of measuring percent oxygen in a gas stream utilizes a unique ceramic sensor shown diagrammatically in Figure 5-38. It consists of a closed-end tube made of ceramic oxide which, when hot, becomes an electrolytic conductor because of the mobility of the oxygen ions. Electrodes of porous platinum are coated onto the outside of the tube and are connected to a meter to complete the electrochemical sensing cell system.

The principle of operation is based on the fact that when two electrodes are in contact with gases having different levels of oxygen partial pressure, a voltage is produced that depends on the oxygen partial pressure ratio. If the

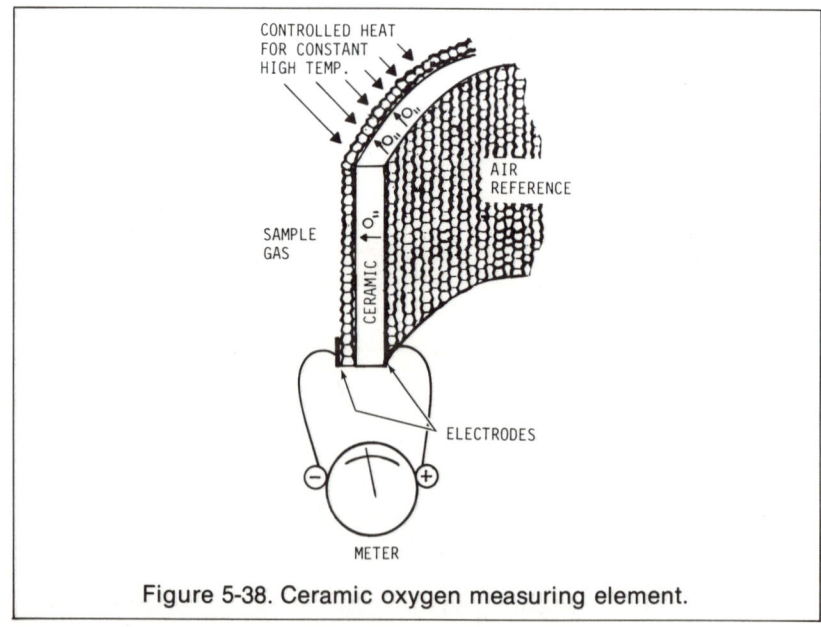

Figure 5-38. Ceramic oxygen measuring element.

oxygen in one gas is known (the reference gas—usually air), that of the other is indicated directly by the emf of the cell.

This technique can also be used to measure combustibles/combustion products because in the absence of molecular oxygen the sensor responds to small amounts of oxygen produced by the dissociation of water and carbon dioxide at the high operating temperature. Since the extent of this dissociation at any fixed temperature is determined by the amount of hydrogen and/or carbon monoxide present, this analyzer will indicate the ratio of water to hydrogen and the ratio of carbon dioxide to carbon monoxide.

Other Oxygen Measurements

Different techniques are used to measure oxygen dissolved in liquid, especially important in measuring conditions of streams near mill sites. One method of measuring dissolved oxygen (DO) is based on the polarographic principle. The sensor in this case is an electrolytic cell enclosed in a polyvinyl chloride housing. Inside the cell are a silver anode, a gold cathode, and the electrolyte. A ptfe membrane is stretched over the gold cathode to keep out water and all undissolved and dissolved solids in the water. Free oxygen in the water can diffuse through the membrane to the cathode where it is reduced. The amount of current flowing is directly proportional to the amount of oxygen in the water. The sensor is temperature compensated so that it reads directly in ppm oxygen.

Another version uses a probe which is essentially a self-generating galvanic cell having a platinum cathode and a lead anode. A ptfe membrane separates an electrolyte solution from the sample being measured. Oxygen diffusion through this membrane causes a galvanic action which develops a signal directly proportional to the concentration of the dissolved oxygen. This signal is temperature compensated, sent to a meter, or used for recording and control.

Thermal Conductivity Gas Analysis

This measurement system makes use of the principle of thermal conductivity—the fact that each elemental gas or gas compound has a characteristic ability to conduct heat at a different rate. The combined effect in a gas mixture is related to the sum of the effects of the components. However, if the thermal conductivity of one is significantly greater than the others, measured changes in thermal conductivity effects can be correlated directly with changes in its concentration, i.e., the gas of interest as a percentage of the mixture.

A very common installation used in paper mills to measure SO_2 content of gases from sulfur burners is illustrated in Figure 5-39. A sample gas is drawn from process line A by an aspirator or pump through a constant head device and through coke filter B to remove any sulfuric acid mist present. It is then drawn through a glass measuring cell, assembly C. An electrically heated platinum filament, located in the measuring cell, forms one arm of a Wheatstone bridge circuit, indicated by M, of the system circuit diagram in

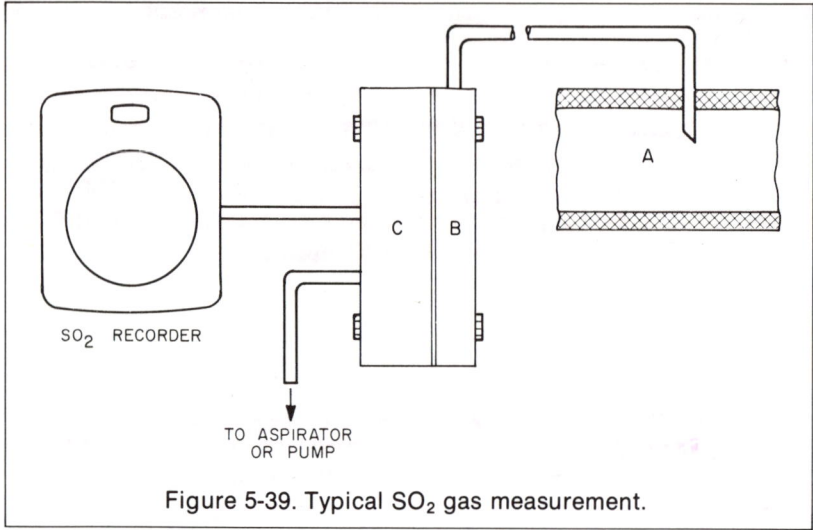

Figure 5-39. Typical SO_2 gas measurement.

Figure 5-40. A similar filament is in the cell with a reference—standard gas R—and forms another arm of the bridge. As the sample gas mixture flows through measuring cell M, the heat loss from the filament will vary with any changes in thermal conductivity, i.e., any change in the concentration of the gas being measured. This, in turn, will change the filament temperature and, as a result, its resistance. The electrical bridge circuit continuously compares the

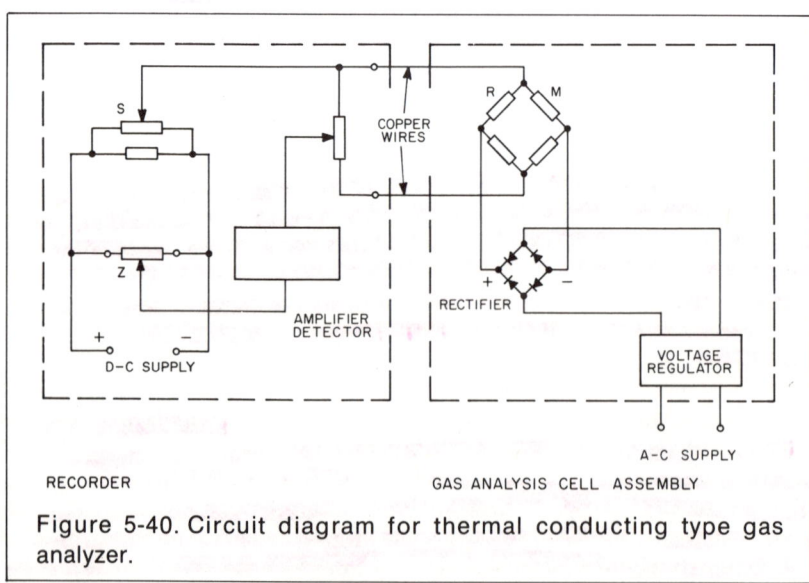

Figure 5-40. Circuit diagram for thermal conducting type gas analyzer.

resistance of M and R and produces a signal which is a function of the concentration of the measured gas.

The electrical circuit in the secondary measuring instrument, shown as a recorder in this case, continuously compares the output signal of the bridge with a standard potential drop across slidewire S. It detects a change in output resulting from a change in resistance of M and automatically repositions the contact on the slidewire to restore null balance and then moves the recorder pen and pointer to a position representing percent gas concentration. Controls SP and Z permit manual adjustment of span and zero. This method can also be used for such other gas analyses measurements as carbon dioxide, oxygen, hydrogen in chlorine, and hydrogen in nitrogen.

Other SO₂ Measurements

When measuring SO_2 from other areas such as mill stacks, different methods are being used. One detector employs a grating spectrometer mounted on one side of a heavy aluminum casting as shown in Figure 5-41.

Light from a tungsten halogen lamp is collimated and then reflected off a flat zero/read mirror (Figure 5-42). Light is directed to the probe mirror which relays energy back to the spectrometer. If sulfur dioxide is present in the probe slot it will absorb energy and regularly spaced bands at 3000 Angstrom degrees. Light passing through the entrance slit is reflected off the modulator mirror to the diffraction grating. The grating disperses the energy, spatially displaying a focused absorption spectra of sulfur dioxide on the exit mask. The optical center of the modulator is tilted at a slight angle with respect to its axis of rotation. Thus, as the mirror rotates the angle at which the light strikes the diffraction grating varies, in turn, causing the absorption spectra to sweep past the exit mask in a circular fashion. The exit correlation mask consists of alternate clear and opaque spaces cut to match the absorption spectra. Sweeping of the absorption bands past the fixed mask lines creates a bank of beat frequencies in light impinging on a photomultiplier tube. As the concentration of gas in the probe slot varies, the depth of the absorption spectra varies and likewise the signal at the photomultiplier tube varies. The rms value of the beat frequencies is directly proportional to the concentration of gas times the slot length or parts per million times meters.

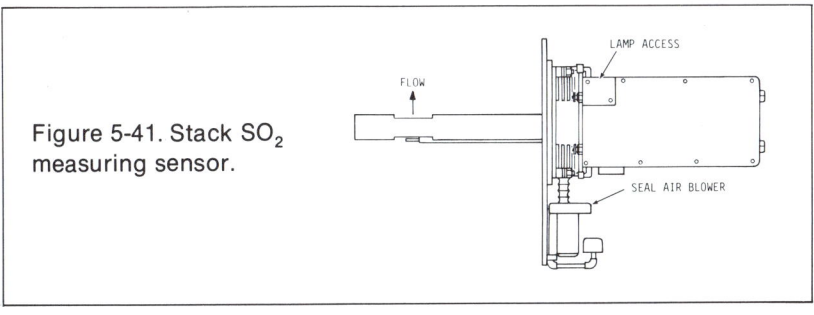

Figure 5-41. Stack SO₂ measuring sensor.

FLOW

LAMP ACCESS

SEAL AIR BLOWER

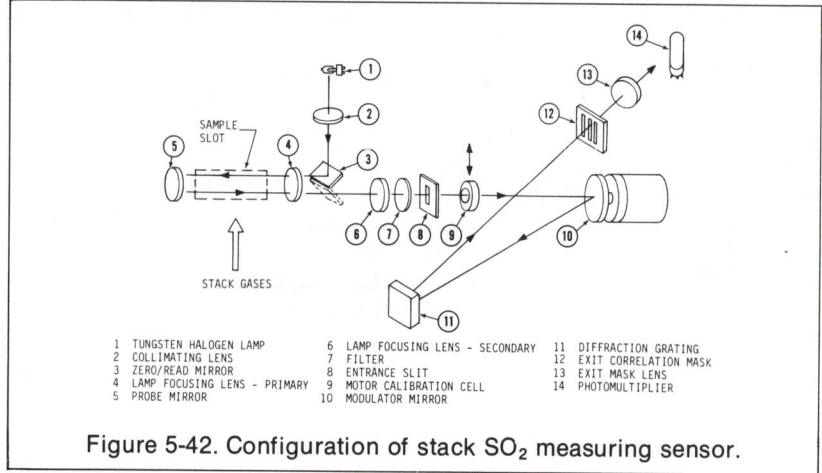

Figure 5-42. Configuration of stack SO$_2$ measuring sensor.

1	TUNGSTEN HALOGEN LAMP	6	LAMP FOCUSING LENS - SECONDARY	11	DIFFRACTION GRATING
2	COLLIMATING LENS	7	FILTER	12	EXIT CORRELATION MASK
3	ZERO/READ MIRROR	8	ENTRANCE SLIT	13	EXIT MASK LENS
4	LAMP FOCUSING LENS - PRIMARY	9	MOTOR CALIBRATION CELL	14	PHOTOMULTIPLIER
5	PROBE MIRROR	10	MODULATOR MIRROR		

The circuitry consists of a nanoammeter which drives an automatic gain control loop and a bank pass filter. The filter selects those frequency components which are created by the presence of the sulfur dioxide spectra and then fed to a precision demodulator. The output is a dc level proportional to the sum of the rms values of the amplified beat signals from the photomultiplier.

Another technique used for the analysis of SO$_2$ in the range of 0.5 to 5000 ppm is based on the principle of *pulsed fluorescence* and utilizes the fluorescent emission of SO$_2$ molecules that have been drawn into a chamber. The SO$_2$ molecules then emit fluorescent radiation in direct proportion to their concen-

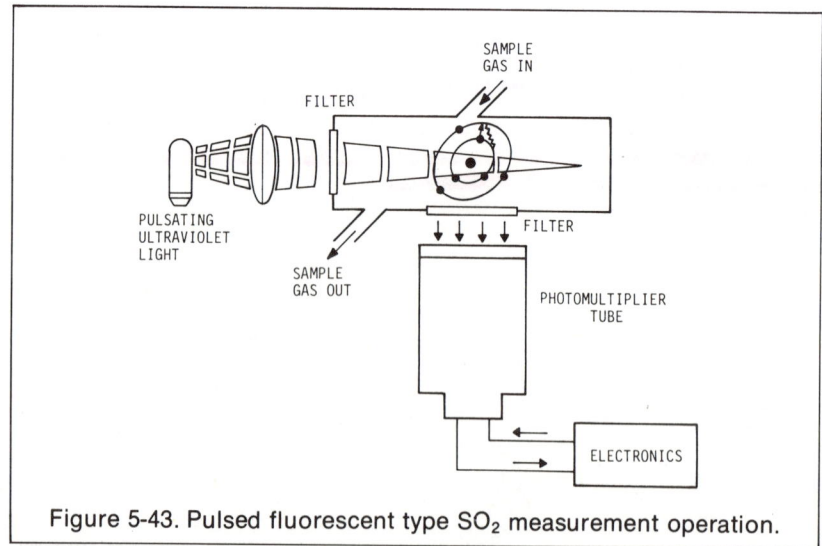

Figure 5-43. Pulsed fluorescent type SO$_2$ measurement operation.

tration. This, in turn, is detected by a high gain photomultiplier tube and is processed through solid-state electronics to be displayed as a direct indication of SO_2 concentration in parts per million.

Figure 5-43 illustrates in block diagram form the general principle of operation of an analytic instrument for the pulsed fluorescent monitoring of SO_2. Pulsating ultraviolet light is focused through a narrow band pass filter into the fluorescent chamber. Here it excites the SO_2 molecules which give off their characteristic decay radiation. A second filter allows only this radiation to fall on a sensitive photomultiplier tube. Electronic signal processing transfers the light energy impinging on the photomultiplier into a voltage which is in direct proportion to the concentration of the SO_2 in the sample stream being analyzed.

Stack Smoke and Dust

Smoke and process dust emission from stacks have been measured by the use of a device which is responsive to energy being radiated from a light source. The light source produces a measuring filament resistance proportional to the detected energy.

Figure 5-44 is a representation of a typical unit. The light source and meter are mounted in permanent alignment on opposite ends of a slotted pipe. A schematic diagram of this device together with the electrical receiving device is shown in Figure 5-45.

The filament of the meter is one arm of an ac Wheatstone bridge measuring circuit. The electrical resistance of the filament changes according to the radiant energy detected. A null/balance amplifier in the receiver continuously positions a servomotor and measuring slidewire to maintain the circuit in balance. The adjacent arm of the bridge circuit is the compensating filament which corrects for normal variations in line supply voltage and ambient temperature changes. An ac signal is sent to the recorder to complete the system.

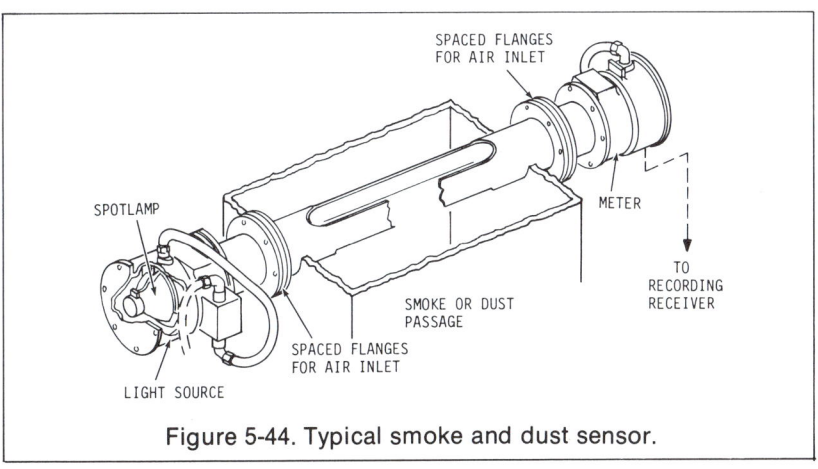

Figure 5-44. Typical smoke and dust sensor.

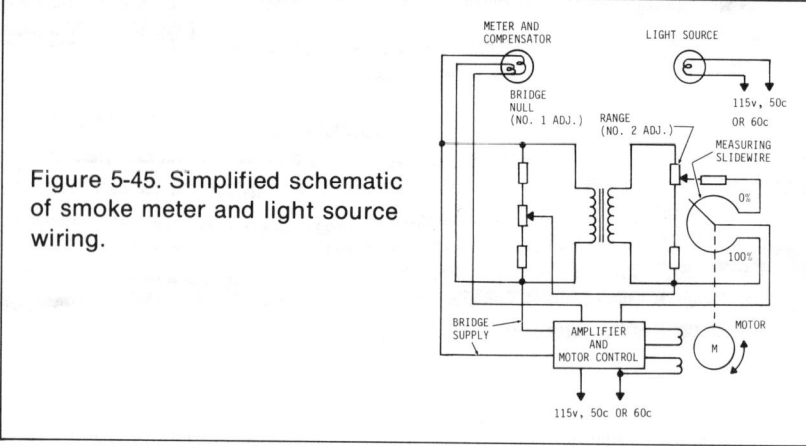

Figure 5-45. Simplified schematic of smoke meter and light source wiring.

BIBLIOGRAPHY

Bates, R. G. *Determination of pH–Theory and Practice*. New York: John Wiley & Sons, 1964.

Conductivity Cells. Foxboro, MA: The Foxboro Company, TI 43-10a, November 1967.

Considine, D. M. *Process Instruments and Controls Handbook*. New York: McGraw-Hill Book Company, 1957.

Durst, P. A. *Ion Selective Electrodes*. Special Publication 314, National Bureau of Standards, 1969.

Fribance, A. E. *Industrial Instrumentation Fundamentals*. New York: McGraw-Hill Book Company, 1962.

Fundamentals of pH Measurement. Foxboro, MA: The Foxboro Company, TI 1-90a, November 1961.

Light, T. S. "Selective Ion Electrodes." In *Analysis Instrumentation*, vol. 5. Edited by L. Fowler, R. G. Harmon, and Dir. Roe. New York: Plenum Press, 1968.

Resistance Dynalog Instrument Use for Electrolytic Conductivity, Foxboro, MA: The Foxboro Company, TI 43-9a, September 1963.

Rhodes, T. J. *Industrial Instruments for Measurement and Control*. New York: McGraw-Hill Book Company, 1957.

Shinskey, F. G. *pH and pIon Control in Process and Waste Streams*. New York: John Wiley & Sons, 1973.

Siggia, S. *Continuous Analysis of Chemical Process Systems*. New York: John Wiley & Sons, 1959.

Von Hippel, A. *Dielectric Materials and Applications*. New York: John Wiley & Sons, 1954.

STUDY QUESTIONS

Indicate whether following statements are True or False by inserting T or F in parentheses:

1. The micromho is a millionth part of the ohm. (T)

2. Conductivity of solutions is affected by concentration and temperature. (T)

3. Conductance is a function of the area of and distance between electrodes of a conductivity cell. (T)

4. Specific conductance in ohms per millimeter is also correctly referred to as a solution's "conductivity." (T)

5. Conductivity cells with low constants should be used for measurements in solutions with high conductivity. (F)

6. Automatic temperature compensation in a conductivity measurement involves the use of a resistance temperature detector inserted in the vicinity of the measuring cell. ( )

7. A direct current is fed to the conductivity measurement circuit to minimize polarization effects on the conductivity cell. (F)

8. pH is a measure of hydrogen-ion concentration of a solution. (T)

9. The strength of a solution increases with a decrease of the dissociation constant of a chemical. (T)

10. If concentration of hydrogen ions (H^+) and hydroxyl ions is the same in a solution, then they each must be $\frac{1}{10}7$. (T)

Select answer or answers which correctly apply to the following statements:

11. The principle of operation of an ion-selective measurement depends on
    a) its equal sensitivity to all ions in a solution.
    b) its high degree of specificity to a single ion in solution.
    c) its high degree of selectivity to a class of ions in solution.

12. Electrode assemblies used for other analytical measurements which can also be considered as ion-selective are
    a) conductivity electrodes.
    b) capacity electrodes.
    c) pH electrodes.

13. A sulfide ion-selective measuring electrode membrane is made of
    a) glass.
    b) silver sulfide.
    c) silicone rubber.

Analytical Measurement   165

14. At 25 C, a tenfold change in ionic activity causes the ion-selective electrode potential measuring a monovalent ion to change by
    a) 59.16 mv.
    b) 29.58 mv.
    c) 19.72 mv.

15. The activity of an ion becomes more proportional to its concentration in solutions that are of
    a) high concentration.
    b) more dilute concentration.
    c) a combination of ions.

16. Oxidation-reduction refers to chemical reactions in which participating chemicals
    a) lose electrons.
    b) gain electrons.
    c) remain unchanged.

17. Typical oxidants are:
    a) hydrogen sulfide.
    b) sodium.
    c) chlorine.

18. The measurement of oxidation-reduction potential is used as a guide to
    a) the rate at which a reaction proceeds.
    b) the degree of completion of a reaction.
    c) the number of elements entering the reaction.

19. Oxidation-reduction potential measurement can be used in the pulp and paper process to
    a) control the manufacture of hypochlorite bleach solutions.
    b) determine degree of cooking in the pulping process.
    c) monitor strengths of chemical additives in the paper mill.

20. The most common ORP electrode system used in the paper industry is made up of
    a) an aluminum measurement electrode and a copper reference electrode.
    b) a glass measurement electrode and a carbon reference electrode.
    c) a platinum measurement electrode and a silver reference electrode.

Indicate whether following statements are True or False by inserting T or F in parentheses:

21. A capacitor consists of two insulators separated by an electrode. The insulators are known as its "plates." (   )

22. If one capacitor charges to three times the number of unit charges as another, when the same voltage is impressed on each, the capacitance of the first is three times that of the second. (T)

23. The material between the "plates" of a capacitor is known as its "dielectric" and affects the electrical size of the capacitor. (F)

24. The dielectric constant of a material used as a dielectric of a capacitor is equal to the ratio of the capacitance produced by the material to the capacitance produced by vacuum in the same capacitor. (T)

25. Turbidity is a physical characteristic of a solution which causes light to be transmitted in a straight line through the fluid. (F)

26. Tyndall ratio is equal to the scattered light divided by the transmitted light. (T)

27. The nephelometer measures only the intensity of scattered light. (F)

28. Color can be measured by optically comparing light reflected from a sample, with light reflected from a standard. (F)

29. The general types of gas analyzers used in industry today include chemical, electrical conductivity, thermal conductivity, paramagnetic and mechanical. (T)

30. Measuring the heat of combustion of a gas is a typical mechanical method of analyzing its composition. (F)

(See Appendix for answers)

6. Density and Specific Gravity Measurement

Frequently during the process of making paper it is desirable to measure density or specific gravity in order to determine the concentration of a solution or mixture so it can be monitored and/or controlled within desirable limits for optimum operation. Measurements of density and specific gravity are often made in other industries on gaseous, liquid, and solid materials. However, in the paper industry they are primarily made on liquids and slurries such as: ore slurries used in the manufacture of SO_2; magnesium hydroxide slurry used in the manufacture of cooking liquors; black liquors from brown stock washers, from evaporators and to recovery furnaces; raw green liquor in smelt tanks; lime slurries to recausticizing process; lime slurries to the lime kiln; white liquor to the cooking process; and clay and starch slurries in paper coating makeup process.

MEASUREMENT THEORY

Density and *specific gravity* are used interchangeably in the pulp and paper industry. Although they characterize the same physical property of a fluid, theoretically the numerical values for the two can be quite different by definition. *Density* is defined as weight per unit volume and expressed in units of pounds per cubic foot or pounds per gallon. Theoretically, density is defined as mass per unit volume. However, for all practical purposes in the pulp and paper industry, it can be considered as weight per unit volume. *Specific gravity* is the ratio of a fluid's density to the density of water at a standard temperature and, therefore, is dimensionless. Density and specific gravity measurements, being intimately related when referred to the weight of an equal volume of water, can be easily converted from one to the other under similar or known conditions. Density measurements are commonly expressed in specific gravity units in the pulp and paper industry.

One of the first specific gravity measuring devices was a float immersed in the solution fitted with a scale that was submerged according to the specific gravity of the solution. Specific gravity being a dimensionless value, the scales were arbitrarily graduated in units convenient to the particular industry using it. The API scale was adopted by the petroleum industry, while Balling became common in the brewing industry, Barkometer in the tanning industry, Brix in the sugar industry, Quevenne in the dairy industry, Richter, Sikes, and Tralles in alcohol manufacturing, Twaddle in the heavy chemical industry. The pulp and paper industry elected to use Baumé units as the scale for many of its measurements.

The National Bureau of Standards has adopted two standard Baumé scales: (1) for liquids heavier than water and (2) for liquids lighter than water. These scales are related to specific gravity in the following manner, with 60°F being the standard temperature:

For light liquids $°Bé = \dfrac{140}{s.g.} - 130$

For heavy liquids $°Bé = 145 - \dfrac{145}{s.g.}$

As the density of a liquid varies inversely with temperature to an amount which differs with different liquids, a specific gravity measurement must be corrected for temperature to be completely accurate in terms of standard reference conditions for density and concentration. For most liquids found in the paper industry, the temperature effect is relatively small and often negligible over narrow temperature spans. Where specific gravity measurement is critical, the temperature must either be controlled at a constant value, corrected to the desired base temperature in the instrument calibration, or else mechanical temperature compensation must be used.

DENSITY MEASUREMENT SYSTEMS

The many methods of measuring density and specific gravity range from a simple hydrometer through fixed volume and differential pressure methods, electrolytic conductivity, and boiling point rise by temperature difference, to the most recent uses of radiation and optical devices. Although all methods can be applied equally well with some minor modifications, the selection depends largely on the physical characteristics of the material under measurement, convenience of adapting to the installation conditions, and the economics of the particular measurement.

The simple hand hydrometer, consisting of a weighted float with a small-diameter stem proportioned so that more or less of the scale is submerged according to the specific gravity, is widely used in pulp and paper mills for making "spot" or off-line intermittent density measurements of process liquids. Continuous measurements are made by fixed volume, differential pressure, boiling point rise, radiation, or optical methods.

Fixed Volume Method

A common continuous density measuring device utilizing the fixed volume density principle is the so-called displacement meter, schematically illustrated in Figure 6-1. Liquid flows continuously through the displacer chamber with the buoyant body, or displacer, completely immersed. A buoyant force which is dependent upon the weight of the displaced liquid, and in turn is a function of

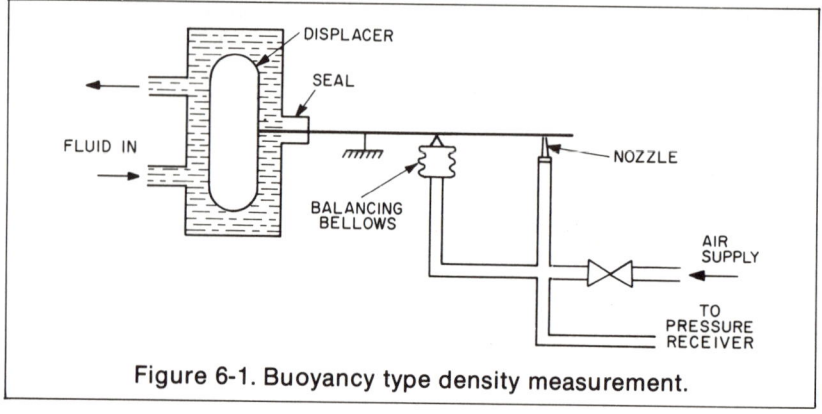

Figure 6-1. Buoyancy type density measurement.

the volume and specific gravity, is exerted on the displacer. If the volume is constant, the force will vary directly with the specific gravity. An increase in specific gravity will produce a greater upward force on the displacer and on the left end of the rigid beam. This causes the right-hand end of the beam to move closer to the nozzle, creating an increase in the pressure at the nozzle that is sensed by the balancing bellows which will expand. As the bellows expands it will force the baffle away from the orifice, causing a reduction in pressure in the pneumatic system. The bellows will move just enough to reestablish a new position of torque balance with somewhat different pressure, which is read on a pressure receiver instrument calibrated in density or specific gravity units. By adding weight to the left-hand end of the beam, some of the buoyant force can be suppressed and the force variations can be read on a calibrated instrument of lower range.

Another fixed volume device that has been used in paper mills to measure density of black liquor, lime mud, clay, and size slurries is based on actual weighing of the solution in a balanced pipe loop, diagrammatically shown in Figure 6-2. In operation, the process fluid flows through a hairpin loop of a tube pivoted on flexures about the horizontal axis. Weight of the tube and its contents is transferred to a weight beam and counterpoised by an adjustable balance weight. A change in fluid density produces a directly proportional change in force on the weight beam which is measured by a pneumatic force-balance device, producing an air signal which is directly related to the weight of the solution of the loop. The secondary indicating, recording, or control instrument can be calibrated in density units.

Differential Pressure Method

One of the simplest and most widely used methods of continuous density measurements is based on pressure variations produced by a fixed height of liquid. As shown in Figure 6-3, difference in pressure between any two elevations below the surface is equal to differences in liquid head between these elevations, regardless of variation in level above the higher elevation. This difference in elevations is represented by dimension H, which must be multiplied by the specific gravity G of the liquid to obtain the difference in head in inches of water, the standard unit for measurement calibration. To measure the change in head resulting from a change in specific gravity from G_1 to G_2, the difference between H inches x G_1 and H inches x G_2 must be calculated:

$$\Delta P = H(G_2 - G_1)$$

where ΔP = differential pressure in inches of water
   H = change in difference in elevations in inches
   G_1 = minimum specific gravity
   G_2 = maximum specific gravity

It is common practice to measure only the span of actual density changes by elevating the instrument ''zero'' to the minimum pressure head to be encountered, allowing the entire instrument working range to be devoted to the differential caused by density changes. For example, if $G_1 = 1.0$ and $H = 100$ inches, the range of the measuring instrument must be elevated H x G, or 100 inches of water. For a $G_1 = 0.6$ and $H = 100$ inches, the elevation would be 60 inches of water. The two principal relationships to be considered in this type of measuring device are then:

Span $= H$ x $(G_2 - G_1)$

Elevation $= H$ x G

It follows then, that for a given instrument range, a lower gravity span requires a greater H dimension.

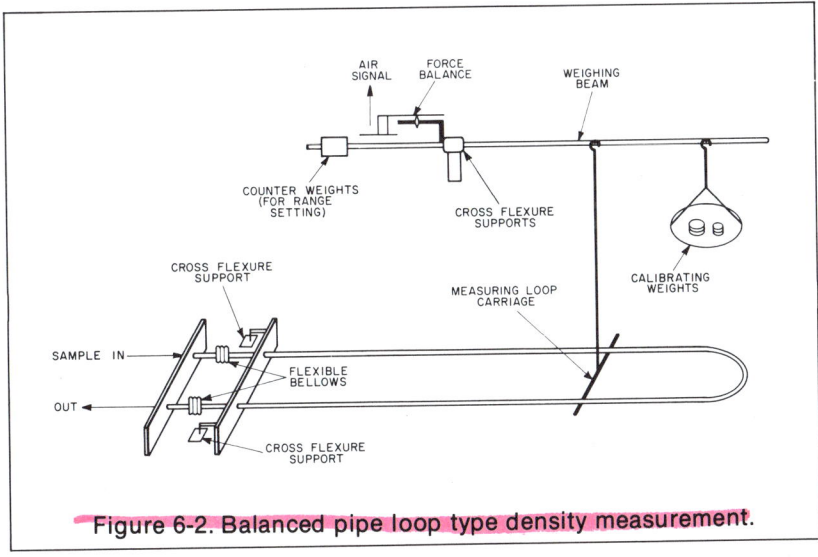

Figure 6-2. Balanced pipe loop type density measurement.

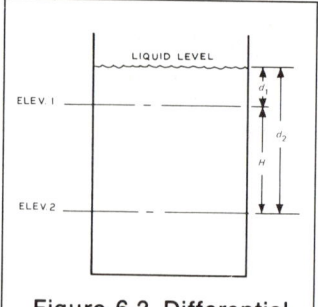

Figure 6-3. Differential
pressure density
measurement method.

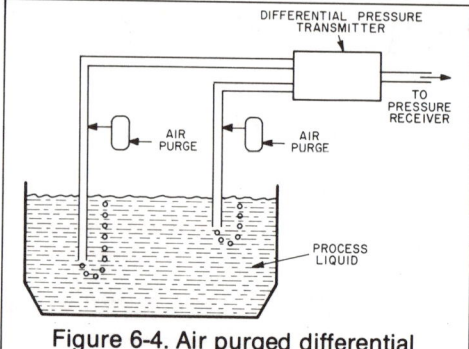

Figure 6-4. Air purged differential
pressure density measurement.

An open tank installation of this type utilizing air purges consists of two bubble tubes installed in the fluid so that the end of one tube is lower than the end of the other, as shown in Figure 6-4. The pressure required to bubble air into the fluid is equal to the pressure of the fluid at the ends of the bubble tubes. Since the outlet of one is lower than that of the other, the difference in pressure will be the same as the weight of a constant-height column of liquid. Therefore, the differential pressure measurement is equivalent to the weight of a constant volume of the liquid and can be represented directly as density, as shown. Weak black liquor, white liquor, and bleach liquor are frequently measured by this method. Span elevation or suppression is achieved by installing a reference chamber filled with a fluid with an appropriate density between the purge unit and the outlet of the low-pressure bubble tube.

This method can also be used to measure density in vertical process pipelines, as shown in Figure 6-5. Two taps at different elevations are installed in the side of the pipe and purged with a reference fluid, usually water, and connected to a differential pressure measuring instrument. In effect, the measured differential pressure is created by two equal columns, one of water and the other of the fluid to be measured. This system has been used to measure the densities of heavy black liquor, green liquor, lime mud, clay, and starch slurries.

Whenever possible, the pulp and paper industry uses the flanged type differential pressure cells, Figure 6-6, in order to eliminate the necessity of purges and bubble tubes. Figure 6-7 illustrates how they are typically installed and used on open and closed tank density measurement applications.

Boiling Point Rise Method

Water boils at a definite temperature for any given set of conditions. Under the same conditions, a solution boils at a higher temperature depending on the amount of material dissolved in it. Therefore, the difference between these temperatures is called boiling point rise. It is directly proportional to the density of the liquid and can be calibrated in specific gravity units. This

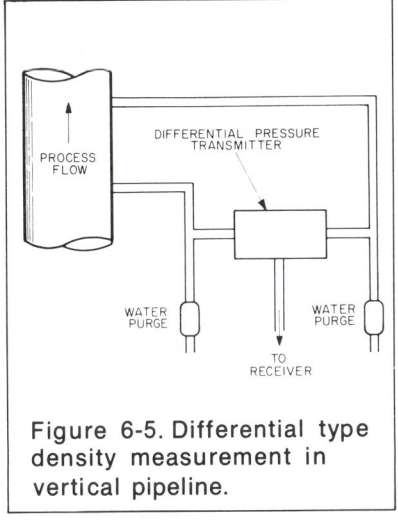

Figure 6-5. Differential type density measurement in vertical pipeline.

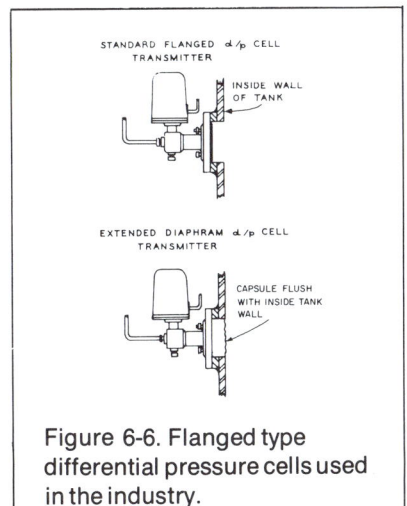

Figure 6-6. Flanged type differential pressure cells used in the industry.

principle is commonly used to measure density of black liquor from evaporators in pulp mills, as shown in Figure 6-8.

To determine boiling point rise, the temperature of the boiling solution is usually measured by a resistance temperature detector, described in the chapter on temperature measurements, and compared to a reference temperature measured in the same manner. The reference temperature is essentially that of pure water boiling at the same pressure as that of the solution being measured.

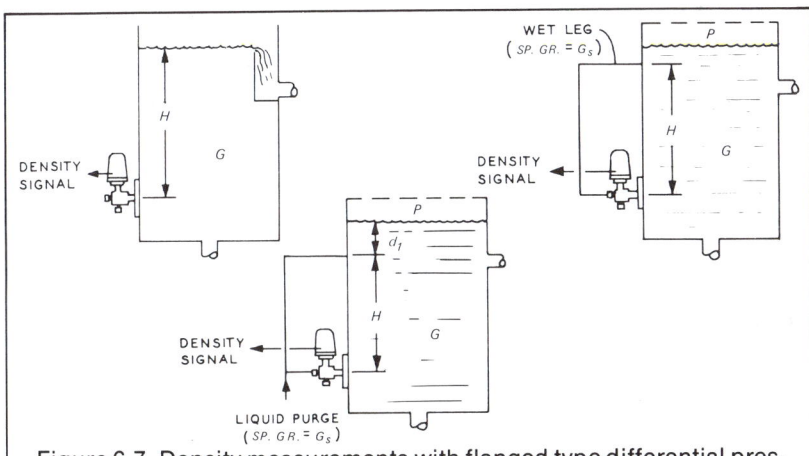

Figure 6-7. Density measurements with flanged type differential pressure elements. Top: left, open tank, constant level; right, open or closed tank, varying level and/or pressure wet leg. Bottom: open or closed tank, varying level and/or pressure, liquid purge wet leg.

Density and Specific Gravity Measurement   173

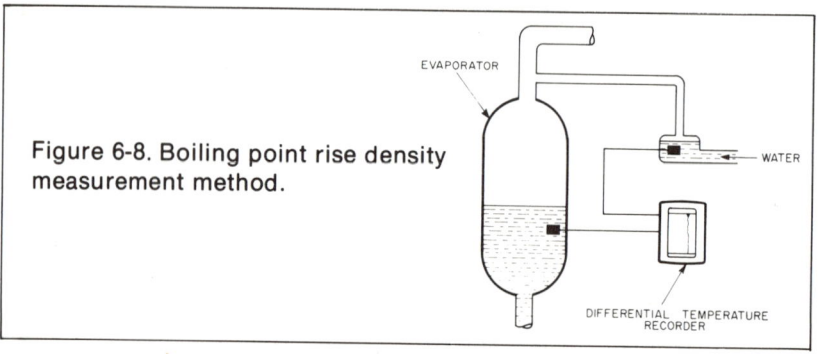

Figure 6-8. Boiling point rise density measurement method.

A simple and convenient method of providing a reference temperature measurement is to install the resistance temperature detector in a chamber located in such a manner that vapors rising from the boiling solution will condense in it. The vapors, being free of materials dissolved in the solution, will condense and form a liquid also free of dissolved material and at a temperature representing its boiling point. Condensed steam and boiling water at solution pressure can also be used for reference temperature measurements. One condensing chamber configuration that has been used for reference temperature measurements is shown in Figure 6-9.

Nuclear Radiation Method

Nuclear radioactive devices, similar to those described in the chapter on level measurements, can also be used to measure density. Their operation is based on the principle that absorption of gamma radiation increases with the mass of the material being measured. The two principal components of a nuclear radiation detecting system are the emitter and detector.

Emitter. Radioactive isotopes, either natural or man-made, emit radiation in the form of particles or waves as they go through the process of decay. The three most common types of emitted radiation are referred to as Alpha (α), Beta (β), and Gamma (γ), and differ in their energy levels. Gamma radiation is used almost exclusively for density measurements because of the higher energy level and the associated penetrating power of the various gamma-emitting isotopes. They are found to be suitable for the relatively thick-walled pipes and vessels used in the pulp and paper processes.

Detectors. There are three types of gamma radiation detectors in common use in density gauges: the scintillation crystal/photomultiplier tube, the Geiger-Mueller tube, and the ion chamber.

In the first type, the scintillation crystal/photomultiplier detector, the incident radiation (gamma waves) strikes an appropriate crystal after penetrating through the material being measured. As shown in Figure 6-10, the crystal is usually sodium iodide (NaI) which then produces scintillations of light. The photons of light then strike the photocathode of a photomultiplier (PM) tube producing photoelectrons which are attracted to the positive charged dynodes.

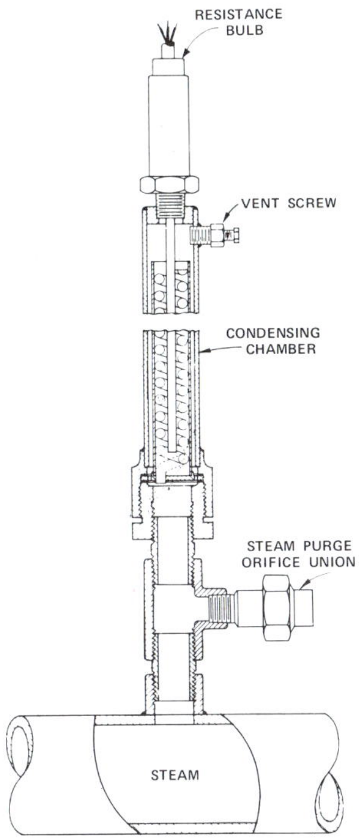

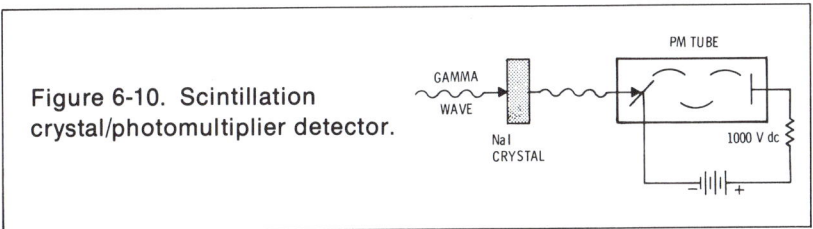

Figure 6-9. Reference temperature condensing chamber.

These, in turn, produce even more electrons by the phenomenon of secondary emission. The resulting current is then a function of the intensity of the incident light which is related to the density of the material under measurement. This current is then transmitted to suitable readout and control devices and converted to appropriate density units.

The second type, the Geiger-Mueller tube, consists of a small-diameter, thin-walled metal tube filled with an inert gas such as argon under pressure (approximately 10 cm/Hg). See Figure 6-11.

The gamma rays transmitted from the material under measurement strike the tube filled with argon. A very thin wire which makes up the center

Figure 6-10. Scintillation crystal/photomultiplier detector.

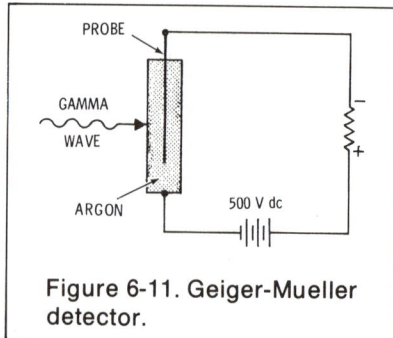

Figure 6-11. Geiger-Mueller detector.

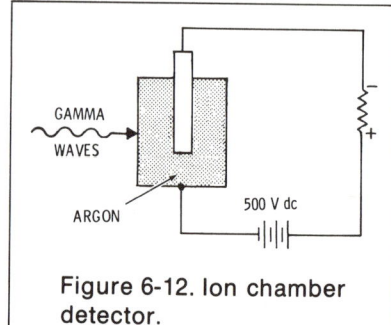

Figure 6-12. Ion chamber detector.

electrode in the tube creates high electric potential stress so that any nuclear particle producing a single ion pair within the gas can initiate an avalanche of electrons by gas multiplication. This avalanche of electrons produces a pulse of current which can either be counted or averaged as a measure of density.

The third type, the ion chamber schematically shown in Figure 6-12, consists of a relatively large-diameter steel chamber with a large-diameter center electrode. It is filled with a heavy inert gas such as argon at a specific pressure of several atmospheres. Using a heavy gas under pressure increases the chances of collision between gamma photons and gas molecules, causing ionization. This ionization of gas results in a continuous current which is proportional to the intensity of the incident radiation that represents a measurement of the density of the penetrated material.

Measuring Assembly. A representative radiation type density primary measuring element assembly is depicted in Figure 6-13. It consists of a constant gamma ray radiation source, which can be of radium, cesium, or cobalt, mounted on the wall of the pipe and a radiation detector mounted on the opposite side. Gamma rays are emitted from the source through the pipe and into the detector. Materials flowing through the pipeline and between the source and the detector absorb radioactive energy in proportion to their densities. The remainder of the radioactive energy is received by the radiation detector. The amount varies inversely with the density of the stream. The radiation detector unit converts this energy into electrical energy which is transmitted to a secondary indicating, recording, or controlling instrument.

A simplified electrical diagram of a basic complete density gauge system is shown in Figure 6-14. The radiation, after passing through the process material of a given density, penetrates into the ion chamber and causes current to flow in the ion chamber circuit. The resulting voltage appears across the "hi-meg" resistor because it is many times greater than the feedback potentiometer (R1). The zero suppression voltage is opposite in polarity to the hi-meg voltage and can be adjusted so that the resultant voltage at the input to the amplifier is zero volts dc. With zero volts dc into the amplifier, there will be zero volts dc out of the amplifier. If the process density increases, then more radiation will be absorbed and the ion chamber current will decrease. This results in less voltage across the hi-meg so that a net positive voltage appears as the amplifier

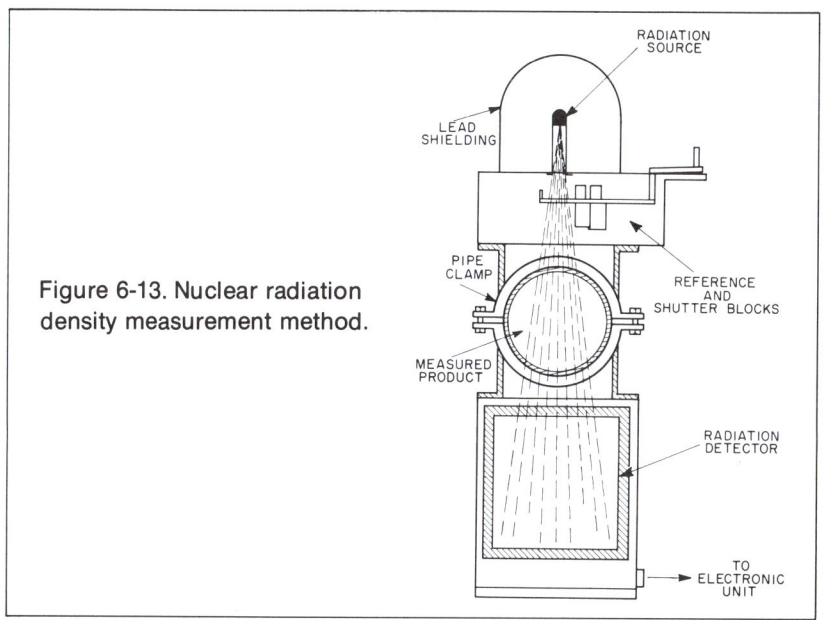

Figure 6-13. Nuclear radiation density measurement method.

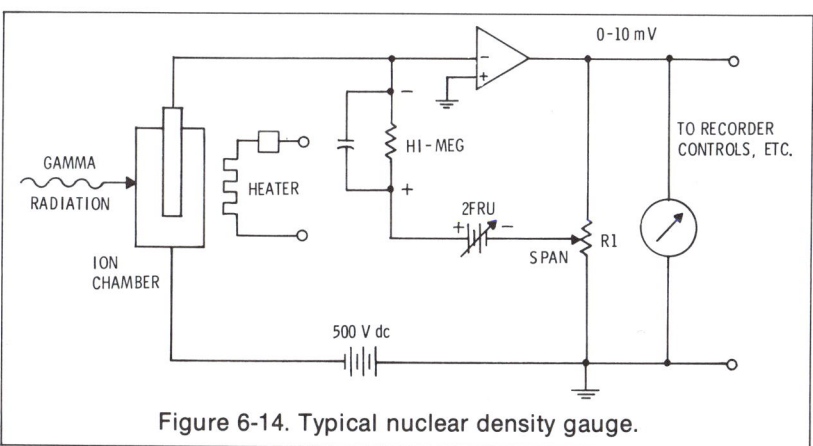

Figure 6-14. Typical nuclear density gauge.

input. Feedback or gain of the amplifier is set by a span potentiometer so the output voltage read by the meter corresponds to the derived density calibration.

Optical Method

Differential Refractometer. A relatively new technique, based on the refractive index of liquids, has recently been used to measure density of black liquor, resins, and starch solutions in the paper industry. The amount that a light beam is bent when directed into a solution is effectively determined by the quantity of solids or density of the liquid. A typical differential refractometer type of density detecting device measures the amount the light beam is bent when the refractive index of a liquid changes. The operation of such a system is illustrated by the schematic diagram shown in Figure 6-15. A light beam from an incandescent lamp is directed through a slit, mask, and lens to the sample cell through which the liquid sample is flowing. The beam passes through the sample liquid and a reference standard to a mirror mounted behind the sample cell. The mirror reflects the beam back through the cell for a second traversal of the sample liquid and reference standard.

At this point, the location of the light beam depends on the relative indexes of the sample and the reference standard. The beam is then divided by a beam splitter mirror. Each half of the beam falls on a photocell which forms two arms of a Wheatstone bridge. If the sample changes, the beam is refracted more or less and the amount of light reaching one photocell changes with respect to the amount of light reaching the other.

The bridge circuit becomes unbalanced and the resultant signal is amplified, causing the balancing motor to drive. The beam deflector plate is geared to the motor and turns in a direction to equalize the light falling on the two photocells. The more the refractor index of the sample deviates from the reference, the greater the unbalance in the bridge circuit and the more the deflector must turn to rebalance the split beams.

The position of the deflector, therefore, is a measure of the refractive

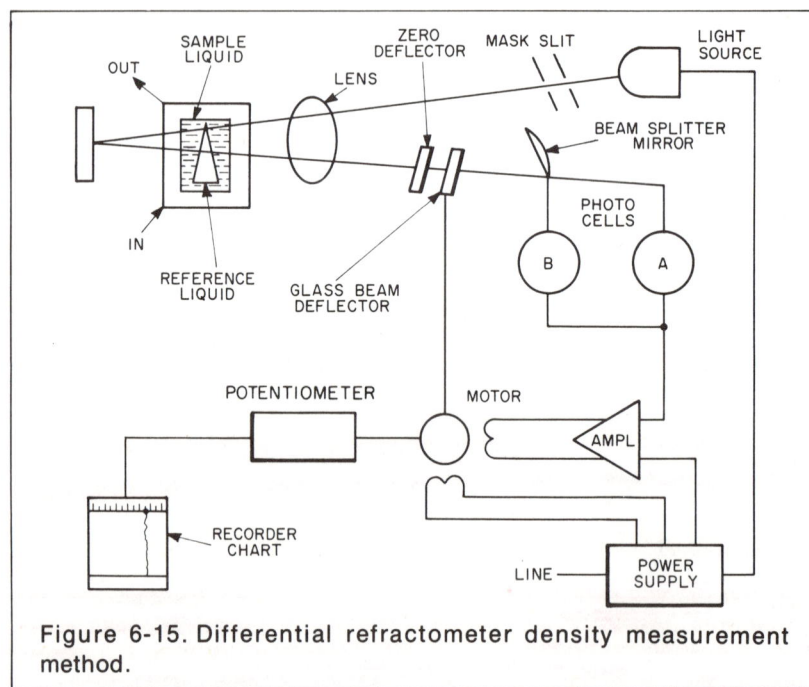

Figure 6-15. Differential refractometer density measurement method.

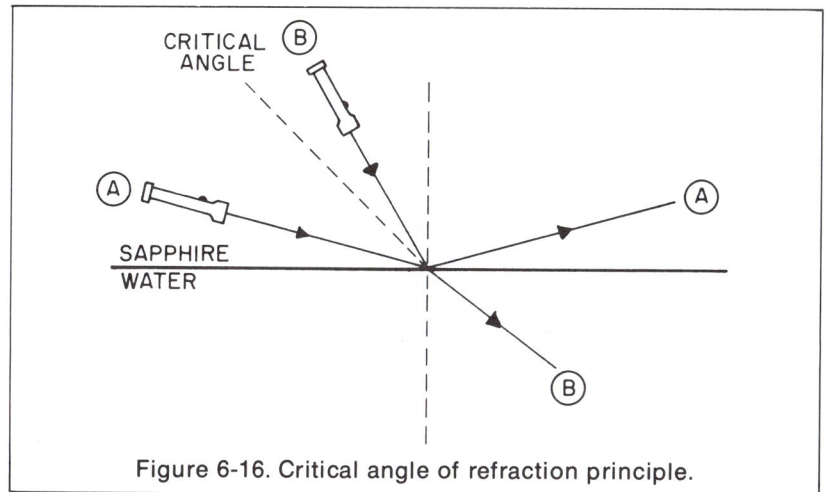

Figure 6-16. Critical angle of refraction principle.

index of the sample. At the same time the motor drives the deflector, it drives a potentiometer across which an electrical voltage has been impressed. The output signal from the potentiometer activates the meter and a secondary instrument which indicates or records the concentration, percent solids, or density of liquids. Of course, this signal can be used for control purposes.

Critical Angle Refractometer. A phenomenon called the *critical angle of refraction* is the basic principle underlying the operation of another optical density measurement method.

If one were to aim a flashlight at a surface of water from a very low angle (position *A*, Figure 6-16), almost all the light would be reflected from the interface or surface of the water.

If the flashlight is raised so as to be more perpendicular to the surface, the beam of light will no longer be reflected, but refracted into the water. Now the angle at which the beam of light transcends from reflection to refraction (the critical angle) is very precisely defined, given the physical properties of the conjugate substances. Therefore, if one could measure this angle it would define the relative indices of refraction of the two substances. For example, if the two substances are sapphire and water, the critical angle (the angle at which a beam of light converts from reflection to refraction) would be:

$$\text{arc sin} = \frac{1.33}{1.70} \text{ or } 52 \text{ degrees}$$

Adding some sugar, salt, sulfuric acid, etc., to the water would raise its optical density and, in turn, change the critical angle.

It has been stated that there is good correlation between refractive index and density, and that measuring the critical angle is a means of determining the refractive index. Figure 6-17 depicts one of the methods of determining the critical angle.

Density and Specific Gravity Measurement   179

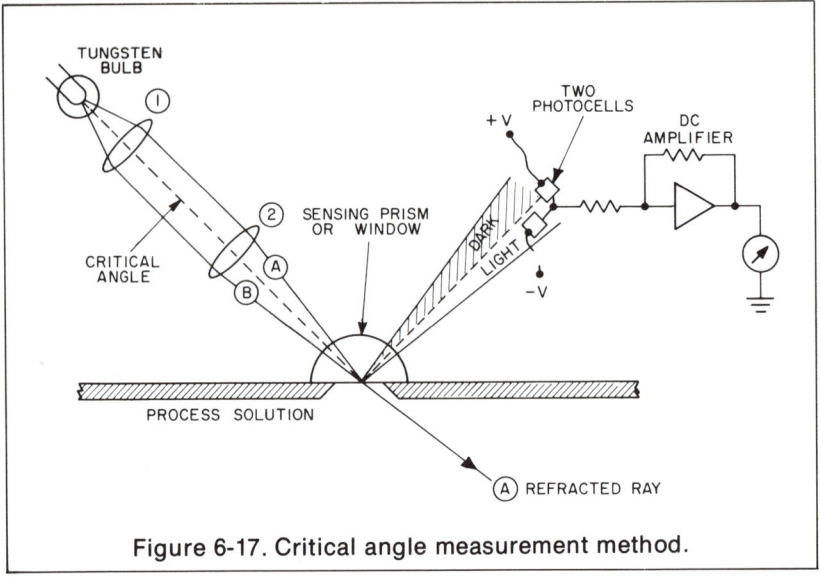

Figure 6-17. Critical angle measurement method.

Light rays from a tungsten bulb centered on the critical angle are collimated (or made parallel) by lens 1. This collimated beam of light is then focused by means of lens 2 on the interface between the sensing window (synthetic sapphire or diamond for abrasion resistance) and the process solution. Now note that the uppermost ray of light in the focused beam impinges more directly on the interface and is indeed above the critical angle as shown by the broken line. Therefore, ray A refracts into the process solution. On the other hand, ray B is below the critical angle and therefore reflects from the interface. All other "in between" rays in the focused beam do likewise, depending on whether they are above or below the critical angle.

This combination of reflection/refraction then produces light and "dark" beams which are detected by *two* photocells placed as shown. (The dark beam is not completely dark as there is always some reflection from an interface.) The signals from the two photocells are then amplified and converted to a meter reading. As the refractive index (hence density) of the process solution changes, the critical angle shifts its position, which in turn produces more or less reflected light depending on which way the critical angle moves. Again, this shift is indicated on the meter.

Another method of meauring the critical angle of refraction is similar to that just described except, instead of flooding the critical angle with a steady beam of light, the angle is continuously scanned by one ray of light (Figure 6-18).

As before, light rays from a tungsten bulb are collimated by lens 1. After collimation, one ray of light is allowed to pass through an aperture, or slit, and then impinges on an octagonal prism rotating at 3600 rpm. Any 45° revolution (⅛ turn) of this rotating prism causes the entering light ray to leave the prism at position A and sweep down to position B—then the cycle is repeated.

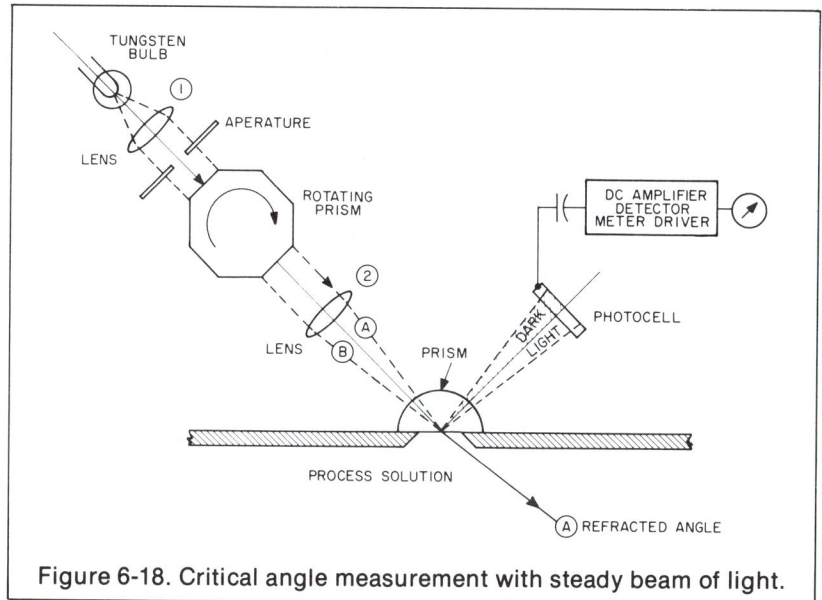

Figure 6-18. Critical angle measurement with steady beam of light.

This sweeping ray of light is then focused on the process solution by lens 2. At this point, note that the focused ray now starts at a high position *(A)* and sweeps down through the critical angle to position *B*. Again the cycle is repeated at a rate of 480 times per second (motor speed of 60 rev/sec times 8 prism flats per revolution). Refraction occurs during the time that the ray of light sweeps from position *A* down to the critical angle. Beyond the critical angle to position *B* we get reflection of the ray.

This action again produces a reflected beam consisting of light and dark regions, but with a very important difference over the previously described method. Here, the light and dark signals are ac, or alternating with time. This allows the use of just one photocell as a detector. The output of this photocell is a square wave (idealized) as depicted in Figure 6-19.

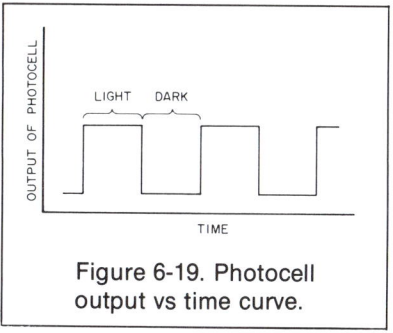

Figure 6-19. Photocell output vs time curve.

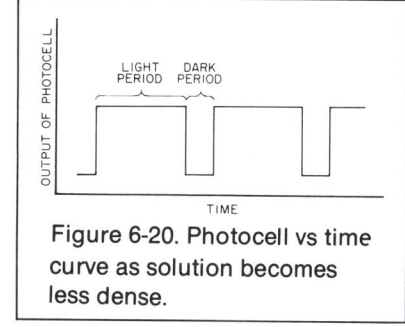

Figure 6-20. Photocell vs time curve as solution becomes less dense.

Here, the width of the light and dark pulses is important, not the amplitude. Therefore, it is of relatively little consequence if the photocell ages, the tungsten bulb changes intensity, or the process solution changes in opacity. And, of course, there is no matching required with only one photodetector.

If the process solution density changes, the relative widths of the light and dark pulses will change. Figure 6-20 shows the effect when the process solution becomes less dense. In this case, the pulse representing reflection is wider with respect to the width of the dark or refracted pulse. This signal is next amplified by ac amplifiers and then detected and measured in terms of density.

BIBLIOGRAPHY

Considine, D. M., ed. *Encyclopedia of Instrumentation and Control*. New York: McGraw-Hill Book Company, 1971.

Considine, D. M. *Process Instruments and Controls Handbook*. New York: McGraw-Hill Book Company, 1957.

Density Measurement Methods for Liquids and Liquid Slurries. Foxboro, MA: The Foxboro Company, TI 1-50a, 1973.

Filbance, A. E. *Industrial Instrumentation Fundamentals*. New York: McGraw-Hill Book Company, 1962.

Holzbock, W. G. *Instruments for Measurement and Control*. 2nd ed. New York: Reinhold Publishing Corporation, 1962.

Lipták, B. G., ed. *Instrument Engineer's Handbook*, vol. 1, "Process Measurements." Philadelphia: Chilton Book Company, 1969.

Stephenson, J. N. *Pulp and Paper Manufacture*, vol. 4. New York: McGraw-Hill Book Company, 1955.

STUDY QUESTIONS

Select answer or answers which correctly apply to the following statements:

1. Density is defined as weight per unit volume and can be expressed in
   a) pounds per cubic foot.
   b) pounds per gallon.
   c) pounds per square foot.

2. Specific gravity is the ratio of a fluid's density to
   a) the weight of an equal volume of water.
   b) the density of water at a standard temperature.
   c) the total head of an equal volume of water.

3. A typical specific gravity scale used in the pulp and paper industry is
   a) Baumé.
   b) A.P.I.
   c) Barkometer.

4. The National Bureau of Standards has adopted standard Baumé scales for
   a) liquids lighter than water.
   b) liquids same as water.
   c) liquids heavier than water.

5. For primary detection purposes, a common fixed volume type density measuring device utilizes
   a) the difference in liquid head.
   b) the boiling point rise.
   c) a displacer.

6. With the differential pressure method of density measurement, it is common practice to measure only span of actual density changes by
   a) expanding the range.
   b) suppressing the range.
   c) elevating the instrument "zero."

7. In measuring density in vertical pipelines
   a) a single purged tap is used.
   b) two taps at different locations are used.
   c) a sealed process connection is used.

8. Boiling point rise method of measuring density is based on the phenomenon that
   a) a solution boils at a temperature that is related to the amount of material dissolved in it.
   b) there is a temperature difference between a boiling liquid and vapor produced.
   c) the boiling point of a solution depends on the total pressure applied to it.

9. The measurement of density by nuclear radioactive devices operates on the basis that absorption of gamma ray radiation is related to the mass of the material measured in
   a) a direct manner.
   b) an indirect manner.
   c) an indiscriminate manner.

10. A refractometer type density detecting device measures the degree to which a light beam passed through a solution is
    a) transmitted.
    b) bent.
    c) absorbed.

(See Appendix for answers)

Viscosity measurement is not as extensively applied in the papermaking process as other measurements. However, its use is important to coating and many makeup operations, size presses, starch conversion processes, and clay slurries.

Viscosity is a measure of the combined effects of adhesion and cohesion of a fluid's molecules which manifests itself as an internal force resisting the flow of the fluid. In general terms, viscosity can be considered a frictional force arising when one layer of fluid is made to move in relation to another layer. Fluids with a high internal resistance to flow are called viscous and will not pour or spread as easily as fluids of lesser viscosity. In many cases, viscosity measuresments can be used as a guide to the rate or degree of a reaction, or an indication of the process end point. Where viscosity can be related to another variable such as concentration, density, or color, it can be used as an indirect measurement of these variables which are difficult to obtain in the more conventional manner.

BASIC THEORY OF VISCOSITY

In his classic mathematical definition of viscosity, Sir Isaac Newton used a "sliding plane" model, similar to that in Figure 7-1, to explain liquid behavior. The liquid under consideration is placed between the two planes with equal area A. The distance between the planes is X. If the planes are at rest or moved in the same direction at the same velocity, there will be no velocity difference $(V_2 = V_1)$. The space between the two boundary layers is filled with a number of layers of the same fluid, each of area A and thickness dX. If a tangential fixed force F is imparted to the top plane with the bottom stationary, a velocity difference is imposed on the system and V_2 is *greater than* V_1, so that $\Delta V = V_2 - V_1$. The uppermost layer of liquid molecules will move at the same velocity V_2 as does the plane. Due to the internal friction or viscosity of the fluid, the second layer of liquid molecules will move at a slower rate, the third layer will move even more slowly, as will the fourth, fifth, and so on. Therefore, since ΔV is constant, the rate of movement or velocity gradient, $\dfrac{dV}{dX}$ is set up in the liquid. Newton assumed that the force per unit area necessary to maintain the constant velocity difference between the adjacent planes was

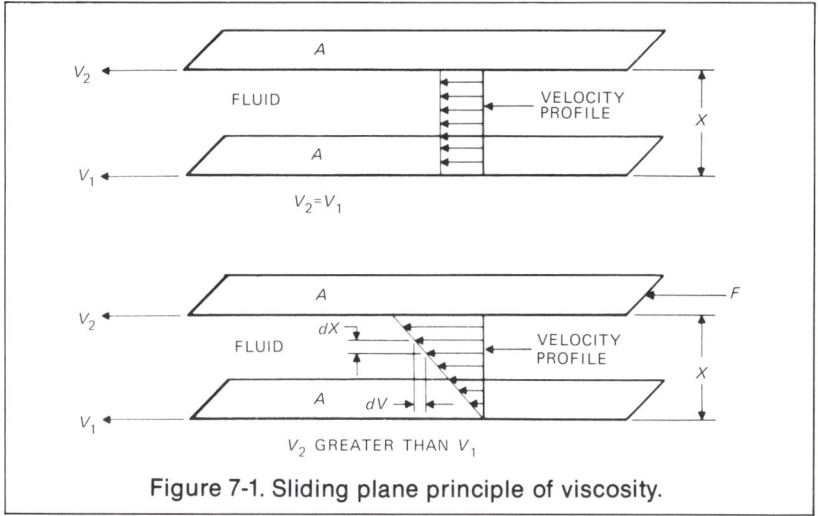

Figure 7-1. Sliding plane principle of viscosity.

proportional to the velocity gradient and could be expressed by the following relationship:

(1) $\dfrac{F}{A} = n\dfrac{dV}{dX}$

in which n represents the proportionality factor and is constant for a given material; n in this case is called the coefficient of viscosity or the fluid's *absolute viscosity*. The velocity gradient $\dfrac{dV}{dX}$ represents the shearing which the liquid experiences and is referred to as the *shear rate*. The force per unit area, $\dfrac{F}{A}$, which imparts this shear to the layers of liquid is called *shear force* or *shear stress*.

If F' represents shear stress $\dfrac{F}{A}$, and S represents shear rate $\dfrac{dV}{dX}$, equation (1) can be restated to define viscosity as:

(viscosity) $\qquad n = \dfrac{F'}{S}\ \dfrac{\text{(shear stress)}}{\text{(shear rate)}}$

Viscosity Types

In his original work, Newton assumed that shear stress was directly proportional to shear rate and, at a given temperature, viscosity would be independent of the shear rate as graphically shown in Figure 7-2. These became known as Newtonian fluids. However, it was later found that many fluids did not maintain a fixed ratio of shear stress to shear rate and that viscosities were not constant at a given temperature, but were a function of shear rate. These

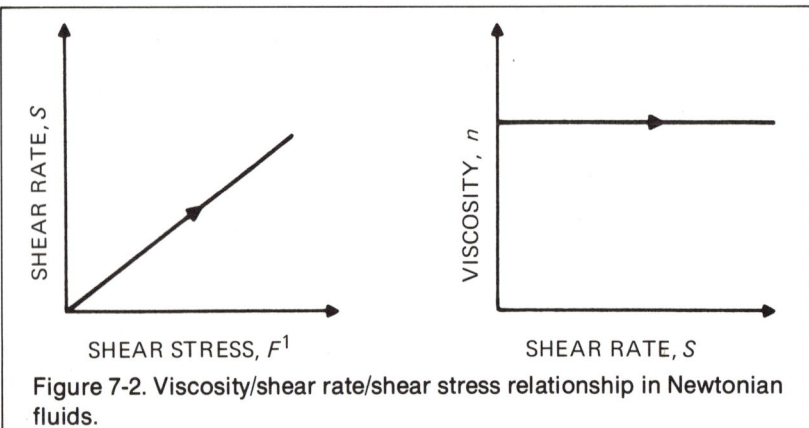

Figure 7-2. Viscosity/shear rate/shear stress relationship in Newtonian fluids.

materials are called non-Newtonians because they do not follow Newton's original theory. It was further noted that not only did the shear stress not bear a linear reaction with shear rate but that viscosity depended on whether the shear rate was increasing or decreasing so that non-Newtonian fluids became classified in accordance with their peculiar characteristic behavior in this respect. Thus, we refer to non-Newtonian fluids as either plastic, pseudo-plastic, dilatant, thixotropic, or rheopectic.

Viscosity Units

The basic unit of viscosity is the poise. Poise is determined by the shear stress in dynes per square centimeter required to maintain a shear rate of one centimeter per second per centimeter in a liquid. The viscosity would be one poise if one dyne were required, 10 poises if 10 dynes were needed, and so on. For convenience, the commonly used viscosity unit is the *centipoise* which is one hundredth of the larger *poise* unit. Viscosity can also be expressed in English units such as pounds (mass) per foot second or pound (force) seconds per square foot, but these units are not commonly used in the paper industry.

A determination of *viscosity* of a non-Newtonian liquid gives a value which is valid only under the exact conditions of measurement and, therefore, is termed *apparent viscosity*, as differentiated from *absolute viscosity*.

Kinetic viscosity is the ratio of absolute viscosity to the density of a given fluid. It is more directly obtainable from standard viscosity measuring systems than absolute viscosity, and can be accomplished by timing the gravity flow of a fixed volume of liquid through a hole or capillary tube with the results reported on a time basis. The basic unit of kinematic viscosity is the *stoke* (or centistoke) which has units of square centimeter per second in the c.g.s. system and square feet per second in English units.

There are other units of viscosity used in industry such as *specific viscosity, relative viscosity, Saybolt universal seconds* (SSU), and *Saybolt ferrol seconds* (SSF), but they are not commonly found in the paper industry.

PRIMARY VISCOSITY MEASUREMENT DEVICES

A wide variety of devices is available for off-line and on-line measurements of viscosity. They can be roughly classified in four categories: (1) rotation types, (2) falling-body types, (3) efflux or flow measurement types, and (4) oscillation types. Only a small percentage is designed for continuous process use in making viscosity measurements, and the first two types are the ones which predominate in papermaking.

Rotation Type

Spindle. The principle of the rotational viscosity device is that viscosity of a fluid is directly proportional to the torque produced by the resistance to shearing stress on a spindle rotating in the fluid (Figure 7-3). By inserting a beryllium-copper spring between the driven shaft and the driving shaft, the driven shaft will "lag" the driving shaft by an angle proportional to the "drag" on the spindle. This spindle drag is proportional to the viscosity and the length and diameter of the spindle. Capacitor plates are mounted on the two shafts and thus measure the lag. The secondary measurement instrument detects changes in capacitance to provide a measure of viscosity for indicating, recording, or control purposes.

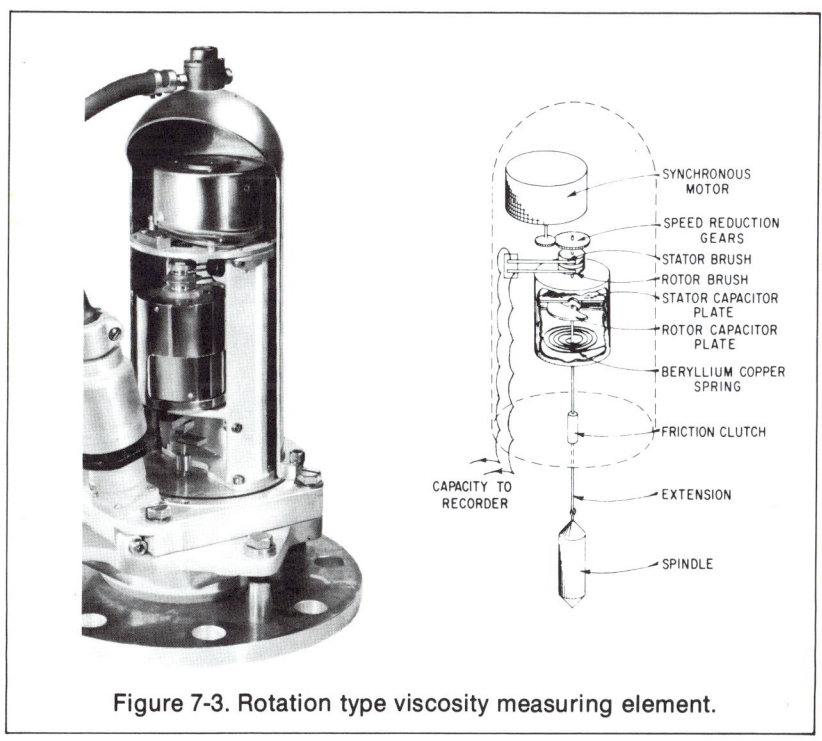

Figure 7-3. Rotation type viscosity measuring element.

Another rotational spindle type viscosity measuring element operates on a continuous sample drawn from the process into a sample cup which is rotated at a constant speed. A spring-controlled angular displacement spindle is immersed in the sample with the resultant torque being measured by a coil attached to the spindle and located between poles of a magnet. Measurement of current needed to restrain the coil is then directly converted to viscosity.

Another method for measuring viscosity, which operates in a similar manner to the rotating spindle type, is based on agitator power. The torque exerted by a process agitator blade is measured by a transmitting wattmeter (thermal converter) or solid-state Hall-effect transducer. It measures the power consumed in driving an agitator in a mixing tank and uses this signal as a measurement of viscosity.

Disc. Another rotation type viscosity measuring element uses two discs and operates according to the shear force principle with one rotating and one stationary disc (Figure 7-4). The disc which rotates at a constant speed is provided with radial slots that have one edge angled outwards. These "wings" constantly draw a supply of the process media at actual viscosity into the gap between two discs. The torque generated between the rotating and stationary discs is proportional to the viscosity of the media and is transmitted to a transducer. The transmitter, by means of a force-balance principle, converts this torque to a pneumatic or an electrical output signal which is proportional to actual process media viscosity. The gap distance is adjusted to change the measuring span, thereby facilitating the measurement of a wide range of viscosities. The generated signal is transmitted to appropriate indicating, recording, or control devices.

Falling Body Type

The principle of operation involved in falling-body types of viscosity measuring elements is based on the rate of fall of a ball or piston through the

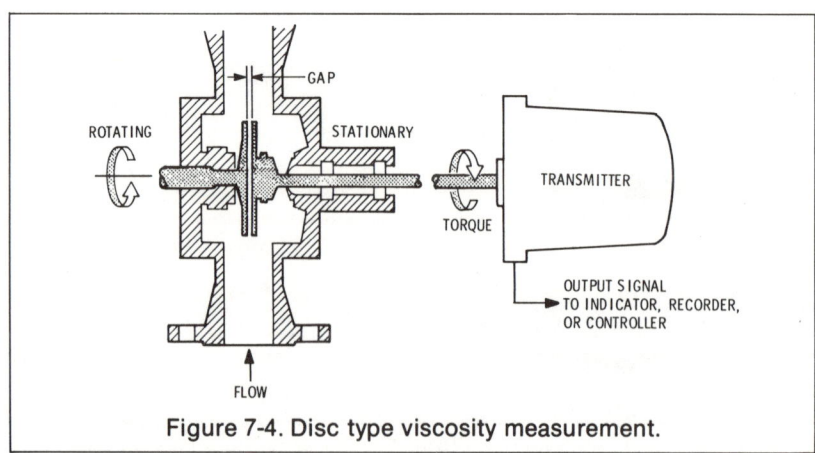

Figure 7-4. Disc type viscosity measurement.

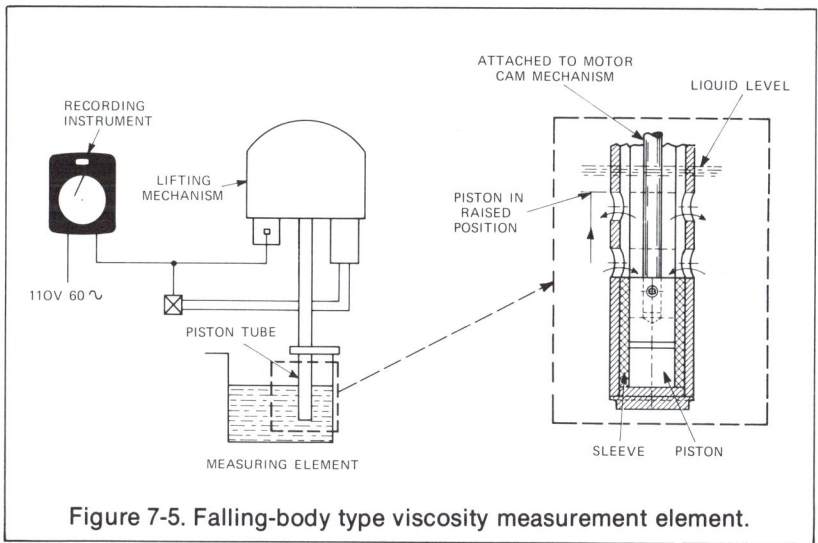

Figure 7-5. Falling-body type viscosity measurement element.

liquid. An assembly that operates on the falling-piston principle is shown in Figure 7-5. The schematic view of the measuring element shows the piston arrangement which is immersed in the liquid whose viscosity is to be measured. The piston is slowly raised by the motor cam mechanism shown in the upper portion of the illustration. It is then allowed to drop by gravity. A sample of the liquid is drawn in through orifices as the piston is raised, and expelled when the piston is dropped. The time required for the piston to drop is a measure of the viscosity, and this measurement is electrically transmitted to an indicator, recorder, or controller where it is converted to desired viscosity units.

A modification of the rotameter, previously described under area flow metering systems, has also been used to measure viscosity and can be considered under this category. For measuring flow, the float is specially designed to be virtually unaffected by viscosity of the fluid. A second float, which is normally sensitive to flow and viscosity, can be installed in the same tube so that the two floats will assume different positions in the flowing liquid. The difference between their responses, therefore, determines viscosity of the liquid (Figure 7-6). In operation, the flow is adjusted to a known constant flow rate indicated by the flow-sensitive float. The position of the viscosity-sensitive float can be calibrated in desired viscosity units for the fluid.

Ultrasonic Type

The ultrasonic type viscometer (Figure 7-7) consists of a probe and an electronic pulse-producing device connected by a coaxial cable. The pulse-producing device sends out short pulses of current to a coil which is situated inside the probe and around a thin blade. The resulting magnetic field excites

Viscosity Measurement   189

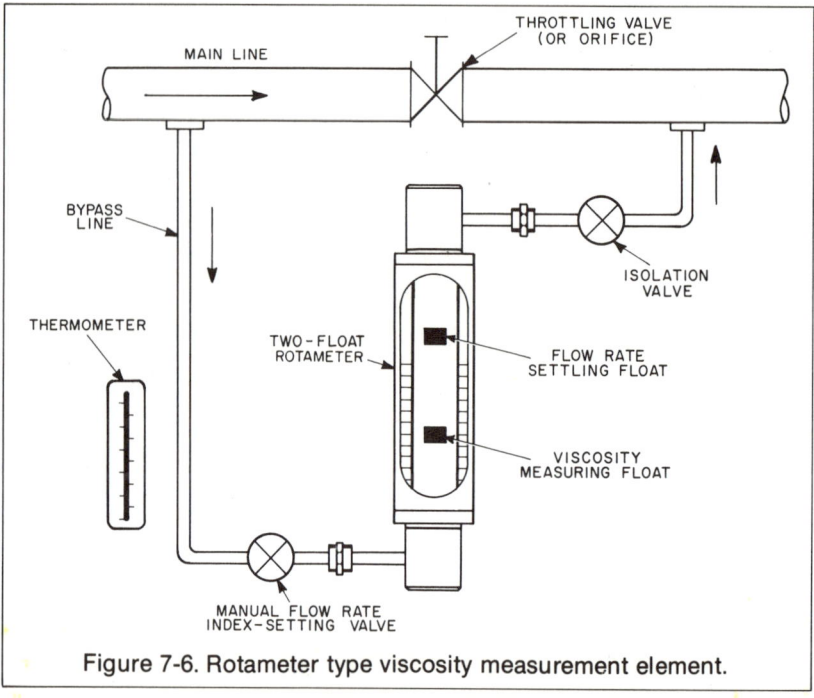

Figure 7-6. Rotameter type viscosity measurement element.

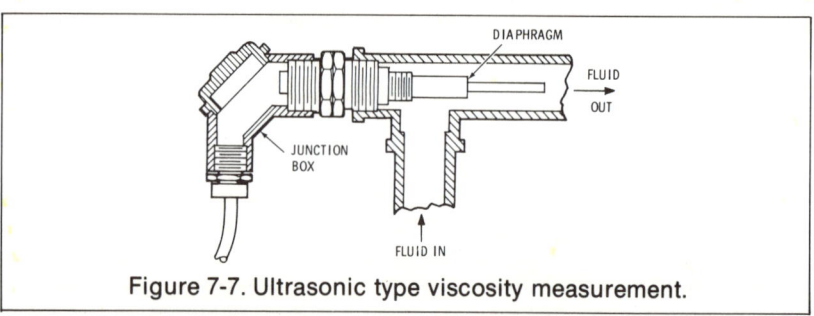

Figure 7-7. Ultrasonic type viscosity measurement.

the magnetostrictive member and causes the blade to vibrate at its own natural frequency which is determined by the length of the blade.

The probe acts as a transducer measuring the damping effect or viscous drag of the liquid. The damping effect on the ultrasonic vibration by the probe is a function of the liquid viscosity and is converted to appropriate values as a measure of viscosity.

Vibrating-Reed Type

The vibrating-reed type viscometer (Figure 7-8) consists of a frequency generator vibrating a spring rod, a probe, and a pickup unit to complete a

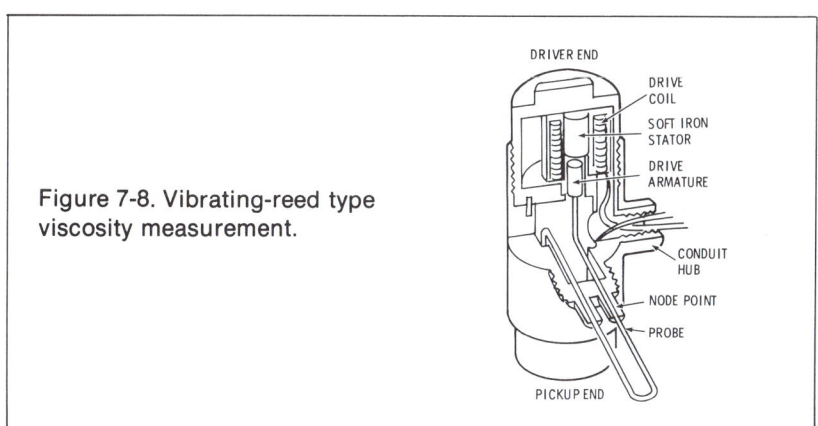

Figure 7-8. Vibrating-reed type viscosity measurement.

measurement loop through the process material.

The fundamental principle of operation is that the amplitude of probe vibration depends on the viscosity of the process media. The resistance to the shearing action caused by the probe vibration increases with an increase in the process media viscosity. This resistance is measured and converted to proper viscosity units for use with indicating, recording, or control instruments.

BIBLIOGRAPHY

A Primer on Viscosity. Foxboro, MA: The Foxboro Company, TI 1-50a, December 1973.

Carrol, G. C. *Industrial Process Measuring Instruments*. New York: McGraw-Hill Book Company, 1962.

Considine, D. M. *Process Instruments and Controls Handbook*. New York: McGraw-Hill Book Company, 1957.

Fribance, A. E. *Industrial Instrumentation Fundamentals*. New York: McGraw-Hill Book Company, 1962.

Holzbock, W. G. *Instruments for Measurement and Control*. 2nd ed. New York: Reinhold Publishing Corporation, 1962.

Lipták, B. G., ed. *Instrumentation in the Processing Industries*. Radnor, PA: Chilton Book Company, 1973.

Rhodes, T. J., and Carrol, G. C. *Industrial Instruments for Measurement and Control*. New York: McGraw-Hill Book Company, 1972.

Indicate whether following statements are True or False by inserting T or F in parentheses:

1. Viscosity is a measure of the combined effects of adhesion and cohesion of a fluid's molecules. (   )

2. Viscous fluids have relatively low internal resistance to flow and will pour or spread more easily than fluids with a higher viscosity. (   )

3. Viscosity can be defined mathematically as
$$\text{Viscosity } n = \frac{S \text{ (shear rate)}}{F' \text{ (shear stress)}}$$

4. "Newtonian" fluids are those in which shear stress is proportional to shear rate at a given temperature. (   )

5. Fluids which exhibit a fixed ratio of shear stress to shear rate, with viscosities being constant at a given temperature as a function of shear rate, are called "non-Newtonians." (   )

6. Poise is the basic unit of viscosity and is determined by the shear stress in dynes per square centimeter required to maintain shear rate of one centimeter per second per centimeter in liquid. (   )

7. The commonly used viscosity unit is the "centipoise" which equals 100 poises. (   )

8. Rotational viscosity measuring devices operate on the principle that viscosity of a fluid is directly proportional to the torque produced by the resistance to shearing stress on a spindle rotating in the fluid. (   )

9. The principle of operation involved in the falling body type viscosity measuring elements is based on relationship of viscosity and the rate of fall of a ball or piston through the liquid. (   )

10. Single-float rotameters have also been used to measure viscosity of liquids. (   )

(See Appendix for answers)

8. Consistency Measurement

Consistency measurements in the pulp and paper industry are primarily made on pulp and paper stock slurries and are considered among the most important and most difficult measurements to obtain. Process efficiency and the ease of equipment operation depend largely on the value and uniformity of consistency of the pulp and stock supply. To date, no practical device has been developed that can be applied to a process stream for continuous measurement of true absolute consistency. A variety of devices is successfully used to inferentially measure consistency of pulp and stock slurries associated with the following operations:

A. Pulp Manufacturing
 1. Continuous digester
 2. Screening
 3. Washing
 4. Bleaching
 5. High-density storage

B. Paper Manufacturing
 1. Refining
 2. Stock blending
 3. Repulping
 4. Saveall
 5. Paper machine wet end

BASIC THEORY OF CONSISTENCY

The term *consistency* is not used in exactly the same sense in the measurement of pulp and stock slurry as it is with the more common fluids. Although consistency is related to density and is sometimes used in the same sense in other fluids, it is not the same in all cases involving stock slurries. Density commonly applies to total weight per unit volume, whereas consistency is defined in the pulp and paper industry as the percentage, by weight, of dry fibrous material in any combination of pulp and water, or stock (pulps and additives) and water. It is calculated by the following formula:

$$C = 100 \frac{F}{W}$$

where C = the consistency of pulp or stock slurry in percent
 W = the total weight of a particular amount of pulp or stock slurry
 F = the weight of fibrous material in that amount

Basic determination of consistency is a laboratory procedure. The test consists of obtaining and weighing a representative sample, removing the water, drying, and weighing the remainder to determine the amount of fibrous material.

Consistency is usually expressed on a bone dry (b.d.) basis, where the percentage of fibrous material is determined by comparing the total sample weight with the stabilized oven dry (o.d.) sample weight. Occasionally, consistency is expressed as air dry (a.d.) with the assumption that the fibrous material contains 10 percent water. However, bone dry is usually assumed in the pulp and paper industry when no other specific reference is given. Consistencies of less than 1 percent are usually considered low; those greater than 6 percent are high. Most consistency measuring devices operate in ranges between those values.

The impossibility of determining a continuous direct absolute measurement of stock consistency is further complicated by the fact that stock is a two-phase suspension of water and fibers which is non-Newtonian and exhibits no well-defined hydraulic properties. Water, on the other hand, is a Newtonian liquid with well-defined hydraulic properties. However, experience has shown that an empirical relationship can be established between certain stock characteristics and consistency. All consistency detecting equipment available today is designed to sense a change in one of these characteristics and relate that measurement to consistency. Under these conditions, and depending on the design of the measuring device, this inferred value called *consistency* is affected by such known variables as velocity of stock flow, type of fibers, freeness, wetness, temperature, pressure, treatment, and broke addition, as well as other variables still not understood. Accordingly, this *inferential* measurement exhibits no long-term repeatable relationship between measurement and consistency and the short-term relationship is neither linear nor the same for different furnishes. Therefore, continuous consistency devices must be designed to selectively measure a characteristic that is closely related to fiber content and not greatly influenced by other variables over acceptable limited spans. These limitations have not prevented the widespread use of this general approach to consistency measurement. The ability of this approach to sense variations in consistency satisfies the requirements of the pulp and paper industry, and accurate measurement of absolute consistency is not, at this time, important.

CONTINUOUS CONSISTENCY MEASUREMENT DEVICES

A general classification of continuous consistency measuring devices is shown in Table 8-A.

They fall into groups determined chiefly by their installations. First, sampling devices operate on a sample stream removed from the main stream. Secondly, in-stream types are installed directly in the main stream with the entire flow passing through or past the primary measuring assembly. They can be further categorized as to whether the measurement point is open to atmosphere (atmospheric or open box types), or enclosed and under positive pres-

TABLE 8-A
CONSISTENCY MEASURING DEVICES

I. Sampling devices
  A. Atmospheric (open box)
    1. Apparent viscosity
      responsive sensor
    2. Force responsive sensor
  B. Pressurized
    1. Apparent viscosity
      responsive sensor
    2. Force responsive sensor
      a) Driven
      b) Stationary

II. Constant in-stream devices
  A. Atmospheric (open box)
    1. Apparent viscosity
      responsive sensor
      a) Driven
      b) Stationary

  B. Pressurized
    1. Apparent viscosity
      responsive sensor
    2. Force responsive sensor
      a) Driven
      b) Stationary

III. Miscellaneous devices
  A. Optical
    1. Transmitted and scattered
      light
    2. Polarized light
  B. Sonic
  C. Vibration
  D. Microwaves
  E. Drainage rate
  F. Electrical resonance

sure (pressurized). Sampling devices are mostly of the atmospheric variety, while in-stream types are common in both configurations. All the devices that have been commonly used differ somewhat in the measurement methods. Viscosity and the dynamic forces of a stock slurry are very closely related and both of these properties are used as the measured variables in inferential consistency measuring devices. Some sensors are primarily responsive to consistency-related apparent viscosity properties and others are primarily responsive to consistency-related forces such as fiber friction, impact against the sensor, and velocity-independent fiber network shear resistance. Elements are designed to isolate or compensate for the effect of velocity on the forces. Some are driven, moving sensors and others are stationary. The more common driven sensors have historically been in the form of paddles, modified propellers, cylinders, spheres, cones, discs, and screws. Common stationary element forms are pinned shaft, rod, and flat blades. Finally, Table 8-A lists the miscellaneous devices, methods that have been tried with some success, but not with general acceptance in industry. Some of these concepts are based on the use of optical methods, ultrasonics, vibration, microwaves, drainage, and electrical resonance.

Sampling Devices—Atmospheric with Driven Sensor

In the past, the most widely used device in the industry was the atmospheric, open box sampling type with electrically driven moving sensors. This device operates on the principle that the torque required to rotate an object in the stock slurry against friction caused by the slurry's apparent viscosity is indicative of the consistency. Figure 8-1 shows a sensor that uses a paddle-shaped feeler agitator suspended in a suitable flow box, and driven by a vertical motor mounted on a ball-bearing swivel base. As the consistency of the stock coming to the sample box changes, a corresponding change in torque is required to rotate the sweeping agitator. The torque is measured and converted to either a pneumatic or electronic signal used to control a dilution water valve. This sytem has been used in the 0.75 to 8 percent consistency range.

Utilizing the same principle based on apparent viscosity, the primary measurement can be made by sensing elements of other shpaes such as a cone (Figure 8-2). The cone is driven by water which is injected under pressure through a special reducing valve and nozzle turbine blades and then into a vessel inside the measuring body. This causes the cone, vessel, and connecting spindle to rotate. The speed of rotation is a function of the friction caused by apparent viscosity on the outside of the sensing cone, and varies inversely with its related consistency. The surface of the accumulated water inside the vessel assumes a paraboloidal shape which varies with the speed of rotation. By using a bubble pipe inside the vessel, the position of the lowest point on the

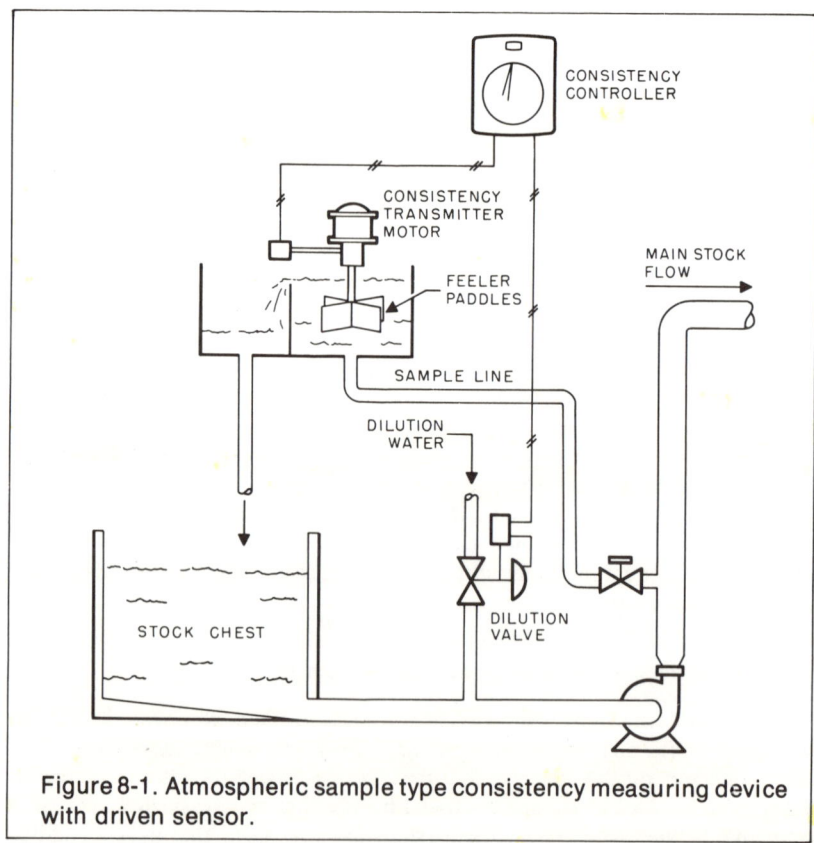

Figure 8-1. Atmospheric sample type consistency measuring device with driven sensor.

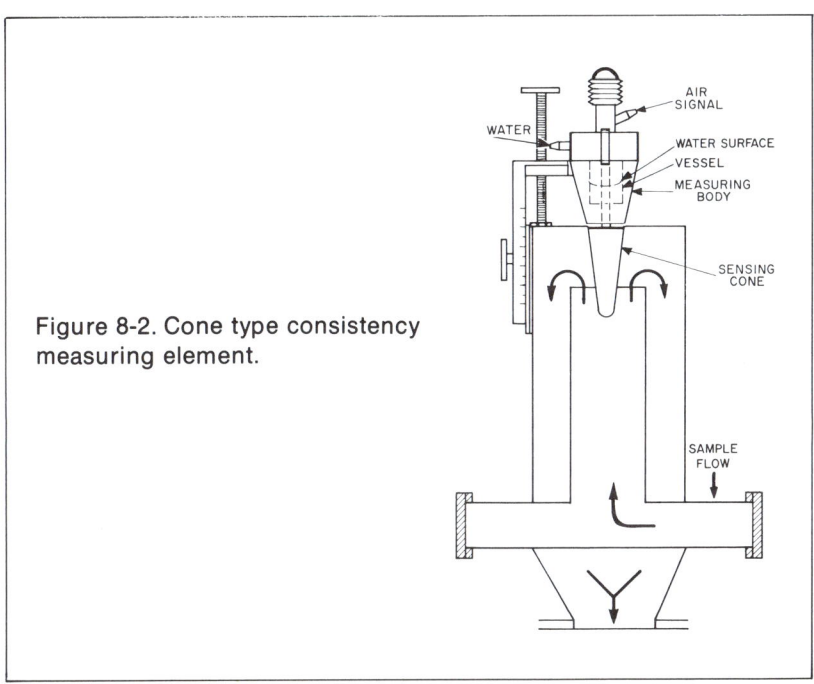

Figure 8-2. Cone type consistency
measuring element.

paraboloid can be gauged pneumatically to give an indication of consistency.
This unit can be used in consistency ranges from 1.3 to 6 percent.

Figure 8-3 shows a photograph and drawing of an atmospheric sampling-type
consistency measuring device using a spindle as the constant-speed driven
sensor. This device is similar to the viscosity measuring unit described in
Chapter 7. The spindle is a plastic cone-shaped element and senses consistency in terms of viscous torque. For a given stock, the sensitivity curve of the
transducer is governed by the physical size of the spindle, speed of rotation,
and calibration of the torsion element. The transducer is calibrated to transmit
either a pneumatic or an electrical signal. The system is normally used in
ranges of 0 to 0.5 percent to 0 to 2.0 percent consistency.

Another variation of the sampled open box measurement using an electrically driven rotating spherical sensing element is shown in Figure 8-4. Its
operation depends on apparent viscous frictional properties which are imparted
to the sensing element immersed in the pulp suspension. When the torque
changes, impulses related to consistency are electronically transmitted to a
servo system for control purposes. This system is used for stock consistencies
between 1.5 and 3.5 percent. By changing the shape of the sensing element to a
cylinder, it can be used in the range of 0.8 to 1.5 percent. By using a shaft fitted
with radial pegs for the sensing element, it can be used in a consistency range of
3.5 to 7 percent. However, this type of element is primarily responsive to
shear forces rather than apparent viscosity and relates the consistency measurement accordingly.

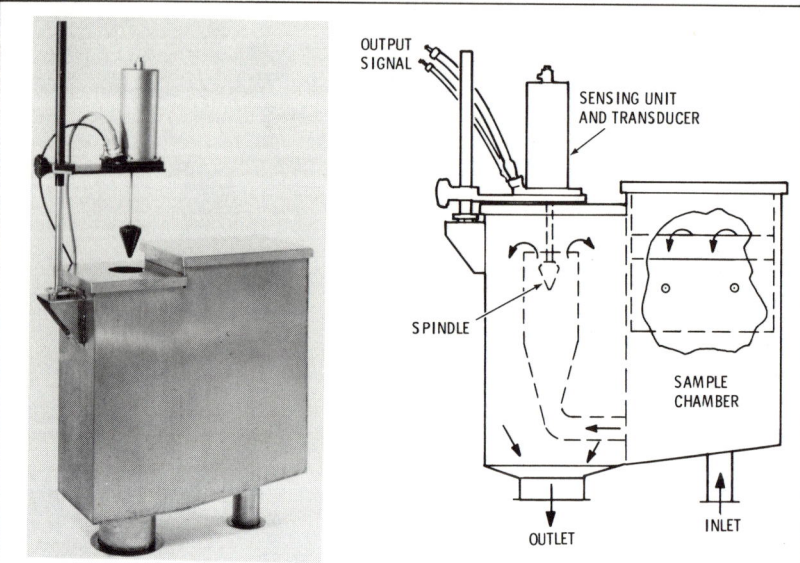

Figure 8-3. Spindle type consistency measuring device. Left, photograph; right, drawing.

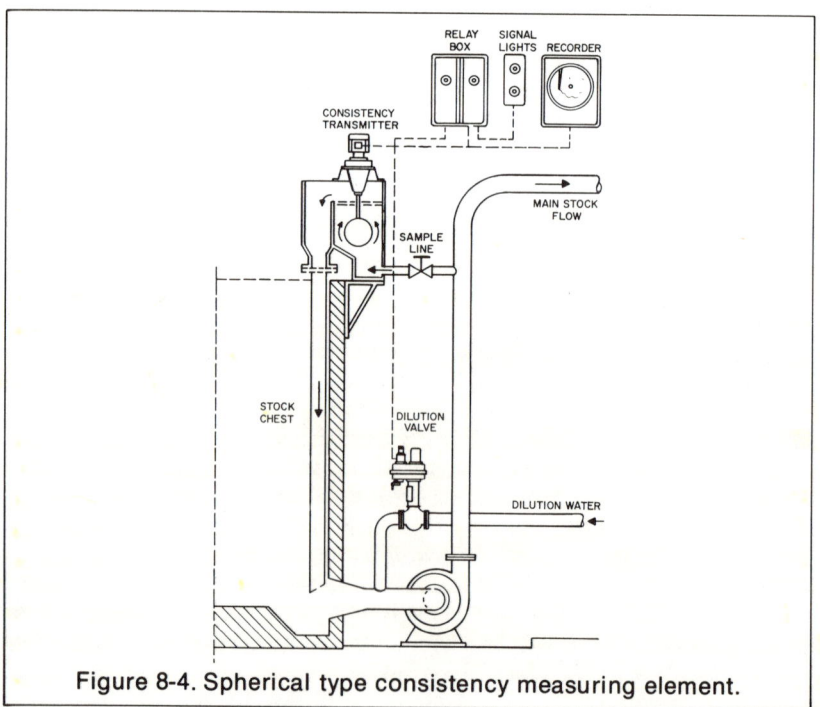

Figure 8-4. Spherical type consistency measuring element.

Sampling Devices—Atmospheric with Stationary Sensor

Most of the consistency detecting devices in this category are primarily responsive to apparent viscosity of the pulp slurry under measurement. Some common types still used in many mills operate on the basis of resistance to flow of stock in a viscosity or friction tube as shown in Figure 8-5. Here, there is a two-compartment box through which a sample is diverted. A constant head in the upper compartment is maintained by a weir and feeds a steady flow via an orifice into the standpipe in front of the box. The consistency-related measurement in this device is the pressure head necessary to pass this flow through a viscosity tube located in the output of the lower compartment. The system is designed to maintain a constant liquid head by increasing or decreasing the supply of dilution water before the sampling point. This type is normally used in the range of 2 to 4 percent consistency.

Another atmospheric sampling device which is apparent viscosity-responsive is shown in Figure 8-6. It operates on a sample of stock diverted through the open measuring box. A constant level is maintained in compartment A. Stock then runs through the measurement channel and returns to the process. The speed at which the stock runs is affected by the apparent viscosity and the level will accordingly rise or fall in the measurement channel. The movement of float B riding on the surface is then related to consistency. This motion is transmitted to an indicating or recording control system which is applied to stocks ranging from 0.5 to 4 percent consistency.

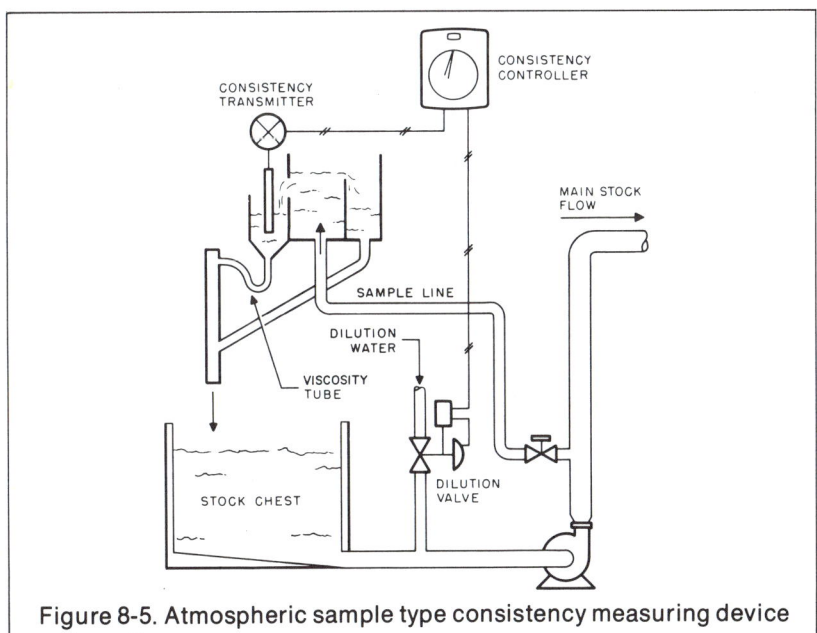

Figure 8-5. Atmospheric sample type consistency measuring device with stationary sensor.

Consistency Measurement   199

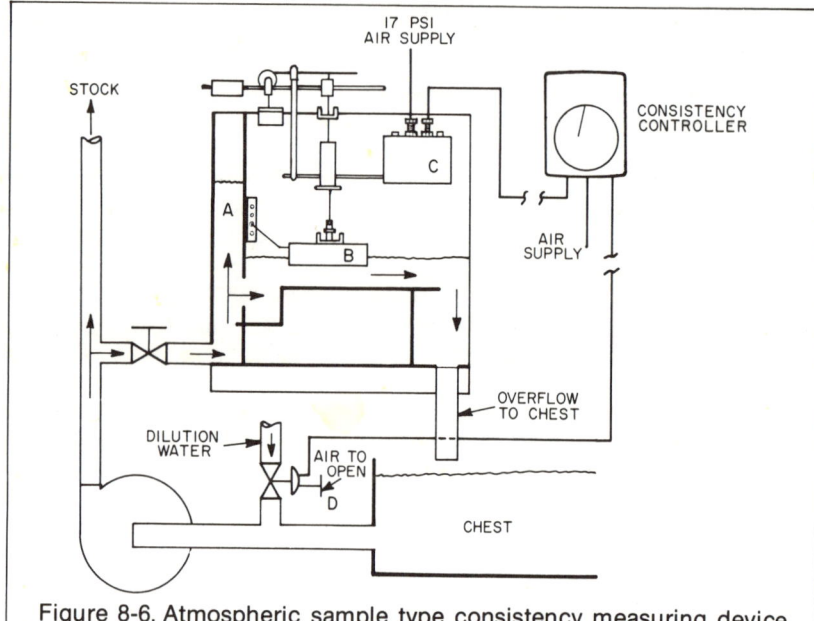

Figure 8-6. Atmospheric sample type consistency measuring device with float sensor.

In-Stream Device—Atmospheric with Driven Sensor

In a typical in-stream open box device utilizing the apparent viscosity of the pulp for measurement, a motor is employed to rotate the paddle-shaped sensing element at a constant speed (see Figure 8-7). The torque necessary to rotate the paddle is related to apparent viscosity and becomes a measure of consistency. This torque is converted to an appropriate pneumatic or electrical signal for indicating, recording, and controlling. Alternative systems measure friction loss in a viscosity tube or other restriction in which the friction head is measured at a constant flow or flow determined at a constant head. Operating ranges for these devices are from 0.75 to 8 percent consistency.

In-Stream Device—Atmospheric with Stationary Sensor

Another device in the in-stream atmospheric category, but with a stationary element, measures consistency by the angle of slope of stock slurry flowing through a trough or channel (Figure 8-8). The natural slope is dependent on apparent viscosity of the stock, thereby providing a means for relating consistency. A measurement of the drop in level between two spaced points in the trough can then be related to consistency of stock flowing through it. A differential pressure measuring unit detects the difference in back pressures due to drop in liquid level. Because the pressure is taken at two points in the same trough, it follows that variation in static head caused by reasons other

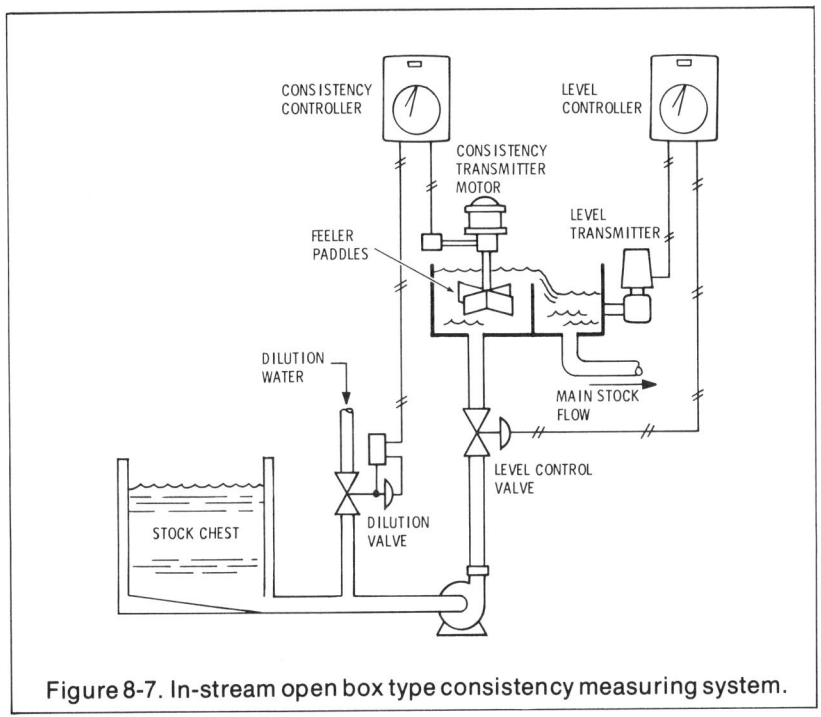

Figure 8-7. In-stream open box type consistency measuring system.

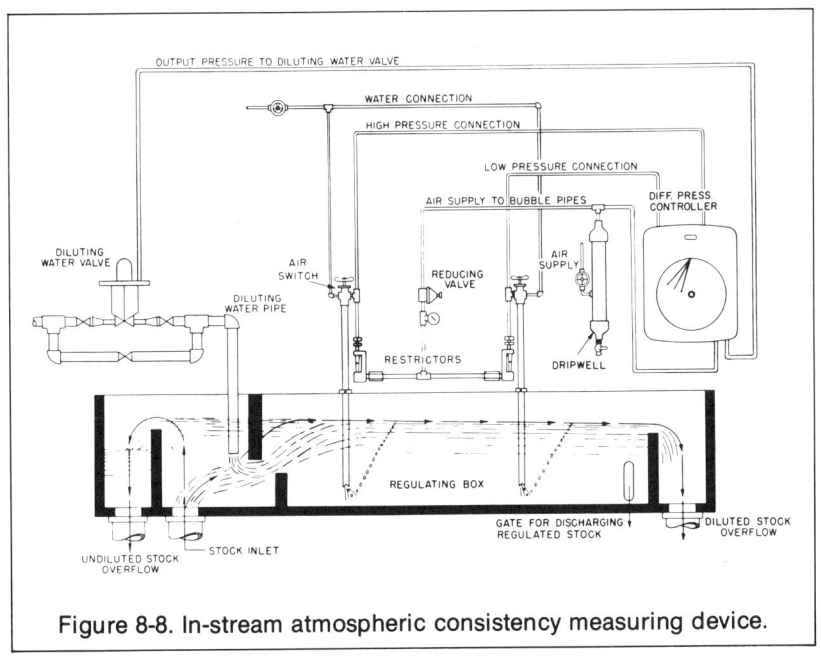

Figure 8-8. In-stream atmospheric consistency measuring device.

Consistency Measurement   201

than consistency changes is the same at two points and will not affect the differential pressure measurement. This system has been used in consistency ranges of 2 to 6 percent.

In-Stream Pressurized Devices with Driven Sensors

Consistency detecting devices of this general category which are primarily force-responsive are usually mounted as complete assemblies directly in the process pipeline. Most of their primary sensing elements are motor-driven but each has its own unique construction which makes it peculiarly responsive to a combination of forces related to the consistency of stock slurry. This force is measured by the amount of energy required to move the sensing element through the fluid surrounding it.

Figure 8-9 shows an in-stream, pressurized consistency detecting device which operates with a paddle-shaped electrically driven sensor responsive to apparent viscosity of the stock. The principle of operation is the same as the atmospheric sample type with moving element (Figure 8-1) except that it is located in the main process line rather than in the bypass sample line. It also operates in the same ranges between 0.75 and 8 percent consistency.

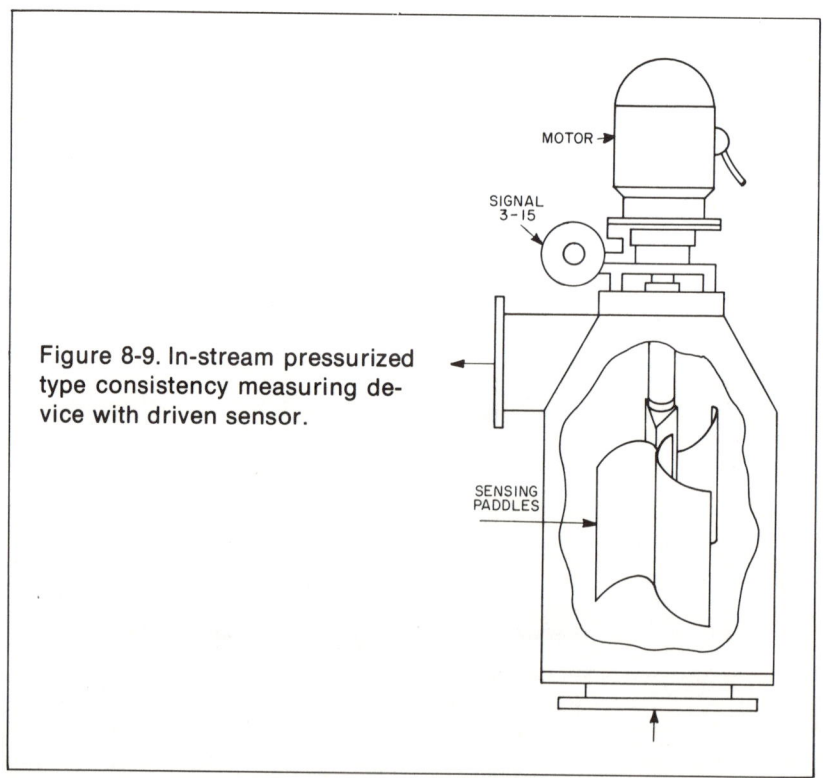

MOTOR

SIGNAL
3-15

Figure 8-9. In-stream pressurized type consistency measuring device with driven sensor.

SENSING PADDLES

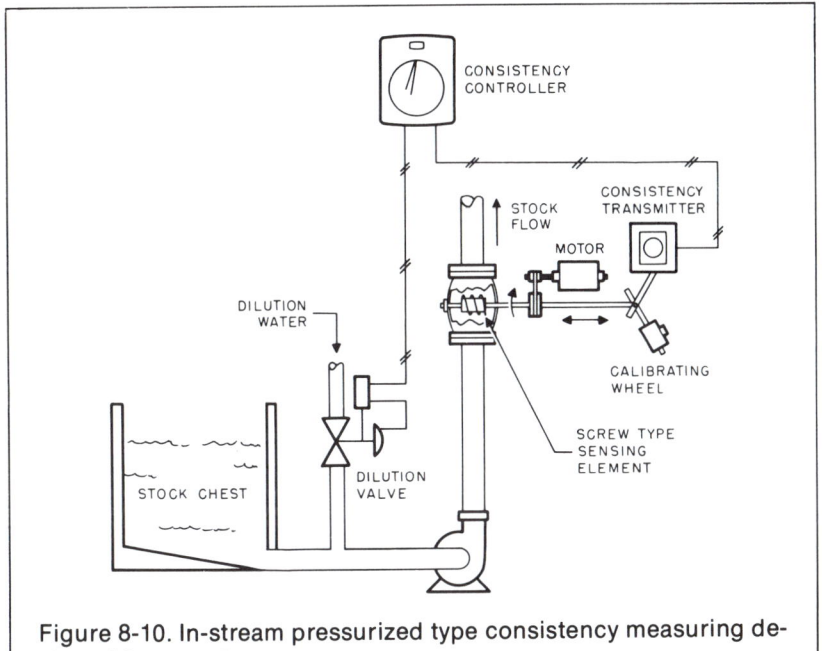

Figure 8-10. In-stream pressurized type consistency measuring device with screw type sensor.

Another version of the in-stream pressurized unit responsive to apparent viscosity has a motor-driven screw-shaped sensor (Figure 8-10). A change in consistency-related viscosity of the stock causes a change in thrust against the rotating screw, which then moves the screw shaft axially. This motion is linked to a position transmitter which produces a signal suitable for the control of consistency ranges between 2.5 and 8 percent.

In a typical measurement unit with a rotating sensing element (Figure 8-11), the rotor is a flat circular plate with involute ribs on the surface. The torque required to drive the rotor in opposition to the shearing force of the stock varies with the consistency. This torque is sensed by either an electronic or pneumatic bridge which balances the torque and produces a signal which is used for control in ranges within 0.75 to 6 percent.

Figure 8-12 is a view of a similar pressurized in-stream consistency detecting assembly with a motor-driven rotating sensor. The primary difference here is in the design of the sensing element and method used in converting the torque measurement to consistency. The rotating member is a flat, smooth motor-driven disc. The torque caused by the shearing forces is balanced against a spring whose motion generates a pneumatic signal related to consistency and used for indicating, recording, and/or control in ranges from 2 to 6 percent.

Still another version of the pressurized in-stream consistency measuring device (Figure 8-13) has a motor-driven primary measurement assembly consisting of a propeller agitator and a disc type measuring element. Both members are motor-

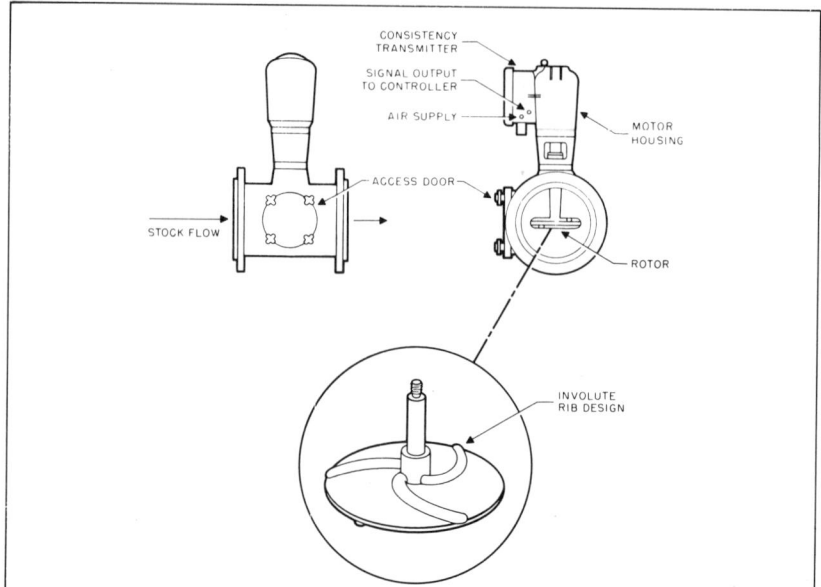

Figure 8-11. In-stream pressurized type consistency measuring device with rotating sensor.

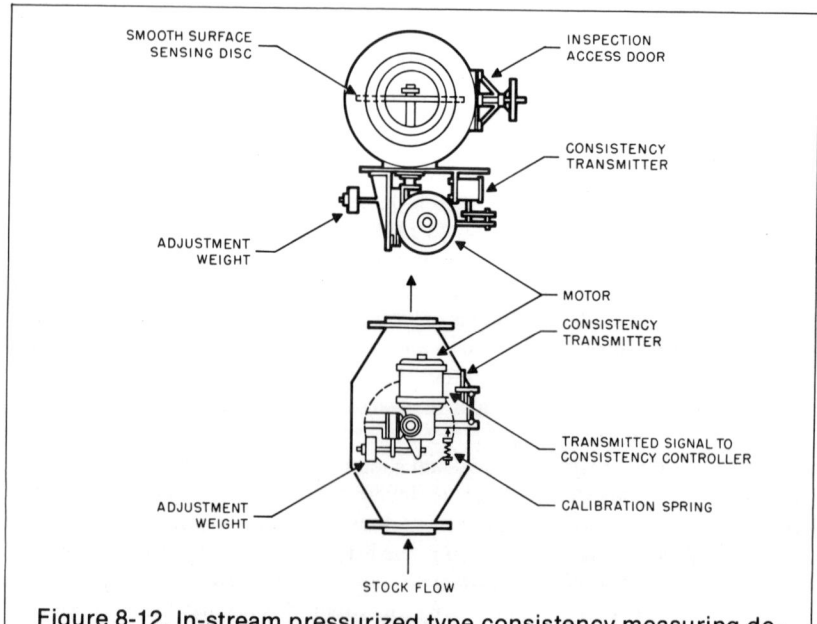

Figure 8-12. In-stream pressurized type consistency measuring device with smooth surface disc sensor.

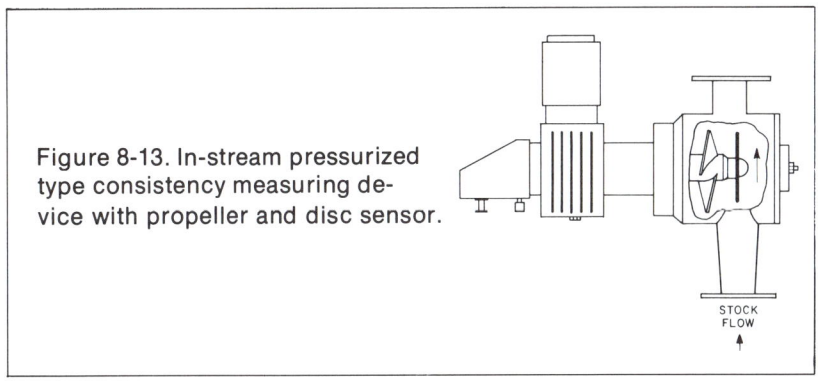

Figure 8-13. In-stream pressurized type consistency measuring device with propeller and disc sensor.

STOCK
FLOW

driven through two concentric shafts. The measuring element rotates in the pulp at the same speed as the propeller. The propeller drive shaft and the measuring element shaft are elastically connected. As the measuring element cuts through the pulp, a moment of force occurs on the measuring shaft as a result of the cutting resistance forces, which then influences the measuring element. The torsion of the measuring shaft in relation to the propeller drive shaft is measured by a pneumatic force-balance transducer, producing a pneumatic signal related to consistency. This device can be used in ranges from 1.8 to 10 percent.

In-Stream Pressurized Types with Stationary Sensor

More recent developments of in-line consistency detecting devices have brought forth an entirely new group which is characterized by specially designed fixed or stationary sensors that are responsive to forces exerted by the fluid passing them. They are available with both pneumatic and electrical signal outputs correlated to consistency.

A unit which employs a special design shear float for the sensing element is shown in Figure 8-14. The float which consists of a cylindrical shaft from which cylindrical fingers project is connected by a spring to one side of a modified pneumatic differential pressure transmitter. Stock flow passing the float exerts a force as it is deformed around the finger. This force, which increases with increasing stock consistency, is applied to the pneumatic transmitter which transmits a signal related to consistency to the control system. Range of operation is 2 to 8 percent consistency.

In an electronic version (Figure 8-15), the detecting element is a stainless steel cylindrical probe mounted directly in the pipeline. It contains four strain gauges forming a resistance bridge. As the forces caused by the deformation of stock flow passing the probe change with consistency, the bridge resistance changes accordingly. This signal is fed into a potentiometer type instrument circuit which measures the force as a function of consistency. Range of operation is also 2 to 8 percent consistency.

A recent development in the field of stock consistency detectors uses a stationary element, shaped like a scimitar blade, that is designed to compensate for

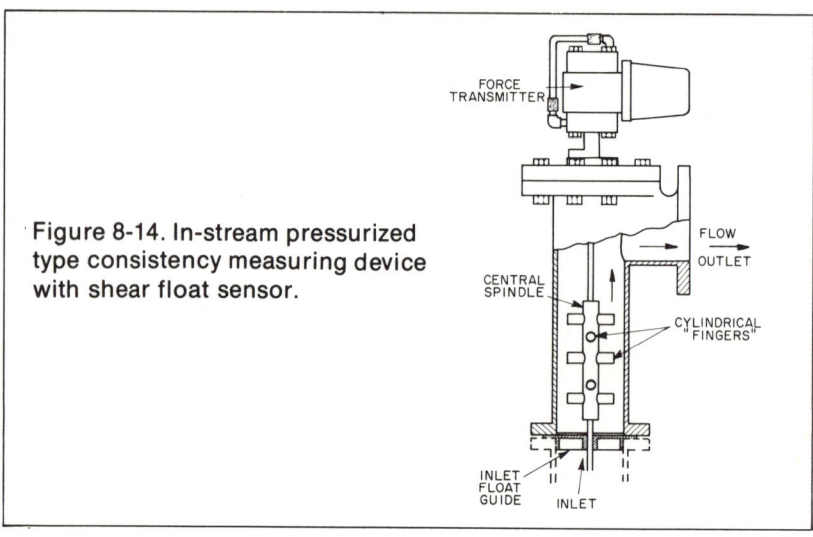

Figure 8-14. In-stream pressurized type consistency measuring device with shear float sensor.

impact forces caused by velocity head changes on the frontal areas and the drag forces due to flow along the sides. Therefore, it is primarily sensitive to fiber network shear, hence to consistency (Figure 8-16). Under normal operating conditions, two kinds of forces act simultaneously on the element. The first (shown as F) results from the drag of the pulp fibers as they flow over the flat sides of the plate. This force is primarily related to consistency variations. The other forces (F_1 and F_1') are due to impact and shearing forces of the stock slurry on the frontal area of the leading edge and on the tail edge respectively. These two forces act perpendicular to their respective surfaces and are primarily sensitive to flow variations.

By geometric proportioning of the blade, the moments of these forces (F_1 and F_1') balance out. The remaining force (F) creates the measured moment about an alloy seal pivot. The resulting moment is converted into a signal output related to consistency and used by a secondary instrument for control within operating ranges of 2 to 6 percent consistency.

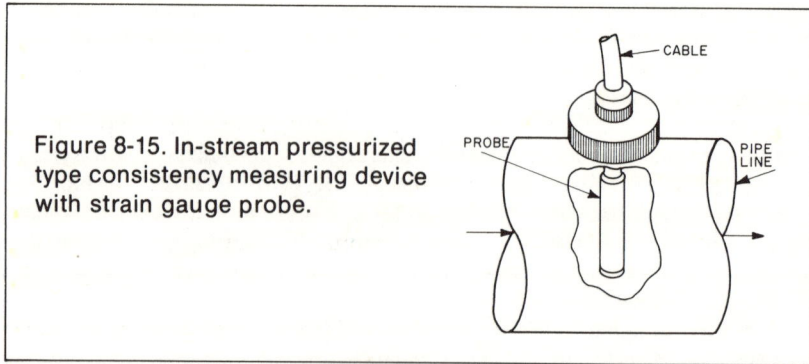

Figure 8-15. In-stream pressurized type consistency measuring device with strain gauge probe.

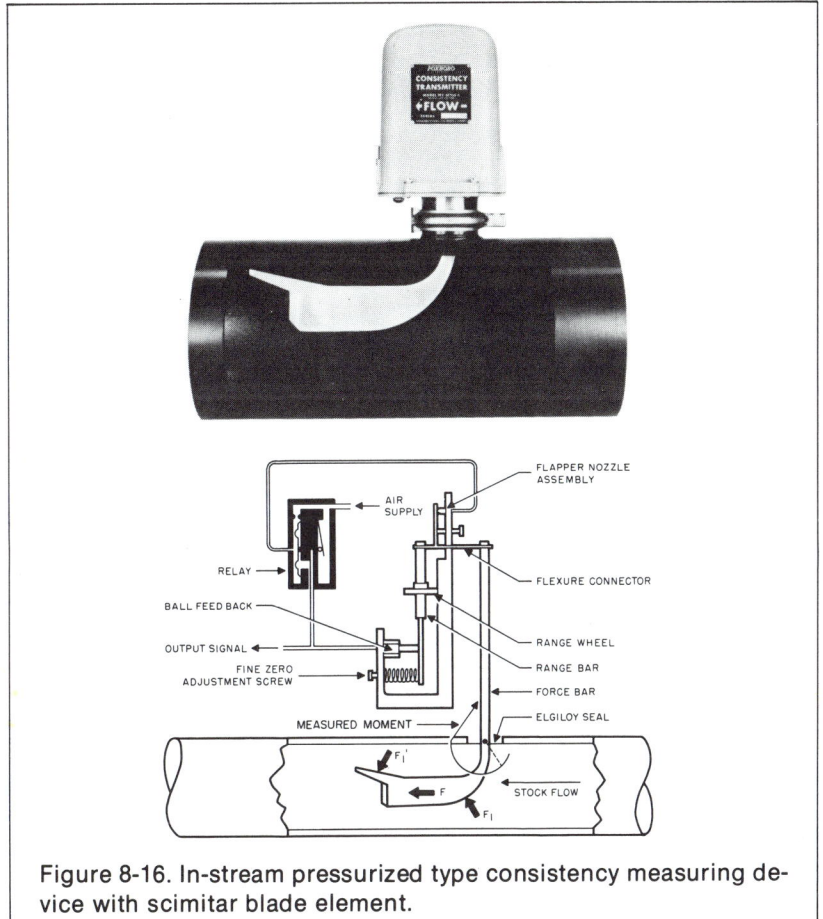

Figure 8-16. In-stream pressurized type consistency measuring device with scimitar blade element.

Another in-line consistency detecting device using a stationary blade type sensor is shown in Figure 8-17.

It depends on the same basic principle of forces for its measurement, but it differs in the way it compensates for velocity effects and in its means for measuring and converting torque from the sensor to a pneumatic output related to consistency. A rib on the sensor compensates for velocity effects. Frictional forces of flow on the rib create an opposite force equal to the force on the leading edge of the sensor. As consistency changes, the sensor and torque arm pivot on a flexure pivot. Torque arm movement is sensed by a pneumatic bridge circuit and converted to a pneumatic signal varying in proportion to consistency changes. This unit can be used to measure consistency changes of stock from 1.75 to 6 percent.

Practically all of the full stream pressurized type consistency measuring devices can be used as pressurized sample types, but this is not common practice. It is done only when the total flow exceeds the capacity rating of the measuring

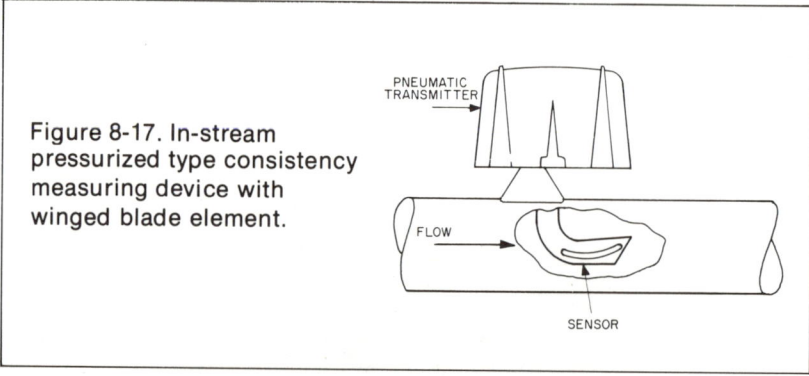

Figure 8-17. In-stream pressurized type consistency measuring device with winged blade element.

assembly. Such installations are made with the measuring device installed around a restriction in the main line and are sized to pass a flow not to exceed the minimum normal main stream requirement.

Miscellaneous Devices

Many other principles and methods of detecting consistency of pulp and stock slurries have been tried with varying degrees of success.

Optical. The light transmission method, illustrated in Figure 8-18, utilizes the principle that a beam of monochromatic light passing through a sample of fiber slurry will change in intensity (in accordance with Beer's law) as the consistency of the stock changes. Beer's law states that in the transmission of monochromatic light through a given thickness of solution sample, the amount transmitted varies exponentially with the concentration of the absorbing species. Therefore, in a beam of light which passes through a flowing stock and is incident to a photo cell, the intensity of the transmitted light can be interpreted as its consistency.

Similar optical consistency measuring devices have used reflected light and the ratio of reflected to transmitted light for the basis of measurement.

Another optical method utilizes polarized light. This technique depends on the phenomena that fibers in suspension will depolarize light and the degree of

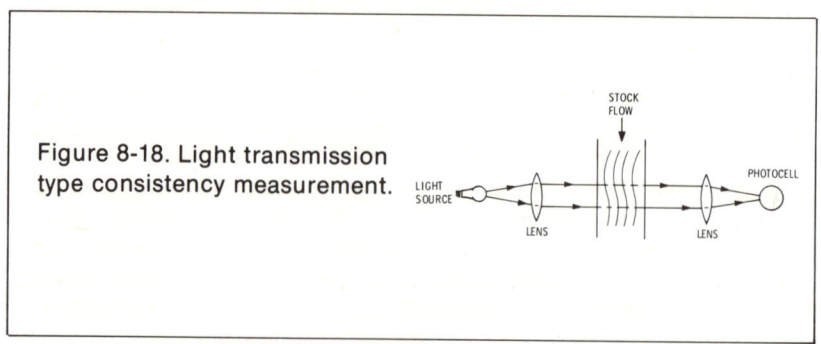

Figure 8-18. Light transmission type consistency measurement.

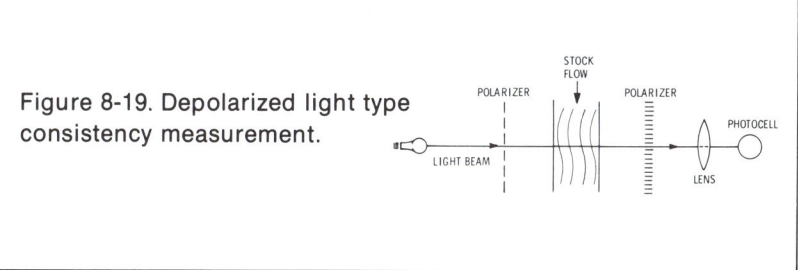

Figure 8-19. Depolarized light type consistency measurement.

depolarization is related to the concentration of fibers; for example, consistency. Cellulose fibers are optically active and thus able to rotate polarized light. If a plane of polarized white light which consists of many frequencies is incident on a series of optically active substances, each frequency will rotate a different amount and thus result in depolarization. A simplified measuring cell with glass windows through which the fiber suspension flows is shown in Figure 8-19.

The cell is placed between two crossed polarizing filters. A parallel white light beam is directed through the system to a photocell. When there is a flow of fiber suspension in the cell, a part of the light will be polarized due to the optical-active nature of the fibers suspended in the cell. The depolarized light will pass the second filter in amounts depending on the concentration of the fiber in suspension.

The optical type consistency measuring devices are particularly adaptable to low consistency ranges of 1 percent and below.

Other Methods

1. Ultrasonics utilize the principle of directing pulses of ultrasonic energy into the fiber slurry and measuring the intensity of the pulses at a fixed distance. This measurement is then converted to a relationship of consistency for ranges of 0.1 to 1.5 percent.
2. Motor load of an electric motor operating an agitator mixer on storage chests has also been used as an indication of consistency of the stock in the tank.
3. Conductivity of pulp slurries after the addition of alum has also been investigated as a means of measuring consistency in the very low ranges of less than 0.5 percent.
4. Flow pressure drop caused by friction loss in a length of existing stock line has been measured with a volumetric differential pressure transmitter and used as an indication of consistency.
5. Other principles, such as vibration, microwaves, drainage rates, and electrical resonance have also been investigated as possible measurements of consistency in fiber slurries.

Research scientists are still trying to find some undiscovered phenomenon of fiber slurries that can be used to obtain a more direct measurement of consistency and provide the pulp and paper industry with a new handle on this most elusive variable.

BIBLIOGRAPHY

Balls, B. W. "Towards Better Understanding of Consistency Measurements." *Measurement and Control*, vol. 1, no. 9 (September 1968).

Canadian Pulp and Paper Association, Process Control Committee, Technical Section. *A Consistency Manual*, June 1967.

Chedomir, G. A. "Electronic Consistency Regulator." *Tappi*, vol. 47, no. 12 (December 1961).

Johnson, B. "How Well Can We Control Consistency?" *Paper Trade Journal*, April 12, 1965.

Lavigne, J. R. "Consistency Control Concepts—Old and New." *Tappi*, vol. 51, no. 1 (January 1968).

Meyn, F. W., Landmark, P., and Aagedal, A. "An Instrument for Continuous Measurement of Low Fiber Concentration Based on the Use of Polarized Light." *The Paper Maker*, pp. 66-69, June 1968.

Morrisey, D. J. "A True In-Line Consistency Regulator." *Paper Mill News*, vol. 86 (1963), no. 33.

Overall, J. E. "A New Concept in Consistency Control—Progress Report." *Tappi*, vol. 42, no. 1 (January 1959).

Ziegenhagen, P.D. "On Machine Low Consistency Indications With Transmitted Light." *Tappi*, vol. 49, no. 10 (October 1966).

STUDY QUESTIONS

Select answer or answers which correctly apply to the following statements:

1. In order to calculate the consistency of a stock slurry, we must know its
   a) total weight.
   b) total volume.
   c) fiber weight.

2. If 1 cubic foot of stock slurry weighs 62.4 pounds and it contains 1.25 pounds of fibers, the percent consistency of 12.5 gallons of slurry is
   a) 10.00%.
   b) 6.24%.
   c) 2.00%.

3. Consistency measurements on stock are difficult because stock slurries
   a) are Newtonian liquids.
   b) are subject to other variables which interact with consistency.
   c) have no well-defined hydraulic properties.

4. Sample type consistency measurement devices
   a) are installed directly in the flow stream.
   b) operate on a portion of the main stream.
   c) exist only as pressurized installations.

5. A common configuration for the driven sensor used by consistency devices is the
   a) paddle.
   b) rod.
   c) blade.

6. Stock properties used by inferential consistency devices include
   a) temperature.
   b) apparent viscosity.
   c) shear forces.

7. A common stationary element form used by consistency measurement devices is the
   a) propeller.
   b) pinned shaft.
   c) flat disc.

8. Most medium-range consistency measuring devices operate within ranges
   a) below 1%.
   b) above 6%.
   c) between 1% and 6%.

9. Impact forces caused by velocity head changes on blade-shaped stationary consistency measuring elements are compensated for by the use of
   a) a rib.
   b) a paddle.
   c) a tail.

10. Optical methods of measuring consistency have utilized the principle that a beam of light passing through a sample of fiber slurry is affected by change in consistency as shown by a change in the beam's
    a) color.
    b) shade.
    c) intensity.

(See Appendix for answers)

9. Humidity Measurement

The use of air is common to all paper mills, and it accomplishes many important purposes in papermaking to eliminate condensation, deterioration of buildings and equipment, to provide comfortable working conditions, and to pick up moisture liberated in drying processes before it is exhausted. Humidity in the air must be controlled to eliminate static electricity, reduce dust, and allow the process to operate to the best advantage. High-grade papers, newsprint, and other papers must be dried under carefully controlled humidity conditions. Humidity control is also important to these products after they are dried.

Humidity measurements are found in: (1) paper machine ventilating systems, (2) paper finishing rooms, (3) paper warehouses and storage areas, (4) paper testing laboratories, (5) press or printing rooms, (6) container and bag plants.

Practically all paper testing is done in laboratories with constantly controlled temperature and humidity because the properties of paper vary with surrounding moisture content. Accurate humidity measurements are important so that ambient conditions in the room can be changed at will and then automatically kept stable.

BASIC THEORY OF HUMIDITY

Humidity is an expression of the amount of moisture in a gas or gases, whether isolated or as a part of the atmosphere. There are two basic quantitative measures of humidity—absolute and relative. Absolute humidity is the amount of water vapor present in each cubic foot or other unit volume and expressed in various units such as dew point, grains of water per pound of air, or pounds of water per million standard cubic feet. Relative humidity describes the ability of air to moisten or dry materials and compares the actual amount of water vapor present with the maximum amount of water vapor the air could hold at that temperature. For example, air which is considered saturated at 50°F (100 percent relative humidity) would be considered quite dry if heated to 100°F (19 percent relative humidity). The graph in Figure 9-1 shows the maximum amount of moisture that can be held by air at various temperatures.

Humidity can be calculated by assuming that the atmosphere to be measured is an ideal gas. The ideal-gas law states that the partial pressure of one of

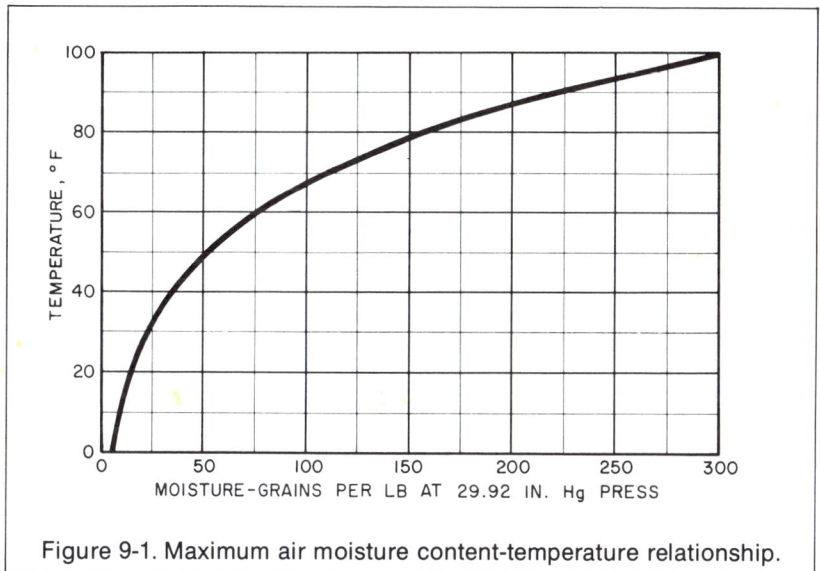

Figure 9-1. Maximum air moisture content-temperature relationship.

the constituents of a mixture is the pressure that would exist if that constituent alone occupied the volume of the mixture at the same temperature. This is expressed mathematically by:

$$W_v = \frac{P_v V}{R_v T} \quad \text{and} \quad W_g = \frac{P_g V}{R_g T}$$

where $W =$ weight of constituent (v – vapor and g – gas)
$\quad\quad\;\; P =$ absolute partial pressure of constituent
$\quad\quad\;\; V =$ volume
$\quad\quad\;\; T =$ absolute temperature
$\quad\quad\;\; R =$ gas constant for a given gas

The absolute humidity (H)—weight of vapor per unit weight of dry gas—is, therefore:

$$H = \frac{W_v}{W_g} = \frac{R_g}{R_v} \times \frac{P_v}{P_g}$$

The density of the water vapor (D)—weight of vapor per unit volume of space, which is the same as the weight of the water vapor per unit volume of dry gas, and is sometimes referred to as absolute humidity—is expressed as:

$$D = \frac{P_v}{R_v T}$$

Humidity Measurement   213

The relative humidity (H_R), which is the ratio of actual partial pressure of the vapor (P_v) in the gas to the saturation partial pressure (P_{sat}), is then:

$$H_R = \frac{P_v}{P_{sat}} \cong \frac{D}{P_{sat}}$$

The value of saturation pressure can be found in steam tables in engineering handbooks.

Dew point is defined as the temperature at which the air or a gas becomes saturated. If the mixture is cooled at constant pressure to the dew point, condensation of vapor will begin.

The dry bulb-wet bulb temperatures are also used to determine humidity of a gas or air mixture. Dry bulb temperature of a gas mixture is measured by an ordinary thermal measuring element. The wet bulb temperature is measured by a thermal element covered by a wick, fully wetted by water. The difference between dry bulb temperatures and wet bulb temperatures is sometimes called the wet bulb depression. This is due to the cooling effect on the bulb of evaporation from the wick.

The relationship between dry bulb, wet bulb, relative humidity, absolute humidity, and dew point can be found in a graph known as the psychrometric chart. This is a convenient device for quickly indicating the condition of an air or gas mixture. The psychrometric chart shown in Figure 9-2 is designed especially for the paper industry and shows air values in the ranges most commonly encountered in air systems.

HUMIDITY MEASUREMENT SYSTEMS

Humidity affects many materials in diverse ways, and the measurement of water present, dew point, and other variables is accomplished by a wide variety of instruments employing quite different methods and principles. Humidity measurements are considered inferred because they depend on differences between two thermometers, expansion or contraction of different materials, temperature at which the water vapor will condense, or the temperature at which certain salt solutions are in equilibrium.

They are classified according to the physical effects on which they are based: psychrometric, hygrometric, dew point, heat of absorption, electrolytic, moisture absorption, vapor equilibrium, and infrared.

Psychrometric

A well-established empirical method of measuring relative humidity is based on psychrometry and involves the reading of two thermometers; one bulb is directly exposed to the atmosphere and the other is covered by a continuously wet wick. Actually, the second bulb measures the thermodynamic equilibrium temperature reached between the cooling effected by evaporation of water and heating by convection. The device used for making this measurement is called a *psychrometer*. The simplest form of wet and dry bulb psychrometer is known as a *sling psychrometer* (Figure 9-3).

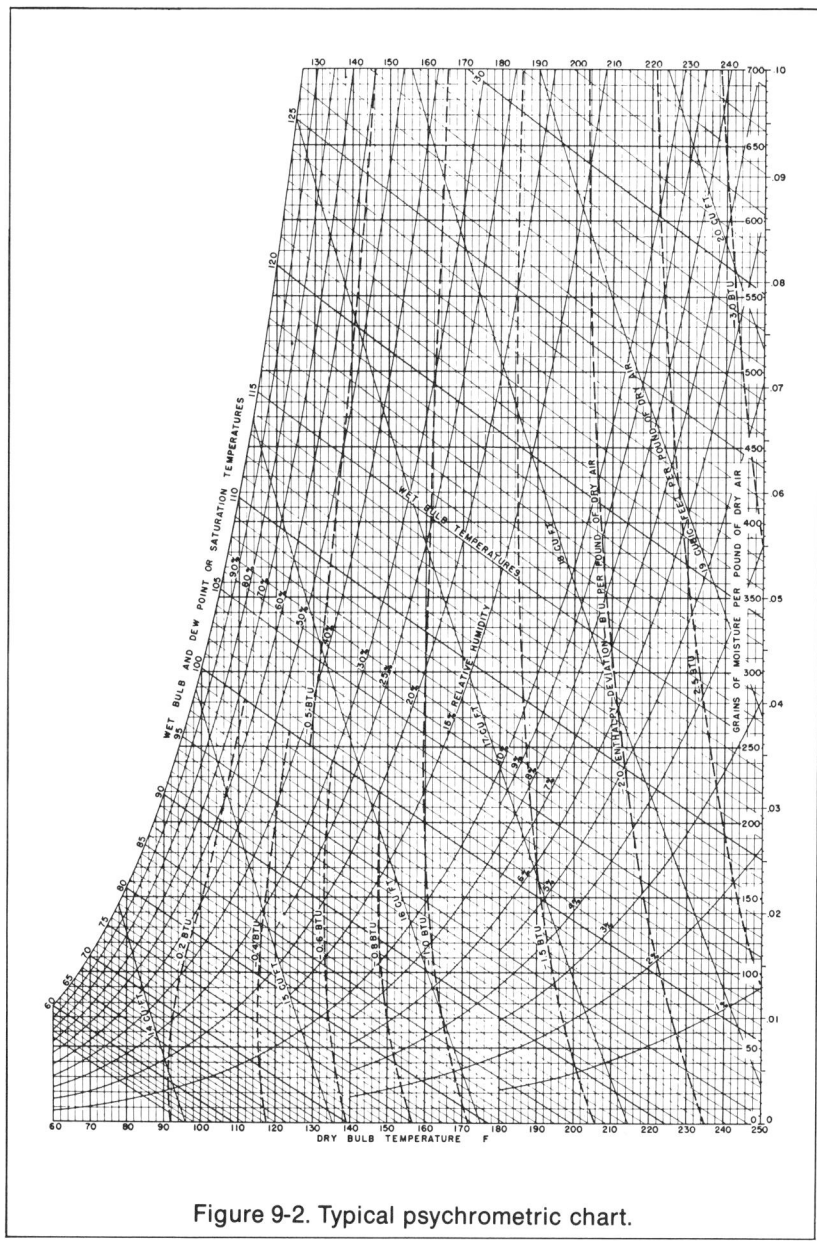

Figure 9-2. Typical psychrometric chart.

The device consists of two mercury-in-glass thermometers mounted on a suitable frame and arranged with a chain handle at one end so that the assembly can be swung rapidly to give proper air velocity. In use, one bulb (the wet bulb)

Humidity Measurement . 215

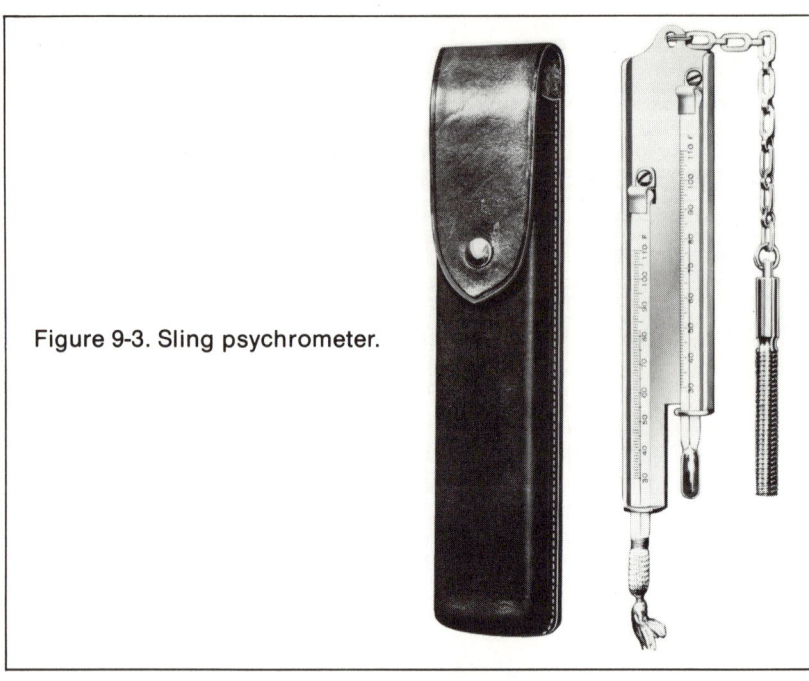

Figure 9-3. Sling psychrometer.

is covered with a wetted wick within a few degrees of room temperature. It is then whirled in a regular circular path for 15 to 20 seconds and readings are quickly made; the wet bulb first before temperature begins to rise. This is repeated until two consecutive wet bulb readings agree. The sling psychrometer is a common checking device for other humidity instruments. Once wet bulb and dry bulb temperatures are known, relative humidity can be determined by tables, special slide rules, or psychrometric charts.

Wet and dry bulb psychrometers for continuous industrial measurement of humidity are available in a wide variety of configurations. A typical system is schematically shown in Figure 9-4. It consists of a pressure spring type thermometer with two separately filled thermal systems. The dry bulb is located in the open while a wick or a porous, ceramic-covered wet bulb is located in the moving air stream. A reservoir or supply of water maintains the proper wetness of the wick.

Resistance type thermal systems are often used instead of the filled-pressure spring system. The wet bulb-dry bulb temperature readings can then be converted to suitable relative humidity values by use of proper scales or charts.

Hygrometric

This method of measurement depends on the change in dimensions of hygroscopic materials such as hair, wood, animal membrane, or paper as the

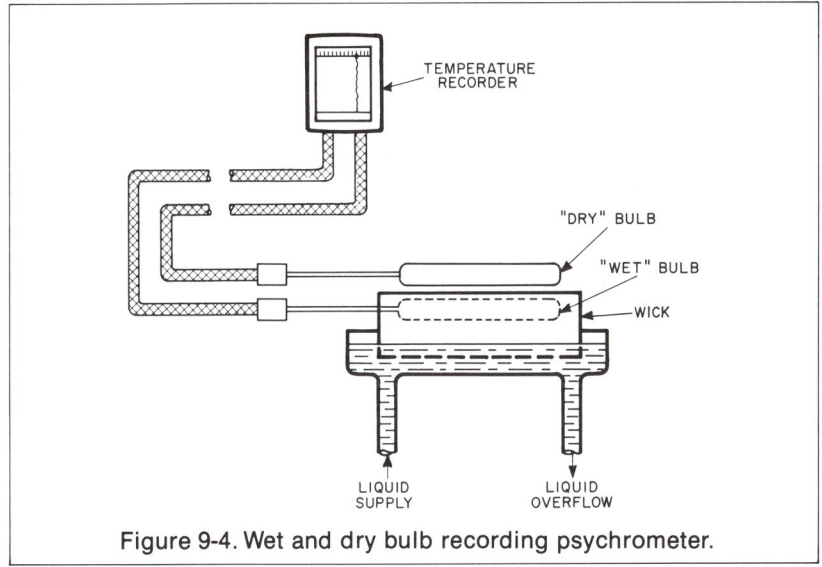

Figure 9-4. Wet and dry bulb recording psychrometer.

relative humidity in the surrounding atmosphere changes. The hygrometer is calibrated to read directly in terms of relative humidity.

Human hair is one of the most common hygroscopic materials used (Figure 9-5). The hair absorbs moisture from the ambient atmosphere in an amount that is a function of the temperature of the hair and of the partial pressure of water vapor in the atmosphere. As water content of hair increases, the hair lengthens with relation to relative humidity. Expansion and contraction of the hair is used to actuate a pointer or pen to provide a continuous reading of relative humidity which can also be used for control purposes. Figure 9-6 shows an actual hair element and where it is mounted in relation to a recorder.

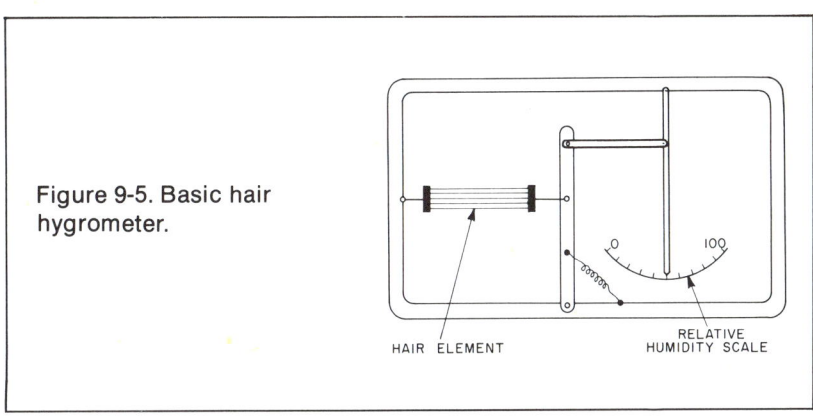

Figure 9-5. Basic hair hygrometer.

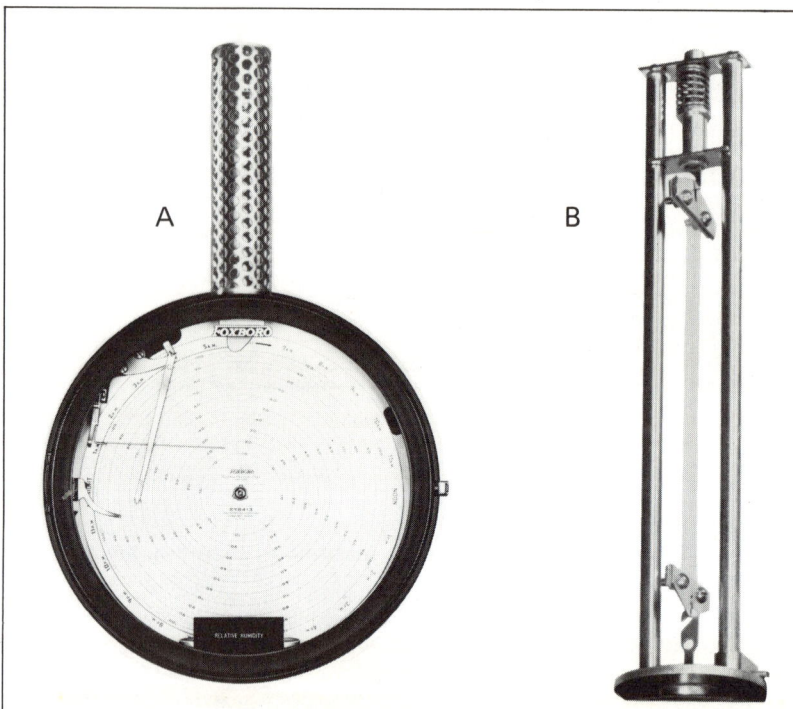

Figure 9-6. Relative humidity measuring and recording system. A, recorder; B, human hair element.

Dew Point

As previously stated, dew point is the temperature at which a mixture of air and water vapor is saturated. The classical method of determining dew point consists of slowly cooling a polished surface until condensation takes place. The temperature of the surface when the first droplet appears is considered the dew point. This method can be used to determine absolute humidity and partial pressure of the vapor.

A widely used approach to continuous measurement of dew point is based on the temperature of vapor equilibrium. The temperature at which a saturated solution of a hygroscopic salt (lithium chloride) is in vapor equilibrium with the atmosphere is measured. The temperature for the salt solution, being much higher than the temperature for pure water, is reached by electric heating.

Structurally (Figure 9-7), a tube containing a temperature measuring element is wrapped with a glass fiber wick wetted with a saturated lithium chloride salt solution. Two conductors are wrapped around the assembly in contact with the wick and are supplied with low-voltage (25-volt) alternating current. Current flow through the salt solution generates heat, raising the temperature. When temperature of vapor equilibrium is reached, water evaporates, reducing current flow and heat input. The temperature cannot go any higher because all the water would evaporate and heat input would cease. It cannot fall because all the salt would then go into solution and too much heat would be generated. Therefore, equilibrium is reached with a portion of the lithium chloride in solution and conductive, and the remainder of lithium chloride dry and nonconductive. Thus heat input is balanced with heat loss. The thermometer bulb when placed inside the metal tube will measure cell temperature or dew point. This is also a measurement of absolute humidity and can be expressed in grains of moisture per pound of dry air, percent water vapor by volume, and other units. Again, dew point readings are easily converted to relative humidity by use of a standard psychrometric chart because one thermal system indicates dew point, while the second reads dry bulb temperature.

Since there is no provision for cooling with this method, its operation is limited to conditions where equilibrium temperature of lithium chloride is above ambient temperature. This corresponds to a minimum of approximately 12 to 15 percent relative humidity over ordinary temperature ranges and is usable up to saturation, or 100 percent relative humidity. Ambient temperatures may vary from as high as 200°F to −30°F. Dew points at higher temperatures must be measured on a cooled sample.

Other Systems

There are many other methods for continuous process humidity measurements in industry, but they are not used as commonly in the paper industry as those discussed above. The basic principles on which some of these methods operate include:

Heat of Absorption. Absorption of water vapor on a solid adsorbent releases heat. A measurement of temperature change when water vapor is

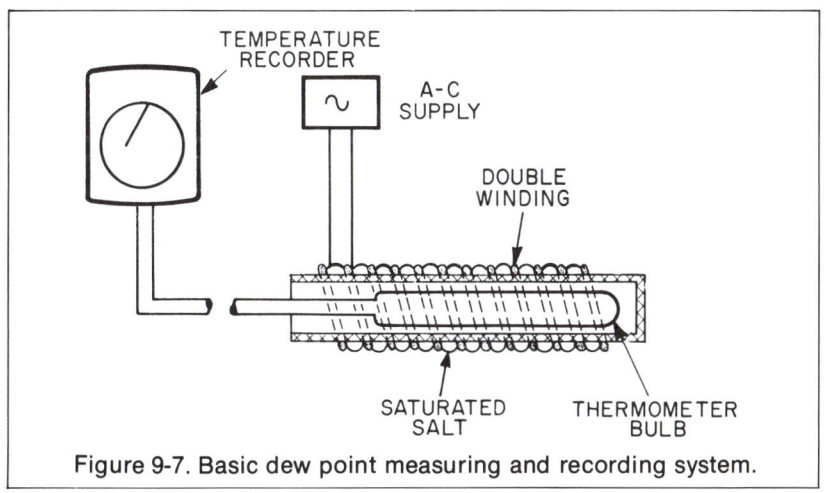

Figure 9-7. Basic dew point measuring and recording system.

alternately adsorbed and desorbed is interpreted as moisture content.

Electrolysis. The moisture in a constant flow of air is absorbed by a thin film of phosphoric acid held between platinum electrodes. A direct current potential applied between the electrodes dissociates or electrolyzes the water into hydrogen and oxygen. This dissociation produces a current which is used as a direct measurement of the concentration of water in the air.

Adsorption. This principle has been used to measure humidity gravimetrically by passing a measured volume of air through a water adsorbing material, such as phosphorus pentoxide, and measuring the gain in weight of the adsorbent as an indication of moisture content. Using this same principle, humidity has been measured by adsorbing the moisture with a quantity of hygroscopic salt and then measuring the change in conductivity due to this moisture by two electrodes in contact with the salt.

Infrared. Water vapor absorbs electromagnetic radiation in certain portions of the infrared region. A measurement of this absorption can be interpreted as moisture content.

BIBLIOGRAPHY

Anderson, N. A. *Instrumentation for Process Measurement and Control.* 2nd ed.

Considine, D. M., and Ross, S. D. *Handbook of Applied Instrumentation.* New York: McGraw-Hill Book Company, 1964.

Dew Point Recording System Using the Dewcel Element. Foxboro, MA: The Foxboro Company, T119-30a.

Eckman, D. P. *Industrial Instrumentation.* New York: John Wiley & Sons, 1950.

Fibrance, A. E. *Industrial Instrumentation Fundamentals.* New York: McGraw-Hill Book Company, 1962.

Lipták, B. G., ed. *Instrument Engineer's Handbook,* vol. 1, "Process Measurements," Philadelphia, PA: Chilton Book Company, 1969.

Pennan, H. L. *Humidity.* New York: Reinhold Publishing Corporation, 1955.

Relative Humidity Measurement—Hair Element. Foxboro, MA: The Foxboro Company, T119-10a.

Tyson, F. C. *Industrial Instrumentation.* Englewood Cliffs, NJ: Prentice-Hall, 1961.

STUDY QUESTIONS

Indicate whether the following statements are True or False by inserting T or F in the parentheses:

1. Humidity can be generally defined as the amount of moisture in a gas or mixture of gases. (T)

2. Absolute humidity describes the ability of air to moisten or dry materials and is expressed in units of grains of water per pound of air. (F)

3. Relative humidity is mathematically expressed as $H = \dfrac{Wv}{Wg}$. (F)

4. The temperature at which a gas becomes saturated is known as its dew point. (F)

5. A common method of determining humidity of a gas is to make a wet bulb and dry bulb temperature measurement. (T)

6. Psychrometric charts are used to determine relationship between wet bulb and dry bulb temperature and relative humidity. (T)

7. The dry bulb of a wet bulb–dry bulb humidity measuring device measures the thermodynamic equilibrium temperature reached between cooling effected by evaporation of water and heating by convection. (F)

8. Hygrometers depend on the change of dimensions of materials sensitive to relative humidity variations for the basis of their operation. (T)

9. The dew point method of determining humidity involves the measurement of the vapor pressure at which equilibrium is established over saturated solution of a hygroscopic salt. (F)

10. The gravimetric principle of measuring humidity involves the measuring of a change in capacitance of a water absorbing material. (F)

(See Appendix for answers)

10. Moisture Measurement

Moisture measurements are used in the paper industry where the moisture content of material must be kept constant or within tolerable operating limits. Moisture differs from humidity in that it is a measurement of liquid adsorbed or absorbed by a solid material and is generally expressed as a percentage of the total weight of the material. Solid material on which moisture measurements are made in the pulp and paper industry exist in a variety of physical shapes and forms. However, the most common forms encountered are: granular materials such as coal, clay, starch; chips as in the wood to digesters; and the most common of them all, the sheet and web.

GRANULAR AND WOOD CHIP MOISTURE MEASUREMENTS

The development of reliable and accurate measurements of moisture in granular materials and wood chips used in the pulp and paper industry has been very difficult. Although no one method has achieved general acceptance, a number of techniques have been used with varying degrees of success. These techniques can be broadly categorized as those depending on electrical properties such as microwave and radio frequencies, and those depending on nuclear properties such as gamma ray radiation.

Microwave Absorption

In Chapter 5 on analytical measurements, under capacity measurement, reference was made to the use of a specially designed capacity measuring element to sense moisture in wood chips and other granular material in the paper industry. Another method used primarily on wood chips operates on the principle of microwave power absorption.

As shown in Figure 10-1, a typical system consists of a primary electrode mounted directly beneath the chip flow and a secondary instrument which converts the signal from the electrode into appropriate moisture units for readout or control purposes. The moisture content of the chips is measured as they flow over the primary sensing electrode. The measurement of moisture by the use of the microwave power absorption technique couples microwave energy to the wood chips and then measures the chips' effect on this energy. The coupling method makes it possible to average the moisture content of the sample in a fixed volume, allowing the unit to discriminate against the natural bulk density changes in wood chips. The end result is a linear calibration which allows direct reading of moisture content. Since the system is sensitive to basic wood density differences, a calibration switch is included for mixtures of different wood species.

Microwave-Nuclear Radiation

Both wood chips' moisture and weight can be measured by use of a combination of microwave and nuclear radiation techniques. Figure 10-2 illustrates a typical conveyor belt installation for detecting chip weight and moisture. Two beams are transmitted through the chip stream for measurement. The first is a microwave beam that is tuned to a particular frequency, absorbed by water molecules but unaffected by other chip components. By measuring the decrease in microwave signal intensity, a direct measurement of chip water content is made. The second is a gamma ray beam which is absorbed in proportion to the total number of molecules present. A division of a properly calibrated water-weight signal by a properly calibrated total mass signal yields percent moisture.

The measuring heads are sized so that microwaves and gamma ray beams scan approximately the same area of chips. Thus, total chip weight and total water weight are measured in a continuous sample of chips flowing on the conveyor. The water-weight signal obtained by the microwave absorption is divided by the total weight signal obtained from the gamma ray penetration to arrive at a percent moisture signal used for indicating or recording purposes.

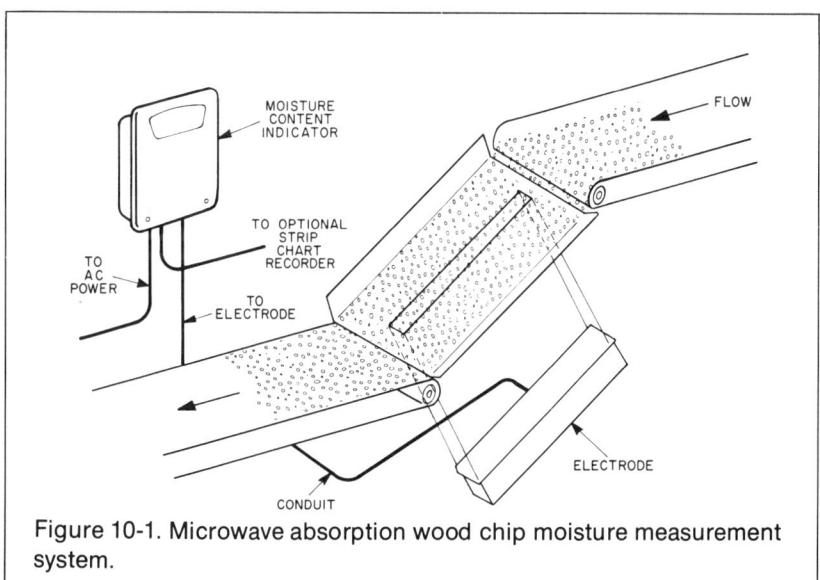

Figure 10-1. Microwave absorption wood chip moisture measurement system.

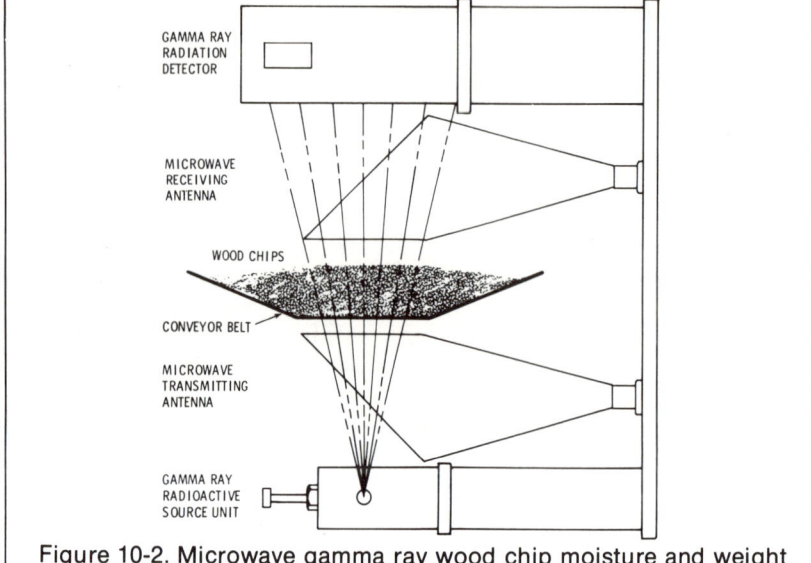

Figure 10-2. Microwave gamma ray wood chip moisture and weight measurement.

Nuclear Radiation

Nuclear radiation has been used to make separate measurements of bulk density (total mass) and total hydrogen in wet wood chips. The resulting signals are electrically combined and ratioed to yield a single signal proportional to percent moisture. Figure 10-3 shows a simplified electrical diagram.

The bulk density channel uses gamma radiation to measure total mass independently of composition. Radiation which penetrates a fixed cross section of material is detected by an ionization chamber that develops a current proportional to bulk density because of the principle of absorption used. The hydrogen channel uses high-energy neutrons to measure the total hydrogen in the chip mass. The fast neutrons lose energy by collision with the hydrogen nuclei which have a mass close to the mass of the neutrons. The number of slow neutrons resulting from these collisions is proportional to the hydrogen content of the material. The slow neutrons are detected by an ionization chamber which develops a current directly proportional to hydrogen density. These signals are then electrically conditioned and ratioed to produce one signal which is representative of the percent moisture and can be used for indicating and recording purposes.

Other Methods

Electrical sensing techniques depend on the measurement of electrical characteristics of wood chips that vary with moisture. Direct current measurements determine conductivity and are useful in measuring moisture in

224   Paper Industry Instrumentation

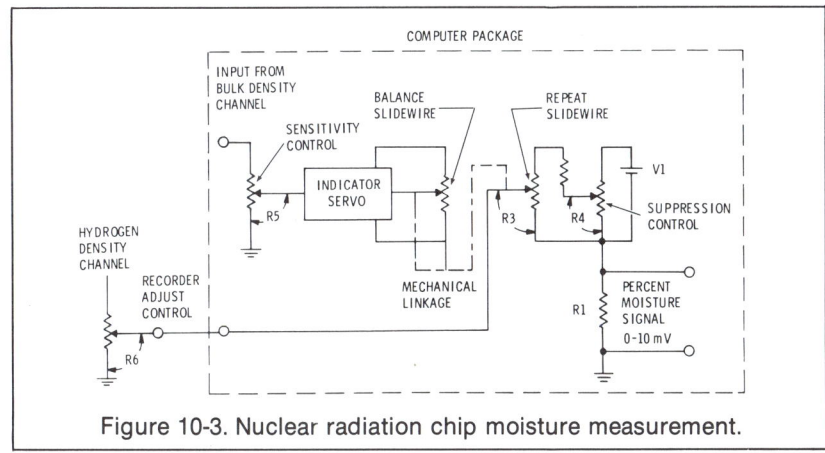

Figure 10-3. Nuclear radiation chip moisture measurement.

low-moisture ranges. Above approximately 20 percent the electrolytic constituents in the wood become dissolved and polarized when dc is applied.

Radio frequencies have been used in higher moisture ranges because, at radio-frequency energy levels, polarization effects are eliminated, the capacitive coupling across individual chips is greatly improved, and the effects due to surface moisture are minimized. The electrical characteristics are measured in terms of capacitance (dielectric constant) and resistance (conductivity). Both variables vary with moisture. Capacitance will vary almost linearly with moisture over ranges typically existing in wood chips.

One system utilizing this technique is essentially a constant-volume sampling capacitance meter operating at radio frequency and has means of measuring temperature and weight of the chip sample. This system is shown in Figure 10-4. The temperature and weight measurements are used to make temperature and density compensations to the measurement of capacitance of a fixed volume of chip samples in order to arrive at the correct percent moisture content measurement of the chips.

PAPER MOISTURE MEASUREMENTS

One of the most difficult and persistent problems is the production of a uniform paper sheet containing the right amount of moisture. According to its degree of moistness, paper can display a number of peculiar physical and mechanical characteristics. If these characteristics are not controlled, undesirable economical and technical consequences can result for both manufacturers and users of paper, particularly in the highly specialized printing industry. Improper sheet moisture manifests itself on the paper machine by:

1. Variations in finish, bulk, and density.
2. Blackening of the paper sheet.
3. Cockles, puckers, and grainy surfaces.

Moisture Measurement   225

4. Excessive breaks on the machine resulting in loss of production time.
5. Nonuniform strength properties such as burst, tear, and fold.

Printers and users also encounter difficulties due to improper sheet moisture:

1. Development of curl.
2. Poor strength properties.
3. Large sheet dimensional change due to shrink or stretch.
4. Poor coating and printing surfaces.

The first essential step toward control of moisture content is its measurement. However, paper web moisture is one of the most difficult process variables to measure. This is because practically all means of continuous moisture measurement are indirect or inferential, and are affected to a greater or lesser extent by other process variables such as temperature, composition, basis weight, and pH. None of the possible methods of measurement is universally applicable to all materials and no material is free of the extraneous effects of other variables. Therefore, the method to use is usually the one that provides the most reproducible results with the materials of interest and at the same time is least affected by other variables known to exist in the process.

Some of the related properties on which a number of the more significant moisture measurement techniques have been developed in the paper industry are classified in Table 10-A.

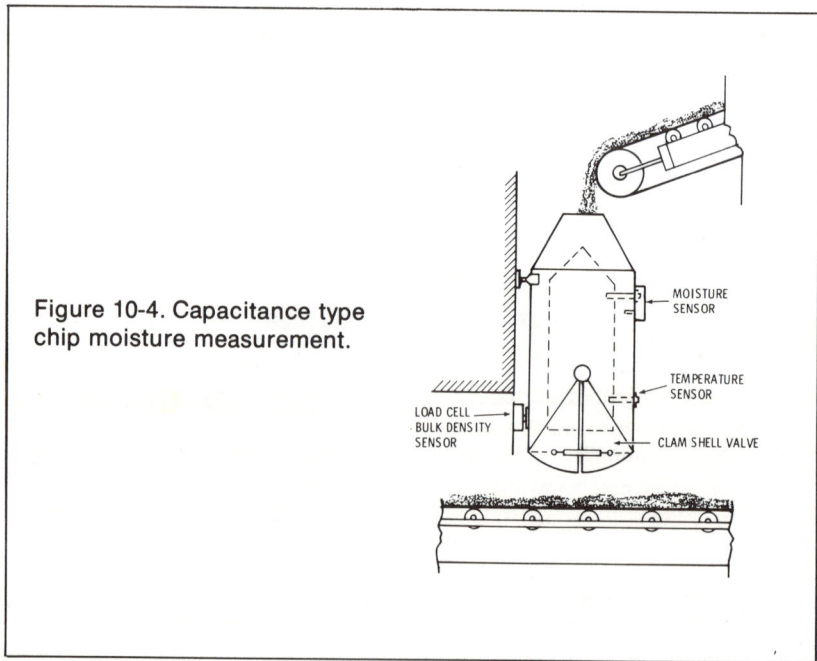

Figure 10-4. Capacitance type chip moisture measurement.

MOISTURE SENSOR

TEMPERATURE SENSOR

LOAD CELL
BULK DENSITY SENSOR

CLAM SHELL VALVE

TABLE 10-A

A. Electrical	D. Temperature
1. Conductivity	1. Temperature
2. Resistance	difference
3. Capacitance	E. Energy absorption
B. Mechanical	1. Microwave
1. Sheet tension	2. Infrared
C. Humidity	F. Pilot dryer
1. Hygroscopic	1. Cascaded control
2. Electrostatic	2. Direct control

ELECTRICAL TYPES

Of the many possible electrical methods of continuously measuring the moisture content of paper, those based on the principles of the electrical properties of paper, such as conductance, resistance, capacitance, and electrostatic effect, have been most extensively used.

The electrical conductivity or resistance method of moisture content is based on the fact that with most types of paper a relationship exists between the moisture content of the material and its electrical conductivity or resistance, which is highly reproducible at a given moisture content.

Conductivity

Paper containing no water is an insulator and produces effectively infinite resistance, or no conductivity. Pure water is also an insulator. However, water in paper is invariably contaminated by electrolytes in solution which make it conductive. If water particles were isolated from one another, the paper would still be an insulator. However, the water particles in moist paper are in intimate contact and present less than infinite resistance, or more conductivity.

Using alternating current, the conductivity of paper can be measured over a range of moisture contents and kinds of paper. The change in conductivity with moisture content is large and the relationship is exponential in form so that a shift of 1 percent in moisture causes conductivity to rise to five times its value at lower moisture content. A change of 2 percent in moisture at lower ranges changes conductivity by 25 times. The block diagram in Figure 10-5 shows a moisture measurement system of this type. The primary measuring electrodes, applied to only one side of the sheet, form a part of a balanced capacity bridge energized with a secondary voltage of about 1000 volts.

A small current, proportional to moisture content, passes through the paper spanning the electrodes and creates a voltage at the input of the first amplifier stage. Grade, temperature, and range controls are used in the next phase to adjust amplifier gain as required. The signal then passes through a second amplifier, then to a phase-sensitive detector, and finally to a secondary instrument which converts the signal to a linear percent moisture response within moisture range limits of 3 percent to 11 percent.

Moisture Measurement   227

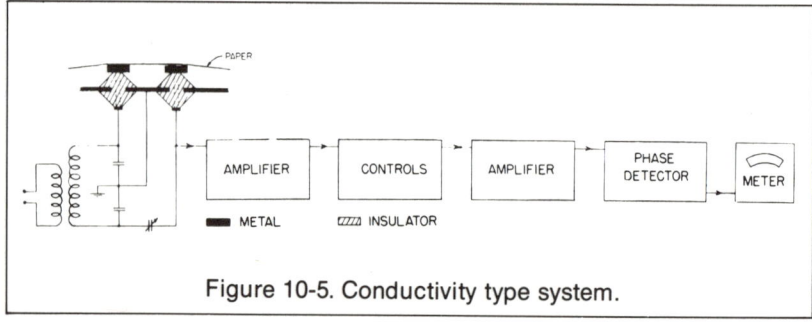

Figure 10-5. Conductivity type system.

Figure 10-6 shows an actual electrode installation on a paper machine. The entire assembly rests on the moving paper by its own weight. A motor-driven vertical shaft provides means for lifting the electrodes off the sheet.

Resistance

The measurement of electrical resistance to current can also be used to measure moisture content of a sheet of paper. Figure 10-7 is a typical system which operates by measuring the electrical resistance through the paper. A detecting roll is installed against the sheet of paper in such a manner that the paper is used as a resistance between the roll and a grounded drying cylinder. By means of a Wheatstone bridge circuit, the resistance across the paper is measured and indicated at a secondary readout or control instrument similar to a resistance thermometer instrument. The resistance of the paper varies from about 25,000 ohms to 50 megohms for various papers with various moisture contents. Calibration in terms of moisture content is established by a calibration curve, which is different for each grade of paper. Satisfactory measure-

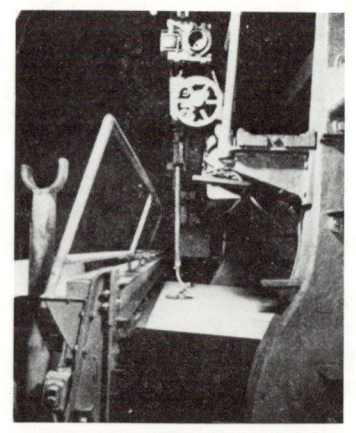

Figure 10-6. Electrode installation.

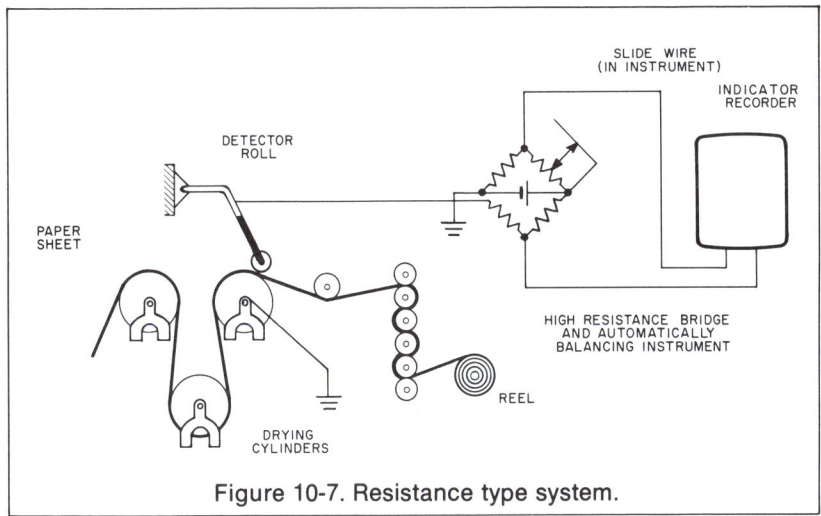

Figure 10-7. Resistance type system.

ments have been made on some papers with moisture contents as low as 2 percent and as high as 40 percent.

Capacitance

This measuring principle, explained in the chapter on analytic measurement, uses the paper web as the dielectric medium of a capacitor which becomes an actual part of the measuring circuit. The paper sheet acts as a variable dielectric material for the capacitor, as shown in the enlarged cross section of one set of the capacitor plates or electrodes in a typical primary measuring element (Figure 10-8). These plates establish an electric field. The broken lines show how the field is intercepted by the moving web of paper.

Electrical capacitance is directly proportional to the dielectric constant of the material in the electric field of the capacitor. Each of the paper ingredients has its own dielectric constants—and the effective dielectric value of the combined ingredients depends on the quantity of each ingredient present. The measuring circuit is designed to be sensitive to capacitance changes caused by the variation in sheet moisture. The dielectric constant of paper is about 3 and that of water is about 80. The dielectric constant of water is so high in comparison to that of other ingredients that a small change in water constant will have a very large influence on the effective dielectric constant of the sheet. It follows that a change in sheet moisture produces a large change in the measured capacitance of the capacitor. The secondary indicating, recording, and/or controlling unit, similar to one described in the chapter on analytic measurement, measures capacitance and displays in moisture percentage.

The actual moisture measuring primary element is a 3-foot-long electrical capacitor containing several sets of electrodes which are located so that they are in continuous contact with, and on only one side of, the running sheet of

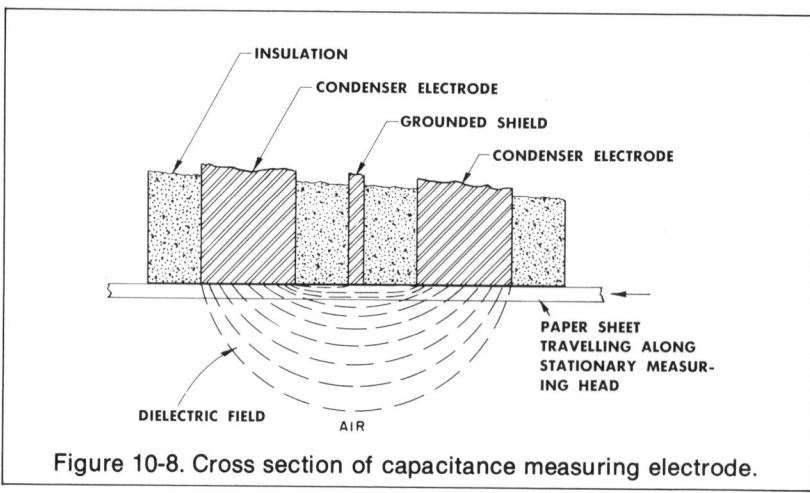

Figure 10-8. Cross section of capacitance measuring electrode.

In the figure, the labels are: INSULATION, CONDENSER ELECTRODE, GROUNDED SHIELD, CONDENSER ELECTRODE, PAPER SHEET TRAVELLING ALONG STATIONARY MEASURING HEAD, DIELECTRIC FIELD, AIR.

paper on the machine (Figure 10-9). Other capacitance types of paper sheet moisture measurement systems utilize primary devices of different configurations. Some are circular, some are square, some mount so that they are on only

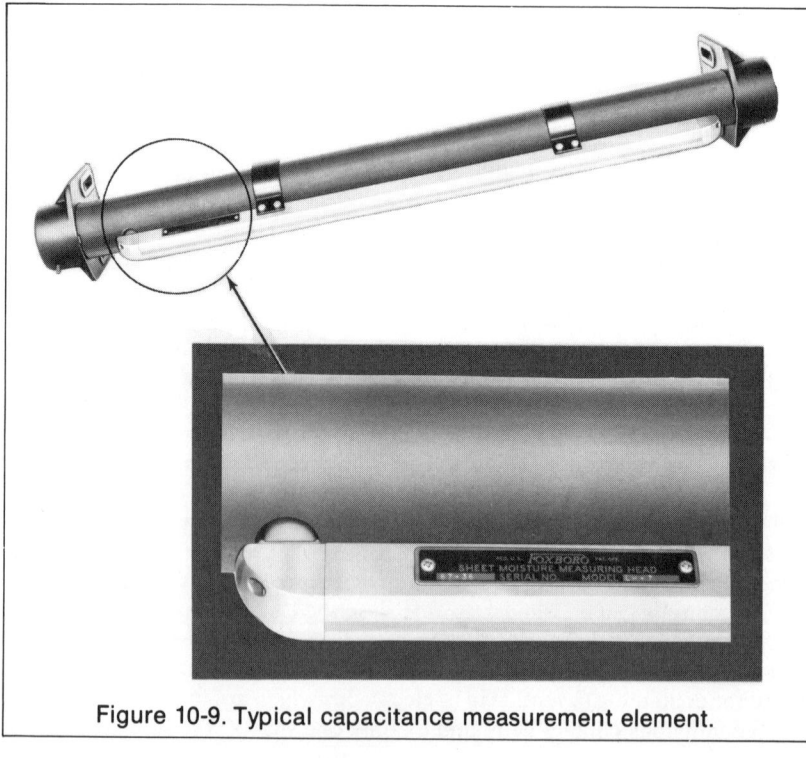

Figure 10-9. Typical capacitance measurement element.

one side of the sheet, and some so they are on both sides. However, they all employ the same basic principle and are normally used in ranges of 4 percent to 8 percent moisture. Some are designed to work with up to 40 percent moisture.

A comparatively new technique which could be classified under this category does not depend on one capacitance measurement alone, but determines total admittance at two frequencies simultaneously. These two values, plus a temperature measurement from a thermistor element, feed an analog computing circuit that produces a temperature-corrected output as a function of dual input channels (Figure 10-10).

The primary measuring element is a fringe field capacitor probe with a large surface containing many strips running in both the machine direction and across the machine. The probe is connected to a bridge circuit that is excited by two different oscillators operating at widely separated frequencies. After linear amplification, the two unbalanced signals from the bridge are filtered, further amplified and rectified. From the resulting dc signals, E_1 and E_2, a simple computing circuit produces a single output voltage E according to the relation $E_1 = \dfrac{(E_1 - E_2)}{E_1}$, which can be read out as percent moisture.

MECHANICAL TYPES

Tension Roll

The theory of this moisture measuring method is based on the fact that a sheet expands and contracts with changes in moisture content. As shown in Figure 10-11, it utilizes a balanced, frictionless tension roll which runs the full width of the sheet averaging out the sheet's moisture content.

The tension roll, which is dynamically and statically balanced, is located at a strategic point in the machine. The roll is made of hard chrome-plated aluminum or steel, and turns on self-aligning ball bearings. The draw of the

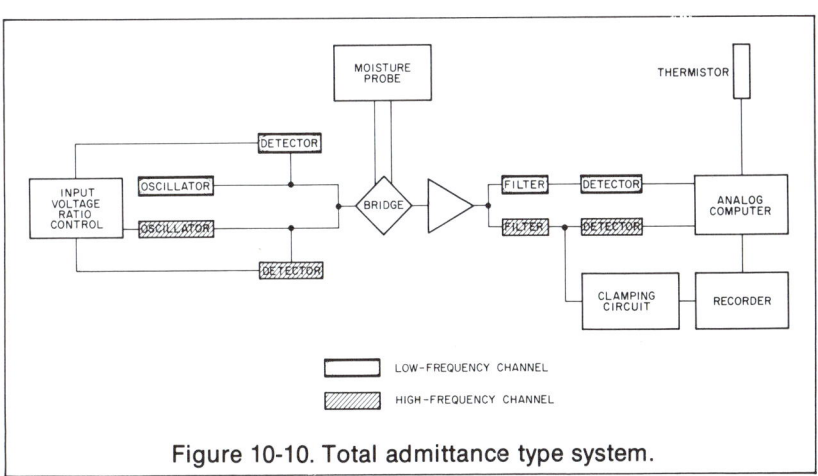

Figure 10-10. Total admittance type system.

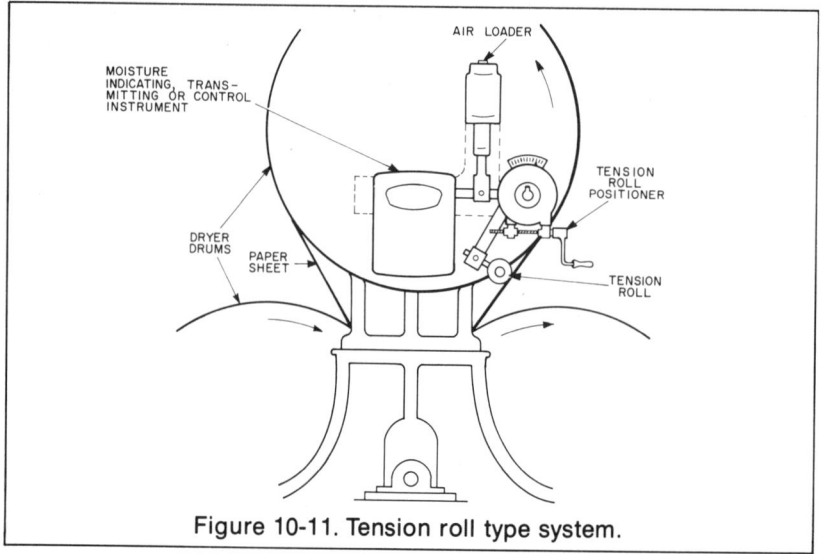

Figure 10-11. Tension roll type system.

sheet due to wetness or dryness actuates the moisture detector according to axial movement of the tension roll. The moisture detector sends out a pneumatic signal proportional to the tension roll's movement. This signal is transmitted to an indicating, recording, and/or controlling secondary instrument.

In older mechanical tension roll systems, the roll was weighed and cable-connected directly to the steam control valve on the machine. Still other types operated small pilot valves which controlled water pressure to cylinder-operated steam valves.

HUMIDITY TYPES

It is a well-known fact that a moving web of paper carries with it a moisture film which is in direct ratio to the amount of moisture in the sheet. This moisture film is in equilibrium with the air surrounding it (see Chapter 9 on humidity). A measurement of the relative air-humidity surrounding the sheet can be used as a measure of the moisture film near the paper web. Based on this phenomenon, the value of sheet moisture can be derived from the value of relative humidity of the air in proximity to the sheet. If a humidity-sensitive detector is placed close to the moving web, an indirect measurement of sheet moisture can be made. Many schemes have been used to accomplish this humidity measurement.

Hygroscopic

One technique based on the hygroscopic principle measures the humidity of the air layer adjacent to the paper web with an element whose resistance changes according to humidity. The primary measuring element (Figure 10-12) consists of a heat resisting hygroscopic rod which changes its resistance rapidly with respect to electric currents as it absorbs more or less moisture from the air film near the paper sheet. The element is housed in a perforated aluminum casing which permits the airstream to flow freely around it. The small variation in electrical resistance causes changes in current which are amplified and transmitted as a percent sheet moisture signal to an indicator, recorder, and/or controller. The element is mounted in a fixed position. A number of additional elements can be fitted, each with its own electronic channel so that cross-sheet moisture variations can be detected.

The secondary indicating or recording instrument includes a single electronic amplifier and two variable resistances used in the calibration; one resistance is for finding one end point of the scale or maximum moisture, and one resistance is for setting the other range for minimum moisture. This permits the scale to be read directly in moisture percentage. When a number of elements are used, the reading represents an average of the measurements. This method has been used to measure paper moisture contents between 1 percent and 25 percent.

Another hygroscopic moisture measuring system, now obsolete for the most part, utilized a special hygrometer for the primary measurement element (Figure 10-13). It was in the form of a hollow shoe or box, tightly placed on the running sheets so that it trapped a sample of air. The relative humidity of the trapped air corresponded to the equilibrium sheet moisture. The humidity of the air was determined by the use of a small capacitor whose dielectric element consisted of a hygroscopic material which gained or lost moisture in relation to the relative humidity of the air surrounding it. The resultant change in capacity was electrically measured and referenced to percent change in moisture content of the paper web for indication and control purposes.

Electrostatic

The moisture of the sheet can also be determined and controlled by electrostatic techniques using the same relative humidity approach. A condenser charged with high voltage is used to develop a static field on the layers of condenser plates to an effective depth of one-half inch. This static field is then utilized to determine humidity. One typical primary measuring head is de-

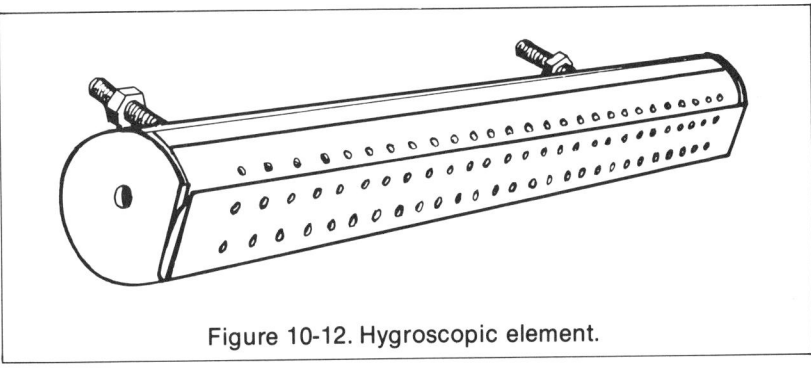

Figure 10-12. Hygroscopic element.

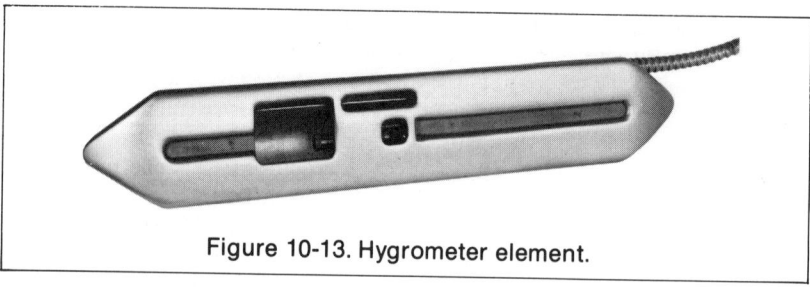

Figure 10-13. Hygrometer element.

signed as a roll and mounted so that it lightly touches the moving web (Figure 10-14). As the humidity in the paper changes so does the static field, thereby varying the discharge of the condenser. The remaining charge in the condenser is used to cause fluctuations in the electromotive force in an electronic circuit which correlates this signal to humidity for display or control purposes. Moisture contents as high as 50 percent can be determined with this system.

TEMPERATURE TYPE

The theory on which this indicating medium works is based on the fact that the temperature of the sheet slowly increases as the moisture content decreases. Thus, a change in moisture content is reflected by a corresponding change in sheet temperature. A sensitive thermocouple primary measuring system is placed on the sheet near the dry end of a paper machine. Any change in temperature can be correlated to a change in moisture content and can be used by an indicating, recording, and/or control system to maintain uniformity in the moisture content of the sheet.

Temperature Difference

A practical method of continuously measuring sheet moisture with this technique involves the measurement of two temperatures—one a reference,

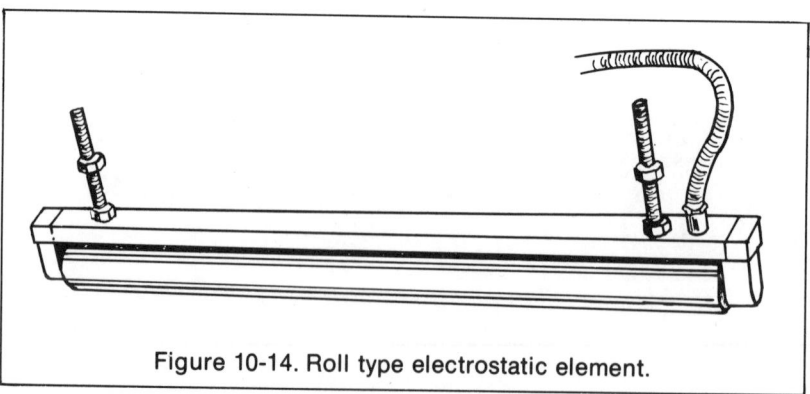

Figure 10-14. Roll type electrostatic element.

the second, the temperature of the sheet after evaporation—and relating the difference in these measurements to the moisture content of the paper. Evaporation, a natural characteristic of the sheet, reveals the percentage of moisture by weight in a sheet of paper as it leaves a drying element.

The drying element, then, can be used as the source of the reference temperature required to make a temperature difference measurement. In the case of a paper machine, a dryer roll, usually the last or next to the last, can be used as the required standard of reference. It is isolated from the other dryers and a precise steam pressure control system maintains its surface at a fixed steam temperature (Figure 10-15).

The sheet, passing over the reference dryer, is heated by conduction and radiation. The surface of the sheet assumes the same temperature as the dryer, but the moisture in the inner portion of the sheet rises only to the boiling point (212°F). In a space of about 8 to 10 inches after the sheet leaves the dryer, the vapor from the inner portion flashes through the surface to produce a sudden drop in temperature. From that point to about 2.5 feet from the dryer, normal evaporation further lowers the surface temperature of the sheet. The total temperature drop is directly related to the percentage of moisture remaining in the sheet. A moisture content of 2 percent will result in practically no temperature drop, while a 15 percent moisture content will produce a drop in the surface temperature of about 75°F.

Each moisture measuring primary element consists of two heat-sensitive heads containing thermocouples (Figure 10-16). One head contacts the dryer to measure the surface temperature, the other lightly touches the sheet to measure its temperature 2.5 feet from the dryer. Both heads are mounted on ball-bearing pivots to follow cylinder eccentricity and sheet flutter. Each head

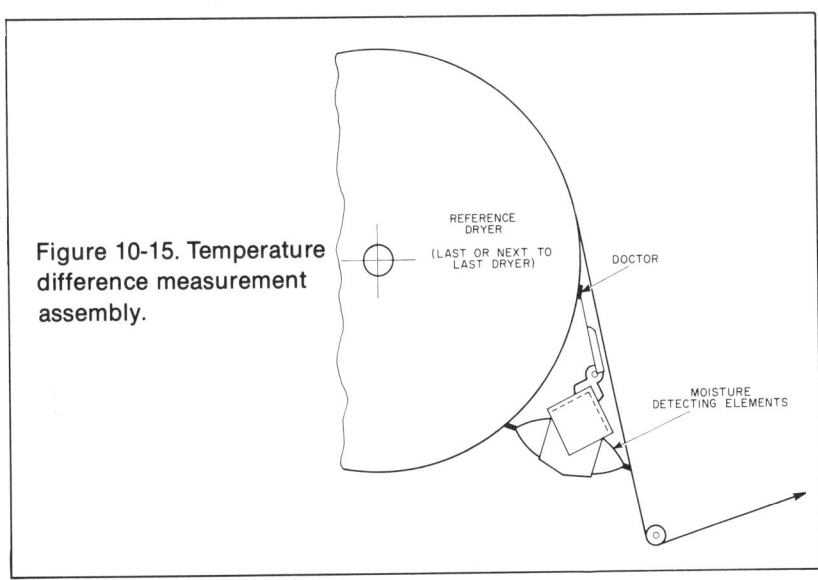

Figure 10-15. Temperature difference measurement assembly.

REFERENCE DRYER
(LAST OR NEXT TO LAST DRYER)

DOCTOR

MOISTURE DETECTING ELEMENTS

Moisture Measurement   235

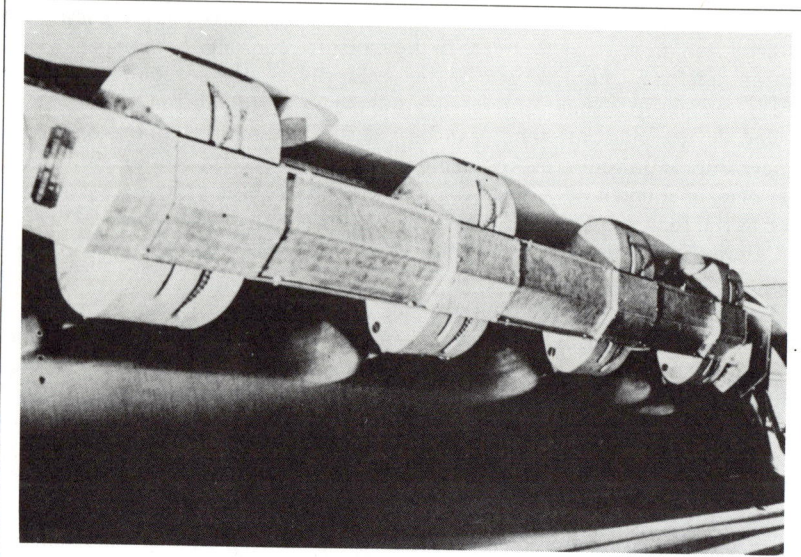

Figure 10-16. Temperature difference element mounted on machine.

has five copper-constantan thermocouples wired in series. A complete primary measuring system is made up of two heads placed in tandem and wired together in series opposition to produce voltages exactly related to the differences in temperature. This difference signal is correlated to the moisture content of the sheet.

ENERGY ABSORPTION TYPES

Recent developments in the field of paper moisture measurements have taken advantage of the fact that water molecules, fiber additives, dyes, fillers, and other material which make up the paper sheet structure will absorb various forms of energy in different amounts. By measuring the energy before and after absorption, the quantity abosrbed by the components in the sheet can be deduced and moisture content derived; several methods utilize this phenomenon.

Microwave Absorption

A moisture measurement system based on the principle of the variable absorption of electromagnetic energy by water molecules is used in several mills. Water molecules are very good absorbers of microwave frequencies. They will absorb several thousand times more of this energy than the same volume of almost any other substance. Thus, the presence of a very small quantity of water may double the microwave absorption factor. The presence of higher levels of moisture may increase the microwave absorption by several thousand times. In this way the system provides a measurement of small moisture variations as well as a measurement of the high moisture level found at the wet end of a paper machine.

The measuring system (Figure 10-17) consists of a microwave transmitter, sensing device, power supply, and an analog computer. The microwave transmitter limits radiation which passes through the sheet. The sensing device located on the opposite side of the sheet receives the variable microwave energy transmitted through the sheet. This energy is electronically compared to the fixed level of radiation at the transmitter to provide a measurement signal. This signal is a dc voltage output which can be used to indicate, record, or control percent moisture of the sheet. It is claimed that this method can be used to measure content up to 80 percent.

Infrared Absorption

The phenomenon of selective absorption of infrared radiation by water molecules has also been used to measure the moisture content of paper. Two measurement techniques have been developed utilizing this basic principle. One operates on the basis of infrared reflectance or backscatter from the sheet and the other on the basis of infrared transmission through the sheet.

The principle of measurement takes into account that there are several narrow bands of frequency in the near infrared spectral region in which the water molecule is excited by a resonant absorption phenomenon. The band of radiation which is absorbed by water is termed the resonant frequency. The energy absorbed is directly related to the energy required to raise the energy state of the water molecule to a higher vibrational-rotational state. The absorption phenomenon at a resonant frequency is, therefore, a measure of moisture content in the path of radiation; the moisture content is determined by the amount of energy absorbed. Radiation other than the characteristic frequencies passes through the molecule unattenuated. Essentially then, if infrared radiation in two very narrow bandwidths is projected onto a specimen of paper

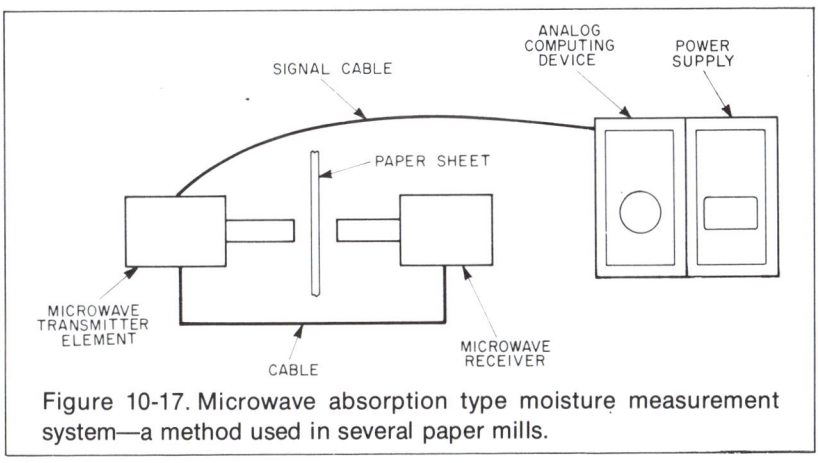

Figure 10-17. Microwave absorption type moisture measurement system—a method used in several paper mills.

and one of these bandwidths is resonantly absorbed by the water molecules while the other is not, the ratio of the reflected or backscattered energies of the two bands is dependent on the amount of water present in the paper within the operating range of 0 to 12 percent.

Figure 10-18 is a block diagram of such a system. In operation, the window in the end of the sensing head is placed against or very near the paper sheet. The source of infrared radiation is an incandescent lamp. Radiation from the lamp is directed by a lens to a dish rotating at 12.5 revolutions per second. This dish contains two narrow bandpass, interference type filters. One of these filters passes radiation at the water absorption band frequency corresponding to a wavelength of 1.91 μm; the other in the adjacent band at 1.80 μm. The filtered, chopped radiation then passes through a tube extending into the integrating sphere through the quartz window of the sphere and onto the specimen as a beam about 0.5 inch in diameter. Radiation reflected or backscattered from the specimen is collected by the integrating sphere and detected by a lead sulfide sensor mounted in the sphere wall. Since the response of the sensor is temperature-dependent, cooling water and a thermostatically controlled heating element are provided to maintain the sensor at constant temperature. The electrical signal from the primary sensor is then given preliminary amplification and routed to the secondary instrument unit. The secondary unit contains the circuitry to power the system and the circuitry for translating the sensing head output in moisture values.

Another method of continuous paper moisture measurement utilizing the basic principle of selective infrared absorption performs its determination on the basis of transmission through the sheet. This fundamental approach distin-

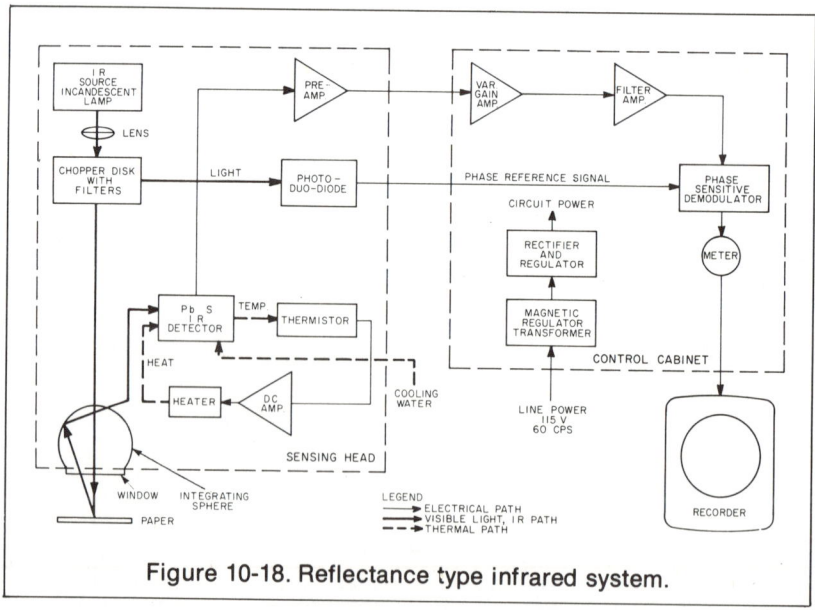

Figure 10-18. Reflectance type infrared system.

guishes the measurement of moisture content of paper by infrared radiation absorption by the same fact that specific wavelengths of infrared radiation are influenced by moisture and cellulose content in the sheet being measured. Direct readout of percent moisture is accomplished by sequential transmission of three discrete wavelengths in the infrared region to solve the equation:

$$\% \text{ moisture} = \frac{(\text{Cellulose} + \text{water}) - (\text{Cellulose})}{(\text{Cellulose} + \text{water})} \times 100$$

A complete transmission type infrared system (Figure 10-19) consists of: a source head containing a tungsten-halogen lamp, optical lens, and three windows in a rotating filter wheel; a detector containing a lead-sulfide-sensitive cell, cooled by a specially designed thermoelectric unit and shielded by a Carrara glass window; and an electronics package which conditions the signal from the primary head/ detector measuring element so it can be used for display, recording, or control.

The source head and detector are located on opposite sides of the paper but do not touch it. The source head sequentially generates infrared energy pulses of three specific wavelengths and associated synchronizing pulse signals. The infrared levels transmitted through the paper are sensed by the detector which provides a continuous train of information pulses. The electronics package processes the detector signal and the analog computer interprets it in accordance with the mathematical equation, providing a continuous direct readout representing 0 to 10 percent moisture.

PILOT DRYER TYPES

This method of indicating moisture is based on the fact that the work done by any individual dryer drum on a paper machine in drying the moist sheet is in direct relation to the moisture content of the sheet as it passes over the dryer. These systems operate on an isolated dryer drum called a *pilot* dryer which is

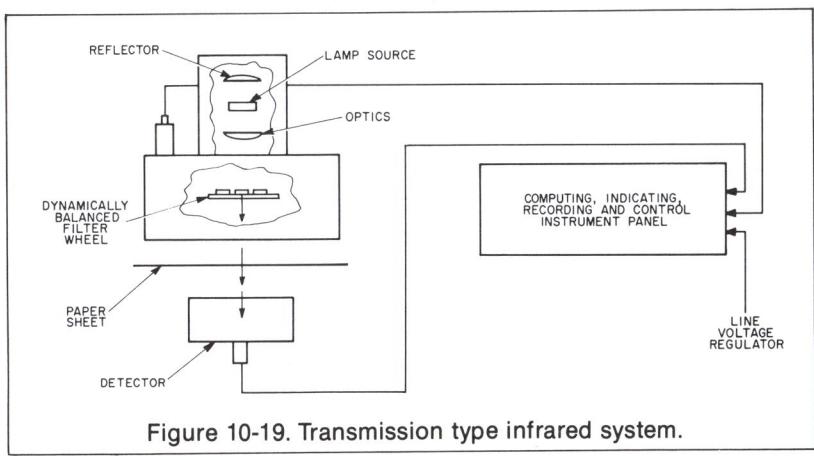

Figure 10-19. Transmission type infrared system.

furnished with an independent steam and condensate system. The rate of evaporation of water from the sheet passing over this dryer increases with the sheet's moisture content. The Btus required for this evaporation come from the steam inside the dryer. If the sheet becomes wetter, more Btus are consumed and more steam is condensed. The dryer pressure would tend to decrease with a fixed steam supply. However, if the pilot dryer is maintained at a constant pressure by automatically manipulating a control valve on the steam supply line to the pilot dryer, the valve opening or position will vary with the moisture content of the paper passing over it. It follows, then, that the signal from a pressure controller can be used as a signal showing moisture variations in the sheet.

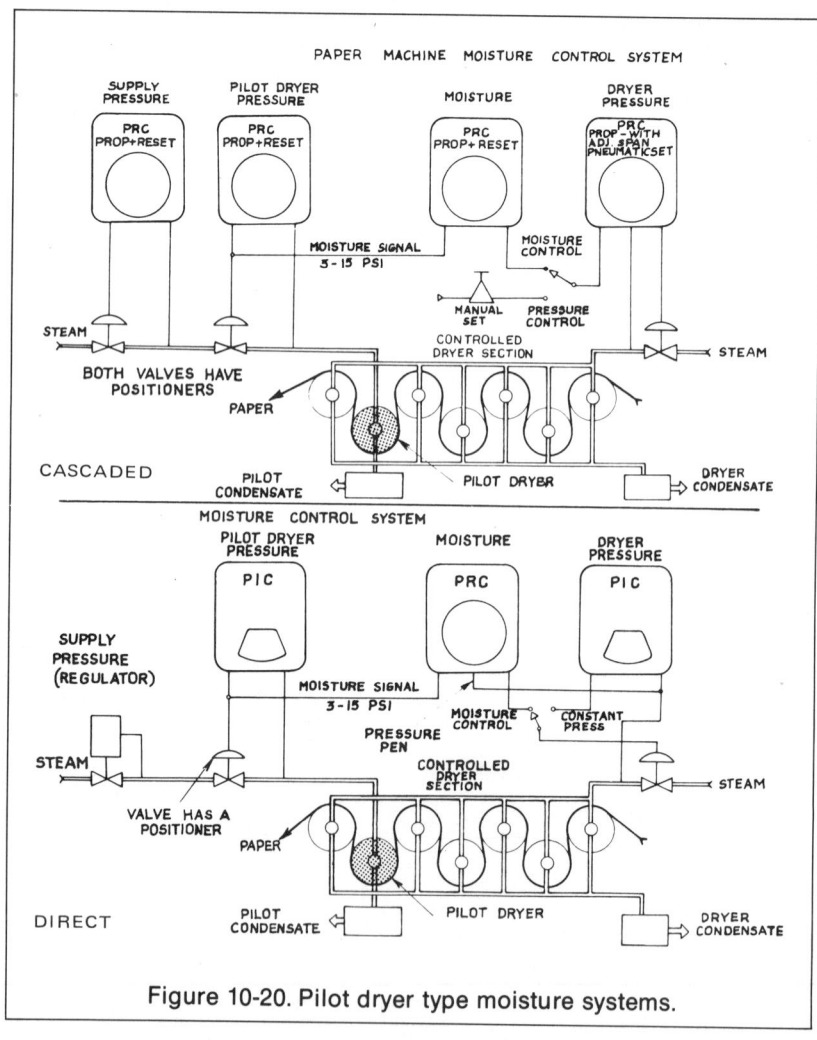

Figure 10-20. Pilot dryer type moisture systems.

Two such systems which have been used extensively in paper mills are shown in Figure 10-20. The upper sketch shows a typical system in which the pressure in the pilot dryer is controlled by a pressure controller. The steam supply pressure is also maintained at a constant value by a second cascaded pressure controller. Any variation in sheet moisture will produce a new condensing rate for the pilot dryer. If the pressure in the dryer tries to change, it is immediately corrected by the pilot dryer pressure controller which sends a new signal to the control valve which, in turn, adjusts the stream flow rate to maintain uniform pressure in the pilot dryer. This variable valve signal is also used as a guide to show moisture deviation of the sheet; is converted to moisture equivalents and retransmitted to a pressure controller on the main steam supply that corrects the steam to the remainder of the dryers in order to maintain constant moisture content in the sheet of paper.

The system shown in the lower sketch (Figure 10-20) operates on the same principle and differs only in the type of instrument used to control the pressure of the steam supply to the pilot dryer, and the elimination of an intermediate controller by operating the main steam control valve directly from the moisture measuring instrument.

BIBLIOGRAPHY

Edenborough, L. D. "Moisture Measurement and the Progressive Papermaker." *Paper Trade Journal*, December 20, 1965.

Evans, J. C. W. *Pulp and Paper Mill Instrumentation*. New York: Lockwood Publishing Company, 1969.

Fishburn, R. E., and Robyn, V.S. "Measurement and Control of Moisture in Paper by the Electrical Conductivity Method." *Tappi*, vol. 41, no. 10 (October 1958).

Hardacker, K. W. "Instrumentation Studies XC, Methods of Measuring Moisture Content of Paper." *Tappi*, vol. 51, no. 5 (May 1968).

Hurm, R. B. "Principles of 'On the Machine' Moisture Measuring Systems—Tappi Project 775." *Tappi*, vol. 44, no. 6 (June 1961).

Lundstrom, J. W. "On-Line Chip Moisture Measurements." *Tappi*, vol. 53, no. 5 (May 1970).

Richesson, M. A. "Continuous Measurement of Chip Moisture." *Tappi*, vol. 50, no. 6 (June 1967).

Skaar, C. *Water in Wood*. Syracuse, NY: Syracuse University Press, 1972.

Smiley, W. O. "Hurletron Moisture." *Tappi*, vol. 48, no. 12 (December 1958).

Thode, E. F. "How Papermakers Measure Moisture. . ." *Control Engineering*, November 1964.

STUDY QUESTIONS

Select answer or answers which correctly apply to the following statements:

1. A typical system designed to measure moisture in wood chips utilizes a
   a) sonic energy absorption technique.
   b) microwave power absorption technique.
   c) microwave power reflection technique.

2. Most continuous moisture measurements used today are
   a) differential.
   b) direct.
   c) unaffected by all other process variables.

3. The use of conductivity to measure moisture in a sheet of paper is made possible because
   a) paper with no water is a good conductor.
   b) pure water is an insulator.
   c) contaminated water is conductive.

4. Capacitance methods of moisture measurement have operated on the basis of
   a) using the paper sheet as the variable dielectric material in a capacitor configuration.
   b) determining total admittance at two frequencies.
   c) electrical resistance to current.

5. The mechanical method of measuring sheet moisture by use of a tension roll relies on the phenomenon that
   a) the sheet will expand as moisture content increases.
   b) the friction of the sheet increases as moisture content increases.
   c) the sheet will contract as moisture content decreases.

6. The measurement of sheet moisture by humidity is possible because
   a) the moisture film carried by the moving web is in direct ratio to the amount in the sheet.
   b) of the pressure equilibrium between moisture and air surrounding it.
   c) the temperature of water and water vapor are the same.

7. The hygroscopic principle of measuring sheet moisture is based on
   a) the vapor pressure of sheet air layer.
   b) the temperature of the air layer next to the web.
   c) the humidity of the air layer adjacent to paper web.

8. The cooling effect of evaporation of water from a sheet of paper is utilized in
   a) the electrostatic method of moisture measurement.
   b) the temperature difference method of moisture measurement.
   c) pilot dryer method of moisture measurement.

9. The microwave absorption moisture measurement system depends on
   a) the variable absorption of electromagnetic energy by water molecules.
   b) the variable reflectance of electromagnetic energy by water molecules.
   c) the selective transmission of electromagnetic energy of water molecules.

10. Moisture measurement techniques presently used on sheets operate on
    a) infrared reflectance.
    b) infrared transmission.
    c) selective infrared absorption.

(See Appendix for answers)

11. Speed Measurement

Rotational speed measurements on motors, pumps, ventilating fans, line shafts, etc., together with linear types made on conveyors and paper machines, are vital to efficient operation of paper process equipment. Early speed measuring devices were referred to as *tachometers*. In many cases, this reference has also carried over into the more modern multicomponent speed measurement systems.

Speed measurement systems can be either mechanical or electrical. They provide final readout information in cycles per second, revolutions per minute, feet per minute, or the ratio of one speed to another.

MECHANICAL SPEED MEASUREMENT SYSTEMS

The first, simplest, and most widely used speed measurement devices in the paper industry were mechanical. They included timed-period revolution counters, centrifugal force indicators, and resonance-operated meters.

Revolution Counter

A basic, hand-held revolution counter (Figure 11-1) must be used with a timing device to determine the number of revolutions in a measured length of time. It consists of a contact point which is pressed against a rotating shaft or

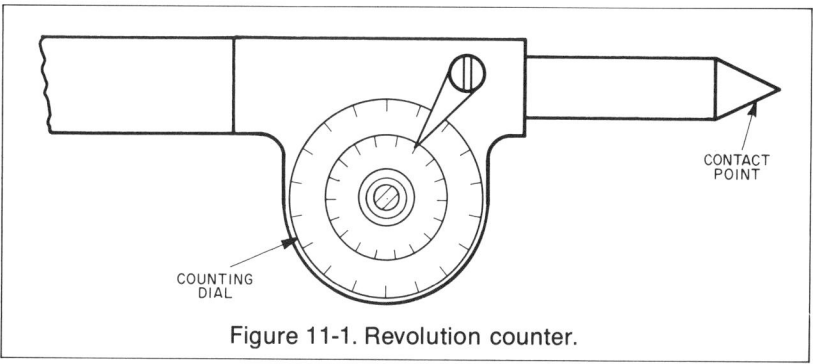

Figure 11-1. Revolution counter.

the source to be measured. The point operates a worm gear that meshes with a spur gear which, in turn, moves a calibrated counting dial indicating total revolutions. A stopwatch started and stopped simultaneously with the counter permits calculation of the average speed in revolutions per unit time. The counter and timer can be combined in one system which simultaneously starts and stops when the contact point is pressed against or removed from the source. This device is normally referred to as a *tachoscope*. A rachet arrangement on the measuring wheel, which frees it for a definite period of time, allows the pointer to indicate directly on a dial calibrated in rpm.

Another version similar to the tachoscope is called a *chronometric tachometer*. It differs in that it automatically repeats its timed period to adjust indication to the average speed for the last timed interval. If these periods are very short the impression of continuous, instantaneous speed measurement is given. With proper attachments, the chronometric tachometer can be mounted on the end of a spindle for measurement of linear speed.

Centrifugal Force Indicator

This type of speed reading device operates on the basis of centrifugal force developed by a rotating mass as a function of speed. This force is applied to a gear and pinion to move a pointer over a graduated scale (Figure 11-2). As the weights rotate, they move outward and, in turn, move a sleeve. This sleeve, or collar, moves a rack and pinion attached to a shaft with a pointer which moves over a graduated scale. Mechanisms of this type can measure up to 40,000 rpm. Other variations of this type employ gravity or pressure to counterbalance the centrifugal force. Indication can be a pointer on a calibrated scale, level of the meniscus of mercury in a manometer, or reading of a pressure instrument.

Resonance Speed Indicator

Vibrating reeds of various lengths mounted on a base with a reference scale in accordance with their natural period of vibration (Figure 11-3) can provide a

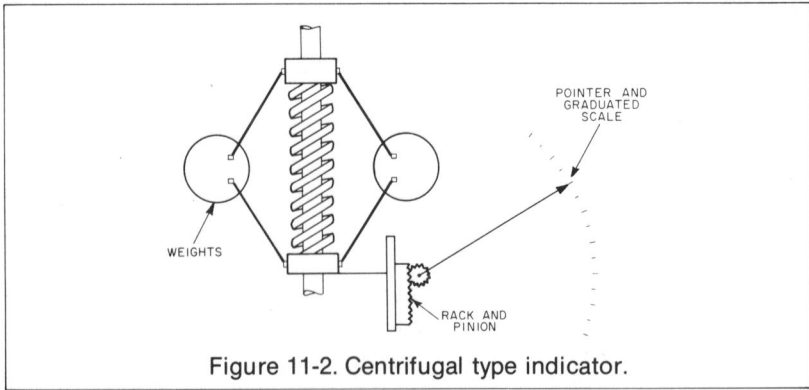

Figure 11-2. Centrifugal type indicator.

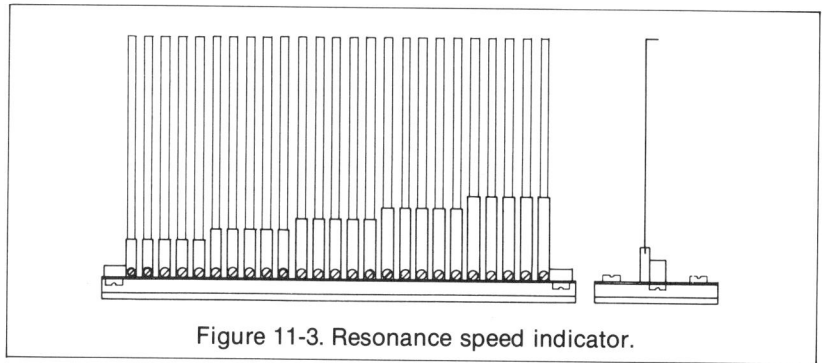

Figure 11-3. Resonance speed indicator.

means to measure the frequency of vibrations or revolutions. By bringing the instrument in physical contact with the member to be measured, mechanical frequencies, oscillations, and rotations may be measured by observing which reed is vibrating. The reed mechanism can be incorporated in various types of housings equipped with calibrated scales so that readings can be made easily.

ELECTRICAL SPEED MEASUREMENT SYSTEMS

Electrical speed measurement systems consist of a primary measuring element or transducer which converts rotational speed into an electric signal output to an indicator or recorder. The primary measuring element produces either an analog signal which can be used for analog indication, or pulses which can be digitally counted in terms of revolutions in a unit time as related to speed.

Magnetic Drag Type

A transducer designed for measuring shaft speeds of compressors, engines, variable speed drives, turbines, blowers, and pumps in paper mills employs a pneumatic signal produced by a magnetic drag torque assembly. The instrument operates on the force-balance principle with magnetic actuation of a pneumatic circuit as shown in Figure 11-4. The input shaft (A) carries an 8-pole permanent magnet (B). A nonmagnetic alloy disc (C) is held in position by flexure mounts (D) near the poles of the magnet. As the input shaft rotates the magnet, the magnetomotive pull on the disc tends to turn in the same direction. This pull positions the force bar (E) attached to the disc in relation to the nozzle (F), causing an increase in back pressure in the airflow through the nozzle. This back pressure, amplified by a relay (G), produces an output air pressure. The output pressure is also connected to the ball feedback unit (H) which rides against the force bar producing a moment to balance the torque on the disc produced by the magnetomotive pull. Because the magnetomotive pull is proportional to the speed of rotation of the magnet, the output pressure is also proportional to the speed.

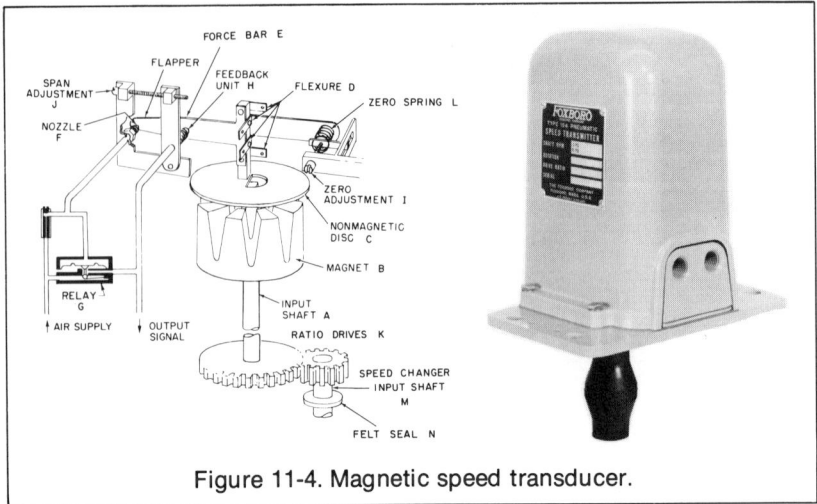

Figure 11-4. Magnetic speed transducer.

Magnetic Pulse Type

The speed of a rotating shaft can also be measured by the use of a magnetic pickup and a digital tachometer as shown in Figure 11-5. In operation, a magnetic pickup is located in close proximity to the teeth of an existing gear on the rotating shaft in the power train of the equipment on which the speed measurement is being made. The magnetic pickup generates an electrical pulse for every tooth on the gear as it moves by the pickup. The pulses are counted by a digital tachometer and displayed as the equivalent engineering units. The time interval over which the pulses are counted is determined by the formula:

$$\frac{60}{\text{rpm x no. of gear teeth}} = \text{count time in seconds}$$

The principle of capacitance has also been used to measure shaft rotation as shown in Figure 11-6. A variable capacitor is formed by vanes attached to the revolving shaft under measurement. Rotation of the shaft alters the capacitance to ground signal. The capacitor forms a part of an oscillator tank so that the number of frequency changes per unit of time is a measure of shaft speed.

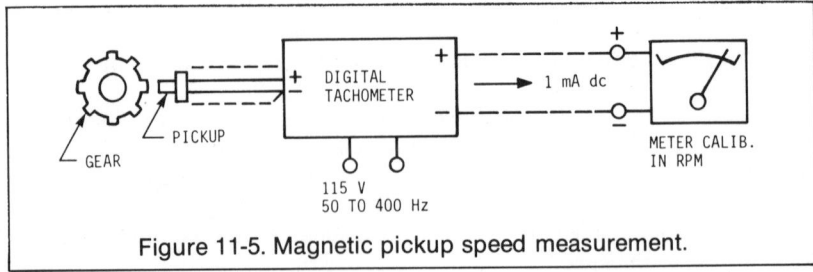

Figure 11-5. Magnetic pickup speed measurement.

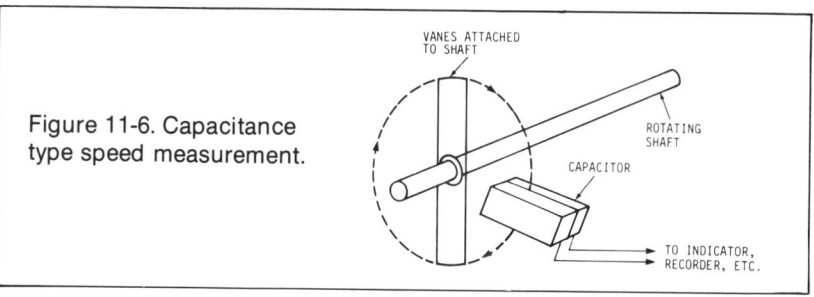

Figure 11-6. Capacitance type speed measurement.

The pulses thus produced go to amplifying, pulse forming, and triggering circuits where they are converted into constant amplitude signals. A dc electrical output signal is thereby produced which is proportional to speed and used for indicating, recording, and controlling purposes.

Electric Generator Types

Many transducers used to convert speed into compatible measurement signals are dc or ac generators which produce voltages used by suitable voltmeter type secondary instruments for continuous display or control.

dc Speed Measurement Systems

A simple schematic diagram of a typical dc speed measurement system is shown in Figure 11-7. The magnetic generator produces a pulsating dc voltage proportional to the speed of the rotor which is driven by the equipment being measured. The generated signal can be read by an ordinary voltmeter or by a null balance type of potentiometer. Polarity of the dc voltage depends on the direction of rotation and can, therefore, be used to indicate this direction by the use of an indicator with its zero point at mid-scale. By using two primary generators on separate pieces of equipment such as on drives in different dryer sections on a paper machine, and the proper readout instrument, the ratio as well as the difference in speeds can be measured. These readings can be

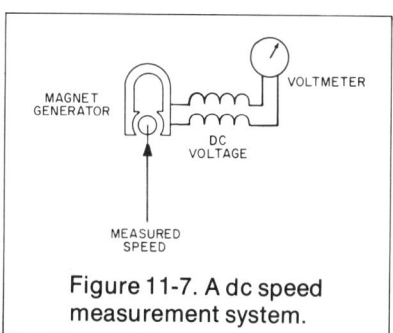

Figure 11-7. A dc speed measurement system.

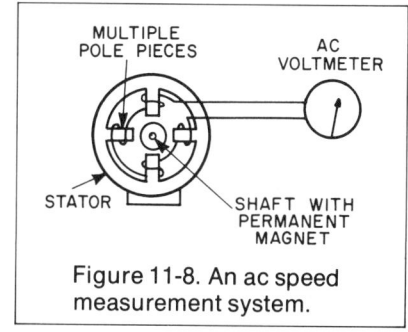

Figure 11-8. An ac speed measurement system.

expressed in terms of percentage and are generally referred to as *stretch* or *draw* of the sheet when applied to a paper machine.

ac Speed Measurement Systems

Similarly, an ac generator can be used to develop a signal that is related to the speed of equipment to be measured. The basic elements involved in such a system are shown in Figure 11-8.

This type has a stator coil with multiple pole pieces; a magnet on the shaft passes the pole pieces. The ac secondary indicating or recording instrument is generally a permanent magnet coil device with a rectifier. This type can also be used to measure difference in speeds from two sources.

Pulse-Shaping Measurement Systems

If the output power of an ac generator is sufficiently high, use of a pulse-shaping device between the primary generator transducer and secondary indicating or recording instrument can improve the accuracy of measurement. As shown in the schematic circuit in Figure 11-9, the saturable transformer acts as a gate during each cycle allowing only a part of the signal to pass to the readout instrument. The transformer confines the balance to the generator itself. The amount of signal appearing on the secondary side of the saturable transformer is constant for each cycle; its value depends on the physical dimensions of the transformer. Thus, readings can be made proportional to the frequency of the generator pulse output rather than the voltage. This method can be used with all ac generators to make readings highly accurate in responding to frequency.

OTHER SPEED MEASUREMENT TYPES

Many other forms of transducers have been used as primary elements for speed measurement systems. Some of these are contactless types. One of them applies the electromagnetic principle utilizing a coil, permanent magnet, and soft iron piece and forms an open magnetic circuit. The circuit is periodically closed by the insertion of a gear tooth or key to produce pulses.

Optical transducers are also considered contactless. One group is designed so that shaft rotation interrupts a beam of light falling on a photoelectric or photoconductive cell. The pulses are amplified and either counted by an

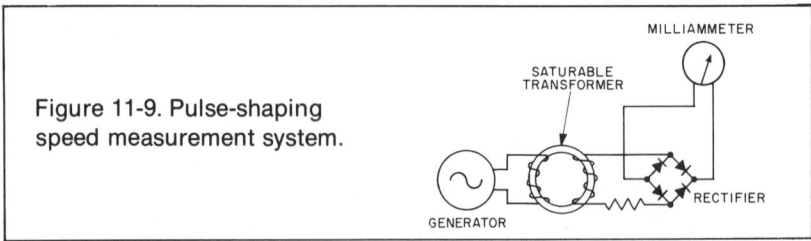

Figure 11-9. Pulse-shaping speed measurement system.

electric counter or shaped to an analog signal before being sent to an indicator or recorder. Another optical method depends on the stroboscopic principle of synchronizing a flashing light with rotation of a shaft, making it appear to stand still. The frequency of the light-producing stroboscope then becomes a measure of rpm without requiring contact with the rotating part.

Other speed measurement systems depend on the frequency of an ac signal produced by the primary device. Common among these are electronic counters, frequency meters, and resonance frequency meters. Systems operating from the primary winding of the ignition coil of an internal combustion engine, on low-frequency pulses and high-frequency oscillations, have also been used for making speed measurements in industry.

There have been reports of the use of laser beam technology in the measurement of rotational and linear speed. Results have been encouraging, but to date the effort has been primarily experimental and the development of a practical and generally acceptable measurement system for this purpose appears to be quite feasible.

BIBLIOGRAPHY

Behar, M. E. The Handbook of Measurement and Control. Pittsburgh: The Instruments Publishing Company, 1951.

Considine, D. M. *Process Instruments and Control Handbook.* New York: McGraw-Hill Book Company, 1957.

Considine, D. M., and Ross, S. D. *Handbook of Applied Instrumentation* New York: McGraw-Hill Book Company, 1964.

Holzbock, W. G. *Instruments for Measurement and Control,* 2nd ed. New York: Reinhold Publishing Corporation, 1962.

16A Series Pneumatic Transmitter-Style C, Foxboro, MA: The Foxboro Company, TI 37-65c, April 1969.

STUDY QUESTIONS

Indicate whether following statements are True or False by inserting T or F in the parentheses:

1. A tachoscope consists of a counter and timer combined in one system. (   )

2. The chronometric tachometer automatically repeats its timer period to adjust indication to the average speed for the last timed interval. (   )

3. The primary measuring elements of the electrical speed measurement systems produce pulses only, which can be digitally counted in terms of revolutions in a unit time as related to speed. (   )

4. Both dc and ac generators are used for making speed measurements. (   )

5. dc generators for speed measurements have a stator coil with multiple pole pieces with a magnet on a shaft which induces voltage in the stator coil as the magnet passes the pole pieces. (   )

(See Appendix for answers)

12. Freeness Measurement

Freeness is a measure of the rate at which water drains from a stock suspension through a wire mesh screen or a perforated plate. This measurement is sometimes referred to as *slowness* or *wetness* and is an indication of stock drainage characteristics.

Drainage characteristics as the sheet initially forms on the wire are an early indication of final sheet characteristics such as tensile strength, burst, bulk, tear, and fold.

In the past, manual freeness tests were utilized to spot-check samples removed from the process at specified intervals of time. Although their use is declining today, these manual methods are still in use in many mills. The most common include the Canadian Standard and the Schopper-Riegler freeness testers which measure the volume of water draining through a perforated plate from a fixed volume of stock beyond a fixed rate of flow, and the Williams precision freeness tester which measures drainage on the basis of time required for a given volume of water to pass through a wire screen from a fixed volume of stock sample.

With modern refining, continuous machine operation, and higher machine speeds, automatic sampling and measurements are required to eliminate the obvious shortcomings of manual testing. A number of automatic freeness measurement systems have been developed to meet this need.

FREENESS MEASUREMENT THEORY

Freeness measurement, like consistency measurement, is somewhat inferential. It is based on the sensing of a phenomenon that is related to freeness and the signal produced is expressed in units of freeness. These units of freeness are usually referred to a measurement standard which corresponds to the method used for making the original manual tests. The values obtained from the Canadian Standard freeness tester are expressed as CSF units, Schopper-Riegler freeness tester results in SR freeness units, and the Williams freeness tester results in Williams freeness units.

Other variables such as consistency, furnish, flow, temperature, additives, pH, dissolved salts, entrained air, etc., have some effect on freeness. Therefore, all these factors must be considered in making a freeness measurement. Stock slurries will drain more freely at high temperature because the viscosity is lower. A 10°F change in temperature can cause as much as a 10 percent variation in freeness in the normal operating ranges (400 CSF). Consistency variations of ±1 percent can cause as much as ±15 percent change in freeness in the same range.

FREENESS MEASUREMENT SYSTEMS

A general classification of some typical automatic freeness measurement systems is shown below:

A. Continuous sample types
1. Level
2. Flow
B. Intermittent sample types
1. Timed level
2. Timed weight

They are either of: (1) the continuous sample type which operates on the basis of a level or flow measurement as the result of processing a sample of stock slurry continuously removed from the process; or (2) the intermittent sample type which operates on the basis of a timed level or weight measurement on discrete samples automatically removed from the process in short timed cycles.

Continuous Sample Types

Level. A typical continuous sample type automatic freeness measurement system which operates on the basis of level measurement is shown in Figure 12-1. In operation, a representative continuous stock sample is continuously diverted from the process and enters the headbox of the measurement system at a constant consistency and temperature. Here it is further diluted to a known consistency of 1 to 2 percent and thoroughly mixed. The diluted pulp flows by gravity to a cylinder vat in which a constant level is maintained by an overflow weir. A sheet is formed on a cylinder covered with a wire screen and then couched off and returned to the stock system. The filtrate water draining through the cylinder screen is held at a constant level by an overflow weir. As a result, a constant head differential is maintained between pulp and filtrate water across the face of the cylinder wire. The filtrate overflow is collected in a standpipe which has a fixed orifice at the bottom, so any change in drainage rate or freeness will vary the level in the standpipe. Hence, change in level in the standpipe corresponds to a change in drainage rate of the pulp. Thus, any change in level is measured and indicated or recorded in arbitrary units which can be converted to common freeness values. This signal can also be used to control grinding and refining operations to maintain freeness of pulp or stock slurries within desired limits.

Flow. Another automatic continuous sample freeness measurement system uses the same principle of drainage through a screen. It differs in that it uses a flow measurement as an indication of freeness changes (Figure 12-2). In

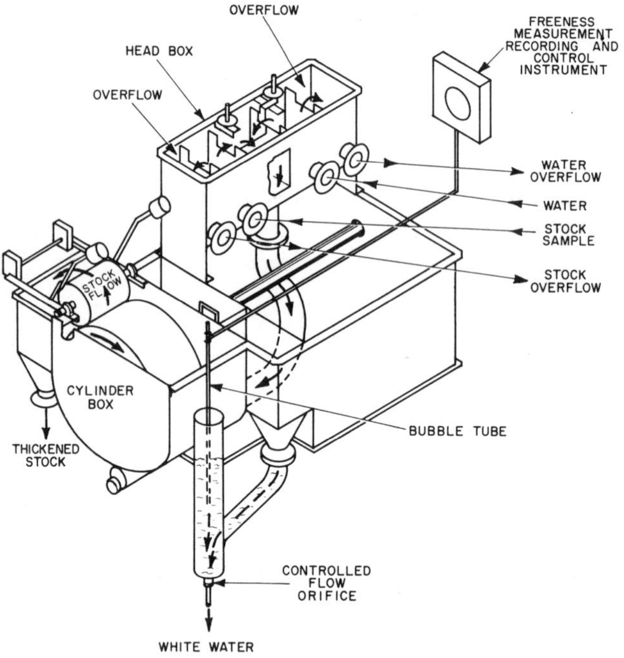

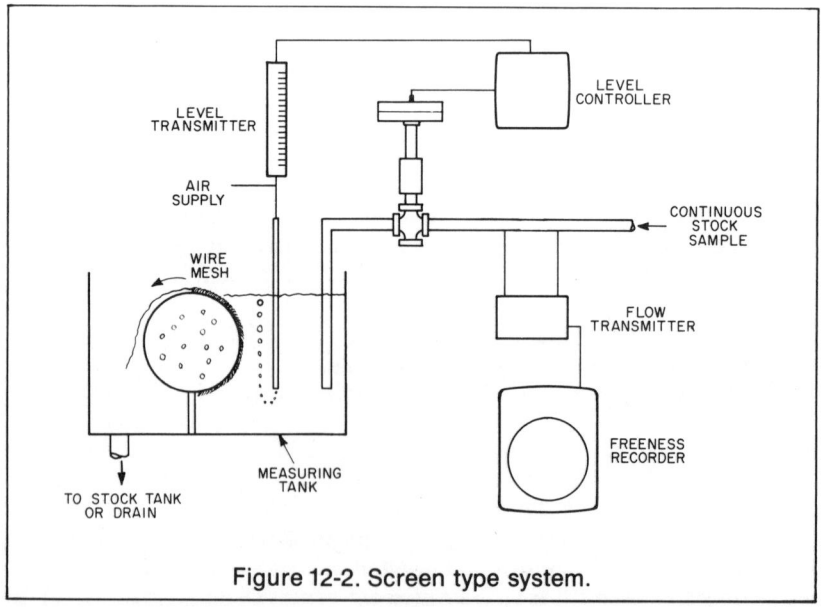

Figure 12-1. Freeness measurement system.

Figure 12-2. Screen type system.

actual operation, a continuous small sample flow of pulp slurry at a constant consistency and temperature is diverted from the process to the measuring tank. The cylinder, which is faced with a wire screen, rotates at a constant speed and is sealed on the bottom and side so that any water drained from the stock must pass through the wire face. A film of stock forms on the cylinder as it rotates. Level of the pulp slurry ahead of the cylinder is measured by a bubble tube and is kept constant by a controller that positions a valve in the stock sample line. The fibers in the slurry flowing to the cylinder are picked up and carried to the downstream side, while the water drains through the fibers and screen. The fibers are washed off the downstream side of the cylinder screen by the water. The rate of water drainage through the fibers is related to the freeness of the pulp. A change in freeness tends to change the level ahead of the cylinder and the bubbler detects this change immediately. The control valve then increases flow accordingly. The flow changes are measured across an orifice plate and recorded in terms of freeness.

Intermittent Sample Types

Level. A typical intermittent sample type automatic freeness measurement system which operates on the basis of a level measurement, after a timed cycle, is shown in Figure 12-3. It consists of a detector, tee, level diaphragm element, ball valve, screen, and transparent plastic measuring column for visual observation, assembled and mounted directly on the stock line on which the freeness measurement is to be made. Shown in Figure 12-3 are the following steps made in intermittent level measurement:

1. Start-Finish. At a given signal, air pressure in the transparent measuring column is reduced below stock line pressure. As the pressure differential is established, a random stock flow develops to the detector unit through the open ball valve. At the screen the fibers separate from the water.
2. Intake. A mat forms on the lower side of the screen while water drains through this mat and enters the transparent detector tube or measuring column. The water column rises uniformly in the tube.
3. Measurement. At a given time signal, a measurement is made of the column of water by a diaphragm level measurement element. This measurement (X) is converted to corresponding freeness units by appropriate calibration of scale or chart in the secondary readout instrument.
4. Exhaust. Air pressure in the detector tube is then increased above stock line pressure so that the water column now reverses its path and acts as a purge to wash the fiber mat from the screen and back into the stock line. At the end of the exhaust cycle all stock has been eliminated from the side connection so that the stock pipeline is operating with no extension of pulp above normal pipeline level. The freeness measurement system then reverts to the first start-finish cycle conditions and repeats the measurement operation about once a minute.

A similar type of intermittent sample freeness measuring device that also operates on the basis of a level measurement after a timed cycle is shown in

Freeness Measurement   253

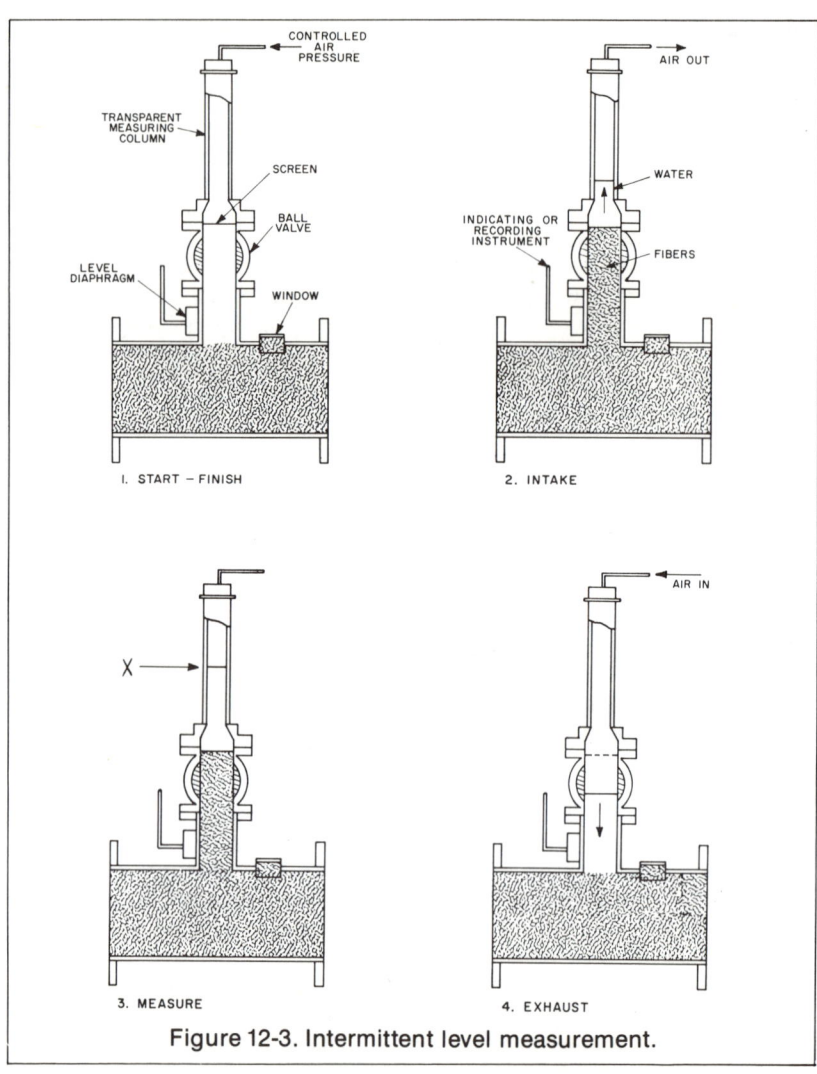

Figure 12-3. Intermittent level measurement.

Figure 12-4. The timed cycle on which it operates is adjustable and determines (A) the sampling phase during which a representative sample of pulp is taken from the pipeline; (B) the measuring phase during which the sample is isolated and drained; and (C) the flushing phase during which the strainer screen is flushed with water.

The operation cycle starts by lifting up the top plate (1) and the mushroom section in which the strainer screen (3) is incorporated into the process pipeline by a pneumatic actuator (6). At the completion of the sampling period (A), the pneumatic actuator is reversed and the pulp sample is drawn into the mushroom-shaped section (2) without changing the pressure or the volume of the sample. The sample is then isolated (*B*) from the pulp line pressure by the top plate and an opening is created between the cover and the upper side of the mushroom section. Water is then fed through this opening through the hollow stem of the mushroom section at a preset pressure. This water exerts pressure on the pulp sample while the underside of the strainer screen is simultaneously opened to atmospheric pressure, thus dewatering and compacting the sample. During this "compression" time or "predrainage" time when the fiber cake is forming on the strainer screen, the drainage goes to waste. This predrainage allows a "tight" fiber cake to build against the strainer. After completion of the predrainage time, the drainage from the fiber cake is collected in the measuring tube (8) for an adjustable preset time (*A*). The collected drainage level is sensed by a bubble tube airflow arrangement and directed to a low-range level transmitter (7) with either pneumatic or electronic output. The drainage volume is a measure of pulp freeness as indicated by the level measurement. Therefore, the output of the transmitter can be converted to freeness units to be used for display and/or control purposes.

After the predrainage and drainage periods are completed, the unit is flushed (*C*) with water and air under pressure. This is directed to the underside of the strainer, thereby cleaning the strainer and discharging the pulp sample, after which the cycle is continuously repeated.

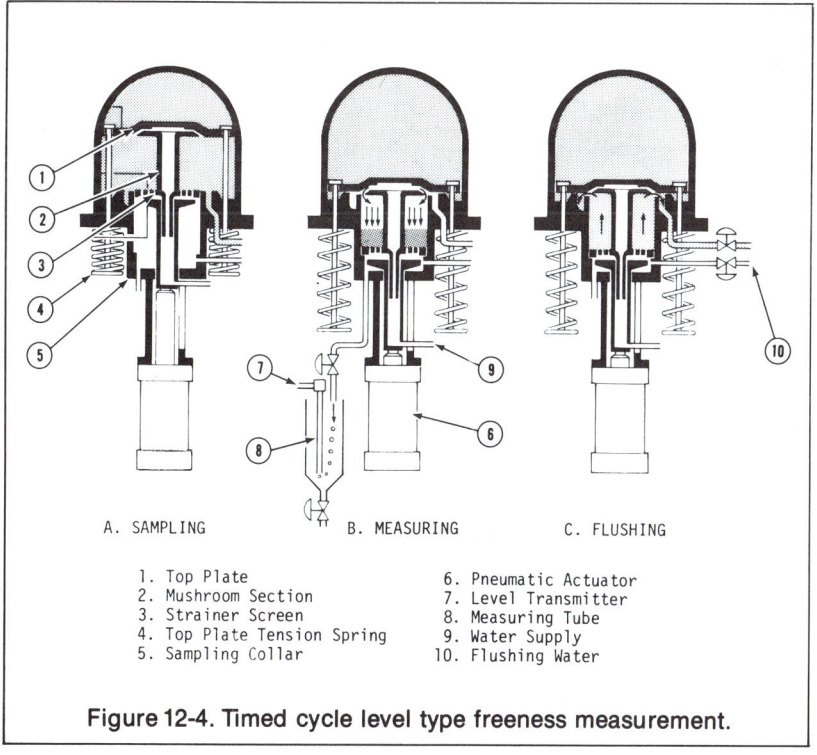

A. SAMPLING   B. MEASURING   C. FLUSHING

1. Top Plate	6. Pneumatic Actuator
2. Mushroom Section	7. Level Transmitter
3. Strainer Screen	8. Measuring Tube
4. Top Plate Tension Spring	9. Water Supply
5. Sampling Collar	10. Flushing Water

Figure 12-4. Timed cycle level type freeness measurement.

Weight. An intermittent sample type automatic freeness measuring system which operates on the basis of a timed weight determination is shown in Figure 12-5. A stock sample of 0.5 to 0.8 percent consistency flows continuously into the overflow tank at the side of the measurement assembly and discharges on the top. The overflow tank has two compartments connected by a U-shaped pipe which is fastened at the bottom of the assembly. A flexible pipe is connected to the drainage container which has an incorporated perforated plate. The top of the container is connected by a small flexible hose to a three-way compressed air valve. The stock slurry enters through the lower flexible hose below the drainage container and rises to the perforated plate. A stock mat is formed on the bottom side of the perforated plate and the drained water rises in the upper part of the drainage container.

The drainage container is suspended on a scale with adjustable counterweights. Depending on the amount of filtrate accumulated in the upper part of the drainage container, the scale inclines and with it the penholder on a recording mechanism. The pen normally does not touch the paper. After a preset drainage time of 40 seconds, a timer opens a compressed air valve which blows sharply on the penholder, setting a mark on the chart. In this way the weight of the filtrate registered after a preset drainage time is recorded in desired freeness measurement units.

The drainage time can be adjusted from 10 to 60 seconds. The pressure differential is constant. The filtrate level in the measuring container rises when the container itself is lowered, keeping a constant pressure difference between the inlet overflow tank and filtrate level.

After the measuring period of one minute, the three-way compressed air valve opens and the drainage container contents are blown back into the stock

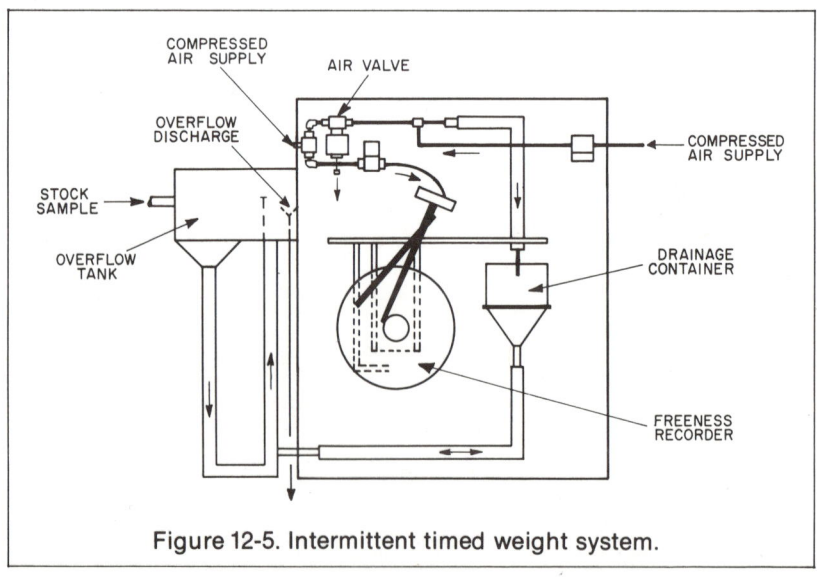

Figure 12-5. Intermittent timed weight system.

line. By a separate solenoid valve, water is injected through an incorporated spray over the perforated plate for complete washup. The cycle repeats every two minutes.

Other Types

A variety of other methods has been used in an attempt to measure pulp slurry freeness. Most of these have been experimental and have not received general acceptance as process measurements. One that has achieved some success as a process measurement operates on the basis of a diluted sample of pulp of approximately 0.8 percent which is passed along a delivery channel where variations in concentration are sensed electronically by the rise and fall of a float-actuated core in a differential transformer. Samples of diluted stock are fed into a one-liter capacity vessel with a bottom consisting of a perforated plate. The base of the pot is sealed for a short time before the stock is allowed to drain freely for a predetermined time. The pot is then emptied, washed, and recharged for the next test. Drainage water is collected in a measuring tank where dc voltages proportional to its volume and temperatures are derived electronically. This volume is a measure of freeness and it is influenced by variations in stock concentration and temperature. The measuring systems for stock concentration and temperature derive dc voltages of amplitude and polarity which are dependent upon degree and voltages of amplitude and polarity which are dependent upon degree and direction of changes from normal standards in either value. These voltages are maintained automatically in the appropriate proportion-to-rate volume. The mean of these values is presented graphically in the form of a continuous recording in corrected standard freeness units.

Paper machine couch vacuum is an indication of drainage characteristics of stock running on the wire and, therefore, has been used as a measurement of freeness. In the same manner, the drainage flow rate from the paper machines's sheet forming wires has been used as a freeness measurement of the stock being run.

BIBLIOGRAPHY

Chedomir, G. A. "A Continuous Freeness Recording Controller." *Tappi*, vol. 44, no. 6 (June 1961).

Evans, J. C. W. *Pulp and Paper Mill Process Instrumentation*, New York: Lockwood Publishing Company, 1969.

Gilbert, H. S. "Drainage Rate Indication and Control." *Paper Trade Journal*, March 11, 1963.

Ottersen, J. S.; Bedsole, J. F.; and Powell, D. T. "New In-Line Freeness Transmitter to Facili-

tate Closed Loop Refining Control." ISA preprints, 2nd International Symposium, Pulp and Paper Process Control, Montreal, Canada. April, 1973.

"Round-Up Report on Three Continuous Freeness Testers." *Paper Trade Journal*, March 11, 1963.

Williams, D. J. "A Continuous Freeness Recorder for Pulp Suspensions." *Tappi*, vol. 43, no. 7 (July 1960).

Indicate whether the following statements are True or False by inserting T or F in the parentheses:

1. Freeness is a measure of stock drainage characteristics. ( )

2. Freeness is basically an inferential type measurement. ( )

3. Freeness is unaffected by other variables existing in a stock slurry. ( )

4. Most systems used for freeness measurements operate on either continuous or intermittent samples. ( )

5. Temperature changes do not cause any variations in freeness, and therefore can be ignored in making a freeness measurement. ( )

(See Appendix for answers)

13. Basis Weight Measurement

One of the most frequently measured physical characteristics of paper or paperboard is basis weight. Paper is usually manufactured and sold within predetermined specifications established for this variable.

In the chapter on density measurements it was noted that weights of various substances are compared in terms of the weight per unit volume expressed in pounds per cubic foot, pounds per gallon, grams per cubic centimeter and similar units. Where the weight of paper is concerned, however, measurement is made on the basis of area and is specified as the weight of a given area or basis weight. Basis weight stems from the so-called ream weight of paper to which it is related but not equivalent. Ream weight represents a given number of sheets of a given size and is expressed in terms of pounds per ream. Although the concept of the ream is accepted by the paper industry, different segments adopted their own standards as to the number and size of sheets in the ream, depending on the grade involved. Therefore, paper having the same basis weight, which is weight per unit area, can have numerous ream weights, depending on the sheet size and the ream count of 480, 500, or 1000 sheets, which have become the most common. The standard sizes most commonly used for various types of paper are:

Writing and printing	17" x 22"
Writing and printing (demy size)	17.5" x 22.5"
Blotting	19" x 24"
Cover	20" x 26"
Tissue (double crown size)	20" x 30"
Cardboard	22" x 28"
Bristol and tag	22.5" x 28.5"
Index	25.5" x 30.5"
News and wrapping	24" x 36"
Book	25" x 38"

Because a "standard" ream that would be universally accepted and used was never established, ream weight had to be further identified by size and number of sheets it referred to. Thus, 20 lb (17 x 22/500) would indicate that a ream consisting of 500 sheets of 17 x 22 inches in size would weight 20 lb and the ream weight of writing and printing paper would be expressed this way.

The industry customarily expresses basis weight as the weight in pounds of 500 sheets, 24 x 36 inches in size, unless otherwise specified in which case the basis weight designation would be followed by size and number of sheets it referred to, as in ream weight and designation. To convert a given weight from one set of conditions to any other is a simple matter. Knowing the number and size of sheets that the weight is referred to, the total area can be determined. By dividing the total area by the weight given, the total weight of one square inch is obtained. This quotient multiplied by the total area under conditions sought gives the weight required. Conversion factors obtainable from published reference tables can also be used.

In determining basis weight manually, it is common to weigh a sheet of paper of known dimensions on a multiplying scale that multiplies the results by 480, 500, or 1000. The weight is then expressed in at least two of the following ways: (1) the equivalent basis weight in pounds for a ream consisting of 500 sheets, 25 x 40 inches in size; (2) the weight in grams per square meter; and (3) the equivalent weight for the ream size commonly used by the paper industry for the particular grade of paper. The weight of paperboard is reported in pounds per 1000 square feet.

Following along with the gradual trend toward the conversion to the metric system of measurements, it is envisioned that basis weight will be stated as grams per square meter (g/m^2).

For example, the metric equivalent of a 20-pound basis weight paper (17 x 22/500) would be expressed as 75 grams per square meter or simply 75 g/m^2. The basis weight is converted to grams per square meter (g/m^2) by multiplying the basis weight by a constant factor (1406.5) and dividing by the number of square inches in the base sheet. Therefore, to convert 20-pound (17 x 22/500), multiply 17 x 22 which equals 374 square inches. Then multiply 20 by constant 1406.5 and divide by 374 which equals 75.2 g/m^2. The 75.2 is rounded off to the nearest whole number which is 75.

BETA-RAY GAUGING

Many unsuccessful attempts have been made to continuously weigh paper during its manufacture by means of sensitive continuous weighing devices. It was not until after radioisotope instrumentation was introduced that beta-ray basis weight gauging became generally accepted as a practical solution.

Beta rays are high-speed electrons emitted by certain radioactive isotopes such as strontium 90, cesium 137, thallium 204, and krypton 85. When a beam of beta rays is directed at a sheet of paper some electrons pass through it and emerge on the other side with sufficient energy to ionize air or gas. Some electrons lose energy in the sheet and become captured and some are sharply deflected from their path and ionize air or gas they contact on the same side of the sheet. The relative number of electrons transmitted through the sheet is an inverse function of its mass (weight per unit area) or basis weight. The number of electrons reflected or backscattered depends directly on the same factor. Therefore, as the paper increases in basis weight, the number of beta rays emerging from the opposite side of the sheet decreases while the number reflected from the sheet increases. Basis weight measuring systems take advantage of both of these phenomena and can be divided into two broad classes—the transmission type and the reflection or backscatter type.

Transmission Types

The principle of the transmission type beta-ray gauges is illustrated in Figure 13-1. A radioactive source is placed below the sheet to be measured and a radiation detector is placed on the opposite side. The detector can be a Geiger counter, scintillation counter, proportional counter, or ionization counter. The ionization chamber is most frequently used. Beta rays, consisting of fast-moving particles of negative electricity (electrons), are emitted from the radioactive source and directed at the moving paper web. The number of electrons that penetrate and pass through the sheet depends on its weight per unit area (mass). These electrons are detected by allowing them to pass into an ionization chamber detector. The ionization chamber is fundamentally an enclosed volume of air or gas, such as argon or nitrogen, with a conductive shell and an insulated collector electrode. Electrons entering the chamber cause a small current to flow through it in direct relation to the degree to which the gas content is ionized. This depends on the amount of beta ray that enters the chamber. The current, which is in picoamperes, is amplified, electronically measured, and displayed or recorded in the desired basis weight units.

An exploded view showing the component parts of a typical beta-ray basis weight primary measurement assembly is shown in Figure 13-2.

On-Machine Basis Weight Measurements

A typical arrangement of a beta-ray system installed on a paper machine to continuously measure the basis weight of a sheet of paper is illustrated in Figure 13-3. The on-machine basis weight measurement can be made with either a stationary or moving head. The former arrangement is sometimes referred to as a *fixed head measurement.* The information resulting from this measurement is used to observe basis weight variations at only one cross-

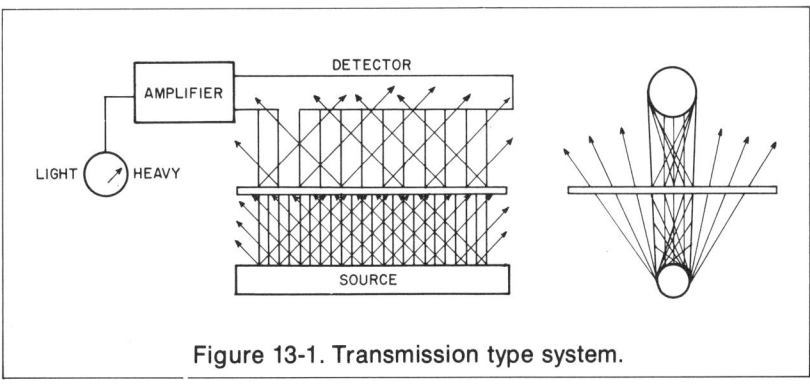

Figure 13-1. Transmission type system.

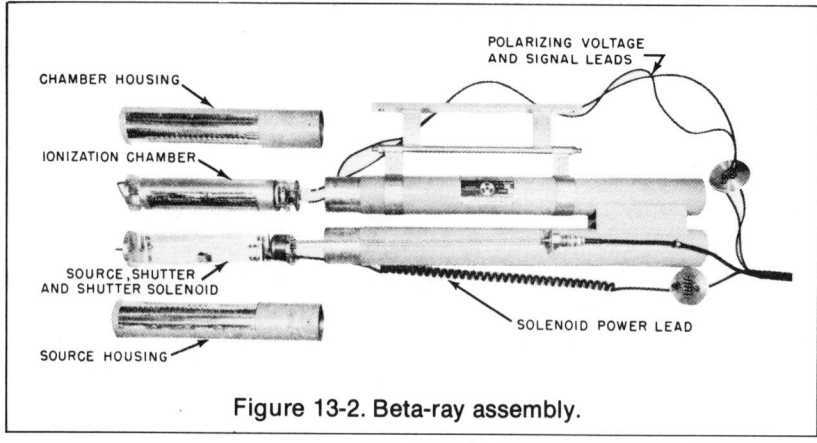

Figure 13-2. Beta-ray assembly.

machine point on the paper web. This measurement is referred to as *machine-direction basis weight*. A *traversing head measurement* is used to obtain basis weight variations across the paper web. It is generally referred to as the *cross-machine basis weight profile*.

Off-Machine Basis Weight Measurements

Cross-machine basis weight profiles are also made off the machine on a sample strip of paper taken from a reel. These units (Figure 13-4), known as *beta-ray sheet weight profilers,* are usually located near machine operators or control laboratory technicians. To obtain a profile record of basis weight variations across the machine, one end of the sample, as obtained above, is inserted into the slot on the front of the cabinet. The strip is automatically fed between a radioactive source and an ionization chamber detector similar to those in on-machine systems. The recorder chart and sample speed are synchronized so that each chart division represents a predetermined amount of sample. The profile record thus produced is used as a guide for making adjustments to level out deviations from the normal configuration.

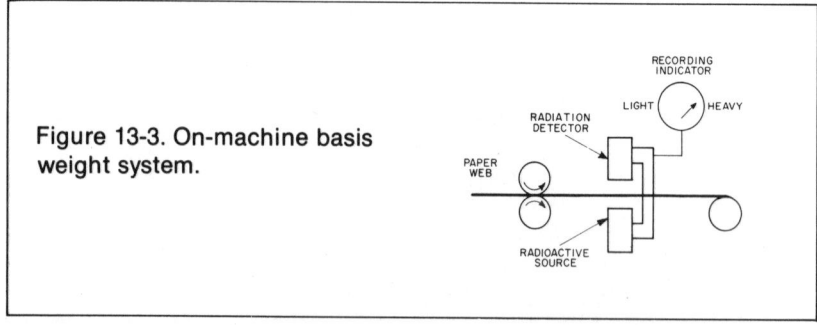

Figure 13-3. On-machine basis weight system.

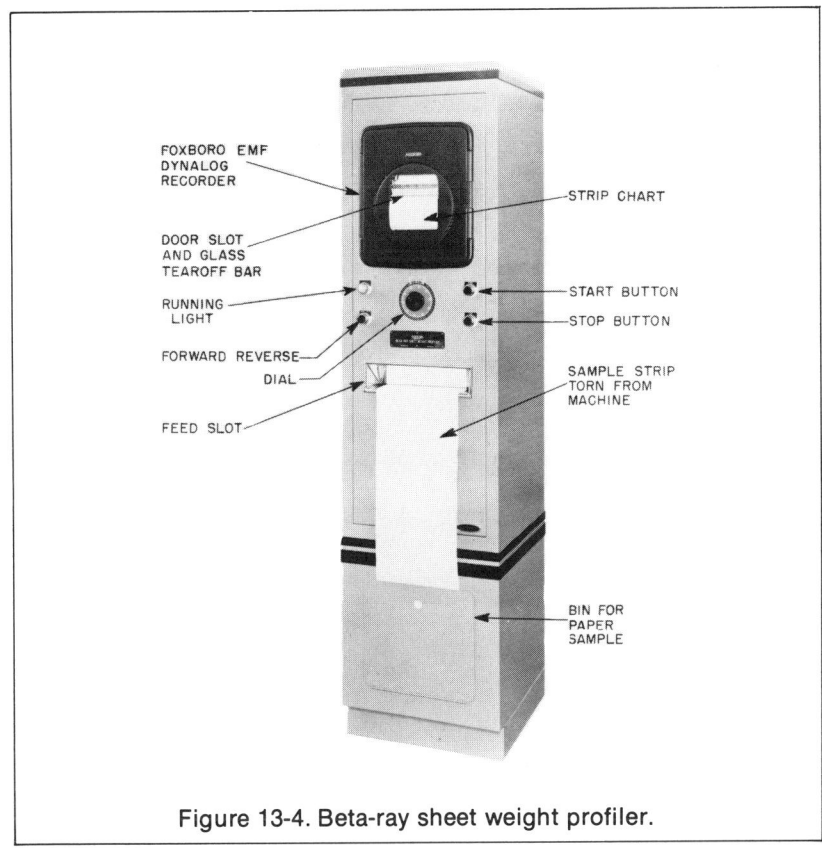

Figure 13-4. Beta-ray sheet weight profiler.

Although uncommon, machine-direction basis weight measurements can be made with the off-machine unit by obtaining from the reel a sample strip of paper in the machine direction and running it through in the same manner as in the cross-machine sample.

Backscatter Types

Basis weight measurement systems that operate on the beta-ray backscatter principle locate both the radioactive source and the ionization detector chamber on the same side of the sheet (Figure 13-5). The measurement is made as the sheet passes over a roll or other backing surface and the source is shielded so that only backscattered radiation reaches the chamber. Thus, when sheet material is placed in the measuring position the ionization chamber picks up diffusely reflected beta rays and the resulting chamber current is converted to the desired basis weight units.

Two factors must be considered when this method is used: the atomic number and weight per unit area of both the material to be gauged and the

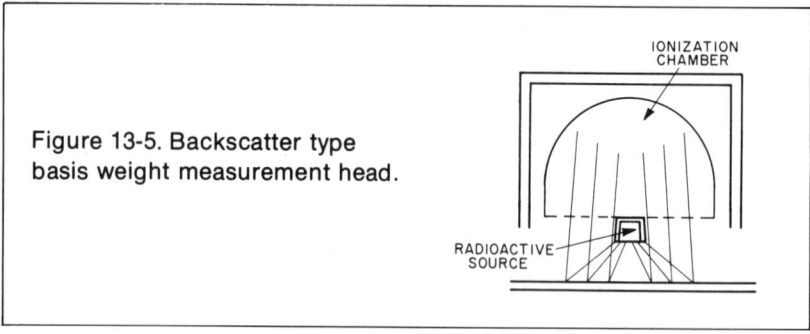

Figure 13-5. Backscatter type
basis weight measurement head.

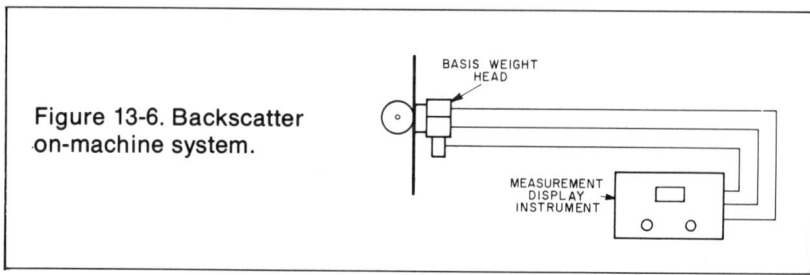

Figure 13-6. Backscatter
on-machine system.

backing material. Ability of a material to backscatter beta rays is related to the electron density within the material, i.e., the material's atomic number. The backing material must be sufficiently thick to ensure that variations in its thickness do not influence the measurement. Figure 13-6 illustrates a typical application of a backscatter type basis weight measurement system.

BIBLIOGRAPHY

Bearer, G. F. "Automatic Control of Basis Weight Using Radioisotopes." *Tappi*, vol. 38, no. 1 (January 1955).

Beta-ray Basis Weight System. Foxboro, MA: The Foxboro Company, TI 9-51c, February 1966.

Brunton, D. C. "The Betameter." *Pulp and Paper Magazine of Canada*, vol. 54, no. 3 (Convention Issue 1953).

Crawford, E. N., and Stram, M. "The Beta-ray Basis Weight Gauge." *Tappi*, vol. 33, no. 4 (April 1950).

Hazelwood, E. *Beta Gauge on Machine*. New York: Vance Publishing Corporation, 1974.

Stephenson, J. N. *Pulp and Paper Manufacture*, vol. 4. New York: McGraw-Hill Book Company, 1955.

Van Den Akker, J. A.; Hardacker, K. W.; and Dearth, L. R. "Instrumentation Studies LXXIX. Beta-ray Gauges." *Tappi*, vol. 40, no. 5 (May 1957).

Select the answer or answers which correctly apply to the following:

1. Basis weight of any paper may be defined as the weight of a given
   a) length.
   b) width.
   c) area.

2. Ream weight of paper is the weight of a given number of sheets of a
   a) specified area.
   b) fixed width.
   c) given length.

3. Book paper with a weight of 30 pounds (25 x 38/480) would indicate that a ream of this paper
   a) consists of 25 sheets 38 inches square.
   b) would have 480 square meters of total area.
   c) would weigh 30 pounds.

4. The effect of beta rays on air or gas is to
   a) ionize.
   b) condense.
   c) neutralize.

5. The relative number of electrons transmitted through a sheet of paper is a function of its
   a) strength.
   b) mass.
   c) formation.

6. Basis weight measuring systems used in the paper industry are of the
   a) transmission type.
   b) reflection type.
   c) backscatter type.

7. On-machine basis weight measurement with a fixed head is used to make
   a) across-machine profiles.
   b) machine-direction measurements.
   c) combination of a and b.

(See Appendix for answers)

14. Thickness Measurement

Thickness, or caliper, affects such physical and electrical properties of paper as strength, bulk, and density. Thus, thickness is important to the paper machine operator in maintaining these properties within specifications.

A form of micrometer is used to manually determine the thickness of paper and paperboard (Figure 14-1). A single sheet is held under a given pressure between two circular, plane, parallel surfaces, one called the *anvil* and the other the *plunger* or *pressure foot*. The precision mechanism of the micrometer indicates the distance between the two surfaces. This is converted and expressed as caliper units in thousandths of an inch. One-thousandth of an inch is commonly called a mil in reference to paper and a point in reference to paperboard. It is anticipated that caliper will be expressed in millimeters (mm) when the metric system of measurements is adapted by the paper industry.

Bulking thickness, or caliper, is the average thickness obtained by measuring a pile of sheets and dividing by the number of sheets; the result is usally not the same thickness as a single sheet because of the packing or nesting effect of surface irregularities.

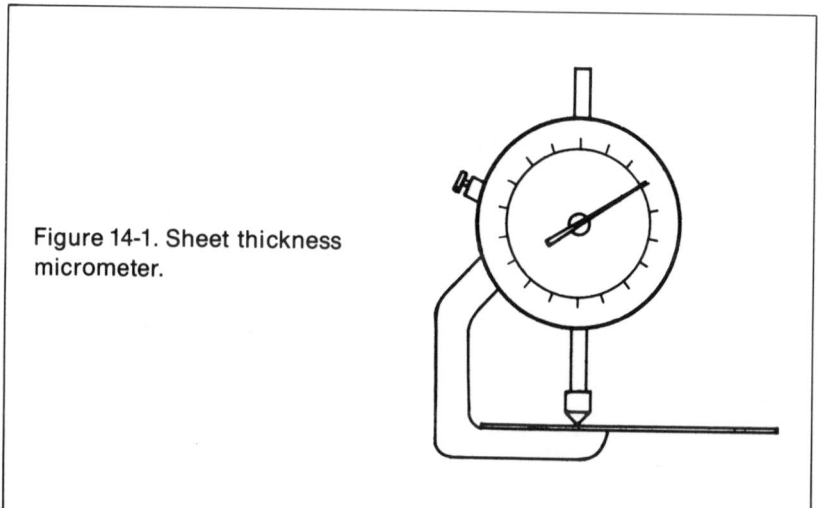

Figure 14-1. Sheet thickness micrometer.

THICKNESS MEASURING SYSTEMS

A number of systems have been developed and used continuously to measure thickness of a paper or paperboard sheet. The caliper measurement systems are classified into two general groups: contacting types and noncontacting types.

Contacting Types

Thickness measuring systems in which the primary sensing element must come into physical contact with the sheet employ electrical, mechanical, or electromechanical techniques in their operation.

Electrical. One of the more successful electrical approaches to thickness measurements involves the principle of magnetic reluctance. These devices sandwich the moving web of paper between the pole faces of an electromagnet and an iron or steel plate. The unit is energized by an alternating current. The paper, being a nonmagnetic material, allows fewer magnetic lines of force to permeate the magnetic core. Consequently, the inductance of the magnet is decreased more or less, depending on the thickness of the web. This is measured by a meter connected in an appropriate circuit. The permeability of iron or steel to magnetic lines of force is about 1000 times that of air or other nonmagnetic materials. As a result, these instruments can be extremely sensitive to changes in the amounts of nonmagnetic interference in their circuits. Figure 14-2 shows a typical electrical schematic and detector design for such a device. Its operation is based upon varying the inductance of a magnetic circuit. The magnetic element in the top detector is positioned by the thickness of the paper web separating the two parts of the variable reluctance circuit. The center leg of the E lamination is surrounded by a coil of wire whose leads are terminated at a source of ac voltage. An ac milliammeter is shown on one lead to read the current of the circuit. The impedance of this circuit is determined by the thickness of the nonmagnetic material separating the E and I laminations. The thicker the material, the greater the milliammeter reading.

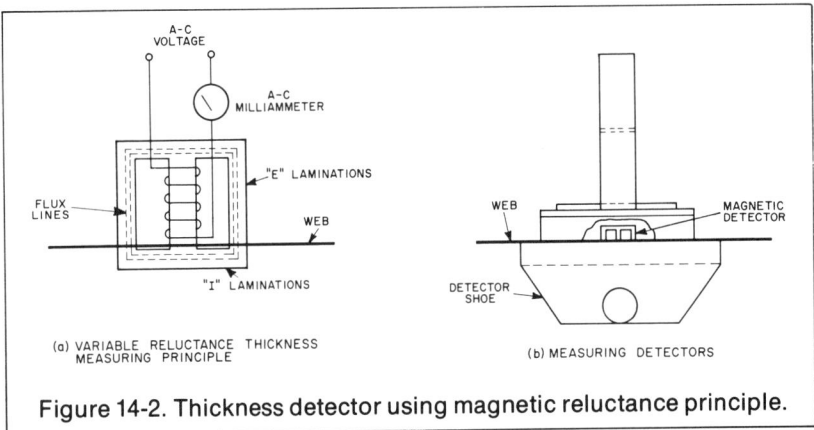

Figure 14-2. Thickness detector using magnetic reluctance principle.

The instrument will read a minimum value when the nonmagnetic material is removed altogether and the two parts of the core are brought into intimate contact. A bottom shoe supports the underside of the web. The top detector contains the induction coil which rides on the web. Thickness measurement essentially involves detecting the change of inductance of the current.

Mechanical. A typical on-machine mechanical sheet thickness measurement system is shown in Figure 14-3. It is normally located between the last press and first dryer of a paper machine and carries two rolls, the bottom fixed in position, the top pivoted on a caliper mechanism. Any change in paper thickness moves the top roll which operates the dial indicating thickness. The bottom roll is held in place against machined stops by tension springs. The dial is mounted on a steel drum on a stud fixed to the caliper housing. The disc at the top of the caliper housing is calibrated in 0.001-inch divisions and this dial is set to the caliper to be run. When the paper caliper is at the specified setting, the dial will stand at zero. An adjustable counterweight is provided to obtain the desired pressure between caliper rolls.

Electromechanical. This type of device uses an ac-excited capacitor mounted on a carriage and arm similar to that shown in Figure 14-4. The sheet fits tightly over a roll made of cast iron. The arm carries the measurement device at a distance of 1.5 mm from the roll's surface. The thickness of the sheet displaces some of the air gap between the device and the roll, and current through the device changes accordingly. A milliammeter is used to measure this current and convert it to appropriate thickness or caliper readings. The measurement carriage can be fixed or it can traverse the sheet for a cross-machine thickness profile.

Sonic. The sonic gauge principle of measuring thickness is diagrammatically shown in Figure 14-5. A quartz or barium titanate crystal oscillator is

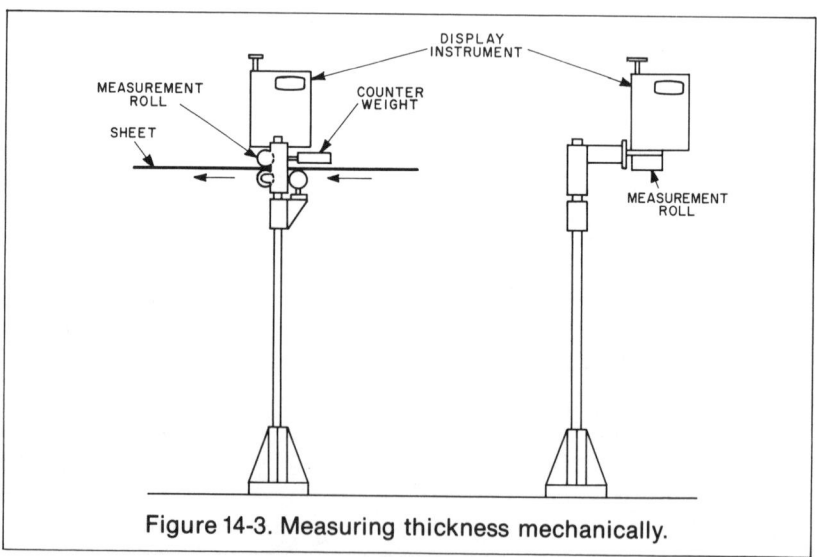

Figure 14-3. Measuring thickness mechanically.

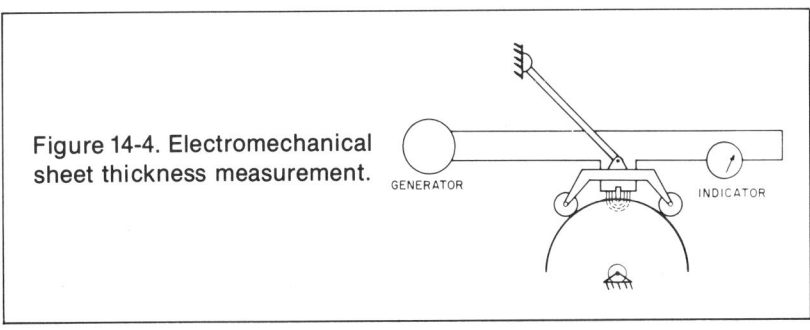

Figure 14-4. Electromechanical sheet thickness measurement.

GENERATOR

INDICATOR

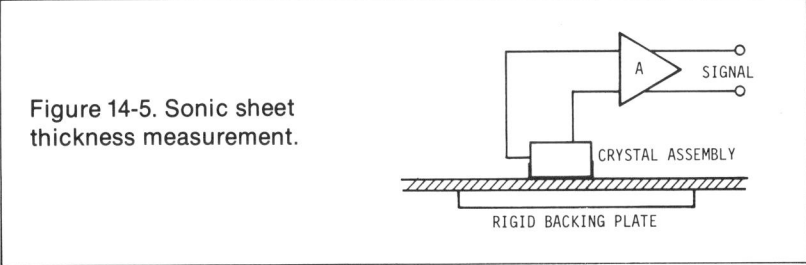

Figure 14-5. Sonic sheet thickness measurement.

A   SIGNAL

CRYSTAL ASSEMBLY

RIGID BACKING PLATE

applied to one side of the sheet with a backing plate on the other side. The crystal is excited until resonance is reached. The resonance frequency is a function of thickness and the speed of sound in the material. This type of measurement is more applicable to the more rigid products such as boards.

Magnetic Comparator. Figure 14-6 shows another contacting type of paper sheet caliper measuring device based on the magnetic comparator principle of operation. As the paper sheet passes over a rigid roll, a small roller wheel rides on top of the sheet moving an armature in and out of a magnetic field as the sheet thickness varies. The output will vary in accordance with the depth to which the armature enters the magnetic field, which is directly related to and a measurement of thickness or caliper of the paper.

Noncontacting Types

Systems that measure thickness without physical contact with the sheet are not as common as the contact types. Although several techniques have been attempted, few have received general acceptance on a practical basis. Optical, fluidic, and beta-ray radiation methods have attained some success.

Optical. Figure 14-7 depicts the concept of a typical optical approach to the measurement of sheet thickness. It consists of a collimated light source positioned in such a way that the center line of the light beam passes tangent to the surface of a precisely machined roll carrying the moving sheet. As sheet thickness varies, an increase or decrease in roll diameter will be noted by detecting the resultant increase or decrease of light transmission. Primarily used as a single point unit, it can also traverse the sheet for profile on a

Thickness Measurement   269

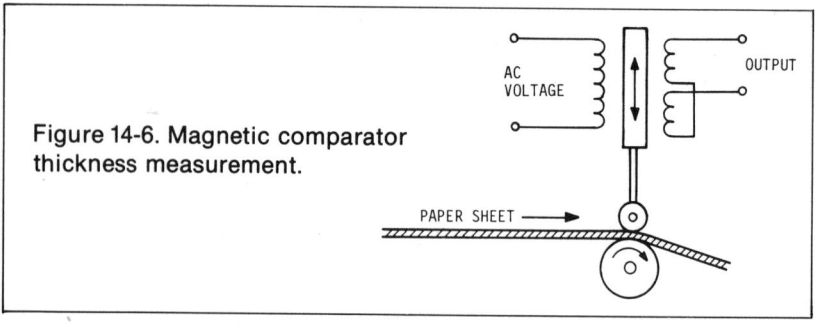

Figure 14-6. Magnetic comparator thickness measurement.

carefully designed mechanism. Other more recent optical methods for thickness measurement involve the use of laser beams. They look very promising toward attaining more accurate results in this area.

Fluidic. A schematic of a fluidic type sheet thickness measuring system is shown in Figure 14-8. This technique employs a jet of air projected against the sheet which is stabilized by a backing plate as part of the sensing assembly. The assembly is designed to operate as a fixed point measuring device and/or traversing device for making cross-machine profile records.

Another fluidic technique employs two opposing air jets: one located below and the other located opposite to and above the first one as shown in the sectional view of the sensor head illustrated in Figure 14-9. Air flows through

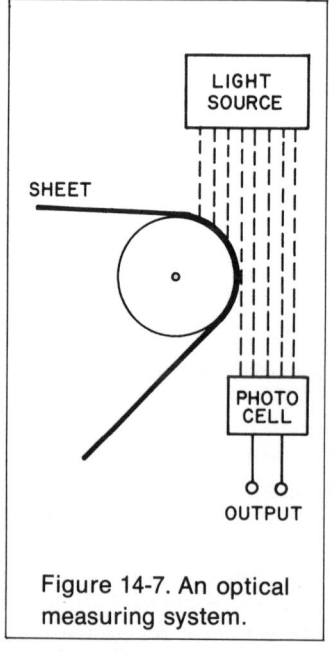

Figure 14-7. An optical measuring system.

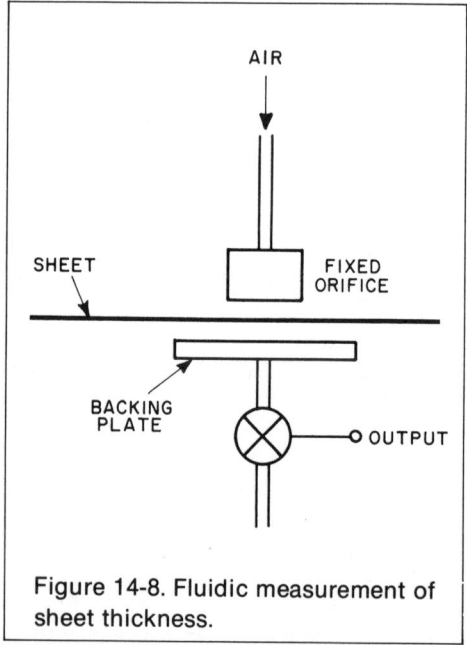

Figure 14-8. Fluidic measurement of sheet thickness.

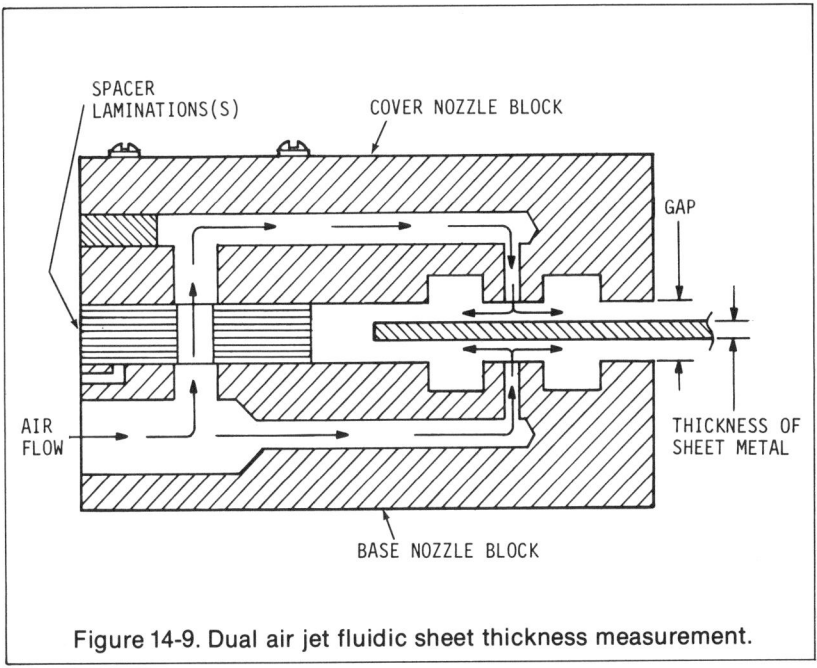

SPACER
LAMINATIONS(S) COVER NOZZLE BLOCK

GAP

AIR
FLOW

THICKNESS OF
SHEET METAL

BASE NOZZLE BLOCK

Figure 14-9. Dual air jet fluidic sheet thickness measurement.

the upper and lower nozzles from a single source. If the sheet material being measured becomes thicker, the flow through the nozzles is restricted and the air pressure in the sensing head increases. As the sheet material becomes thinner, the flow through the nozzles becomes less restricted and the air pressure in the head decreases. These pressure changes are proportional to the change in thickness of the material passing through the sensing head. The pressure changes are transmitted to appropriate display instruments and used as a continuous measurement of changing sheet thickness or caliper.

Beta-ray Radiation. Beta-ray gauges similar to those used for basis weight measurements have been used for making sheet thickness or caliper determinations under certain specified and fixed conditions. Because of this, the beta gauge is commonly misnamed a thickness gauge. A beta gauge is a mass gauge and for most practical purposes gives a direct measurement of the mass per unit area of the sheet. Therefore, it measures thickness or caliper only when the density of the sheet is constant. This is generally the case in measuring metal, rubber, or plastic sheets, but is frequently untrue for paper. On many grades of paper, basis weight and thickness are so closely related that the beta gauge can be used to measure caliper; however, it will not measure thickness directly.

Off-Machine Sheet Thickness Measurement

Cross-machine sheet caliper profile measurements are also made in an off-machine manner utilizing a unit that is similar to the basis weight profile measuring unit in appearance and in the method of operation. However, the internal mechanism is designed to make a continuous thickness measurement on a sample strip of paper taken from a reel of paper after manufacture and in the cross-machine direction. The design is usually based on a mechanical-electric or mechanical-pneumatic detection system.

A schematic of a typical mechanical-pneumatic off-machine caliper profiler is shown in Figure 14-10. The paper sample is fed into the paper chute and travels between the parallel faces of the anvil and sensing button, being drawn through the instrument by pull-through rolls. The anvil forms the button of a micrometer screw to which is attached the edge-reading micrometer dial calibrated in 0.001-inch (1-mil) graduations. The sensing button is attached to a support which is mounted on parallel flexure strips to permit motion perpendicular to the paper-contacting faces while maintaining anvil and sensing button always parallel. The sensing button support is forced into contact with

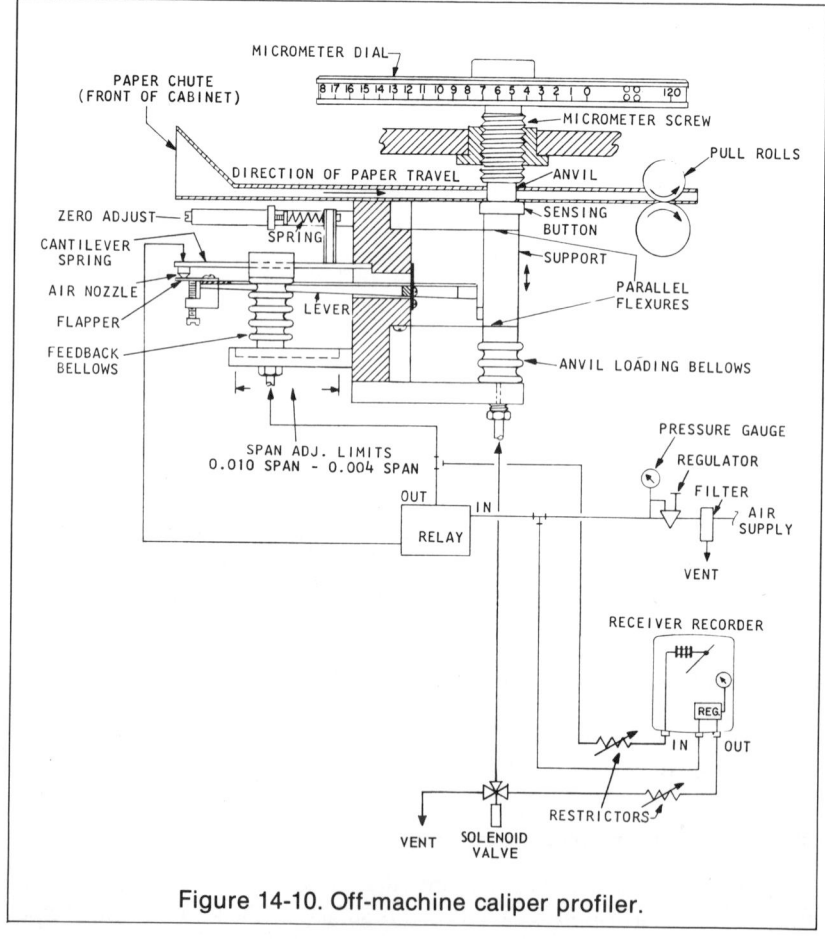

Figure 14-10. Off-machine caliper profiler.

the paper by air pressure applied to the anvil loading bellows and then it moves vertically with any change in thickness.

As the sample is drawn through the profiler detecting mechanism, the sensing button and support follow variations in caliper. Motion of the sensing button support is communicated by a small flexure strip to a lever which, in turn, pivots on crossed flexures. A flapper on the far end of the lever covers and uncovers the air nozzle positioned immediately above it. As the sample becomes thicker, the sensing button moves downward and the flapper moves upward covering the nozzle. The increased back pressure caused by the covering of the nozzle is fed to the air relay which increases the pressure signal output. The output pressure from the relay goes to a feedback bellows which exerts an increased force against the cantilever spring.

When the bellows flexes the cantilever spring, the nozzle is pushed back slightly to reestablish a throttling relationship between flapper and nozzle. If the motion of the flapper is great, the change in pressure will have to be great to force the cantilever spring back far enough to allow the nozzle again to bleed within the throttling range. Thus, the output pressure of the relay becomes a function of the motion of the flapper and, hence, of the sensing button.

The relationship between sample thickness and relay output pressure is adjusted for the desired span by moving the feedback bellows along the cantilever spring. When the sample has been engaged in the pull rolls and the anvil pressure reapplied, the micrometer dial is: (*A*) set to bring the recorded air output pressure to the zero center on the chart, or (*B*) set to the specification thickness of the sheet. On pressing the start button, the automatic thickness profile measurement is made across the sheet sample as it is being drawn through the unit.

BIBLIOGRAPHY

Daniel, A. B. "Continual Recording of Comparative Paper Caliper by Air Gauging." *Paper Trade Journal,* May 14, 1954.

Foxboro Caliper and Weight Profilers Companion Instruments. Foxboro, MA: The Foxboro Company, TI 9-51b, October 1959.

Hart, J. A., and Welsl, D. J. "Magnetic Caliper Gauges for Papers." *Pulp and Paper Magazine of Canada,* November 1951.

Inseerd, P. J. "Caliper Control," *Pulp and Paper,* December 28, 1964.

Kahoun, J. B.; Zurovitch, W.; and Newby, L. Jr. "Systems for On-Machine Measurement and Control of Paper Caliper." *Tappi,* vol. 48, no. 7 (July 1965).

The Foxboro Caliper Profiler for Calipering Paper or Board. Foxboro, MA: The Foxboro Company, TI 9-51a, February 1959.

"Thickness of Paper and Paperboard." *TAPPI Testing Procedure T411.*

Van Munn, P. H. "On-Line Caliper Measurement and Control." *Tappi,* vol. 53, no. 5 (May 1970).

Select the answer or answers which correctly apply to the following statements:

1. The average thickness of a pile of a known number of sheets of paper is generally referred to as its
   a)   sheet thickness.
   b)   caliper.
   c)   bulking thickness.

2. Contacting thickness measuring system types include
   a)   electrical.
   b)   mechanical.
   c)   pneumatic.

3. Noncontacting thickness measuring system types include
   a)   hydraulic.
   b)   pneumatic.
   c)   beta-ray radiation.

4. The off-machine caliper profiler can be used to measure
   a)   cross-machine thickness variations.
   b)   machine direction thickness variations.
   c)   cross-machine basis weight variations.

5. The difference between the off-machine and on-machine sheet thickness measurement is
   a)   the principle of operation.
   b)   a paper sample must be removed from the paper machine to make a measurement with the off-machine device.
   c)   the on-machine device makes the measurement continuously on-line as paper is being manufactured.

(See Appendix for answers)

15. Quality Measurement

Measurements discussed up to this point are primarily those made on variables found in the manufacturing process of pulp and paper, and they can generally be classified as process variables. They are characterized by the fact that they usually can be regulated to specified values or within desired limits by the manipulation of another variable in the process.

There is another class of measurements which are made on the products produced; for example, pulp from a pulp mill and paper from a paper mill. These measurements are made on variables that are representative of the final characteristics or physical properties that determine the product's quality which must be maintained within specified limits. These measurements can generally be classified as Quality Measurements. They are characterized by the fact that usually more than one interrelated process variable must be regulated and/or may require operator adjustment of process equipment to control variables within the desired parameters. Therefore, these measurements are usually prominently displayed so they can be used as a guide for operator manipulation of the process or they are fed into some type of digital computer which evaluates them. Based on the results of the evaluation, the computer automatically initiates a number of preprogrammed and interactive manipulations to maintain the quality measurement of the product within specified limits.

These quality measurements are interrelated and can be grouped on the basis of the property they represent, such as optical, physicochemical, strength, structural, and surface properties. They can be listed in the following categories:

A. Optical Properties

1. Brightness
2. Color
3. Gloss
4. Light transmission
5. Opacity
6. Reflectance

B. Physicochemical Properties

1. Liquid permeability
2. Moisture content
3. Sizing
4. Water vapor permeability

C. Strength Properties

1. Bending	7. Hardness
2. Burst	8. Softness
3. Tearing	9. Puncture
4. Tensile	10. Stiffness
5. Compressibility	11. Stretch
6. Folding endurance	

D. Structural Properties

1. Air permeability	3. Formation
2. Caliper	

E. Surface Properties

1. Erasability	3. Smoothness
2. Fuzz	4. Surface bonding strength

These are the typical quality measurements made on pulp and paper. Many other special measurements are made and depend on the type and grade of the product and its intended end use. Also, not all of the above measurements are made in all pulp and paper mills. The desired measurements are selected on the basis of the product and its final use.

Most of these measurements are made on samples taken from the process to laboratories and are normally referred to as *lab tests*. However, the laboratory technology of some of these tests has been recently extended into the field of on-line continuous monitoring for manual or automatic process control. There is an increasing demand on technology to supply the amount of information which can be used productively by on-line digital computers, and more and more attention can be expected to be focused on quality-type measurements for this purpose. A number of these lab tests have never emerged out of the experimentation trial stage while others have realized relatively more application success. Some of these are reviewed in this chapter.

BRIGHTNESS

Although the measurement of brightness on white pulp and paper is sometimes considered synonymous with color, strictly speaking it is not a colorimetric quality measurement. Brightness measurements are made on pulp during the bleaching process to control the amount of chlorine added during the chlorination stage and to measure the degree of whitening in order to control the addition of bleaching chemicals such as hypochlorite and chlorine dioxide in subsequent bleaching stages.

Dual Wavelength

One measurement system based on the dual wavelength principle of operation consists of an illuminating source, an optical sensor, a window to provide access to the pulp in the process, and an electronic analog receiver-computer to resolve the reflectance information and convert it to an electrical output signal for recording and control purposes.

The sensor, as applied in the control of chlorination, views the pulp slurry in the stock line shortly after the addition of chlorine. The basic elements and the principle of operation of the sensor are shown in Figure 15-1. The process material (pulp) to be measured is illuminated with white light by the illuminating lamp. A flexible skirt around the viewhood occludes the ambient light. The reflected light energy is accepted by a lens and the amount of energy seen by the optical device is controlled by a variable iris. The light energy passes into an integrating sphere, which has an internal coating of a high-reflectance white material. Its purpose is to reflect and distribute the light energy equally over its inner surface. This method averages or integrates the energy received from light or dark spots caused by surface texture or mosaic effects of the pulp slurry. The result is an average value of light energy on the walls of the integrating sphere. The sensors do not see the process directly, but "see" the averaged light energy on the opposite wall of the integrating sphere. A colored filter, red on one end and blue on the other, defines that part of the spectral energy that is allowed to pass to each tube. This light energy is converted into electrical currents representative of the red channel and the blue channel, which are connected to the main chassis of the instrument for processing. The instrument is connected in a manner which indicates the relative difference between the two resulting currents. Thus, the difference in light energy at the two wavelengths is displayed or recorded in values of brightness.

The sensor can also be adapted to look at and measure the brightness of the pulp cake or sheet on the washer drums when applied to bleaching stages later in the process.

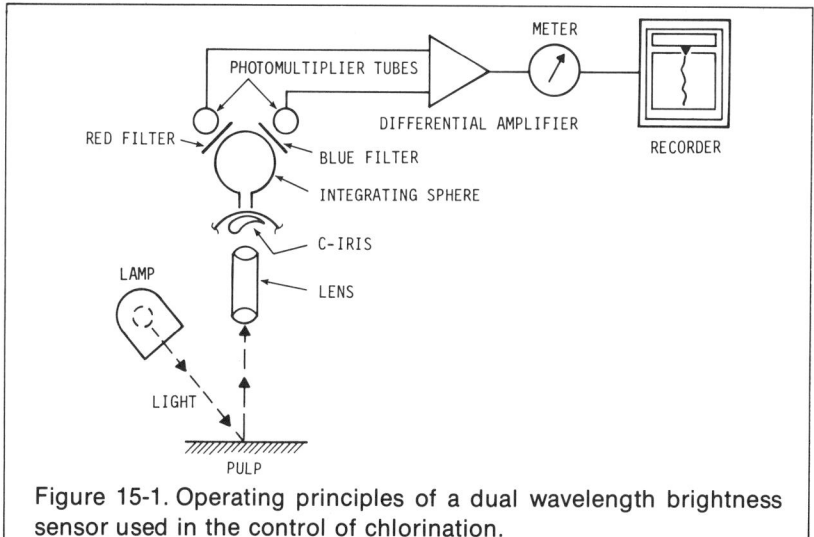

Figure 15-1. Operating principles of a dual wavelength brightness sensor used in the control of chlorination.

Single Wavelength

Another type of sensor that has been used to measure brightness of the pulp sheet formed on the washer drum during the bleaching process utilizes a single wavelength of light as the basis of its measurement.

Figure 15-2 shows schematically the operation of such a sensor. Light from flood lamps B_1 and B_2 illuminates the pulp sheet traveling under the sensor. A small portion of the light is diverted into a compensator where it automatically adjusts the current flow through a slidewire resistor. A fixed value filter permits light of only this wavelength to enter the phototube and depends upon the amount of light energy falling from the cathode structure as well as the wavelength of the light energy. Once the current flowing in the slidewire resistor is established, it will remain at that level until the aperture is readjusted or the characteristics of the light change. The current flow, therefore, through leg 1 is held constant. The current flow through leg 2 of the bridge varies with the amount of light reflected from the surface of the pulp. As the pulp brightness varies, the current in leg 2 varies causing a change in voltage drop across R_3. This causes an unbalance in the bridge voltage between A and B which are the inputs to a servoamplifier. The servoamplifier operates the servomotor in the direction needed to rebalance the bridge voltage. This is done by the servomotor adjusting the slidewire on the opposite leg of the bridge. The position of this slidewire represents pulp brightness and is used to transmit an electrical signal to an indicator, recorder, and control system.

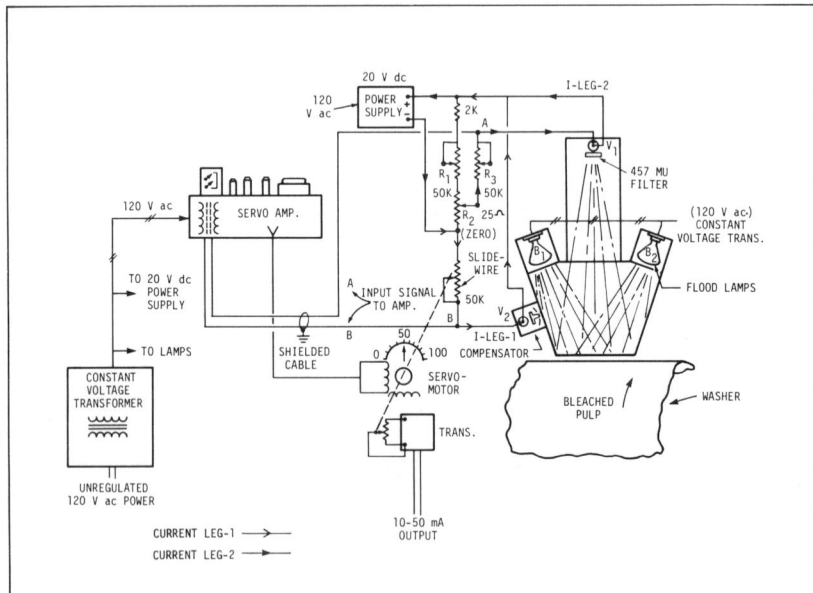

Figure 15-2. Operating principles of a single wavelength brightness sensor used to measure brightness of the pulp sheet during bleaching.

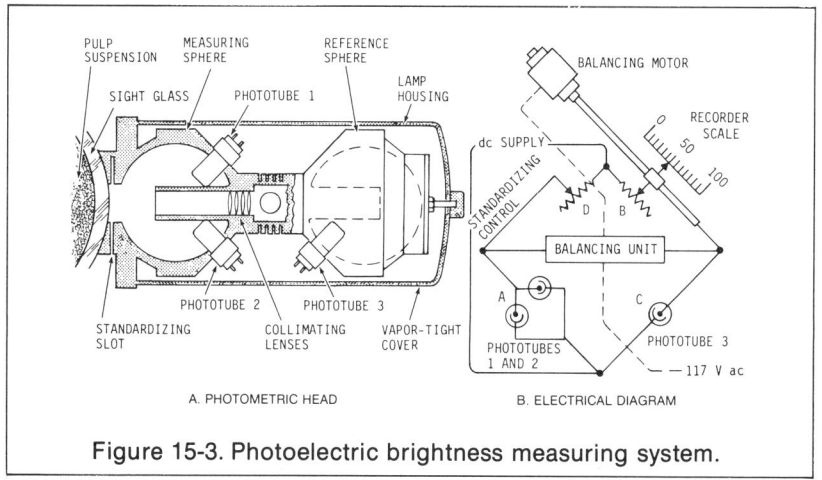

Figure 15-3. Photoelectric brightness measuring system.

Other Brightness Sensors

Another photoelectric device that has been used to measure reflected light or brightness of a pulp suspension in a pipeline is shown in a cross-sectional view in Figure 15-3A. The photometric head consists of a light source, dual light projector systems, and dual light-integrating spheres, all enclosed in a vapor-tight cover mounted on a sight glass installed in the pipeline.

The simplified electrical diagram of the measuring system, illustrated in Figure 15-3B, shows that it consists of a self-balancing Wheatstone bridge in which phototubes 1 and 2 are connected in parallel in arm *A,* and a linear slidewire included in the opposite arm (arm *B*). Phototube 3 is included in the adjacent arm *C* and a standardizing control resistor is in arm *D.* The recorder pointer is coupled to the slidewire slider so that its reading corresponds directly to the position of the slider.

The balancing unit operates the motor to keep the bridge in a state of balance; the motor acts through a suitable mechanism to change the position of the slider and pointer. A change in brightness of the pulp suspension produces a change in light to phototubes 1 and 2, proportionally affecting the resistance of arm *A.* With resistance of the other arms of the bridge constant, the resistance of arm *B* will vary in inverse proportion to that of arm *A* in order to keep the bridge balanced. This resistance will, therefore, be directly proportional to brightness which results in a corresponding movement of the pointer on the readout scale of the display instrument.

Other brightness sensors have been developed to measure brightness of pulp sheets by use of an electrically chopped fluorescent light source, an electronic two-channel spectrometer from which the ratio of two reflected color bands is determined and related to brightness.

Work has also been done in the field of fiber optics to transmit and receive the light in the design of a brightness measurement system for pulp during the bleaching process.

Quality Measurement   279

PAPER QUALITY MEASUREMENTS

The other quality-type laboratory measurements which have been developed for on-line applications are those primarily concerned with the paper sheet. Like the brightness sensors, they are, in most cases, optical devices.

Color

Color measurements of liquids and gases were described in Chapter 5 on analytical measurements.

In the paper mill, color measurement is primarily made in the control laboratory on paper samples taken at convenient times during the operation of the machine. Although there has always been a high level of interest in making this measurement continuously and on-line (and attempts to accomplish this were made at least as early as 1948), it is still a relatively new and rather limited application. Like many other on-machine optical measurements, color has been very slow to come into use because of the mill environment which is quite hostile to optical instruments. The environment makes accuracy, precision, and prolonged dependability difficult to achieve.

It has been difficult to decide exactly what to measure. One color measurement system optically compares light reflected from the sample point with light reflected from a standard. The resulting differential light is beamed through a blue, then a green, and then an amber filter. Sensing the light differential energy, an electronic phototube converts it to an alternating electric current which is amplified, demodulated, and can be used for indicating, recording, and/or control purposes. A separate reading is taken with each of the three filters. In combination, the three readings "register" a color in somewhat the same way one's own eyes and nervous system perceive color. Behind this measuring system is the theory that any color the eyes can see can be described in terms of some combination of three hypothetical primary colors. The proportions in which these primaries would need to be mixed to match a given color are called X, Y, and Z—or tristimulus values. To measure the three values directly is to measure color itself. This is only one of several techniques under investigation and development. It is expected that other methods and concepts will evolve as the technology of color measurement advances.

Gloss

The terms *gloss, glare, finish,* and *smoothness* are sometimes used almost interchangeably. *Gloss,* however, refers particularly to the polished surface of paper as produced by calendering or coating, and also to the property a surface has of reflecting light specularly, that is, similarly to a mirror. A typical system for making this measurement is shown in Figure 15-4. This scheme uses three beams of light and gives as its output the gloss value difference between the paper being produced and a master or standard selected as the desired gloss. The output from one incandescent lamp is broken up into three identical beams. Beam 1 falls on the paper being produced, while beam 2 falls on the standard. Beam 3 is transmitted directly to a light sensor. The specularly

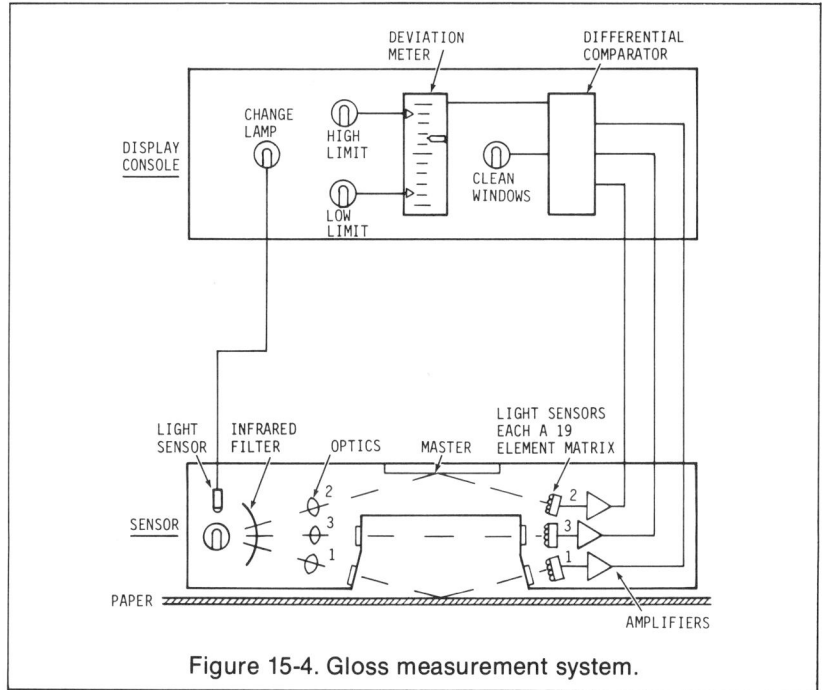

Figure 15-4. Gloss measurement system.

reflected light is photosensed by separate photocell matrices, and the outputs are amplified and fed to a differential comparator which takes the differences and outputs this value. This differential output is read out on a deviation type readout instrument which is scaled with a zero value at the center.

This concept displays the measurement not in terms of absolute gloss units but rather in terms of the difference between desired gloss value and the actual gloss of the paper being produced. A measurement system that displays the measurement in absolute gloss units is schematically shown in Figure 15-5. The sensor consists of a light source (E), a test photocell (TP), a comparison

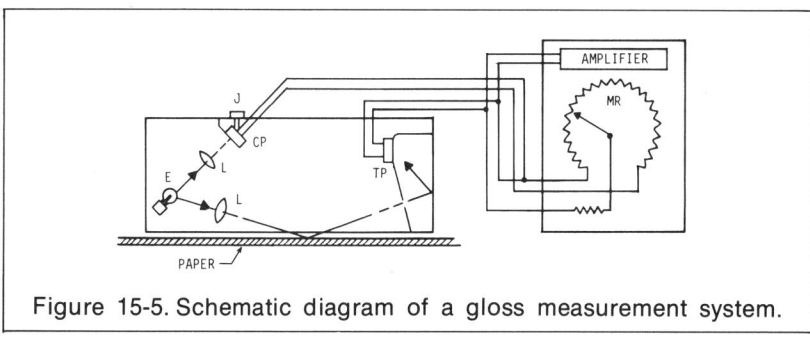

Figure 15-5. Schematic diagram of a gloss measurement system.

photocell (*CP*), a lens system (*L*), and mirrors which direct light from the source to the surface of the paper, all enclosed in an exposure head. The photocells are oriented in the exposure head in such a manner that they receive light from the lamp. The comparison photocell (*CP*) is brightly and constantly lighted through a heat-absorbing lens (*L*). The current generated in this photocell is sent to the display instrument where it passes through the measurement rheostat (*MR*). Between the end and sliding contact of this rheostat is a variable potential. This variable potential is combined with a high resistance to supply a relatively small current that may be varied to balance that from the test photocell. Because the resistance facing the variable potential is high, the balancing current is small relative to the total current and is proportional to the rheostat setting. The amplifier responds to any lack of balance between current from the test photocell and the balancing current. It drives a reversible motor which positions the rheostat contact with the pen and pointer linked to it so that it moves as the current needed to balance that from the test photocell varies. Potential across the test photocell is zero at balance and the test current is a function of the amount of light reaching the test photocell. Therefore, gloss is proportional to the rheostat setting and the pen and/or pointer is scaled in these values on the readout display instrument.

Opacity

Opacity is the property of a sheet which obstructs the passage of light and prevents one from seeing objects through the sheet on or in contact with the opposite side. It is usually measured in the laboratory by the contrast-ratio method in which a sample is uniformly and diffusely illuminated with white light. A black cavity is placed behind a portion of the sample. A white reflecting surface (usually magnesium carbonate) is placed behind another portion of the sample near that covered by the black cavity. The sample is then viewed in a direction perpendicular to the surface of the sample. If the sample is not completely opaque, that portion which lies over the black cavity will appear darker than that which lies over the white surface. The brightness ratio of the portion over the black cavity compared to that of the portion over the white surface is the contrast ratio and a measure of the opacity of the paper sample.

On-Line Contrast-Ratio. A schematic diagram of an on-line opacity measuring system using the control-ratio principle is shown in Figure 15-6. It measures contrast ratio on a sequential basis. As the web moves along, measurements are made alternately on black and white backed paper and the average value read out on an indicator, recorder, or other display and/or control system. The light from a tungsten-iodine bulb passes through two apertures and falls on the paper web at two separate places, shielded from each other. A mirror (white body equivalent) and an empty hole (black body equivalent) are located below the paper. The light reflected from the two illuminated areas on the paper is collected separately by two bundles of fiber optics. The two bundles of fiber optics carrying the light from the two illuminated areas are placed 180° apart on a 180° shutter. The output of one is cut off while the other is on. The two beams passing through the shutter are then square-wave mod-

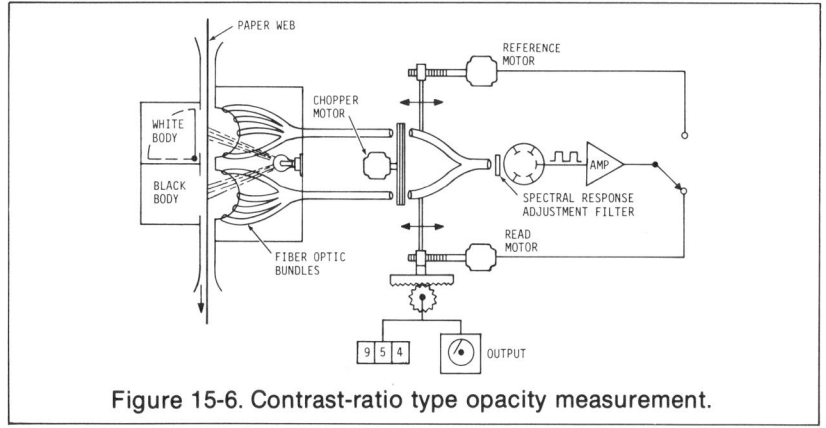

Figure 15-6. Contrast-ratio type opacity measurement.

ulated. The ratio of the two light signals is determined by the use of an electro-optical-mechanical system which attenuates the more intense of the two signals until it equals the weaker. The two fiber optic bundles bringing the light signals from the paper are passed through the shutter to two compartments of the metal box. In the compartment are lead screws which drive carriages that move a second pair of receptor fiber optic bundles. These two fiber optic receptor bundles receive the two chopped optical beams. Only one of these is moved in normal operation. The two bundles merge and illuminate a photomultiplier. The output of the photomultiplier is an ac wave signal, which is the error in the optical intensity balance. This signal, after amplification, drives a servomotor. The motor moves a lead screw and changes the distance between the fiber optic receptor and the fiber optic bundle, bringing the more intense signal (white body) from the light pickup via the chopper. When the two intensities falling on the receiving fiber optics are equal the carriage stops moving. At this point the position of the carriage is measured. Therefore, the position of the carriage is a function of the opacity which is transmitted to suitable output display instrumentation.

On-Line Light Transmission. Other on-line opacity measuring sensors utilize the light transmission technique as a basis of operation. A typical sensor, shown in Figure 15-7, consists of a tungsten lamp and power supply which holds the color temperature steady with minimum intensity variations, a mechanical light chopper to produce a modulated signal at the detector independent of environmental light level, a detector and correcting filter which closely match the spectral response of the human eye, a preamplifier which prepares the signal for transmission to the interface and display instrumentation, and internal standards used during automatic standardization.

Essentially, opacity is measured by transmission of a broad band of visible light approximating the spectral response of the human eye. Standardization is accomplished automatically by the insertion of a neutral density filter. The sensor is usually mounted in a head that scans the width of the paper sheet just before it leaves the paper machine.

Quality Measurement   283

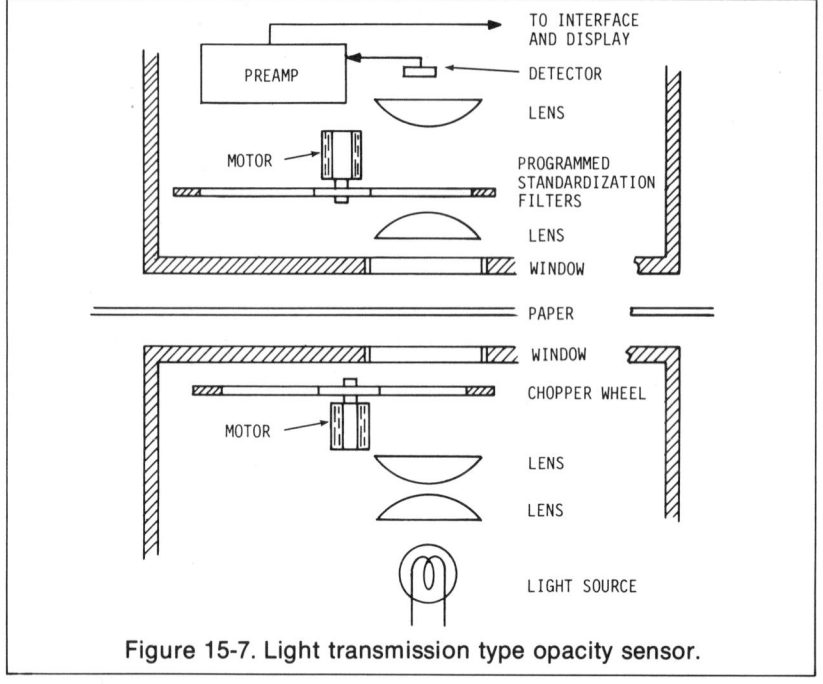

TO INTERFACE
AND DISPLAY

PREAMP

DETECTOR

LENS

MOTOR

PROGRAMMED
STANDARDIZATION
FILTERS

LENS

WINDOW

PAPER

WINDOW

CHOPPER WHEEL

MOTOR

LENS

LENS

LIGHT SOURCE

Figure 15-7. Light transmission type opacity sensor.

Ash

The measurement of ash is used as an indication of the filler content of paper. It is made in the laboratory by igniting a sample in such a way that the combustible and volatile components are removed and the percentage of resulting inorganic residue determined.

The measurement of ash has been made on-line on the paper machine by the use of a principle based on the absorption of radiation passing through the paper web. Using an appropriate radioactive source, it is possible to determine the amount of radiation absorbed by the elements of higher atomic number. The absorption coefficient is proportional to the third power of the atomic number, and since the ash content has a higher atomic number than the pulp fiber the absorption is primarily determined by the ash content. One of the radioactive sources that has been used is the gamma emitter Fe-55.

The operation of a typical ash measurement system is depicted schematically in Figure 15-8. Radioactive radiation rays emitted from the source are absorbed by the fillers present in the paper sheet as it passes between the source and the detecting receiver. The radiation not absorbed is detected by the radiation receiver, amplified and transformed by a frequency-to-voltage transducer into a dc analog signal.

A calibrating unit which contains scaling circuits and a manual adjustment for basis weight converts the signal into a direct reading digital or continuous analog display calibrated in percent ash content.

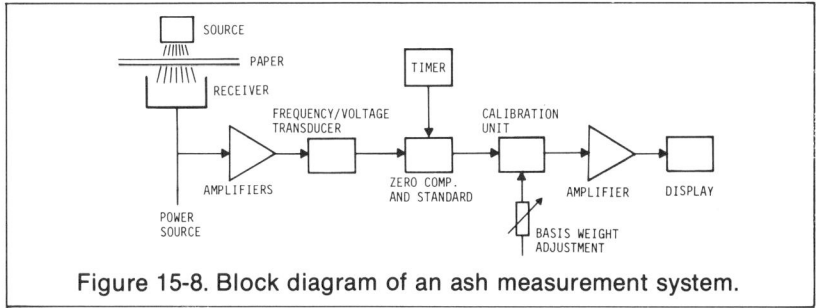

Figure 15-8. Block diagram of an ash measurement system.

Modulus

Although there has been a growing interest in measuring strengths of paper on-line, and although much work has been done in this area, it is still considered to be in the early stages of development. The modulus of elasticity of a paper sheet is related to both strength and bending stiffness. However, with proper correlation it can be used as a measurement of strength only. There are two factors involved in measuring the modulus of elasticity: sound velocity and sheet density. A proposed method of the elastic moduli of paper involves the use of ultrasonics. One such design using this method utilizes wheel type transducers which measure ultrasonic velocity in the sheet material by contacting the sheet. The wheels consist of a ring with a slot cut out to accommodate a piezoelectric element which generates and receives the sonic waves. A pair of identical wheel transducers are used to measure longitudinal sonic wave beam velocity and another pair are used for shear wave beam velocity. These signals are then converted into a measurement of modulus of elasticity which is translated into strength by an instrumentation system shown in Figure 15-9.

Another proposed technique for measuring modulus of elasticity by the use of ultrasonic energy does not require direct contact with the paper sheet. This idea is based on the use of electrostatic transducers radiating ultrasonic energy into air to irradiate the paper. The vibration developed in the paper sheet radiates sound into the air on the other side which is used as a measurement of the velocity of sound in paper. This, in turn, is related to modulus of elasticity and converted to a strength reading.

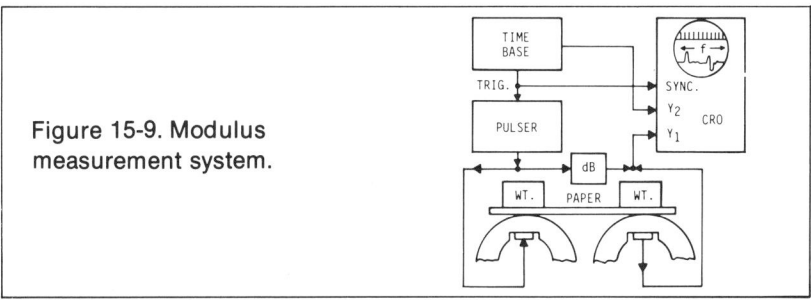

Figure 15-9. Modulus measurement system.

Hole Detection

Although hole detection can technically be considered more as a measurement of paper defect than of paper quality, its immediate detection and identification when they occur can be read out and displayed to aid the machine operator to make machine corrections to eliminate the cause.

There are a number of techniques which have been developed and used to accomplish this. One method uses ultraviolet light to automatically detect holes in paper. The detecting system, diagrammatically shown in Figure 15-10, consists of two rows of ultraviolet lamps encased in a housing, which illuminate the sheet of paper through a slit extending across the width of the sheet. When a void in the sheet passes over the ultraviolet lamps, light reaches one of the photoelectric sensing heads located approximately two feet from the paper on the reverse side. This activates an alarm on the readout station. Signal lights can also be lit depending on where the void is detected. These signal lights are used by the machine operator to locate the hole in the sheet and take corrective action to eliminate the cause. The hole count can be recorded and totalized by instrumentation on the readout station. You will note that the ultraviolet technique does not require any contact with the paper sheet.

Another method used to measure, locate, and record holes requires contact with the sheet and utilizes flexible fine metallic feeler brushes as sensing elements. The brushes are mounted in such a way that they sweep over the surface of the sheet of paper which is riding over a grounded roll, plate, or bar. Whenever a hole in the sheet passes between the feeler brush and the grounded element under the sheet, a low-voltage pulse is transmitted to an indicating readout device for operator use. The operator can then take appropriate corrective action to eliminate the cause of the hole.

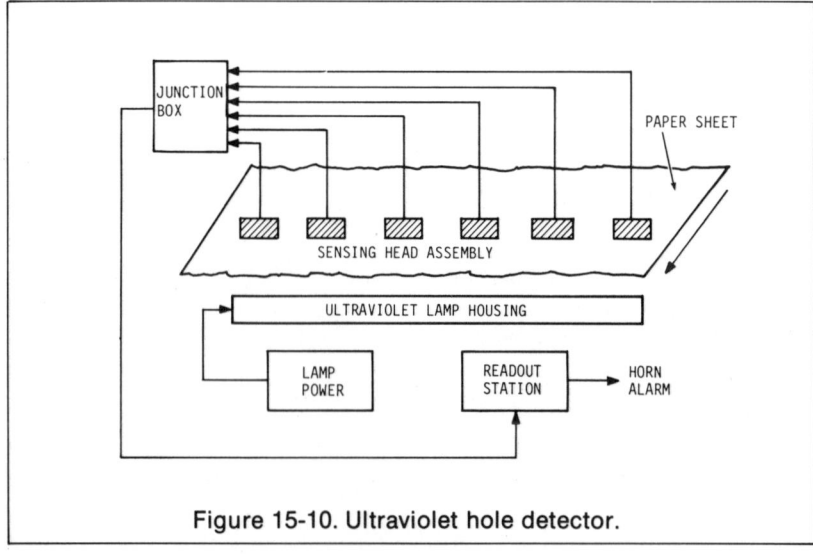

Figure 15-10. Ultraviolet hole detector.

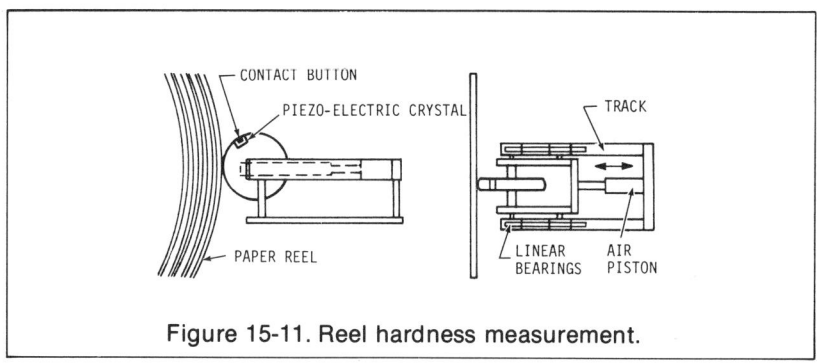

Figure 15-11. Reel hardness measurement.

Reel Hardness

In the production of a reel of paper on the paper machine, it is important to keep the hardness uniform across the reel width. Traditionally, hardness has been measured by the operator using his hand and/or a "billy stick"; from this type of observation he made adjustments to the cooling air showers arrayed across the calender stack. This operation compensated for nonuniformities in the hardness profile.

A sensor has been developed to make this measurement continuously on-line. The simplified drawing in Figure 15-11 shows that this device consists of a disc mounted on a shaft which is air-loaded to contact the reel and is fitted with a contact button on the surface of the disc, flush with the disc surface and mechanically linked to a piezoelectric crystal. The sensor registers against the rotating paper reel. Each time the contact button passes through the region of contact between the reel and the disc, the pressure profile which exists in the contact region is measured.

The measurement is essentially a solid-state measurement in that the piezoelectric crystal has a modulus of elasticity greater than the disc in which it is mounted. There is no relative movement between the crystal and the disc. Other devices which have a method of operation based on impact have been developed and used to make reel hardness measurements.

Off-Line

On-going investigations and experimentations are being conducted on many of the laboratory measurements which are being made off-line in an effort to adapt them into continuous on-line methods of operation.

Formation. Formation is one of the properties which is determined by the degree of distribution uniformity of the solid components of the sheet with special reference to the fibers. This property is very important not only because of its influence on the appearance of the sheet but because it influences the values and the uniformity of the values of nearly all other properties. Therefore, in the manufacture of certain grades of paper, it is one of the more important measurements which would be advantageous to perform on-line.

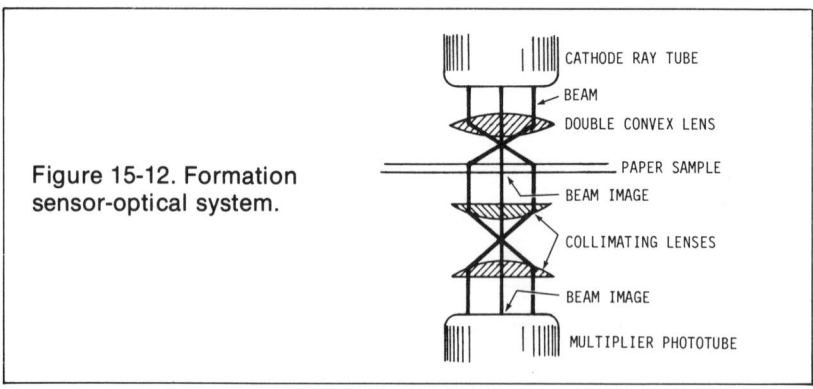

Figure 15-12. Formation sensor-optical system.

To date, this measurement has primarily been made off-line by the use of an optical sensor. This sensor consists of a system made up of a cathode-ray tube, a multiplexer phototube, and a set of appropriate lenses arranged as shown in Figure 15-12. As the sensor sweeps a paper sample in a prescribed pattern, a spot of light is generated on the face of a cathode-ray tube. The electrical output of this phototube is fed into an electrical measuring system which compensates for the opacity of the paper sample, converts the output, and displays it in values representing relative formation in the sheet.

Bursting Strength. This measurement, which is the ability of a sheet to resist rupture when pressure is applied to one of its sides by a specified instrument under specified conditions, is another one that is done off-line. It is largely determined by the tensile strength and extensibility of the paper. Although the on-line measurement of this property has never shown any feasibility of being practical, systems have been built that make a series of measurements continuously and automatically on samples of paper. Figure 15-13 illustrates one such arrangement. In operation, when the operator pushes the "start" button the motor-driven paper-pull rolls run for an interval as set on a timer. When the

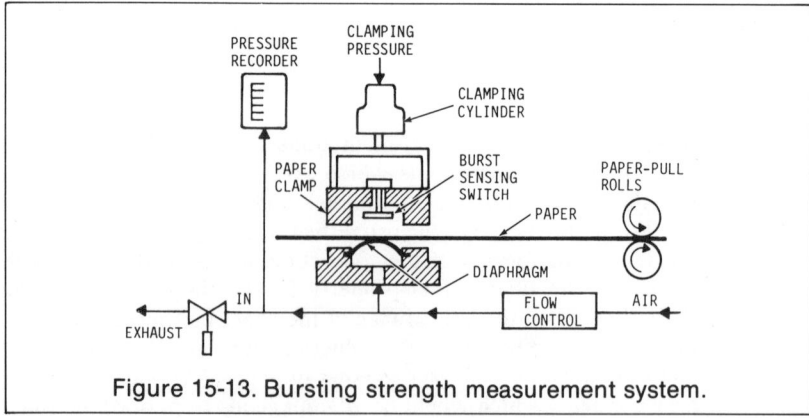

Figure 15-13. Bursting strength measurement system.

timer runs out, its contacts start a pulse-generating motor and open a solenoid which applies air to a piston. The piston, in turn, forces the clamping ring down on the sample in the test position. Air, flowing at a controlled rate, is fed continuously to a pressure chamber under a rubber diaphragm and is vented to atmosphere through a solenoid valve. After a suitable, brief interval as determined by a second timer, the solenoid valve closes to seal off the pressure chamber permitting test pressure to build up behind the diaphragm. Simultaneously, contacts connect the output from the pulse-generating motor to a totalizing counter. The flow rate to the pressure chamber is controlled to produce a uniform rate of pressure rise in the pressure chamber. The rate of pressure rise, in psi, is synchronized with the pulse rate produced by the pulsing motor and registered on the totalizer so that one count is registered for each psi rise in pressure. When the pressure has risen to the point where the paper sample breaks, the diaphragm actuates a burst detector switch in the clamping assembly. Actuation of the burst detector switch allows for a suitable interval for deflation of the diaphragm and then resets the primary timer to start the cycle again.

General

This is by no means a complete coverage of quality measurement instruments that have been investigated or should be studied. The selection of properties that determine quality and the development of instruments for its measurement have been influenced by the interest expressed in industry, the state of the technology, and the extent of available information on the property. Due to these factors, it has not always been expedient to study properties and instruments in order of their apparent importance to the industry. Therefore, it should not be judged that the neglected properties are the ones least desired to be measured by instrumentation. On the contrary, some of the unmentioned measurements are highly important but have been delayed by presenting very stubborn and involved problems in their development.

BIBLIOGRAPHY

Brewster, D. B., and Robinson, W. I. "Paper-making Sensors: The Key to Successful Computer Control." *Pulp and Paper*, March 1974.

Casey, J. M. "On-Line Opacity Measurement and Control." *Pulp and Paper Magazine of Canada*, vol. 75, no. 8 (August 1974).

Cherewich, H., and Walker, O. J. "Automatic Reel Building." *American Paper Industry*, vol. 56, no. 3 (March 1974).

Cook, A. J. "On-Line Appearance Measurement—A Status Report." *Tappi*, vol. 56, no. 2 (February 1973).

Jamison, W. "New Approach to Paper Formation." *Paper Trade Journal*, vol. 140, no. 41 (October 8, 1956).

Karjalainen, P., and Matti, O. "Pulp Brightness Controlled With New Coram Analyzer." *Pulp and Paper International*, vol. 14, no. 11 (November 1972).

Obenshain, D. N. "Black Widow Continually Measures Brightness of Pulp Flowing in a Pipe." *Paper Trade Journal*, vol. 142, no. 4 (February 10, 1973).

Papadakis, E. P. "Ultrasonic Methods for Modulus Measurement in Paper." *Tappi*, vol. 56, no. 2 (February 1973).

Schaffzin, R. A. "Continuous Gloss Measurement Employs Comparison Technique." *Pulp and Paper International*, vol. 12, no. 11 (November 1970).

Strom, J. "Improved Control by Color Measurement." *Tappi*, vol. 56, no. 7 (November 1973).

STUDY QUESTIONS

Indicate whether the following statements are True or False by inserting T or F in parentheses:

1. Quality variables can generally be regulated by the manipulation of one other process variable. (   )

2. Properties which are represented under Quality Measurements can be classified as optical, physicochemical, strength, structural, and surface. (   )

3. Brightness is strictly a colormetric quality measurement. (   )

4. Brightness measurements can be made only by the use of dual wavelengths of light. (   )

5. Gloss refers to the property of paper related to the specular reflection of light from its surface. (   )

6. Opacity is the property of the sheets which allows light to penetrate in order to permit visibility of objects in contact with the opposite side. (   )

7. Ash is a measurement which is related to the amount of filler in a sheet of paper and is best made by the contrast-ratio method of light comparison. (   )

8. Although the modulus of elasticity of paper can be used as a measurement of strength, it is also related to its bending stiffness. (   )

9. Modulus of elasticity of paper can be determined by measuring ultrasonic velocities in the sheet. (   )

10. Hole detection in a paper sheet can be accomplished by the use of optical and electrical methods. (   )

11. Optical, electrical, and mechanical techniques have been used successfully to make on-line reel hardness measurements. (   )

12. Formation is a measurement made to determine the distribution of solid components in a sheet of paper. (   )

13. The on-line measurement of formation can be done by the use of optical and electromechanical methods. (   )

14. Bursting strength is a measurement of the ability of a paper sheet to resist rupture. (   )

15. The measurement of bursting strength is best made on-line by the use of optical and mechanical means. (   )

(See Appendix for answers.)

16. Signal Transmission

The instruments described in Chapters 1 through 15 measure process and product quality variables. These measurements can be visually monitored on indicating scales or recorded on charts by recording devices. They provide the operator with information he can use to control the process by making necessary adjustments himself.

The signals from these measuring instruments are also used in instrument systems which perform adjustments automatically in accordance with control objectives or set points established by the operator. The typical automatic control system (Figure 16-1) employs a primary measuring means (A) which detects changes in the process variable; a transmission device (B) which sends the signal to a controller (D) which compares the measurement signal with the desired value; and a final control device (E) which acts on the control signal to keep the process variable on target. A recording device (C) might also be used in the loop.

When instrumentation was first introduced in the pulp and paper industry, all sensing devices were locally mounted in closely coupled loops. As control systems became more sophisticated, indicators, recorders, and controllers were consolidated on control panels and in control rooms. They were connected by pneumatic lines to the primary measuring devices on the process. As the centralization trend continued, the demand and utilization of auxiliary means of signal transmission over longer distances increased. This transmission is accomplished by coupling the measuring device directly to a transmitting mechanism (transmitter or transducer) which produces a pneumatic or electrical signal proportional to the variable being measured. Electrical transmission signals can be generated as continuous milliamp or millivolt, or discontinuous impulse-duration signals. Today, most measurements in pulp and paper mills are transmitted with continuous pneumatic or electronic signals to remote-mounted indicators, recorders, and controllers fitted with proper mechanisms to accept these signals; these mechanisms are called *receivers*.

TRANSMISSION SYSTEMS

Few primary measuring devices are capable of producing a signal strong enough to be transmitted unamplified over long distances. Regardless of the primary device, the transmission medium (pneumatic or electronic) or the

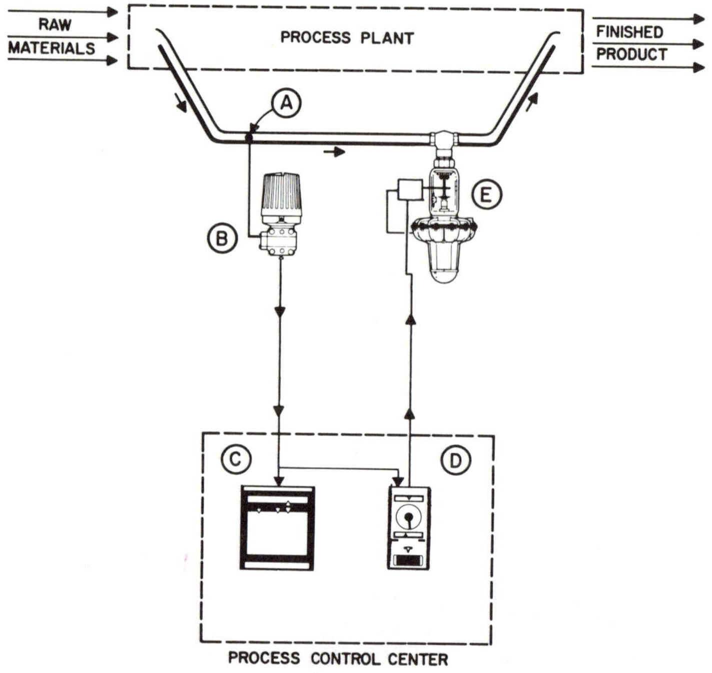

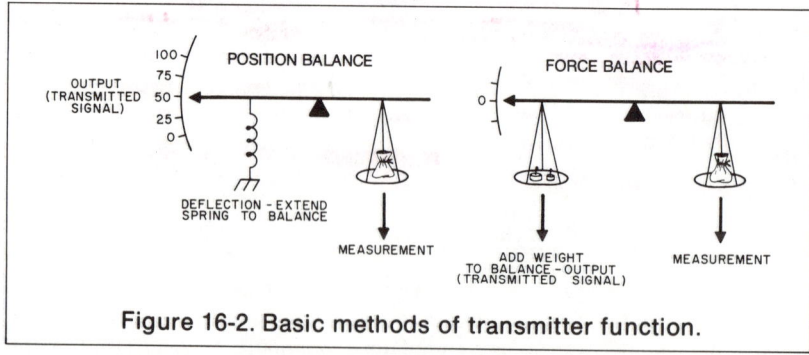

Figure 16-1. A typical control loop.

medium being measured, two basic methods are used to produce a signal strong enough to be transmitted. One type is referred to as *position balance* (often called motion balance in the past) and the other as *force balance*. The difference in the principles involved is illustrated by the simplified sketch in Figure 16-2. In the force-balance type, there essentially is no measure of position and the mechanism need have no net motion. Therefore, it provides

Figure 16-2. Basic methods of transmitter function.

no direct, inherent mechanical indication of measurement. The position-balance type is calibrated against a spring so there is a direct, inherent indication of the measurement.

Position Balance

Through the position-balance type transmitter, the primary measuring element for variables such as flow, pressure, level, and temperature causes motion of a lever-linkage system calibrated against a spring resulting in a position proportional to the variable. A pointer and pen can be incorporated in the lever-linkage system for local indication or recording the position of the lever in the proper measurement units. A pneumatic or electronic system can be used to detect this position and to generate a proportional pneumatic or electronic signal which can be transmitted to remotely located receiver type indicators, recorders, or controllers.

Position-Balance Pneumatic Transmission. To measure and pneumatically transmit pressure by position balance, the simple Bourdon tube primary element is subjected to stress by the gauge pressure. The tube deflects and moves a lever-linkage system (Figure 16-3) which is connected to an indicating pointer and to the flapper of a pneumatic detecting system. Increasing measurement causes the flapper to approach the nozzle. This action increases the back pressure on the pilot side of a relay which, in turn, increases the output of the relay. This output is fed to the feedback bellows and spring assembly and is also transmitted to the receiver instrument. This effect causes the nozzle to back away from the flapper until the pressure relationship balances the new measurement position in accordance with the spring calibration. This transmitted pressure is, therefore, proportional to the position of the element and lever system.

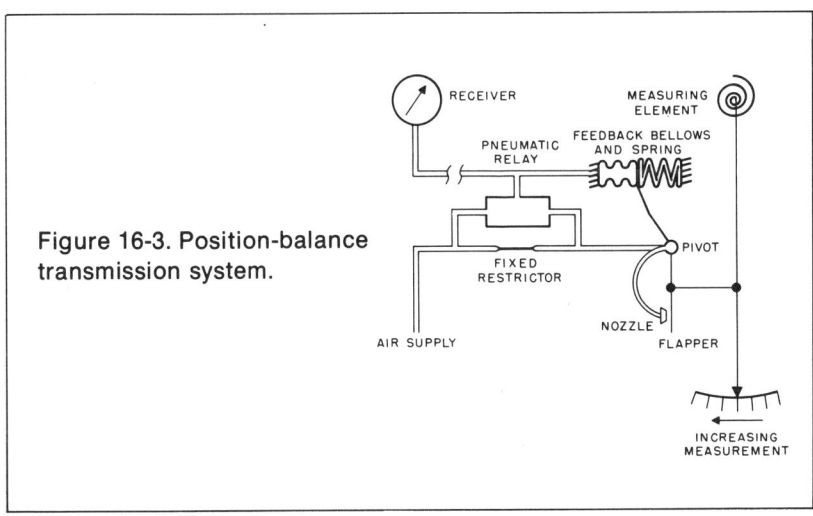

Figure 16-3. Position-balance transmission system.

Relays. Pneumatic relays used in the detecting system are of two basic types—bleed and nonbleed. The principal difference between the two lies in the internal valve design. An example relay would be the bleed type shown in Figure 16-4. It includes the fixed restrictor or reducing tube as part of the mechanism. Back pressure on the nozzle causes pressure to be exerted on the diaphragm. This pressure partially unseats the ball valve, permitting air to pass through the output port from the supply port. Likewise, a decrease in mesurement and a decrease in nozzle back pressure causes the ball valve to partially seat, an action that vents the excess air pressure to atmosphere. Air pressure is thereby continuously throttled between output and vent.

Pneumatic Transmission Signal. The very early pneumatic instruments operated on a vacuum which meant one or two psia to atmospheric pressure (14.7 psia at sea level). When gauge pressure came into use, approximately the same pressure span was used and one of the first common ranges was 2-14 psig. The 2-psi value on the low end of the scale was adopted to provide an elevated instrument zero for an active or live measurement at this point. Although other ranges have been used and are still being used today, a 3-15 psig signal, proportional to a zero to 100 percent measurement scale, was generally adopted as a standard. This relationship is shown in Figure 16-5. A typical 3-15 psig position-balance type pneumatic transmitter with the front cover removed to show components is shown in Figure 16-6.

Position-Balance Electrical Transmission. In order to obtain an electrical transmission signal, an electronic system is used to detect the position of the primary element. If the primary element is the Bourdon tube used to measure a pressure variable, the stress causes the tube to deflect, moving the lever-linkage system (Figure 16-7) which is connected to an indicating pointer and to the core of a variable core transformer which is part of an electronic circuit. Changes in position of the transformer core change the current flow in the two sides of the transformer. This change in output is proportional to the change in the process variable being measured. The indicating, recording, or controlling

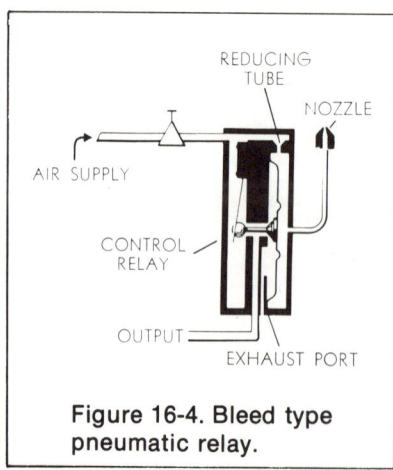

Figure 16-4. Bleed type pneumatic relay.

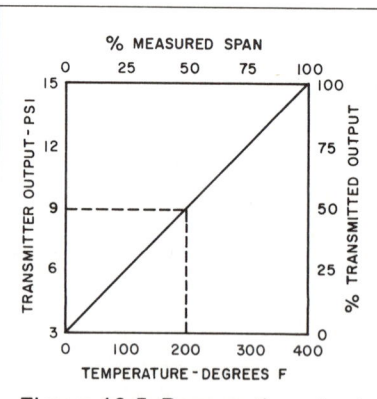

Figure 16-5. Pneumatic output calibration for temperature.

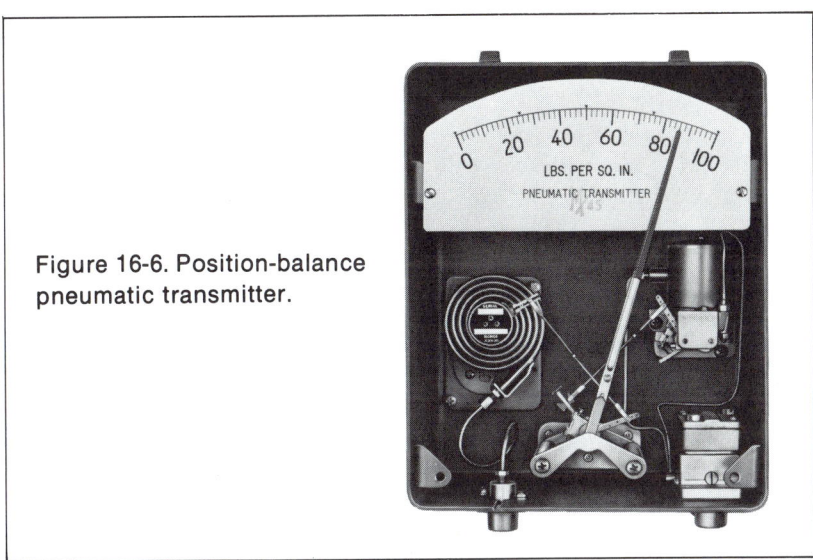

Figure 16-6. Position-balance
pneumatic transmitter.

receiver instrument is actuated by a meter movement operated from the transmitted electrical signals.

Electrical Transmission Signals. Although the signal most commonly used for transmission is current because it is unaffected by power line characteristics and immune to induced voltage, there has been no universally accepted standard as in pneumatic transmission. Basically, there are two categories of electronic signals and a number of accepted ranges within those categories. The two categories are the so-called voltage and current systems. The significant difference between the two is that a *voltage* system is one in which voltage at the output terminals is proportional to measurement. It delivers relatively

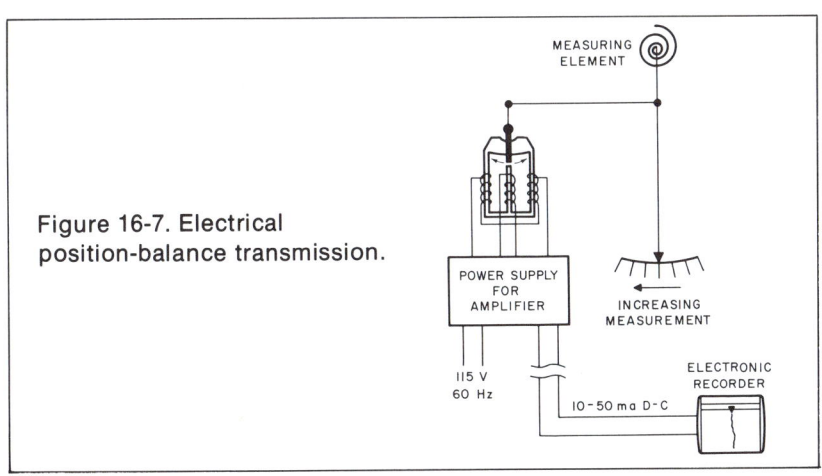

Figure 16-7. Electrical
position-balance transmission.

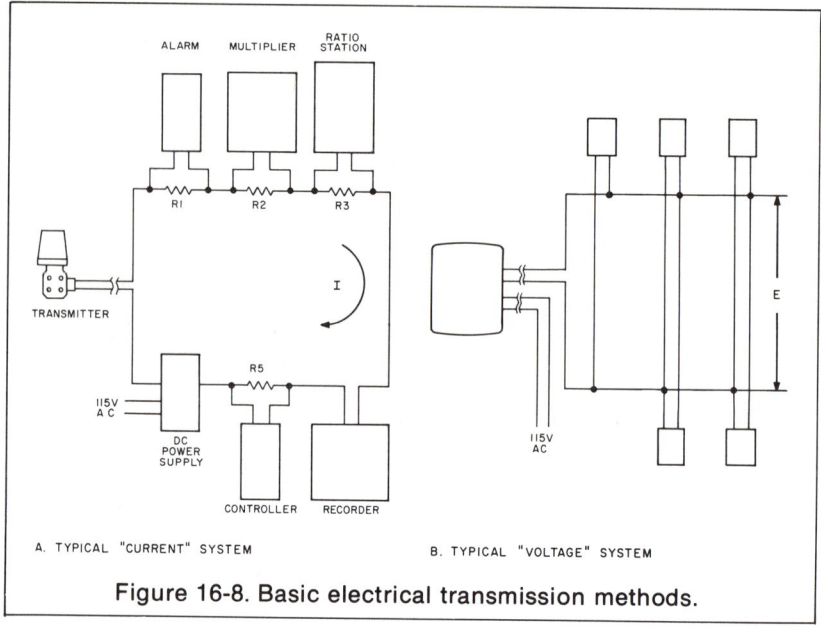

Figure 16-8. Basic electrical transmission methods.

low power and feeds high-impedance receivers in parallel. A *current* system, on the other hand, is one in which current in the loop is proportional to measurement and which has relatively high power output with regulated current delivered to low-impedance receivers in series. Figure 16-8 illustrates these fundamentals schematically. Typical examples of the two categories are shown; there can be variations of these basic systems. Table 16-A lists some typical values of signal, span, impedance, and power developed. Note that while standardization *per se* is not a reality, the industry has generally agreed

TABLE 16-A

TYPICAL VALUES FOUND IN ELECTRONIC TRANSMISSION

VOLTAGE SYSTEMS				CURRENT SYSTEMS			
Span, Volts	Signal, Volts	Resistance, Ohms	Max. Watts	Span, Milliamps	Signal, Milliamps	Resistance, Ohms	Max. Watts
4	1-5	100K	.0002	4	1-5	2500	.06
8	2-10	200K	.0005	16	4-20	800	.3
20	5-25	500K	.0012	40	10-50	600	1.5
20	−10 to +10	500K	.0002				

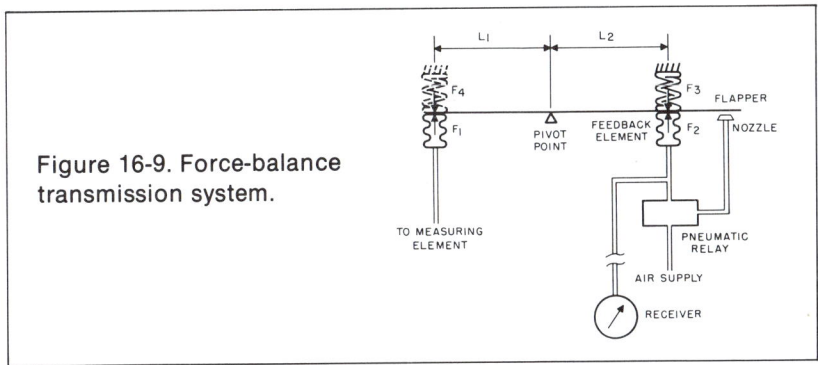

Figure 16-9. Force-balance
transmission system.

on a standardized ratio of 1:5 minimum to maximum signal values, which is the
same as with the pneumatic transmission systems.

Force Balance

With the force-balance type transmitter, energy is added to rebalance that
produced by the measurement of the process variable. Therefore, a change in
this rebalancing energy becomes a measure of the change in measurement and
can be used as a basis for producing a pneumatic or electrical transmission
signal. Essentially, there is no motion in the force-balance mechanism.
Therefore, no inherent mechanical indication through the lever-linkage system is practical. Readout, whether local or remote, must be performed by a
separate receiver mechanism.

Force-Balance Pneumatic Transmission. Figure 16-9 shows a schematic
diagram of a typical pneumatic force-balance transmitter. The energy or force
of the measurement is exerted against one end of a lever pivoted at the center.
Opposite the pivot on the same side of the lever is a feedback element
actuated by the pneumatic relay output to provide the rebalancing energy or
force. The output of the relay is controlled by a flapper-nozzle relationship, the
lever acting as the flapper. In operation, an increase in measurement pressure
F_1 causes the flapper to approach the nozzle. This increases the relay output
pressure and increases the feedback force F_2 until the forces again balance.
The bias spring ($F_3 = 3$ psi) provides the initial output of 3 psig at zero
measurement. Moving the pivot point changes the span of the device. To
decrease the span, L_1 is increased. To suppress a portion of the span, as is
common in transmitting level in elevated tanks, a suppression spring is added
in opposition to F_1 as represented by broken lines in Figure 16-9. The force-
balance equation then becomes:

$$L_1 (F_1 - F_4) = L_2 (F_2 - F_3)$$

Typical Force-Balance Transmitters. The most widely used pneumatic
force-balance transmitter measures differential pressure. Figure 16-10 is a
schematic representation of such a transmitter. Pressure is applied to the high

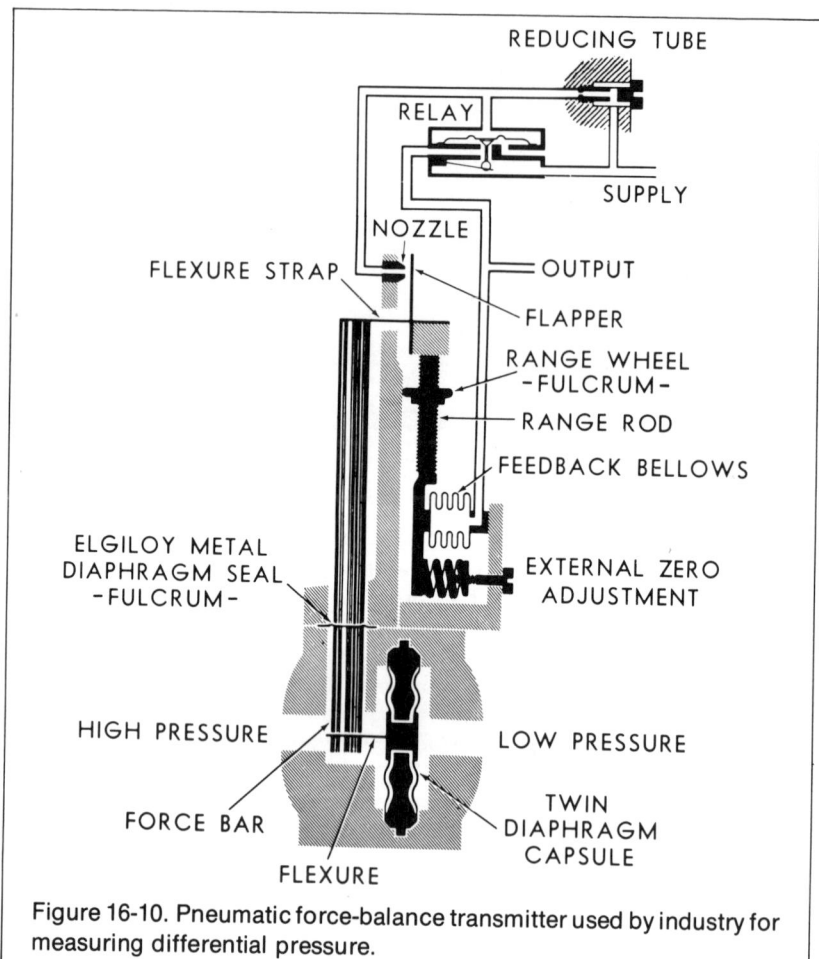

Figure 16-10. Pneumatic force-balance transmitter used by industry for measuring differential pressure.

and low sides of the twin-diaphragm capsule. The difference between these two pressures creates tension in the flexure, and a pulling force is exerted on the lower end of the force bar. The resulting moment at the other end of the force bar causes the flapper to tend to move toward the nozzle. The output of the relay is increased, and the force of the feedback bellows on the range rod creates a moment that will balance the force of the differential pressure across the capsule with its resultant moment on the force bar. This balance of forces results in an output pressure change which is proportional to that in input differential force.

Another type of force-balance transmitter is used to measure temperature and is shown in Figure 16-11. In operation, as the temperature rises, the pressure increases on the thermal system capsule, thereby causing a force on the lower end of the force bar. The moment exerted against this end of the bar tends to rotate it about the flexure fulcrum. The opposite end of the bar, acting as a flapper, tends to move toward the nozzle. The output pressure of the relay is increased and applied to the feedback bellows and also as output transmission to receiver instruments. The pressure change in the feedback bellows produces a moment of force opposing that of the measuring element, and the force bar is balanced by a new output pressure which has changed by an amount proportional to the change in measured temperature.

Electrical Force-Balance Transmitters. The operation of an electrical force-balance transmitter (Figure 16-12) is similar to that of its pneumatic counterpart. Changes in force produced by the sensing element are applied to the bottom of force bar *A*, which pivots through flexure *B*. The laminated core (*C*) at the opposite end of the bar, moves within the air gap of a differential transformer detector. The detector core movement is only 0.005 inch for full-range shift in output. Variations in detector output are rectified, amplified, and then fed through the force-balance feedback coil (*E*) in series with the power supply and receiver. The feedback coil applies a force that is equal and opposite to that of the sensing element, maintaining the system in continuous force balance.

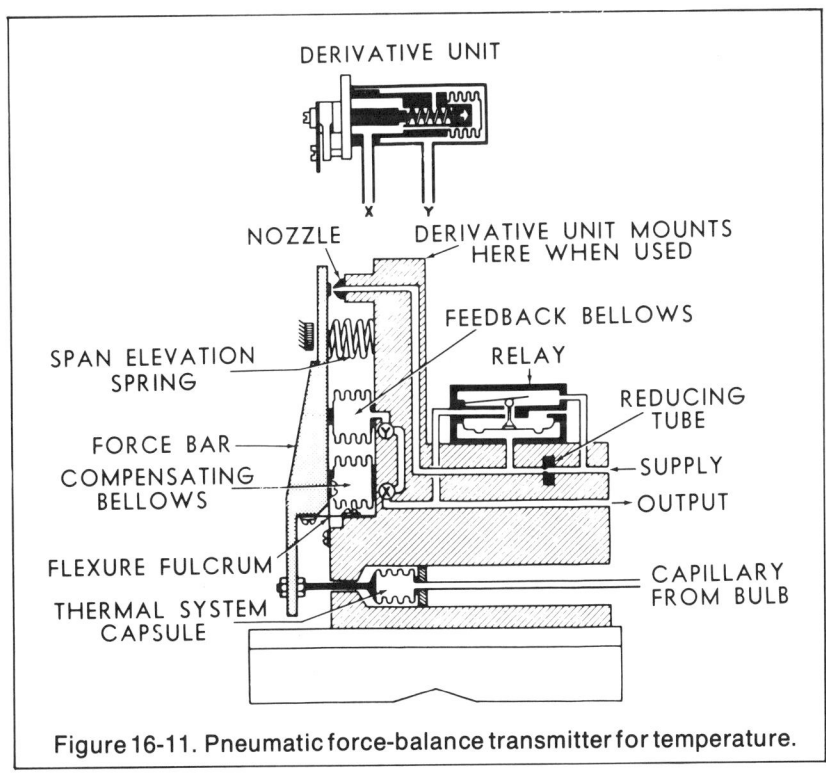

Figure 16-11. Pneumatic force-balance transmitter for temperature.

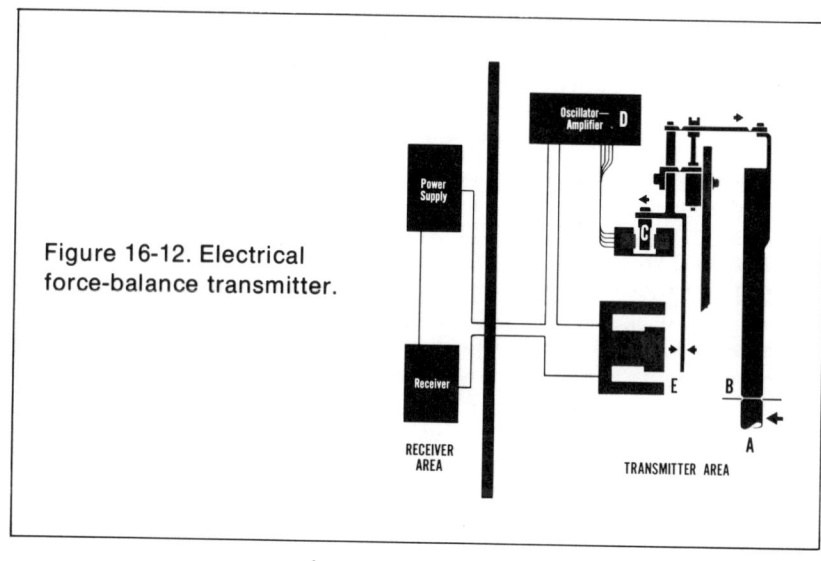

Figure 16-12. Electrical force-balance transmitter.

BIBLIOGRAPHY

Anderson, N. A. *Instrumentation for Process Measurement and Control*. 2nd ed. Philadelphia: Chilton Company, 1972.

Babcock, R. H. *Instrumentation and Control in Water Supply and Waste Disposal*. New York: The Reuben H. Donnelly Corporation, 1968.

Drinker, P. H. "Some Fundamental Considerations about Electronic Controls." At Fifth National Power Instrumentataion Symposium, Fort Worth, Texas, May 1962.

Fribance, A. E. *Industrial Instrumentation Fundamentals*. New York: McGraw-Hill Book Company, 1962.

Holzbock, W. G. *Instruments for Measurement and Control*. 2nd ed. New York: Reinhold Publishing Corporation, 1962.

Instrument Diagrams with Explanatory Notes. Foxboro, MA: The Foxboro Company, Form 2329D, August 1965.

Lavigne, J. R. "Electronic Control—How Can It Help the Papermaker." *TAPPI*, vol. 47, no. 7 (July 1964).

———. "Getting the Most Out of Automation When Expanding or Modernizing." *Paper Mill News*, November 26, 1962.

M/45 Indicating Pneumatic Transmitter Series. Foxboro, MA: The Foxboro Company, TI37-40a, January, 1960.

Tivy, V. V.; Drinker, P. H.; and Kessel, M. C. "Application and Design Factors in Differential Pressure Transmitters." *Machine Design*, October 13, 1960.

STUDY QUESTIONS

Indicate whether following statements are True or False by inserting T or F in the parentheses:

1. Long distance transmission of measurement signals has always been very common in the pulp and paper industry, even in the early days when instrumentation was just beginning to be used for process control. ( )

2. Most common measurement transmission signals found in the pulp and paper industry are either electrical or pneumatic. ( )

3. The two basic methods used to produce a transmission signal are by motion balance and force balance. ( )

4. The transmission medium and the method used to produce it depend on the primary element used to make the measurement. ( )

5. In a pneumatic force-balance transmitter, the transmitted pressure must be directly proportional to the distance between the flapper and nozzle and the motion of the element and lever system. ( )

6. Voltage, current, and power are generally used for electrical transmission signals. ( )

7. A 1:5 ratio of minimum to maximum signal transmission values has been a generally accepted standardization throughout industrial process control. ( )

8. For all practical purposes, a force balance mechanism does not involve any motion. ( )

9. Readout of the measurement on a force balance transmission system can be conveniently made with a pointer incorporated in the lever mechanism for local indication. ( )

10. Although the medium is different, the operation of electrical force balance transmitters is not dissimilar to the operation of pneumatic force balance transmitters. ( )

(See Appendix for answers)

17. Automatic Control

In previous chapters, it has been shown how process measurements are made in the pulp and paper industry so they can be used to maintain the process at predetermined values considered to produce optimum operating conditions. How this is accomplished depends on the nature of the control methods required to achieve these levels of variable values, the degree of disturbance that the process can tolerate, and the cost of providing desirable control. There are some processes in the paper industry that require only minor and infrequent adjustments to correct or limit the deviation of the measured value from some selected reference to maintain satisfactory stability. Under such conditions, an indication or record of the measured variable is used as a guide by the operator to make manual corrective adjustments. In most cases, however, it not only becomes impractical and uneconomical, but also impossible to manually control the process to within safe and acceptable tolerances. Therefore, it becomes necessary to resort to automatic methods to accomplish these more difficult control operations. The complexity and nature of this automatic control is determined by the degree of difficulty of the control problem.

Automatic control can then be considered as automatically balancing the supply and demand of a process over a period of time; it is accomplished by an automatic controller which measures the value of a variable quantity or condition and operates to correct or limit the deviation of the measured value from selected reference. It follows, then, that an automatic controller is made up of three basic parts: (1) a measurement or sensing section that observes the level of the variable to be regulated; (2) a reference source that indicates the desired source of operation; (3) a mechanism for causing the level of the manipulated variable to be changed so that the measurement signal is made to come closer to, or equal to, the reference point (set point).

CONTROL LOOPS

There are many kinds of equipment used to implement automatic control, which employ mechanical, pneumatic, electrical analog, and digital techniques. Although the methods and hardware used may differ, the basic theory of operation is the same and the steps through which the complete cycle of operation is executed are known as the *control loop*. There are two basic types of control loops: open loop and closed loop.

Open Loop Control

Figure 17-1 is a block diagram of a typical open loop control system. Note that the open loop control involves making an estimate of the form or quantity of action necessary to accomplish a desired objective. There is no feedback of information from the process to the controller and, therefore, the measurement is not present in the loop. In a sense, the controller varies the supply without knowing the demand. Speeds of filters, washers, savealls, paper machines, and other similar process equipment are controlled in this manner. Speed targets are set by rheostats, but normally there is no feedback of speed measurement fed into the control system to automatically make any necessary readjustments to maintain balance if there is an upset in original conditions.

Closed Loop Control

Most of the automatic control in the pulp and paper industry is accomplished with the closed loop configuration (Figure 17-2). Here a measurement of the variable to be controlled is made and is compared to the set point. If a difference exists between the actual (measurement) and desired (set point) values, the automatic controller will take necessary corrective action.

Because measurement of the controlled variable is made and sent back to the controller where it is compared with a reference or set point to determine necessary corrective action, this type of closed loop control is referred to as *feedback control*. Another type of control where the measurement is taken in the process feedstream, compared to a reference point, and a predicted amount of corrective action is taken thereon, has more recently been referred to as *feedforward control*.

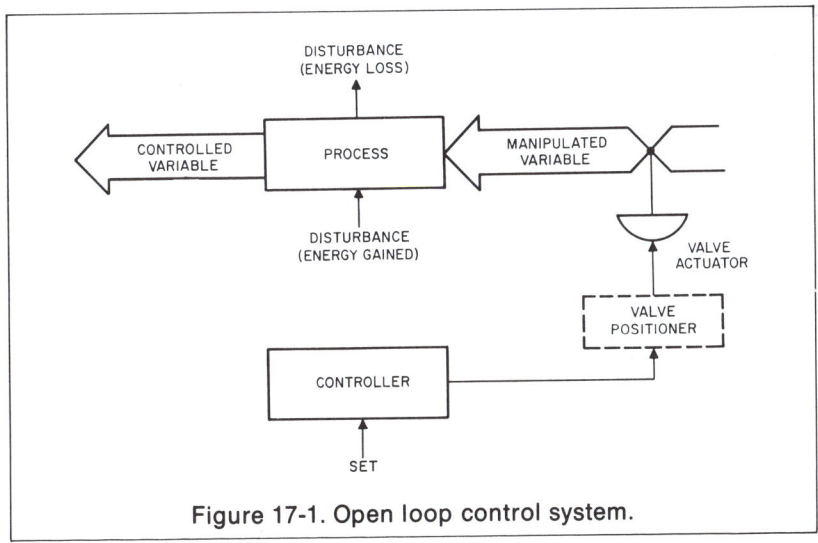

Figure 17-1. Open loop control system.

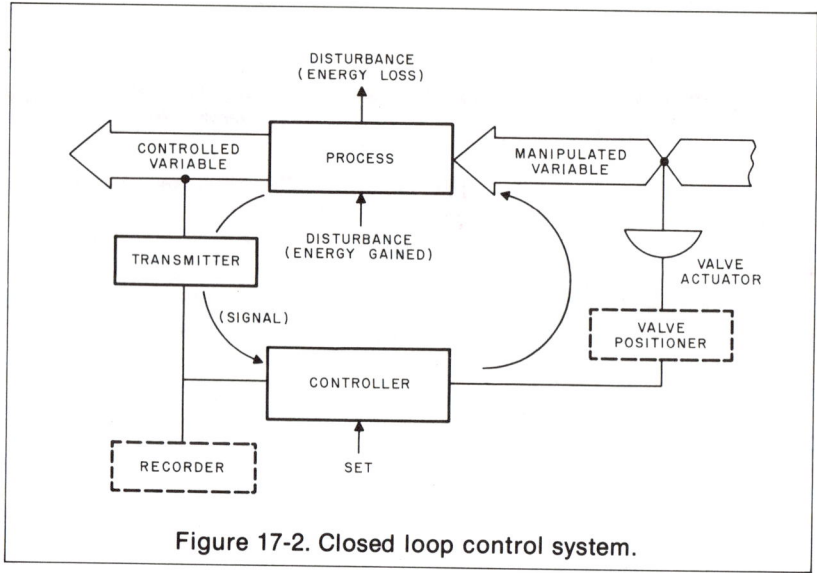

Figure 17-2. Closed loop control system.

CONTROL MODES

The basic automatic control modes used in the pulp and paper industry for process control can be generally classified as either two-position or throttling. With two-position control, the valve or final control element operator is always either in the open or closed position. The controlling unit will never maintain the operator in an intermediate or throttling position. There are several specific methods of control that can be classified as two-position: (1) on-off, (2) differential gap, and (3) time proportioning. On-off is the most common two-position control used in the paper industry. Others are used on special occasions.

Any type of control in which the final operator is purposely maintained in intermediate positions between open and closed is generally known as a throttling control mode. Depending on their action, throttling control modes can be further categorized as proportional, integral, and derivative. Various combinations of these control modes are available for process control; however, those most commonly used in the pulp and paper processes are: (1) proportional, (2) proportional-plus-integral, (3) proportional-plus-integral-plus-derivative. The integral mode is also referred to as the *reset mode*.

Position Control Modes

On-off. With the on-off control mode, as soon as the measured variable differs from the desired control point (set point), the final operator is driven from one extreme to the other (Figure 17-3). For example, as soon as the measured variable exceeds the control point, the final operator is closed. It will remain closed until the measured variable drops below the control point, at which time the operator is fully open. With this type of control the measurement is always cycling. However, when properly applied, the amplitude of the cycles about the control point is so small that the measurement record is very nearly a straight line and satisfactory control is obtained. Thermostatically controlled electric heaters, found in many parts of the plant, and heating of white water are examples of on-off control systems. A typical control device which accomplishes on-off control pneumatically is shown in Figure 17-4. It consists, basically, of a measuring element and a means of recording or indicating changes in the measured variable; a setting knob to set the desired control value and a means of indicating this setting; an air control relay which is an air amplifier and switching mechanism for switching a regulated air supply to a control valve; and a flapper-nozzle assembly which is essentially an error detector that detects deviations of the measurement from the set point.

In operation, if the pen or indicating pointer were below the index and an increase in measurement caused it to move up scale past the index by a slight amount, it would actuate the flapper through a link and lever system, uncovering the nozzle. This reduces the back pressure air signal from the nozzle to the relay and allows the ball valve in the relay to close. The change in the relationship of flapper to nozzle for a 3-15 psi relay output change is less than 0.001 inch, while the amount of pen movement to produce this change is approximately 0.15 percent of full scale. The control valve assumes either a fully open or fully closed position depending on its design, and it remains in that position until the measurement changes and the pen moves down scale past the index. When the measurement signal decreases, it causes the flapper to cover the nozzle, thereby increasing the air signal to the relay. This causes the ball valve of the relay to be lifted from its seat, thereby allowing full air supply pressure to actuate the control valve in the opposite direction.

Differential Gap. Differential gap is similar to on-off control except that a band exists around the control point. A typical response curve as well as the corresponding final operator action is shown in Figure 17-5. When the measured variable exceeds the upper boundary of the gap, the final operator is

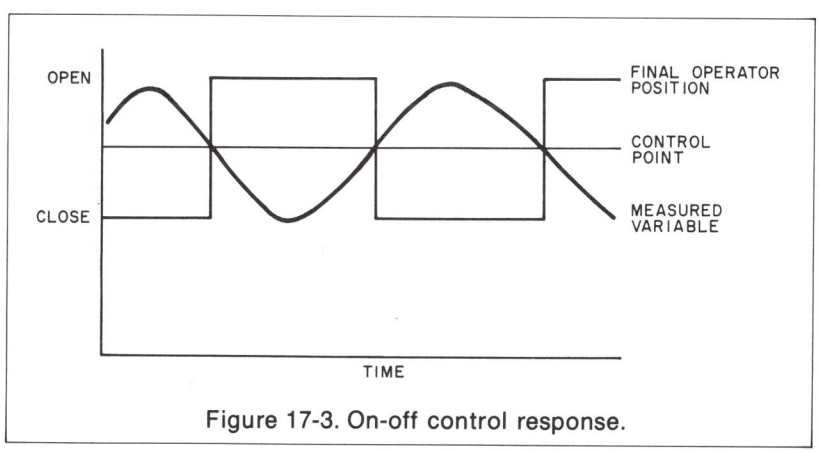

Figure 17-3. On-off control response.

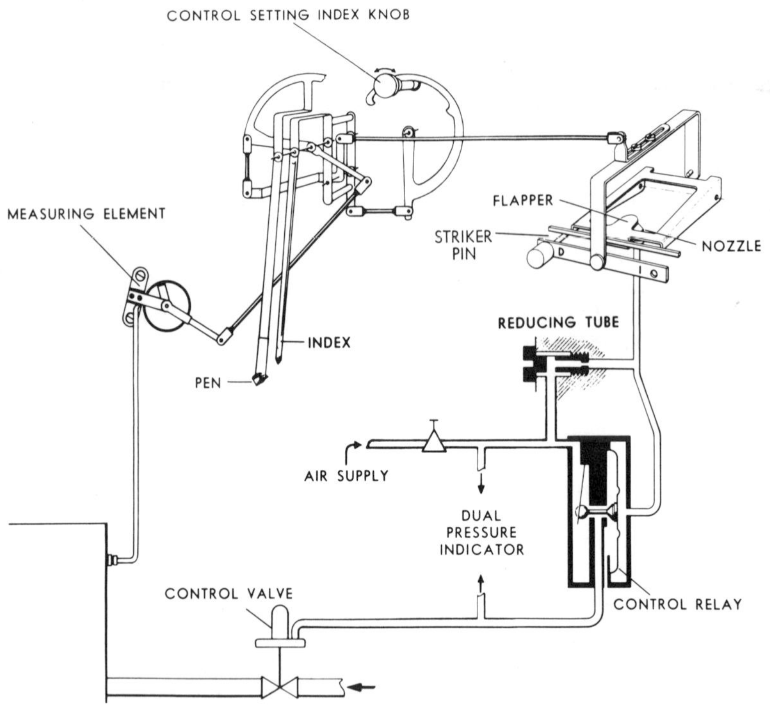

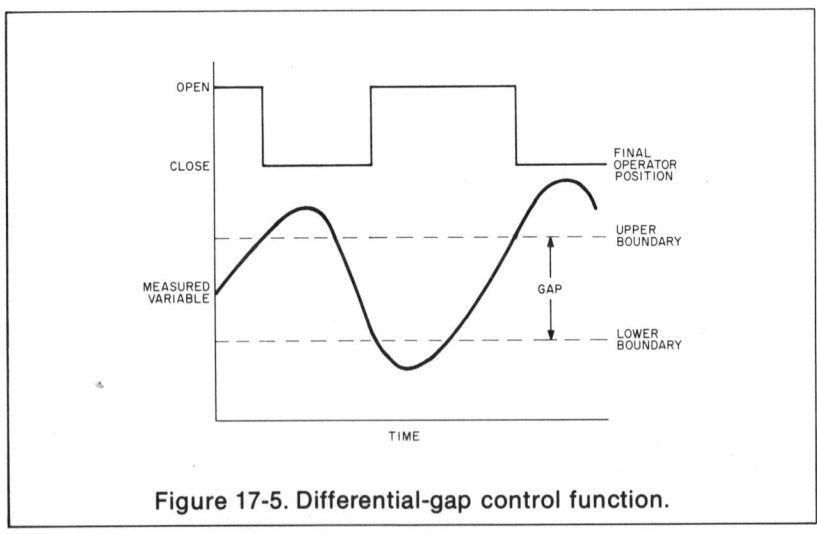

CONTROL SETTING INDEX KNOB

MEASURING ELEMENT

FLAPPER

STRIKER PIN

NOZZLE

INDEX

PEN

REDUCING TUBE

AIR SUPPLY

DUAL PRESSURE INDICATOR

CONTROL VALVE

CONTROL RELAY

Figure 17-4. On-off controller.

OPEN

CLOSE

FINAL OPERATOR POSITION

UPPER BOUNDARY

MEASURED VARIABLE

GAP

LOWER BOUNDARY

TIME

Figure 17-5. Differential-gap control function.

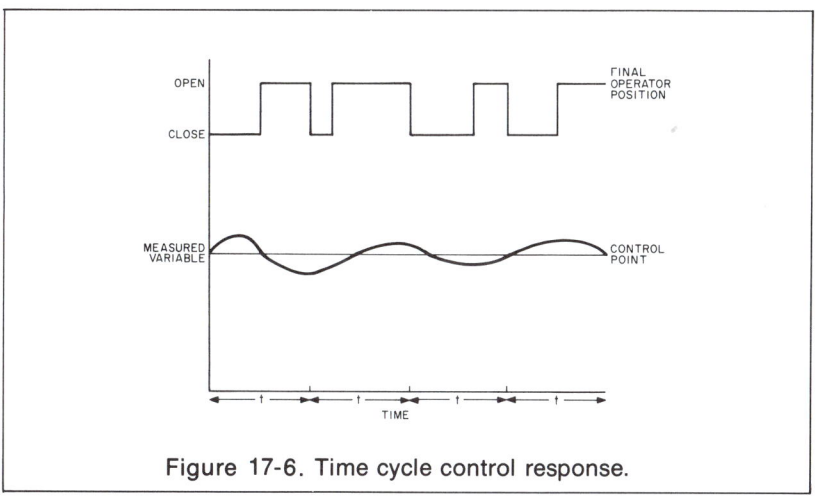

Figure 17-6. Time cycle control response.

closed. It will remain closed until the measured variable drops below the lower boundary. The final operator will remain open until the measured variable again exceeds the upper boundary.

Time Proportioning. In time proportioning control, a time base is established. During this time period, the final operator is closed for a certain percentage of the time and open for the rest. The ratio of the closed-to-open time is determined by the relationship between the measured variable and the control point (Figure 17-6). A time proportioning controller is normally set up so that when the measured variable equals the desired control point the final operator will be open for half the time cycle and closed the other half. As the measured variable drops below the control point, the final operator will remain open longer than it is closed.

Throttling Control Modes

Proportional Control Mode. In proportional control, a throttling action is provided so that the final element (a valve) balances the process input with the process demand by being positioned somewhere between fully opened and fully closed. This throttling action continuously actuates the valve so that the desired process control balance is maintained under varying conditions. The valve position depends on the relation of the measurement to a set point. The set point is the point at which it is desired to maintain the measured variable. It is adjusted by manually positioning the control index. The control point is the value of measurement variable that the controller is maintaining. The term *offset* refers to the amount of deviation between the set point and control point.

Proportional action may be described as output pressure change proportional to actuating signal change. A proportional control mechanism changes valve position in proportion to the change in measurement. This type of control can only keep a process variable within certain limits and, therefore,

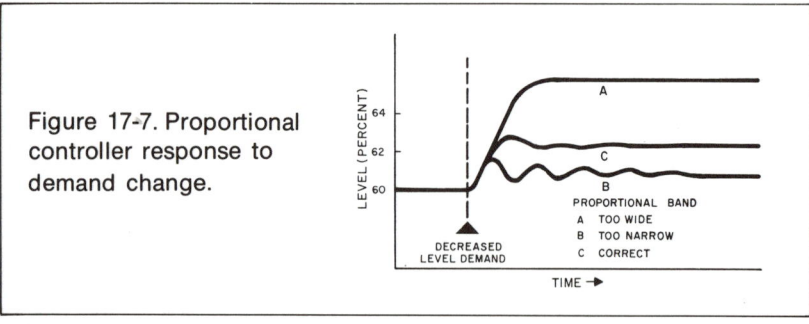

Figure 17-7. Proportional controller response to demand change.

LEVEL (PERCENT)

64
62
60

A

C

B

PROPORTIONAL BAND
A TOO WIDE
B TOO NARROW
C CORRECT

DECREASED
LEVEL DEMAND

TIME →

cannot be used to hold a definite or exact value. The amount that the measurement must change to move the valve from fully opened to fully closed is referred to as the *proportional band.* In a sense it is the ratio of the measurement change to output change, expressed as a percentage of the total instrument range, which is adjustable to obtain stable control under differing process conditions.

The measurement can stabilize at any point within the proportional band which can be narrowed so that a greater output change will result from the same measurement change. However, the series of controller response curves depicted in Figure 17-7 shows that there is a minimum below which the proportional band cannot be narrowed without causing the control mechanism to cycle and it is determined by process conditions.

The relation between measurement change (pen position) and valve stroke is graphically shown in Figure 17-8. Curve A shows that the pen must travel the entire scale to change the valve from a fully opened to a fully closed position. It also shows that for any position of the measurement pen, there is a corresponding position of the valve in terms of its stroke. For example, with the setting index at mid-scale, if the pen is at 20 percent of scale, the valve is at 20 percent of stroke; and if at 80 percent of scale, the valve is at 80 percent of stroke. Because the pen must travel 100 percent of scale for the valve stroke to be 100 percent, proportional band is referred to as 100 percent. Curve B shows that the pen position must change between 25 percent and 75 percent of scale to obtain 100 percent valve stroke. Curve B, then, represents a 50 percent proportional band. Curve C represents 25 percent proportional band, Curve D, a 10 percent proportional band.

An alternative term for proportional action is *gain*, which is defined as change in output/change in input. It is a dimensionless number and is the reciprocal of proportional band. To convert percent proportional band to gain, the formula, gain = 100/P.B., is used.

Proportional Control Mechanisms. As with transmission mechanisms, proportional control mechanisms employ either position-balance or force-balance methods and the medium of operation is predominantly pneumatic or electronic. Figure 17-9 shows schematically the working of a pneumatic proportional position-balance controller. It differs from the on-off controller in that a bellows and an opposing spring stabilizing assembly have been added. In

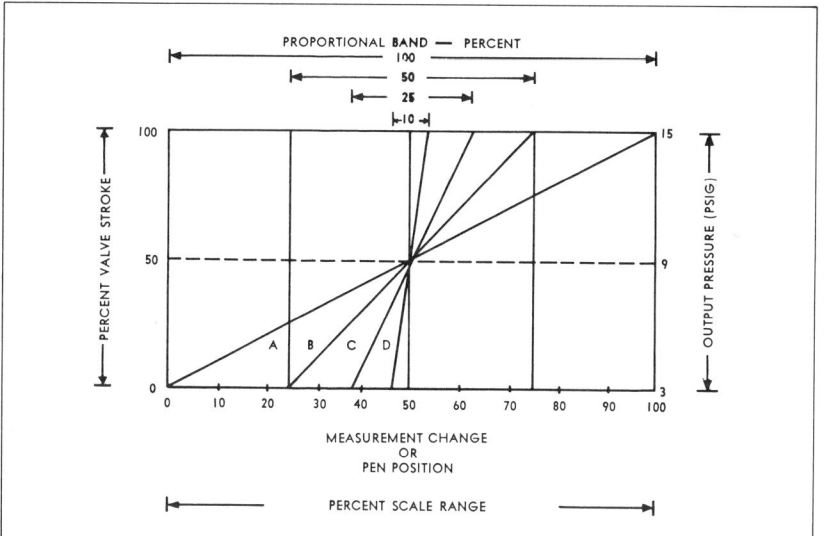

Figure 17-8. Proportional action measurement to valve stroke relationship is shown graphically.

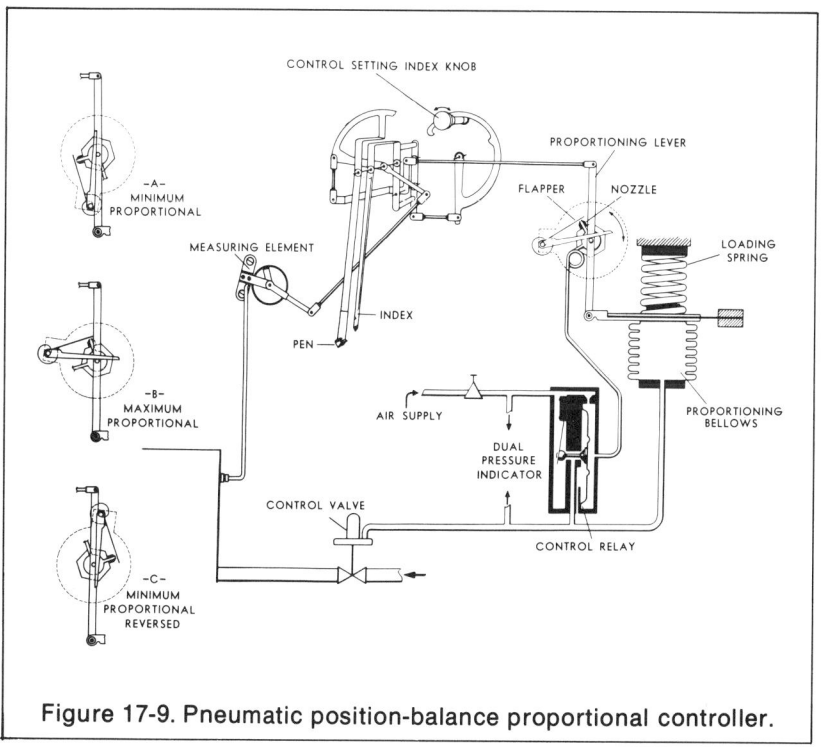

Figure 17-9. Pneumatic position-balance proportional controller.

operation, increasing measurement moves the top end of the proportioning lever to the right through a system of links and levers causing the flapper to move closer to the nozzle. This, in turn, causes a proportionately higher back pressure to be applied to the relay, which increases the output pressure to the control valve and, at the same time, to the proportioning bellows. The control valve either opens or closes in proportion to the increased output pressure, depending on its design. As the bellows receives this increased pressure, it pushes upward against the loading spring. This upward motion repositions the proportioning lever until the flapper moves away from the nozzle. The system is again balanced but at a higher output pressure proportioned to the higher measurement. Thus, the valve is maintained in some position between fully opened and fully closed. In this manner, the controller is in balance and the measurement is kept within the proportional band limits.

The schematic of a pneumatic force-balance proportional controller is shown in Figure 17-10. In this type of controller a balance of forces rather than position balance governs the operation. The controller contains a measurement bellows that receives the measurement signal (usually 3-15 psi) from the measurement transmitter. It also contains a set point bellows that receives an adjustable 3-15 psi signal. The set point and measurement bellows oppose each other and act on a lever. The lever, which rests on a pivot, is also the flapper-nozzle detecting system. Changing pivot position changes the relative lengths of the lever arms, adjusting the proportional band. The feedback bellows and proportional spring between the pivot and the flapper-nozzle act on the lever to reposition the flapper within the throttle band of the nozzle, similar to the position-balance mechanism. The remainder of the mechanism,

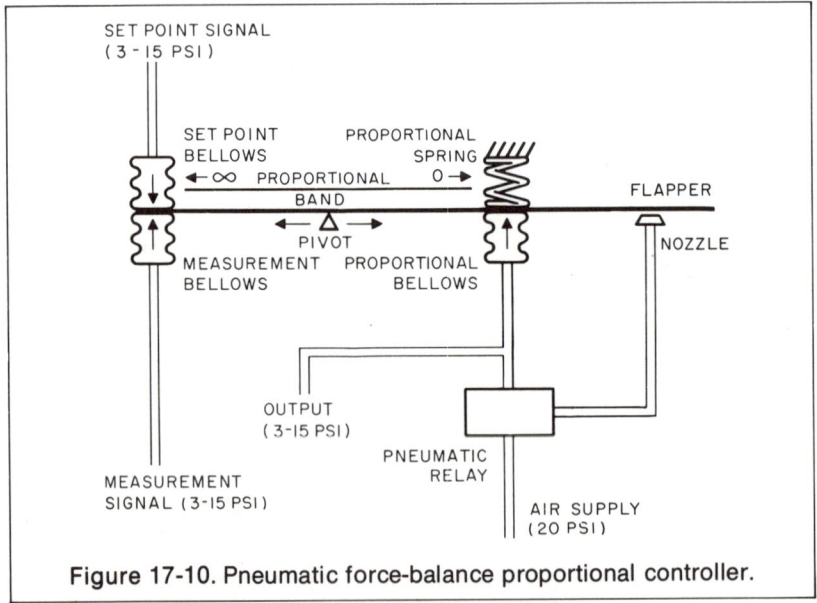

Figure 17-10. Pneumatic force-balance proportional controller.

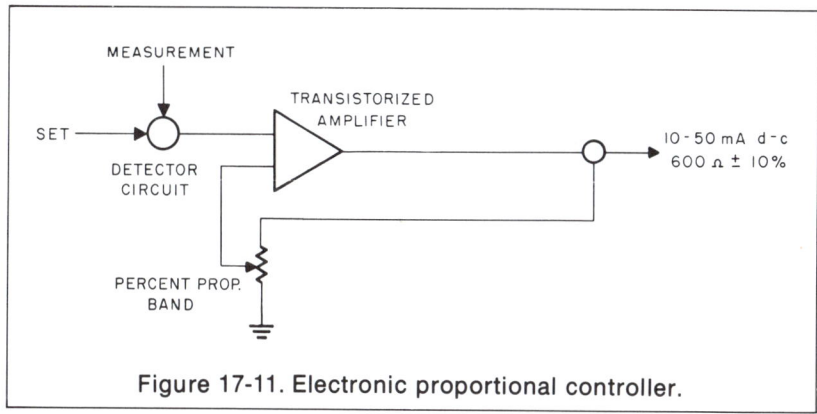

Figure 17-11. Electronic proportional controller.

including the pneumatic relay, etc., is essentially the same as for the position-balance controller.

Electronic control mechanisms perform the same functions as their pneumatic position- and force-balance counterparts. At one time, practically all electronic instrumentation was characteristically designed around slide-wires, vacuum tubes, transformers, and other comparatively large-sized components. Today, practically all modern electronic controllers use solid-state components such as diodes, transistors, magnetic amplifiers, etc. As previously explained in the chapter on transmission, a 10-50 or 4-20 milliampere dc signal proportional to measurement is commonly transmitted to electronic controllers. Controller outputs in corresponding dc signals must be converted to air signals when used with pneumatic final control element operators.

A simplified block diagram of an electronic proportional controller is shown in Figure 17-11. The set point is fed into a detector circuit where it is compared with the measurement. Output of the detector circuit, which is proportional to the difference between measurement and set point, is fed to a transistorized amplifier. Output of the amplifier is used for operation of the valve and for a feedback signal. An adjustable potentiometer in the feedback circuit is the proportional band adjustment.

Proportional-Plus-Integral Control Mode. When a process becomes more difficult to control, the proportional band must be increased in order to eliminate cycling. When this zone in which the measurement will stabilize becomes so wide that poor control results, it then becomes necessary to introduce a control function which will maintain the measurement at a precise point instead of just within the band. This control mode, known as integral, is added to the proportional control mode. Although this function can be accomplished manually, it is inconvenient and practically never done in process control. The combination proportional-plus-integral control generally assumes that the integral function is automatic and part of the control mechanism. With the integral control, the valve movement is a function of the magnitude and duration of the measurement deviation from the set point, whereas with the straight proportional controller it is a function of the deviation magnitude

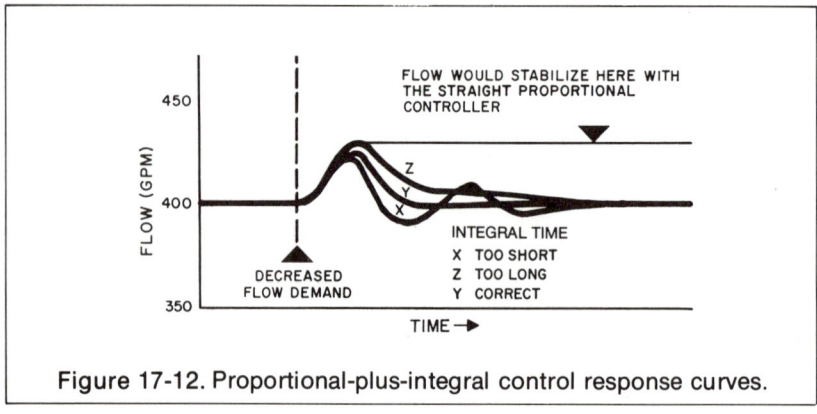

Figure 17-12. Proportional-plus-integral control response curves.

only. Therefore, instead of having a specific valve position for each measurement deviation, the proportional-plus-integral controller can position the valve anywhere from fully opened to fully closed for a given measurement. The valve is continuously positioned as necessary to keep the measurement at the control point. The correct integral rate is determined by the process just as is the proportional band. This rate (time) is set in accordance with the natural reaction time of the process. The proportional-plus-integral control response curves (Figure 17-12) show that when integral time is set correctly the valve is stroked at a rate at which the process can respond. If the integral time is set too fast, the valve will move faster than the measurement and cycling will result; if it is set too slow, the process will not return to the set point quickly enough.

Proportional-Plus-Integral Control Mechanisms. Figure 17-13 shows schematically the operation of a pneumatic position-balance proportional-plus-integral controller. It differs from the proportional controller in that the opposing spring has been replaced with an integral bellows and an integral resistance or restrictor has been added. In operation, if the pen tends to move up scale from the set index causing the proportioning lever to be moved to the right, the flapper, acting through a system of links and levers, will approach the nozzle and the air pressure to the valve will be increased as with the proportional controller. This same pressure is applied directly to the proportioning bellows and through the integral resistance to the integral bellows. The function of the resistance is to control the integral time or rate at which the pressure in the integral bellows approaches the pressure in the proportioning bellows. When the pressures are equal the controller is in a balanced condition. If a process upset or load change takes place causing a deviation between measurement and set point, the pressure to the control valve and the proportioning bellows will increase or decrease. Simultaneously, the pressure in the integral bellows changes in proportion to the magnitude and duration of the deviation. As long as the deviation is other than zero there will be a difference in pressure between the two bellows, allowing the corrective action to continue. This resetting action by the integral bellows causes the controller to be in balance at whatever valve output pressure is necessary to keep the measurement at the control set point. The integral resistance is adjustable to permit matching of the controller response to the process characteristics.

In the pneumatic force-balance proportional-plus-integral controller shown in Figure 17-14, an integral bellows replaces the proportional spring and an integral capacity tank and restrictor are added to the instrument. These are essentially the same modifications made in the position-balance controller.

In the electronic proportional-plus-integral controller (Figure 17-15), a resistance and a capacitance, which are electrical equivalents of a restrictor and a capacity tank, are added to the proportional controller. Integral is achieved by placing an integral capacitor in the feedback line and an adjustable integral resistor in parallel with the proportional band potentiometer.

Proportional-Plus-Integral-Plus-Derivative Control Mode. There are processes in the pulp and paper industry with long time lags, some of which involve the measurement of temperature and pH, when the proportional-plus-integral control mode response may not be sufficient to stabilize itself fast enough on the occurrence of an upset. Under these conditions an additional control mode, *derivative*, is used to improve the control. It has been shown that the proportional mode produces a valve response proportional to the deviation of the measurement from the set point, and integral response is proportional to the time that the measurement is away from the set point. Derivative response is proportional to the rate at which the measurement departs from the set point and is sometimes considered an *anticipatory* action. Because it is rate-sensitive, the derivative control mode when added to the

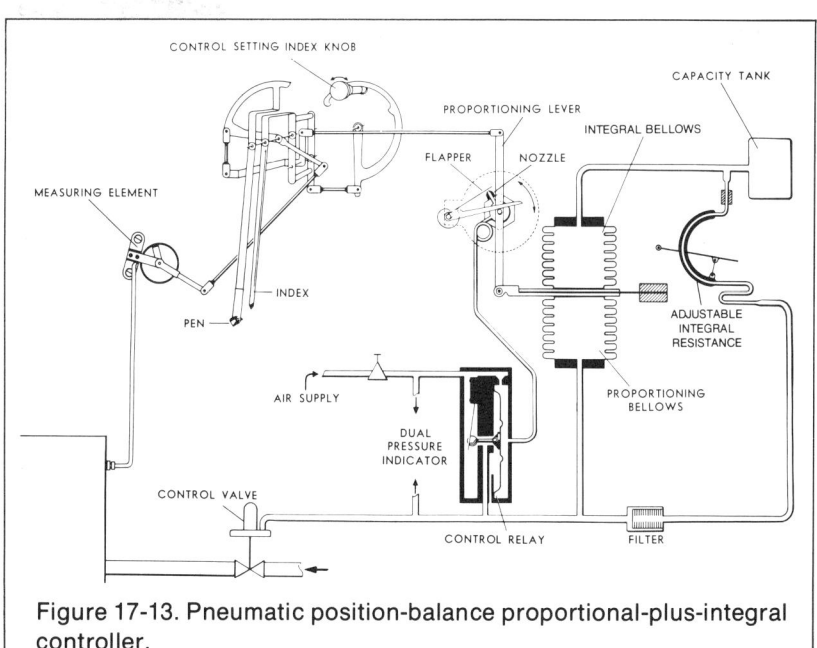

Figure 17-13. Pneumatic position-balance proportional-plus-integral controller.

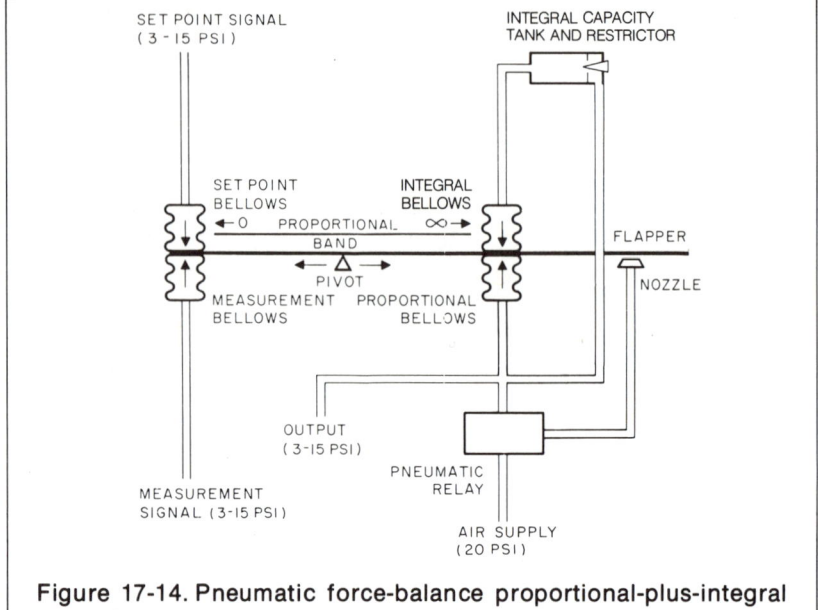

Figure 17-14. Pneumatic force-balance proportional-plus-integral controller.

proportional-plus-integral-plus-derivative control mode permits the use of a narrower proportional band, thus reducing the amount of possible measurement deviation and overshoot on process upset. In effect, it functions to reposition the valve sooner in order to stabilize process upsets. The integral action continues to reposition the valve until any offset or deviation is eliminated, resulting in fast stabilization of process at the set point.

Proportional-Plus-Integral-Plus-Derivative Control Mechanisms. A typical pneumatic position-balance proportional-plus-integral-plus-derivative

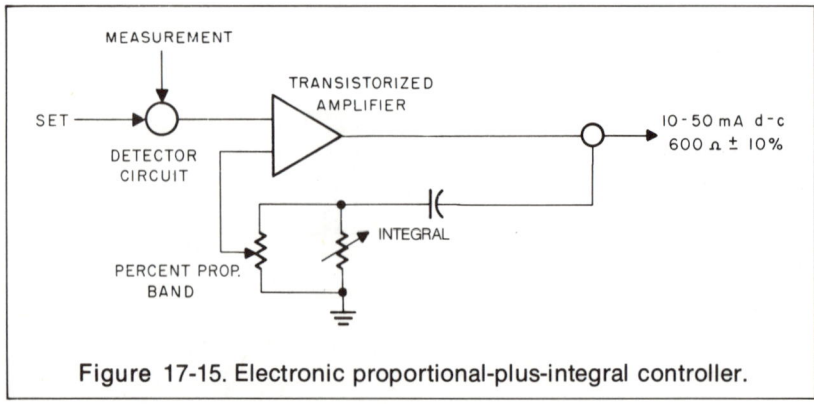

Figure 17-15. Electronic proportional-plus-integral controller.

controller is shown schematically in Figure 17-16. It differs from the proportional-plus-integral controller previously described in that an outer derivative bellows is added to the proportional bellows; an adjustable derivative resistance (restrictor) and derivative capacity tank are also added to the feedback circuit. With the measurement at the set point, pressures in the proportional, integral, and derivative bellows are all equal to the output pressure of the controller. If the measurement starts to increase at a uniform rate, the derivative bellows immediately senses the change, increasing controller output by an amount proportional to the rate at which the measurement is changing. This change takes place immediately. During the duration of the change, a constant differential pressure is maintained across the derivative restrictor and there is a constant flow through it into the proportional bellows. Simultaneously, this flow establishes a differential pressure and, thus, a flow across the integral restrictor which increases linearly as measurement continues to increase. When measurement stops increasing, derivative action decreases control ouput by an amount proportional to the rate of change of the measurement. By design, this rate of change is equal to the initial rate of change; therefore, output changes equal amounts in the opposite directions. In this way, immediate response to measurement changes is obtained. Normally, derivative time is somewhat shorter than integral time but both restrictors must be adjusted to match process dynamics.

Similarly, the pneumatic force-balance proportional-plus-integral-plus-derivative controller (Figure 17-17) is designed by adding a derivative capacity tank and restrictor to the force-balance proportional-plus-integral controller.

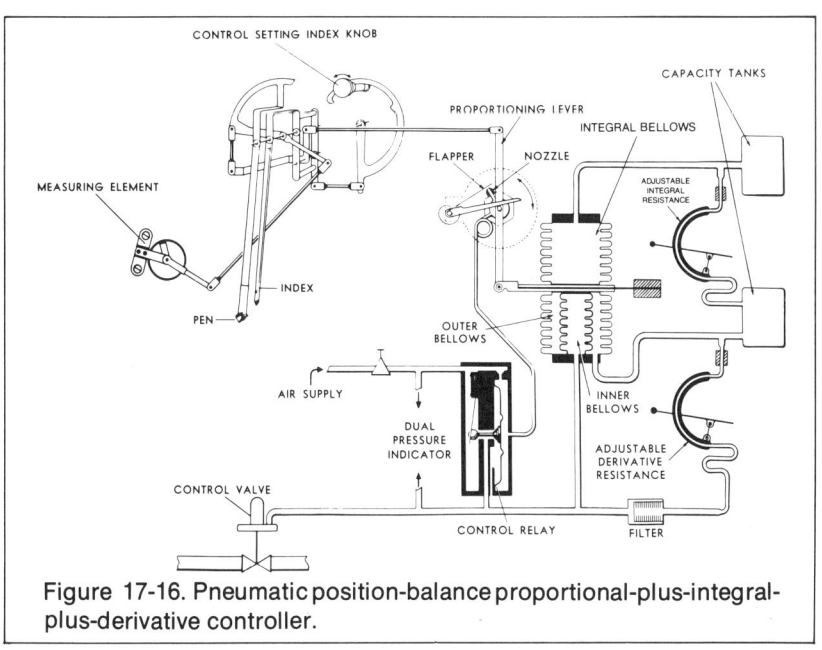

Figure 17-16. Pneumatic position-balance proportional-plus-integral-plus-derivative controller.

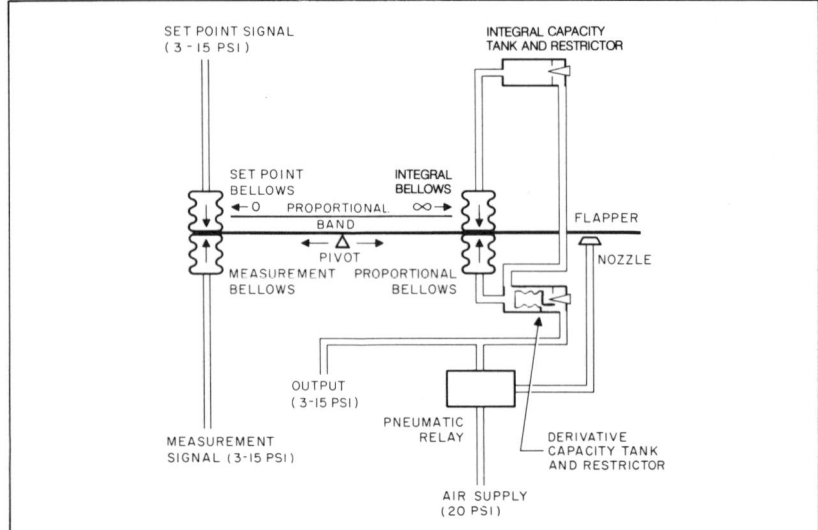

Figure 17-17. Pneumatic force-balance proportional-plus-integral-plus-derivative controller.

The output signal goes through the derivative restrictor and into the derivative tank. From there the signal feeds to both the proportional bellows and to the integral tank and restrictor. This provides an output that responds to rate of departure of the measurement from the set point.

In the same manner the electronic proportional-plus-integral-plus-derivative control schematic circuit is developed from the proportional-plus-integral schematic by placing the derivative resistor in the feedback line and the derivative capacitor in parallel with the two other adjustable proportional band and integral resistors (Figure 17-18).

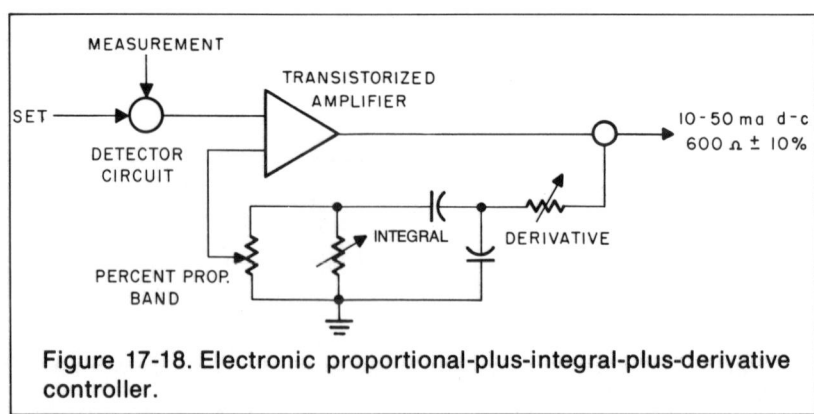

Figure 17-18. Electronic proportional-plus-integral-plus-derivative controller.

To date, the most common controller design is to have all components, set point adjustments, measurement display, control modes, and control mode devices mounted together in one box. More recent designs have employed the split-architecture concept in which the set point adjustment and measurement displays are mounted separately and remotely from the control mode devices.

A simplified block diagram of an electronic proportional-plus-integral-plus-derivative controller using this type of design configuration and which operates on a 0-10 volt dc base electrical signal is shown in Figure 17-19. With this arrangement, all measurement signals are converted to 0-10 volts dc by a suitable converter. The 0-10 volt dc output signal is also converted to suitable current signals to be used by field devices. In operation, the converted field measurement signal and set point signal generated by the set point station are fed into the differential input stage of amplifier A. It sums the measurement and set point signals and provides a ± 10 volt dc error signal to succeeding controller stages. An increase/decrease switch is provided ahead of this stage to enable reversing the action of the controller. This switch simply exchanges the measurement and set point inputs to amplifier A, reversing the polarity of the error signal.

The measurement is fed in parallel to a separate derivative circuit. The output of this derivative circuit is proportional to the rate of change of the measurement signal. The derivative time setting is controlled by a time-modulated tuning circuit. The output from the derivative circuit is summed with the error signal from amplifier A in the variable gain stage of amplifier B. A step change in either of these signals will cause a proportional change in the output of amplifier B. This output passes to the proportional-plus-integral output stage. The output stage circuit functions as an integrator and changes the output of a rate determined by another signal generated in the integral

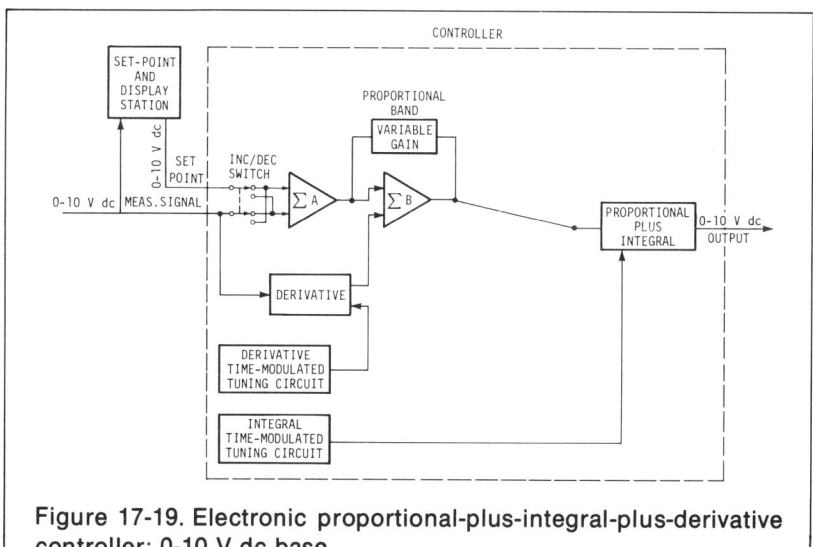

Figure 17-19. Electronic proportional-plus-integral-plus-derivative controller; 0-10 V dc base.

time-modulating tuning circuit. The resultant 0-10 volt dc output from the controller is converted to proper current signals (usually 10-50 mA dc or 4-20 mA dc) used by final control devices in the field.

Controller Tuning

The purpose of tuning a controller is to match the gain and time functions of the controller with the rest of the elements in the control loop (process, transmitter, valve, etc.). There are a number of practical and theoretical approaches to controller tuning. Mathematical equations can be used to predict ideal proportional, integral, and derivative settings for a given process. The predominant practice in the pulp and paper industry is to tune controllers by practical experience and/or by trial-and-error method.

With a proportional controller, the following procedures may be followed:

1. Place the controller on manual.
2. Adjust the proportional band to maximum.
3. Place the controller on automatic.
4. Make a step change in controller set point.
5. Observe the resulting measurement cycle.
6. Reduce the proportional band and repeat steps 4 and 5.
7. Repeat steps 4, 5, and 6 until an amplitude cycle and optimum recovery to stability are observed.

With a proportional-plus-integral controller:

1. Place the controller on manual.
2. Adjust the proportional band to maximum.
3. Set the integral to maximum time.
4. Place the controller on automatic.
5. Adjust the proportional band to an optimum setting as with the proportional controller.
6. Decrease the integral time in steps until cycling begins.
7. Increase the integral time until cycling disappears.

A practical adjustment procedure for tuning a proportional-plus-integral-plus-derivative controller is:

1. Place the controller on manual.
2. Adjust the proportional band to maximum.
3. Set the integral to maximum time.
4. Set the derivative to minimum time.
5. Place the controller on automatic.
6. Adjust the proportional band to an optimum setting as with the proportional controller except that a slight cycle should remain in controller.
7. Increase the derivative time until the cycle stops.

8. Narrow the proportional band until the cycle starts again.
9. Repeat steps 7 and 8 until further increases in the derivative time fail to stop the cycle.
10. Widen the proportional band to stop the cycle.
11. Set the integral time equal to the derivative time.

Combination Control Systems

Some of the more common multivariable control systems encountered in the pulp and paper industry which employ two or more measurements or measuring elements in a control loop are: (1) cascade, (2) ratio, (3) auto-selector, (4) relation, and (5) feedforward.

Cascade Control. In cascade control the output of one controller adjusts the set point of another controller. A cascade control system consists of one controller (primary or master) controlling the variable which is to be kept at a constant value and a second controller (the secondary or slave) which controls some other variables that can cause fluctuations in the first variable. The primary controller positions the set point of the secondary and it, in turn, manipulates the control valve. The objective of the cascade control system is the same as that of the single loop controller in that its function is to achieve a balance between supply and demand and, thereby, maintain the controlled variable at its required constant value. The secondary loop is introduced to reduce the effect of lags, thus stabilizing inflow to make the operation more accurate. The secondary controller can be considered as the final control element being positioned by the primary controller in the same way a single controller would position a control valve.

The secondary variable is not controlled in the same sense as the primary; it is manipulated just like any control medium. One example of simple cascade control systems is where the output signal of a level controller is used to set the set point of a flow controller on the liquid flowing into a storage tank (Figure 17-20). Another example is the use of the output signal from a heat exchanger temperature controller to set the set point of a steam flow controller to the heat exchanger. Cascade control is used to improve control of processes which normally involve upsets and long time constants with which single three-mode controllers cannot cope satisfactorily.

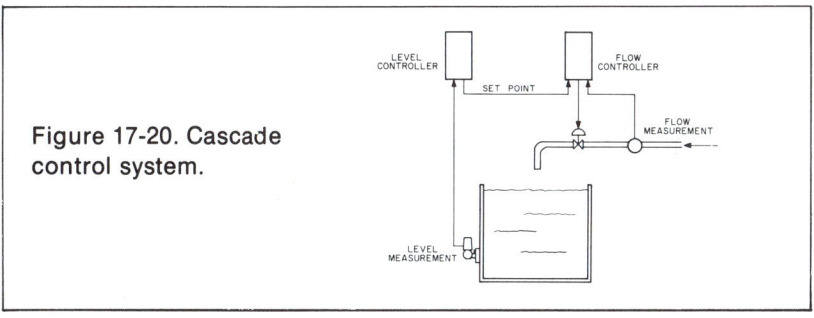

Figure 17-20. Cascade control system.

Ratio Control. When it is desired to maintain one measurement variable, known as the controlled or secondary variable, in a preset ratio to another variable, referred to as the uncontrolled or primary variable, ratio control is used. A simple flow ratio control loop is shown in Figure 17-21. In this system it is desired to control secondary flow B in preset ratio to primary flow A. Flow transmitter A senses the primary flow. The ratio relay multiplies the 0-100 percent output of the flow transmitter by a manually set factor:

(Flow output A) x (Preset factor) = Output of ratio relay

This output then becomes the set point of the controller regulating controlled flow B. At equilibrium, flow B equals the set point of the controller, such that

Flow B = (Flow A) x (Ratio factor)

or

Ratio factor = Flow B/Flow A.

Continuous stock blending control systems used so extensively in today's paper mills utilize this principle of control.

Auto-selector Control. Auto-selector control systems are used when a single final operator is to be manipulated to prevent any one of several process variables from exceeding a preset limit. For instance, it may be desirable to operate a drying process at a certain pressure but there is a more important or overriding limit on the amount of steam that can be used without upsetting the operation of the source of steam. To meet these requirements a pressure controller, with set point at the desired pressure, is used on the dryer; and a flow controller, with set point at acceptable flow rates, is used on the steam flow. Outputs of both controllers are sent to a selector relay. This relay will

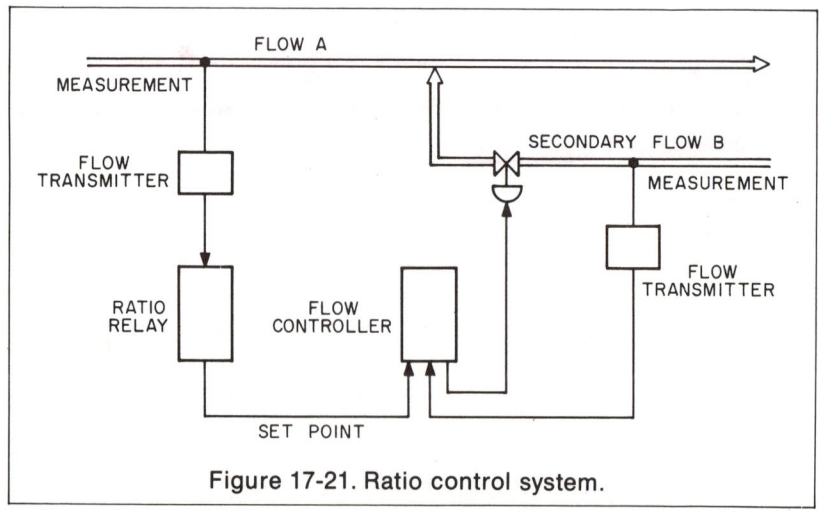

Figure 17-21. Ratio control system.

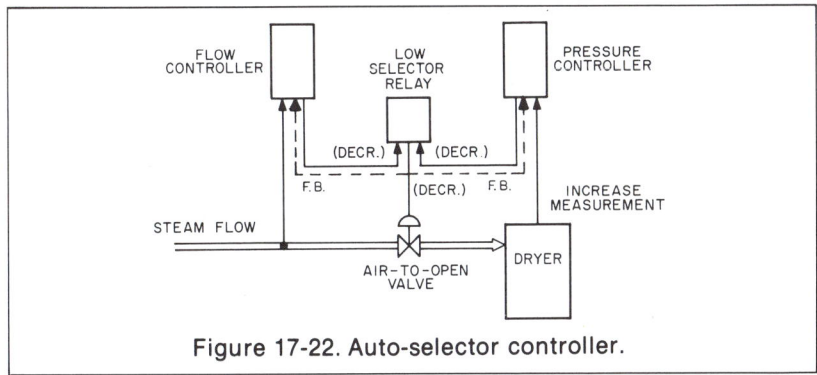

Figure 17-22. Auto-selector controller.

allow the output signal from the pressure controller to be transmitted to control as long as the steam flow is below the safe set point of the steam flow controller. As the steam flow approaches the set point of the steam flow controller, the selector transfers the valve-actuating signal from the pressure controller to the flow controller and will maintain this control until the steam flow drops safely below the set point and the selector relay transfers control back to the pressure controller. The operation of such a system is diagrammed in Figure 17-22.

Relation Control. The relation control system is similar to the ratio control in having two inputs and one output, but it controls one in a fixed relation with changes to the other rather than a strict ratio. For instance, one measurement element may increase an uncontrolled variable such as temperature and is directly connected by a linkage to the control index of the controlled variable such as pressure. The temperature controller then acts as a master and adjusts the set point of the pressure controller to maintain a preset difference or relation between temperature and pressure. In this way, temperature can be used as a measure of saturated steam pressure, a principle used quite often to control relief of batch digesters and density of heavy black liquor from evaporators.

Feedforward Control. Feedforward control relies on the principle of predicting the amount of required corrective action in input change to a process and when the correction should occur. It is used to improve control of processes which have long time delays such as found in bleach plants, evaporators, paper machines, and the like. Under these conditions, the feedback controller suffers from the disadvantage that it is working from a process signal which does not represent the true condition of the process. Figure 17-23 compares a typical feedforward control loop with a typical feedback loop. With feedback, load changes are detected only after they have affected the controlled variable. Thus, correction is developed later than the load change with the possibility that it could occur when none is needed if the load changes had since been eliminated.

The feedforward control loop accepts the process load change as its input and corrects the manipulated variable immediately to prevent the effect of the

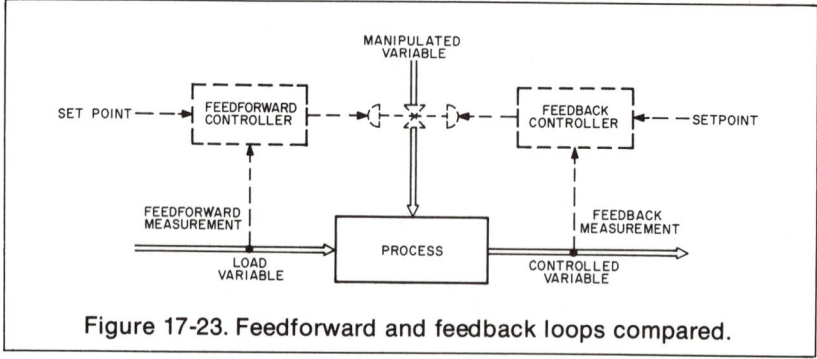

Figure 17-23. Feedforward and feedback loops compared.

load change from reaching the controlled variable. Theoretically, feedforward control is capable of perfect control in that the controlled variable never need deviate from the set point. The feedback controller, on the other hand, cannot function until the controlled variable has in fact deviated from the set point. Thus, perfect control is not theoretically feasible with a feedback controller.

In any process there are a number of kinds of load changes possible. Some can be measured and used for feedforward control. Others are unknown or measurements for them do not exist but they do affect the controlled variable even if only in a minor way.

Compensation must also be made for these latter disturbances. This can be done by using a feedback loop to trim the feedforward control loop based on a measurement of the controlled variable as shown in Figure 17-24. This solution of utilizing the advantage of both feedforward and feedback control will provide the immediate control system response for major load changes—unobtainable with feedback control alone—and the ability to provide for long-term drifting of the controlled variable caused by minor unmeasured process load changes. In Figure 17-24 major measured load changes are shown being fed to the feedforward controller whose output is changing the manipulated variable to maintain the controlled variable at the set point. The minor

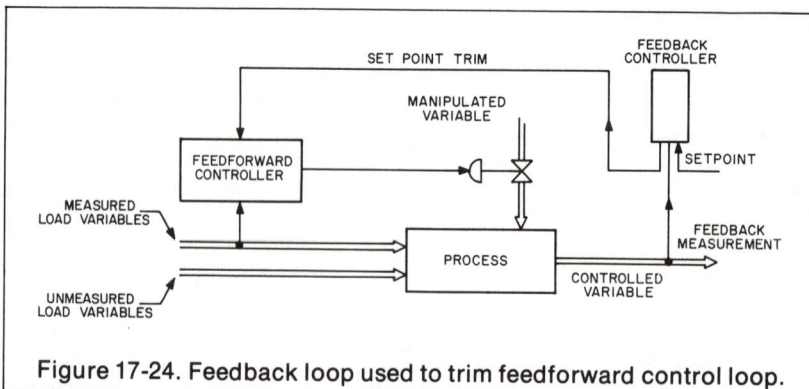

Figure 17-24. Feedback loop used to trim feedforward control loop.

load changes are, however, acting on the process to cause changes in the controlled variable which the feedforward controller is not capable of handling. The controlled variable is fed to the feedback controller which will, in cascade arrangement, cause the feedforward set point to change, thereby compensating for minor deviations in the controlled variable.

BIBLIOGRAPHY

Anderson, N. A. *Instrumentation for Process Measurement and Control.* 2nd ed. Philadelphia: Chilton Company, 1972.

Automatic Control Terminology. Foxboro, MA: The Foxboro Company, TI 3-1a, 1955.

Babcock, R. H. *Instrumentation and Control in Water Supply and Waste Water Disposal.* New York: The Reuben H. Donnelley Corporation, 1968.

LaJoy, N. H. *Industrial Automatic Controls.* Englewood Cliffs, NJ: Prentice-Hall, 1954.

Process Control Instrumentation with Explanatory Notes. Foxboro, MA: The Foxboro Company, Pub. 105A, 1970.

Shinskey, F. G. Process Control Systems. New York: McGraw-Hill Book Company, 1967.

Soule, L. M. "Basic Control Modes." *Chemical Engineering,* October 20, 1969.

Types of Control Action Selection Determined by the Process. Foxboro, MA: The Foxboro Company, TI 3-10a, 1973.

STUDY QUESTIONS

Select the answer or answers which correctly apply to the following statements:

1. That portion of an automatic controller that indicates the desired level of operation is called
   a) the measurement.
   b) a reference source.
   c) output manipulator.

2. In open loop control there is
   a) no feedback.
   b) a measurement.
   c) knowledge of demand.

3. In closed loop control there is
   a) no feedback.
   b) a measurement.
   c) knowledge of demand.

4. With proportional control mode, the final control element is normally
   a) wide open.
   b) closed.
   c) throttling.

5. Control mechanisms employ methods which are
   a) position balance or force balance.
   b) pneumatic or electronic.
   c) opposing or supplementing forces.

6. When there is a difference between set point and measurement, the integral action of a proportional-plus-integral control mode
   a) will continue to drive the final operator in the direction to move the measurement to the set point.
   b) will throttle the final operator to a fixed position relative to the difference.
   c) will throttle the valve in order to maintain this difference.

7. The derivative response is proportional to the
   a) deviation of measurement from set point.
   b) time the measurement is away from set point.
   c) rate at which measurement moves from set point.

8. In tuning a controller, the first adjustment is usually made to
   a) the derivative time.
   b) the proportional band.
   c) the integral time.

9. In ratio control
   a) the output of one controller adjusts set point of another controller.
   b) the secondary measurement variable is controlled at a preset percentage of the primary measurement.
   c) the secondary measurement variable is controlled at a fixed relation to the primary measurement.

10. Feedforward control involves the
    a) measurements in the feedstream.
    b) prediction of correction action.
    c) detection of load changes after they have affected the controlled variable.

(See Appendix for answers)

18. Final Control Elements

The final control element in a loop is the mechanism through which adjustments are made to maintain a required balance between the supply and demand of a process. Generally, this is accomplished by varying the flow of energy or material to a process by manipulating the final control device in response to a signal from the controller to maintain level, pressure, temperature, or some other process condition. Since the controlled medium in the paper industry is usually a fluid of some sort, the final control element is usually a valve. However, depending on the method by which control is accomplished, the final element may also be a damper, louver, variable-speed pump, rheostat, switch, or any such unit. As with transmitters and controllers, the operating signal for final control elements may be pneumatic, electric, hydraulic, or mechanical. There are many kinds of final control elements. For purposes of automatic control, there are two main categories: (1) those designed for on-off control or intermittent (batch) type operations and (2) those designed for throttling service on continuous processes that require smooth and even adjustments to maintain balanced operation.

CONTROL VALVES

A control valve is often defined as a variable orifice in a fluid flow system. It can be adjusted manually by handwheel or lever, or it can be operated automatically by controller output signals. The final operator in a control loop directly changes the value of the manipulated variable by changing flow rate.

A complete automatic control valve consists of three major components:

1. The actuator transforms the controller signal into motion, providing power to vary the orifice.
2. The valve body assembly consists of a pressure-tight fitting that is threaded, flanged, or welded in a fluid flow line and contains one or more internal orifices through which fluid flow is controlled.
3. A plug, damper, or louver is positioned in the orifice by the actuator to control the pressure drop and rate of flow.

Figure 18-1 shows a typical complete pneumatic automatic control valve.

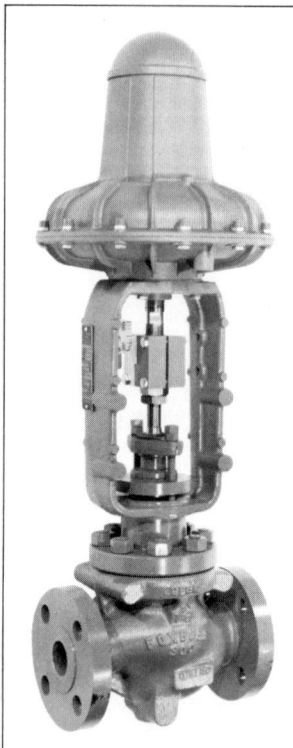

Figure 18-1. Typical complete pneumatic automatic control valve.

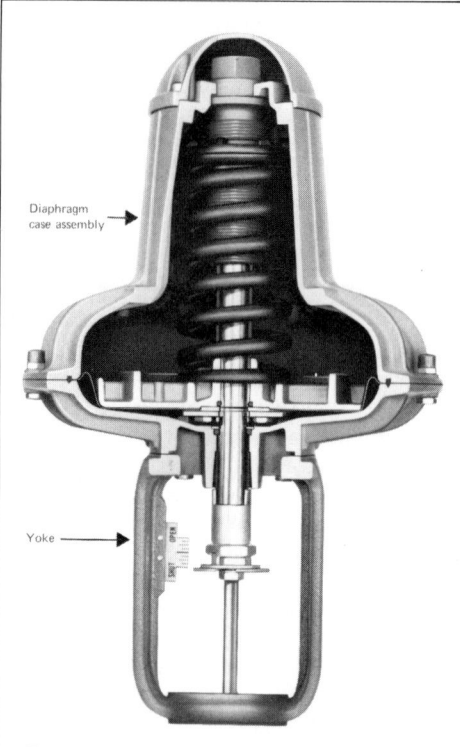

Figure 18-2. Diaphragm type valve has two main parts—actuator yoke and diaphragm case assembly.

ACTUATORS

Depending on the input signal, control valve actuators may be pneumatically, electrically, hydraulically, mechanically, or manually operated. Pneumatic operation is the most widely used actuation method in the pulp and paper industry. Some electric, hydraulic, electrohydraulic, and mechanical actuators have been used in areas where no operating air is available, or where low ambient temperatures create problems of water freezing in the air lines. The two major types used are the diaphragm type and the cylinder type. The diaphragm type may be either spring or springless while cylinder types are usually springless.

Diaphragm Type Actuators

Diaphragm actuators (Figure 18-2) consist of two main parts: the yoke (or superstructure) and the diaphragm case assembly.

The yoke can be a single casting or an assembly of steel, cast iron, or other

326   Paper Industry Instrumentation

suitable metal that will not yield, causing stem misalignment. Diaphragm cases are made of cast iron, steel, aluminum, or other suitable metal. Normally, these actuators use low operating air pressure and perform either *air-to-lower* or *air-to-raise* action.

The purpose of the diaphragm is to act as a seal between the upper and lower diaphragm chambers. The conventional type diaphragm consists of a flat piece of flexible material of circular shape. Its diameter is a function of the required stem thrust. The diaphragm is often premolded in order to prevent crimping or stretching and to give better effective area for stroke characteristics. In the spring type pneumatic diaphragm actuator (Figure 18-3) with air-to-lower action, the controlled air signal is admitted on the upper side of the diaphragm, moving the stem down in opposition to the spring. The actuator stem moves upward with decrease of signal air pressure. This action is sometimes referred to as *direct acting;* but this can be confused with controller action and, therefore, this use is not preferred. When air signal pressure is introduced below the actuator diaphragm moving the stem up, it is referred to as *air-to-raise* action. It is also referred to as *reverse*, but to avoid confusion with controller action, the former is also preferred here. Diaphragm actuators have been designed to be used for a specific action. However, actuators such as those shown in Figures 18-2 and 18-3 can be used for either action by removing the cap, turning the actuator over, and replacing the cap. This type of actuator is referred to as *reversible*.

The springless versions of the actuator (Figures 18-4A and 18-4B), like the spring type, can be made to provide stem position proportional to signal air pressure. The signal air pressure is directly impressed on one side of the diaphragm and an independently regulated air supply is fed to the opposite side to provide an on-off operation. Since the actuator does not have a spring to provide positioning, some type of valve positioner which receives the control signal from the controller is necessary to provide for accurate intermediate

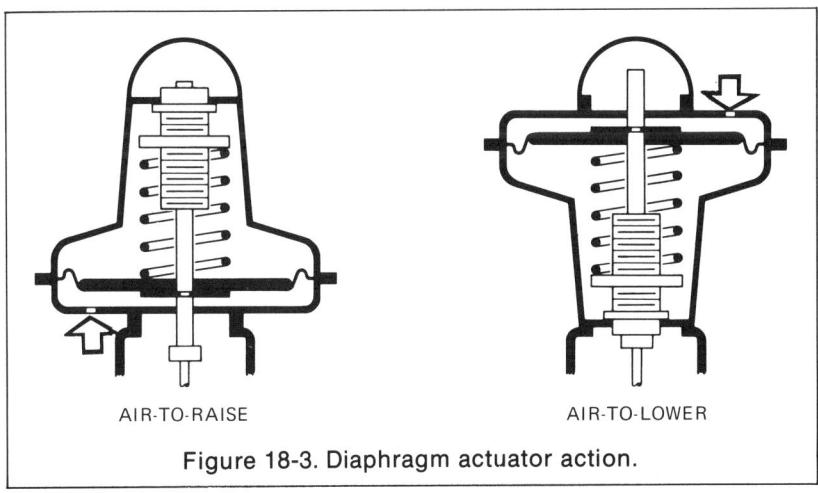

Figure 18-3. Diaphragm actuator action.

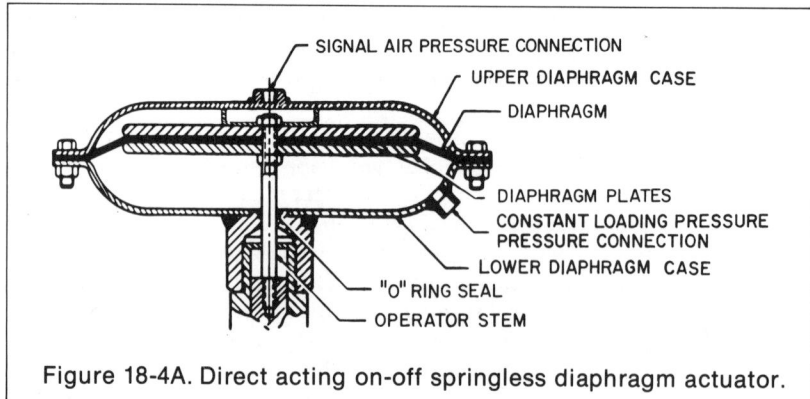

Figure 18-4A. Direct acting on-off springless diaphragm actuator.

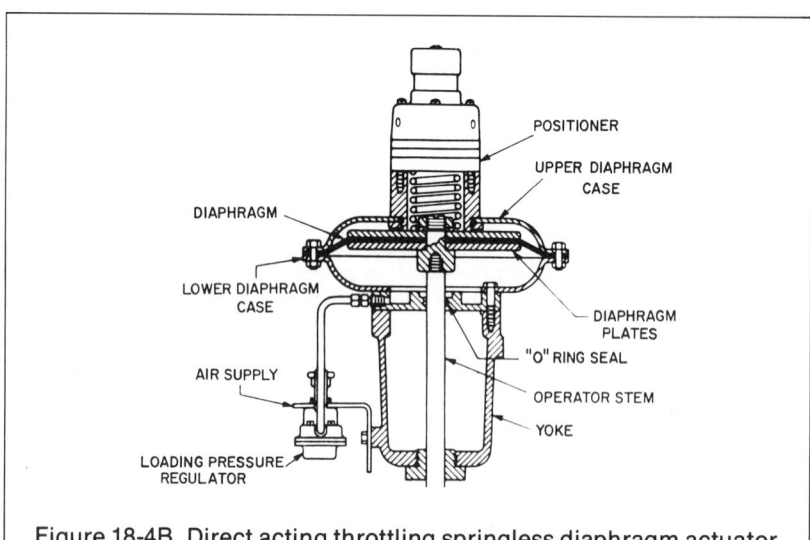

Figure 18-4B. Direct acting throttling springless diaphragm actuator.

stem positioning in throttling applications. When air-to-lower action is required, the regulated air supply is fed to the lower diaphragm chamber and the signal air to the upper chamber. For air-to-raise action, the connections are reversed.

Cylinder Type Actuators

Cylinders are either pneumatically or hydraulically operated. The most common pneumatically operated cylinder type actuator (Figure 18-5) consists of a piston operating in a pressure-tight cylinder. The pistons are made from several metals, such as steel, alloy steel, cast iron alloy or high-tensile aluminum alloy, and are ground to very close tolerances.

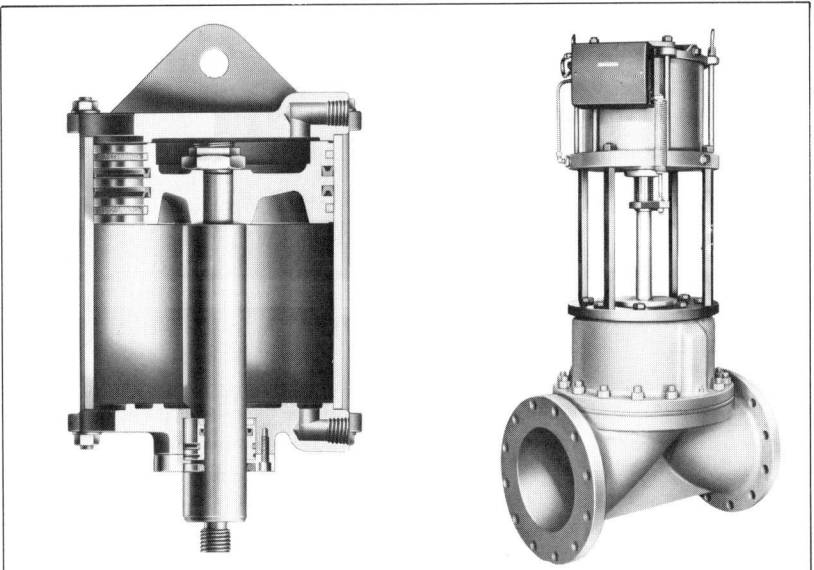

Figure 18-5. Simple cylinder actuator, left; cylinder actuator on valve, right. Unit consists of a piston in a pressure-tight cylinder.

Similar materials are used for the cylinder housings. The inner diameter of the housing is precision-honed and the piston rod is held to close tolerances and carries a high-grade surface finish. The close tolerances and high finish are required in cylinder operation in order to minimize breakaway friction, preventing possible cycling. Seals between the piston and the cylinder walls, and the piston rod and end cap, are used to prevent leakage of air out of the cylinder housing.

As in the springless diaphragm actuator, the air signal is impressed over the upper side of the piston and a lower air pressure is admitted to the lower side of the piston for air-to-lower action and the reverse arrangement is for air-to-raise action. When used on a throttling type final control element, a positioner which receives the air signal from the controller is used to produce the actuating pressures to the cylinder for positioning the stem.

When the cylinder is hydraulically operated it does not require a positioner for throttling because hydraulic pressure is supplied in proportion to the deviation of the controller variable, causing the piston to gradually float in one direction or the other. Other cylinder actuator arrangements have used four-way on-off valves to load and unload pressure above or below the piston, depending on the action desired. Some cylinder type operators have also been equipped for spring loading.

Cylinder actuators can use higher operating air pressures than diaphragm type actuators and, therefore, they can handle greater final control-element positioning loads such as on a valve, damper, or louver.

Electric Actuators

Electric actuators may be classified as solenoid operated or motor operated. Solenoid actuators essentially consist of an electromagnet and a moveable armature which is linked to, or is an integral part of, the final control element (see sketch of solenoid unit on valve in Figure 18-6). These actuators are generally used to provide two-postion (on-off operation) only and they are not adaptable to proportional positioning or to operation with large valves or high pressure drops.

Electric motor actuators consist of a suitable driving motor, reduction gearing, and an output shaft or lever arm, all in a common housing. These have been used where proportional positioning of the final control element has been required. However, they have not been widely used because of the cost and relative difficulty in adapting them to automatic control. To meet the demand for an electrically operated actuator which could be used for true modulating of final control elements, the electrohydraulic actuator was developed. In a typical design (shown schematically in Figure 18-7) the input signal from the controller positions a linear differential transformer located in the feedback loop from the actuator piston to a pilot valve which establishes true linearity of plug position to signal. A three-section pump is used to supply 50 psig hydraulic pressure to operate the control system and two 500 psig streams for cylinder operation. A flapper positioned by a force motor in the controller circuit tilts to cover and uncover two nozzles. Restricting the nozzle flow allows pressure to be transmitted to the side of the piston affected by the restriction. A change in piston position creates a change in feedback spring tension. When the spring force equals the force motor effect, the flapper assumes a neutral position. On loss of electric power, the cylinder shutoff valves lock up the pressure in the cylinder to maintain existing conditions. A bypass valve between cylinder chambers allows pressure equalization for manual operation with a handwheel.

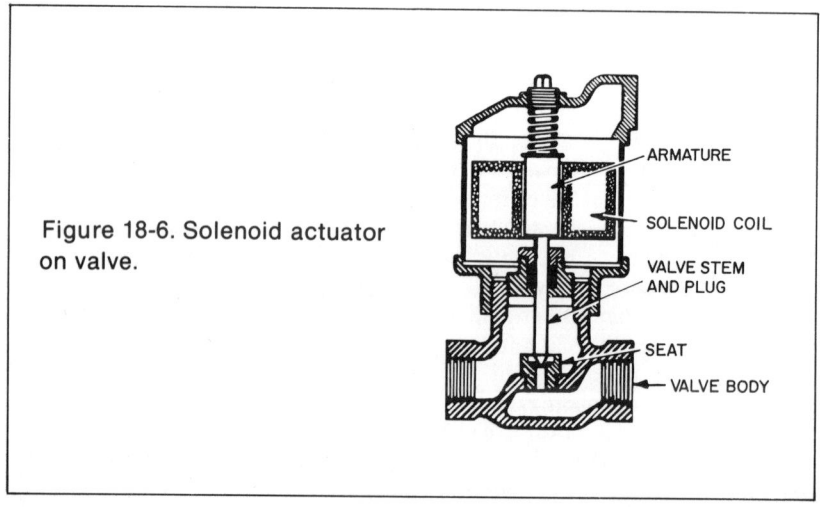

Figure 18-6. Solenoid actuator on valve.

ARMATURE

SOLENOID COIL

VALVE STEM AND PLUG

SEAT

VALVE BODY

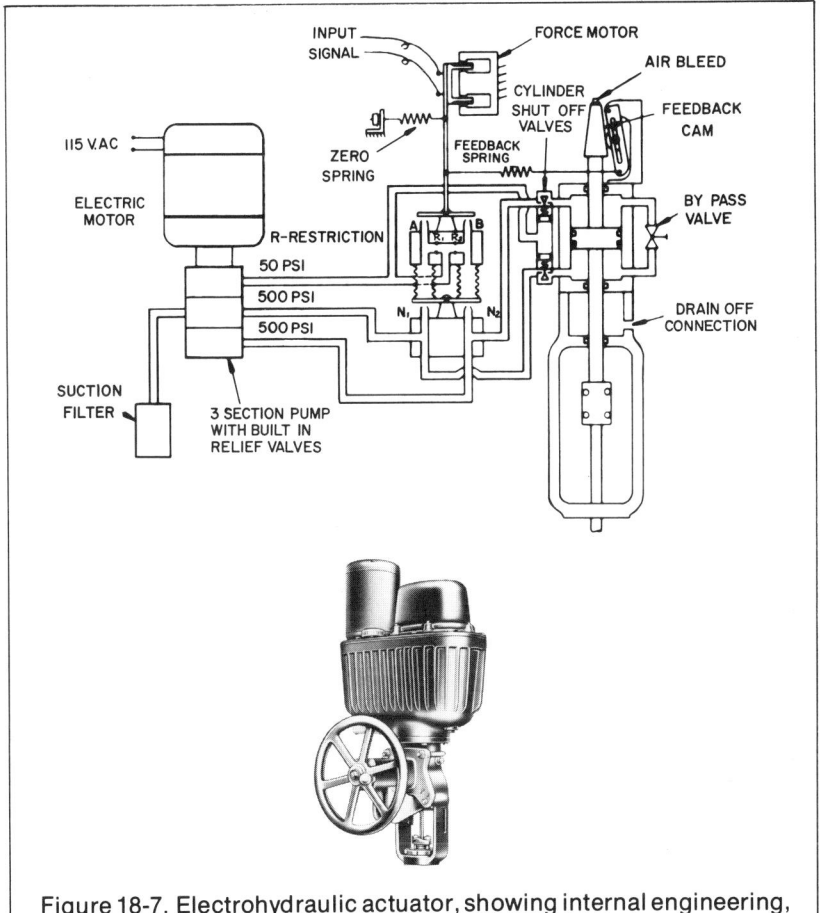

Figure 18-7. Electrohydraulic actuator, showing internal engineering, top, and casing, bottom.

VALVE BODIES

The valve body consists of the outside housing, valve seats or ports, and the stuffing box. It contains the valve plug or inner valve which moves into the valve seat or port. Many varieties of valve bodies are available, and those most common to the pulp and paper industry are:

1. Globe type
2. Saunders patent type
3. Butterfly type
4. Ball type
5. Rotary plug type
6. Three-way type

Final Control Elements   331

Control valves are essentially pressure vessels and they are available with end connections to meet a wide variety of applications. They are manufactured using a wide range of materials including iron, bronze, steel, stainless steel, low alloy steels, special alloys, as well as plastic materials.

Valve Plugs and Characteristics

The valve plug is that movable part of the body assembly which provides the variable restriction to flow and is the principal functional part of the "trim" portion of the valve body. Trim consists of those parts of the valve that come in direct contact with the process fluid. The design of the plug varies according to the specific set of flow characteristics desired to be imparted to the valve. When the percent of the valve lift is plotted against percent of maximum flow, significant curves are formed. Matching these characteristics with those required by the process and control loop enables correct process control.

Valve plugs are normally designed for either two-position (on-off) or for throttling control. With the on-off type used in many control applications in the paper industry, the valve plug is positioned by the actuator at one of two points within its travel. These plugs are available for *lower-to-close* or *lower-to-open* action. The selection depends, among other things, on the actuator used and the fail-safe position required (valve to close or open on air failure). The shape of the on-off valve plug varies. A typical plug used for this kind of service is the *beveled-disc* shown in Figure 18-8. This is a single-seated, lower-to-close, air-to-open valve assembly, sometimes called a *poppet valve*.

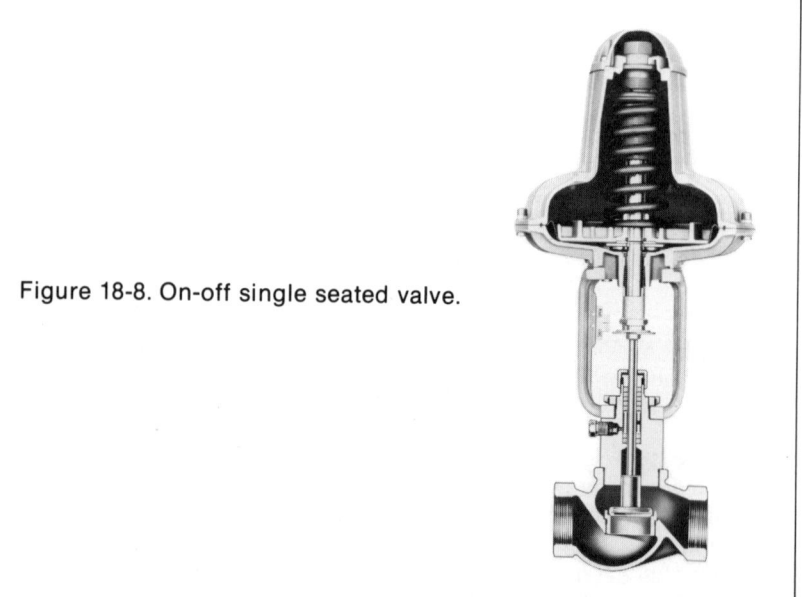

Figure 18-8. On-off single seated valve.

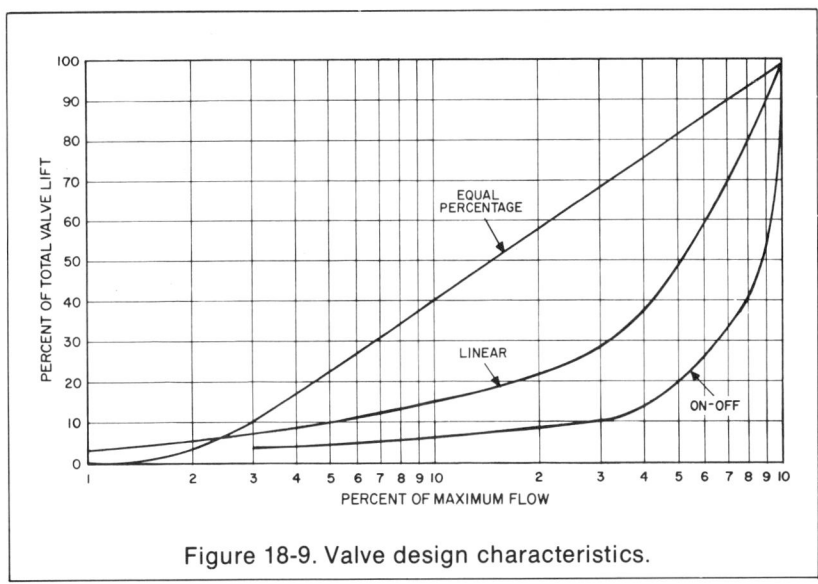

Figure 18-9. Valve design characteristics.

Valve plugs used in throttling control are positioned at any point within their travel as dictated by process requirements. These, too, are available in a variety of shapes which determine the characteristics of operation. At one time, the most widely used plug style was the equal percentage characteristic. That is, for all positions of the plug relative to the port opening—for example, percent of valve lift—equal increments of lift produced equal percentage changes in flow under constant pressure drop across the valve port. A later development was valve plug design with linear characteristics; for example, a plug with a linear relationship between plug-to-port opening position and flow rate. These design characteristics are summarized in Figure 18-9 with semilog plots.

Typical characterized valve plugs used for throttling control are of the so-called parabolic and V-port configurations. The parabolic type is used to produce approximately linear characteristics. When equal percentage characteristic valve response is desired, both the V-port and contoured plugs can be used. Figure 18-10 shows a typical equal percentage V-port plug in a valve body along with the parabolic plug used for the linear characteristic. The equal percentage V-port configuration is a common type used for general applications in the paper industry today, while the parabolic plug is employed to minimize erosion when used under conditions of turbulent flows at high velocity and pressure drops.

On small flow rates a contoured plug called a *needle* has been designed to produce an equal percentage flow characteristic (Figure 18-11). This type of plug ensures precise control even at the lowest flows and is generally used in the control of acid, caustic, or alum on pH systems and dye solutions in stock blending systems.

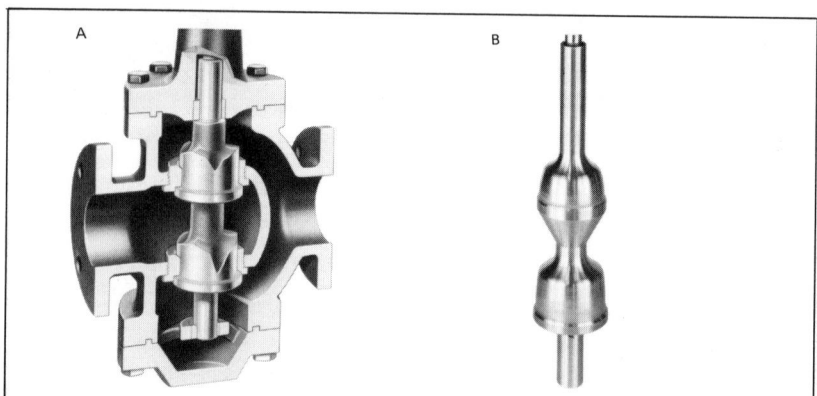

Figure 18-10. Conventional globe valve plug: A, equal percentage V-port; B, linear parabolic.

Process Dynamics

The dynamic response of the process and the transmitter/primary device combination is an important consideration in choosing the proper valve characteristic. To exactly determine the response of a process requires a

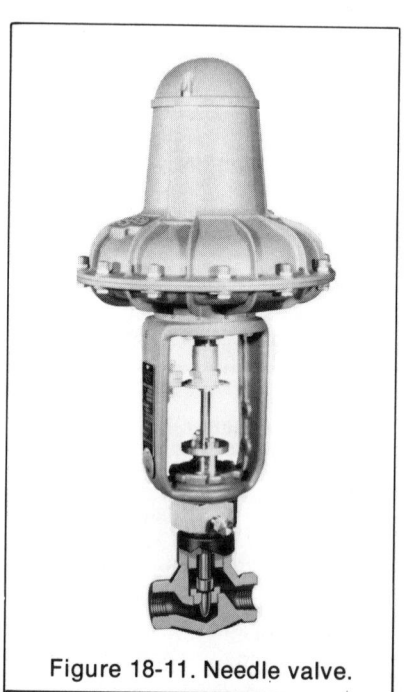

Figure 18-11. Needle valve.

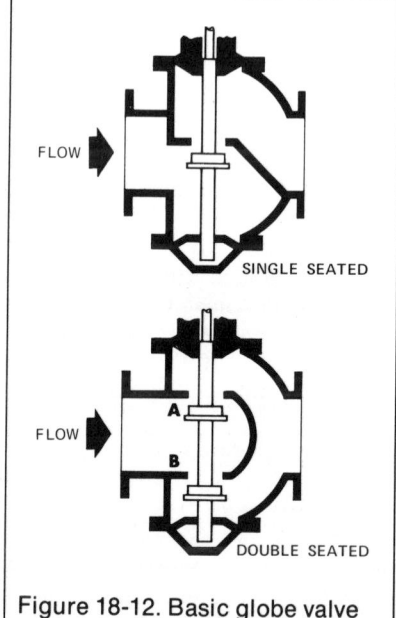

Figure 18-12. Basic globe valve bodies, single and double seated.

complete dynamic analysis for each control loop. This is difficult and usually impractical. However, it is possible to establish general guidelines based on experience. Among the normal applications encountered in paper industry instrumentation are flow, pressure, temperature, level, and pH control. The suggested valve characteristics for these applications are shown in Table 18-A.

A misapplied equal percentage valve results in increasing valve sensitivity at high flow rates and can cause instability unless the controller's proportional band is adjusted at the high flow rate. The control loop will then tend to be over-damped at the low flow rates, with corresponding sluggish response.

A misapplied linear valve exhibits the opposite effect. A controller properly adjusted at the high flow rate would have too narrow a proportional band setting for stability at low flows. The proportional band would have to be adjusted at low flow rates and the control loop would be over-damped and sluggish at higher flow rates.

Globe Type Valves

Globe valves are the most popular type and are available in two general styles. The older style conventional globe valves found in the paper industry are either single-seated or double-seated. Single-seated valves have only one orifice through which the fluid must pass and the valve plug has only one seating surface and a single valve seat. A double-seated valve has two orifices and the plug has two seating surfaces and two valve seats. Double seating offers a balancing action not available in the single-seated conventional designs. Note that in Figure 18-12 (upper sketch), the plug must open against full line pressure effects, requiring a larger and more expensive actuator. In Figure 18-12 (lower sketch) the line pressure tends to open the plug at the lower port

TABLE 18-A

VALVE CHARACTERISTIC SELECTION

Applications (factor controlled)	% of System Drop Across Valve	Characteristics
1. Flow—linear w/differential pressure	<20	Equal percentage
2. Flow—linear w/flow	<40	Equal percentage
3. Flow—linear w/differential pressure	>20	Linear
4. Flow—linear w/flow	>40	Linear
5. Pressure	≈100	Linear
6. Pressure	<50	Equal percentage
7. Liquid level	<40	Equal percentage
8. Liquid level	>40	Linear
9. pH	<50	Equal percentage
10. pH	>50	Linear
11. Temperature	>50	Equal percentage

and close it at the upper port. With this counterbalancing of forces, the actuator is never acting against full line pressure, allowing the use of smaller and less expensive actuators. However, the double-seated conventional style valve should not be used where tight shutoff is desired because temperature variations create unequal expansion between the inner valve parts and the body, making it impossible for both upper and lower ports to be tightly closed at the same time. If tight shutoff is required, the single-seated conventional valve, or cage style, discussed next, could be used.

Cage Style Valves

In a more recent development of globe style valves, the characteristics have been designed in the plug guiding portion of the valve trim called the *cage*. In the balanced trim design, ports of various configurations producing either linear or equal percentage characteristic response are cast into the cage. The plug moves past these apertures to produce the desired response (Figure 18-13).

These cage style valves have been designed to replace the conventional style globe valves. The unbalanced cage trim replaces single-seated valves and the balanced cage trim replaces the double-seated valves. Some balanced cage trim designs serve double duty since they provide tight shutoff, which double-seated valves cannot provide.

Newer materials have been selected by most manufacturers of cage style valves for use in the valve trim. As standard, most are some type of hardenable

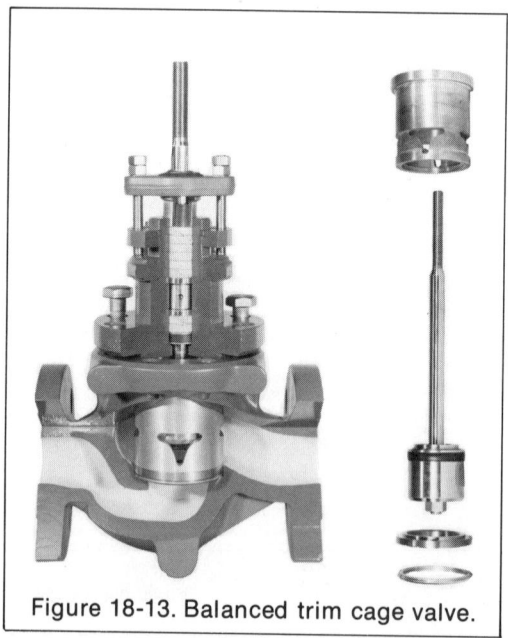

Figure 18-13. Balanced trim cage valve.

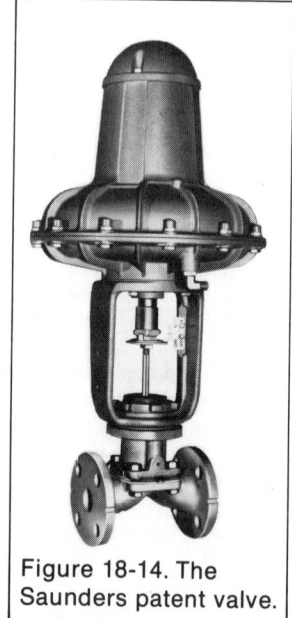

Figure 18-14. The Saunders patent valve.

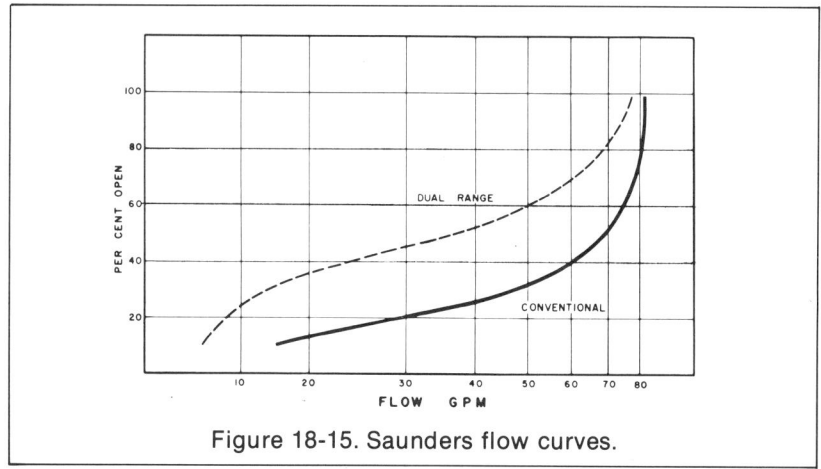

Figure 18-15. Saunders flow curves.

stainless steel such as 17-4PH or 416SS. This provides greater resistance to erosion and wear caused by high-velocity fluids or fluids with suspended solid particles.

Generally speaking, the streamlined bodies of these newer style valves provide higher flow capacities than the older style conventional valves.

The construction of cage style valves permits only lower-to-close action and any reversing of operation must be performed in the actuator.

Saunders Patent Type Valves

The Saunders type control valve body was originally designed for manual operation and later adapted for use in automatic control (Figure 18-14). In operation, the valve actuator raises or lowers a flexible diaphragm to effect process flow control by seating on a weir. The diaphragm isolates all moving parts from the process fluid, providing smooth, virtually pocketless passage not found in globe type valve bodies, which makes them attractive for use on corrosive fluids, slurries, and pulp stock. The major deficiency of a Saunders type valve has been its inability to offer good throttling characteristics. Consequently, its use has been limited to on-off and narrow-range throttling control systems. This has been especially true for the high- and low-flow ranges of the flow curve (Figure 18-15).

To correct the poor throttling characteristics at the low end of the flow curve, the one-piece diaphragm follower is redesigned to provide dual-range response. This modification consists of a two-piece compressor assembly that provides independent control over two areas of the valve diaphragm. The first increments of stem travel raise only the inner compressor from the weir (Figure 18-16). This allows flow through a contoured opening in the center of the valve rather than through a slit across the entire weir as offered in the conventional style. The outer compressor is still held firmly seated by the upper spring in the valve bonnet. With further stem travel, the inner compres-

Final Control Elements   337

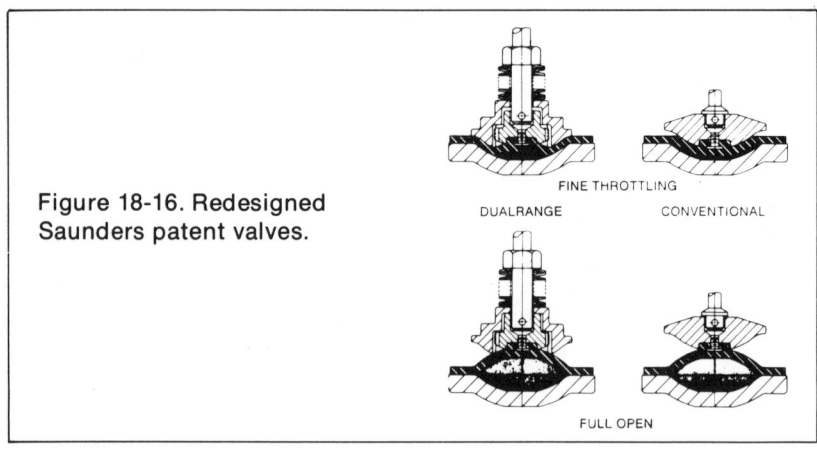

Figure 18-16. Redesigned
Saunders patent valves.

FINE THROTTLING

DUALRANGE   CONVENTIONAL

FULL OPEN

sor contacts the outer compressor and both are now raised together by the valve actuator until the regulated capacity of the valve is achieved.

The construction of the Saunders type valve permits only lower-to-close action and any reversing of operation must be done in the actuator.

Butterfly Type Valves

Butterfly valves are used for low line pressure, large line sizes; for pulpy or semisolid materials which would foul plug type valves; for gas and air flow control and where a large-size, low-cost valve is desirable for such applications as controlling large flows of water, liquors, and stock slurries up to 2 percent consistency. They can be found in practically all areas of a pulp and paper mill controlling such flows as cooling water to heat exchangers, makeup water to hot water storage tanks, condenser water, green liquor to heaters, interstage filtration to brown stock washers, filtration to repulpers, hot water to washers, condensate, and white water. Other butterfly valve applications in the pulp and paper mill include level control in seal tanks and control of flat box vacuums.

A butterfly valve (Figure 18-17) consists of a cylindrical body with a disc or vane that rotates on a shaft installed perpendicular to the axis of the cylinder. The body may be flanged at both ends or made in the form of a solid ring. The latter type is known as a *wafer* valve. When this type valve opens, the disc actually extends into the pipeline.

Butterfly valves can be lined with rubber or other materials for either corrosion resistance or tight shutoff. Generally, the control characteristic is not as good as with the globe control valves.

Torsional force on the shaft of a butterfly valve increases as the valve is opened, reaching a maximum value at approximately 70 degrees from perpendicular to the pipeline, after which it tends to diminish. For maximum stability in throttling positions, this valve is usually not rotated beyond this reversal point, limiting the maximum opening to about 70 degrees from vertical. There are some butterfly valves made so that the vanes strike the body and effect closure at 15 degrees from vertical, resulting in a maximum of about 60 degrees recommended rotation in operation. This metal-to-metal seating may be expected to leak about 0.5 to 1 percent of total capacity in the closed position, a condition which can be improved by a rubber liner.

Rotary Plug Type Valves

Recently the rotary plug type family of valves has gained increased acceptance for controlling the flow of pulp stock and other difficult process fluids in the paper industry. In these valves, the plug is rotated inside the valve housing or body. The shape of the housing depends on the shape of the plug, which may be cylindrical or spherical. The two major types of rotary plug valves are the plug cock style and ball valve.

As illustrated by the schematic drawing of the plug cock in Figure 18-18, the rotary motion uncovers a part of the opening through the plug. The plug may have a rectangular port or a V-port with characteristics the same as corresponding globe valves. With rotary plug cock valves, characterization is obtained by means of a linkage arrangement between power unit and plug cock, a cam in the positioner, and/or configuration of the port opening in the plug cock. A typical rotary plug cock valve with a positioner containing a cam for characterization is shown in Figure 18-19.

Like the rotary plug cock valve assembly, the ball valve plug is designed with either a V-notch or straight-through bore and characterization is obtained

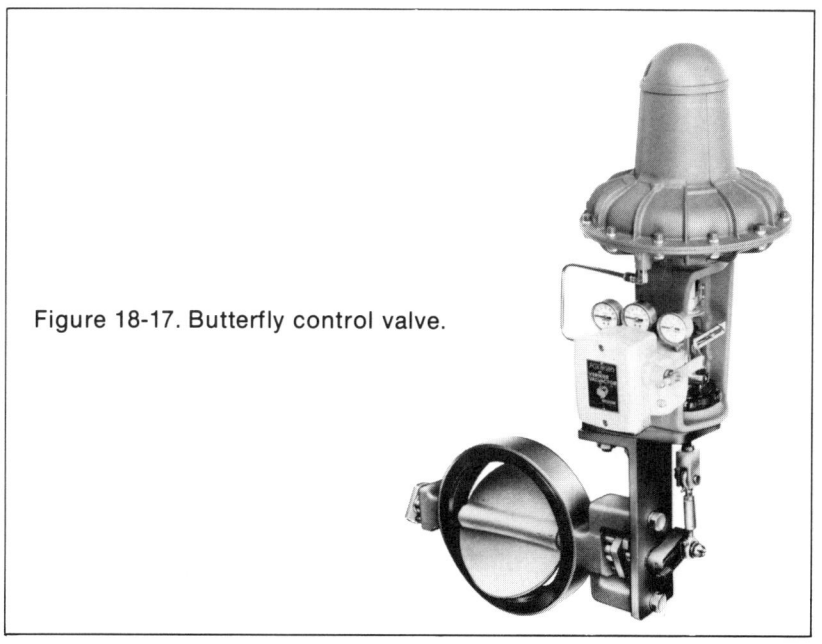

Figure 18-17. Butterfly control valve.

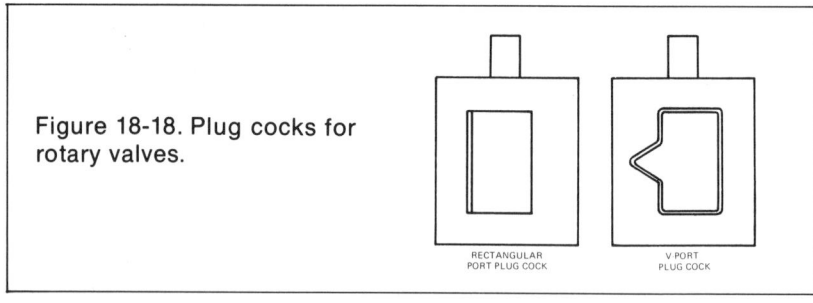

Figure 18-18. Plug cocks for rotary valves.

RECTANGULAR
PORT PLUG COCK

V PORT
PLUG COCK

in a similar manner. The V-notch ball results in the same flow characteristic response as the modified parabolic plug in the globe style valves.

On the other hand, the straight-through flow pattern ball plug (Figure 18-20) with a double elliptical orifice configuration provides inherent equal percentage characteristics. During throttling (Figure 18-21) the double orifice reduces the effect of velocity erosion by splitting the total pressure drop across both orifices; allows the valve to regulate pulp stock slurry flow without dewatering.

The seal between plug and body of a rotary plug valve is either metal-to-metal or to some other material of construction, such as rubber or plastic, to obtain tight shutoff conditions. Figure 18-22 shows where PTFE (Teflon) is used as a seating material. The flexible lip exerts continuous sealing pressure, compensating for wear.

Three-way Control Valves

All of the control valves discussed so far are used to regulate the flow of fluid and are characterized with one inlet and one outlet and are sometimes

Figure 18-19. Typical V-port rotary plug cock valve.

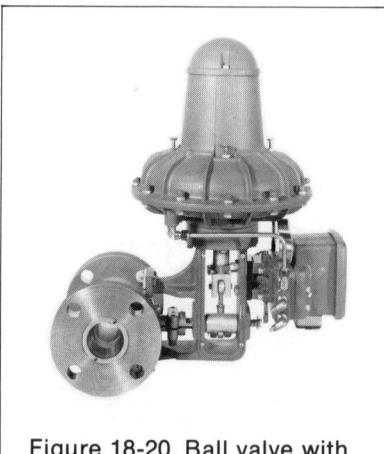

Figure 18-20. Ball valve with straight-through flow pattern.

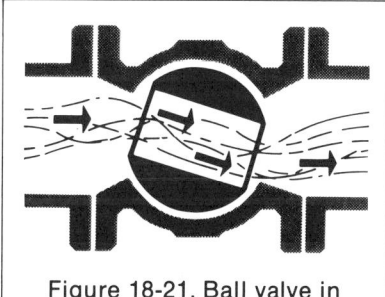

Figure 18-21. Ball valve in throttling position.

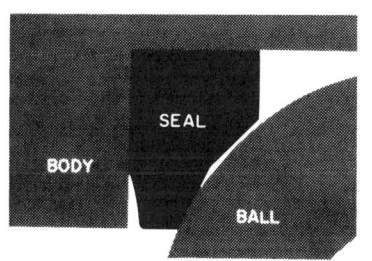

Figure 18-22. PTFE seating on ball valve.

referred to as *two-way valves*. There is another group of valves which is a modification of the two-way valves obtained by the addition of a third flow opening, making them suitable for splitting or blending flow streams. Three styles of three-way control valves have been developed to handle such paper industry applications as throttling split flow control, throttling blending control, and on-off diversion control. In considering the application of a three-way valve, it should be understood that as the stroking of the inner plug alters the area relationship at one seat ring, it conversely changes the exposed area at the other seat ring. Therefore, the total capacity of a three-way valve is a summation of the two discharge or input ports, depending on the direction of flow.

Split Flow. Figure 18-23 illustrates an astroid ported three-way valve. It is designed to duplicate the performance of two throttling type control valves. The upstream pressure entering the side port is proportionally split in two directions against an area-balanced inner plug so that the dynamic hydraulic forces acting in opposite directions are nullified. This null balance effect allows the pneumatic actuator to perform a smooth, steady control action which is necessary to throttling control systems.

Blending The three-way valve shown in Figure 18-24 is used to proportionally combine two flowing streams and incorporates two seats and double

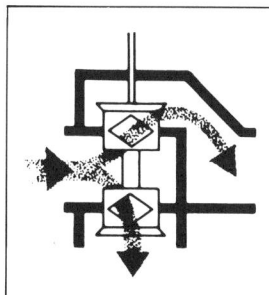

Figure 18-23. Split flow three-way valve.

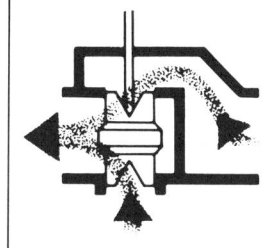

Figure 18-24. Blending three-way valve.

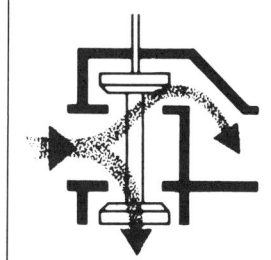

Figure 18-25. Diversion three-way valve.

V-ports. The left-hand port is the common discharge port and the two inlet streams to be blended are admitted from the right-hand and bottom ports. The inlet pressure forces acting on a single disc oppose each other, canceling most of the dynamic hydraulic effect. Again, this counterbalancing of forces allows the pneumatic actuator to deliver a smooth action.

Diversion. Figure 18-25 is a three-way, double-seated control valve with two poppet style beveled discs. This style is recommended for on-off service only where the full flow is directed through one port at a time. For on-off diversion service, the inlet port is the left-hand port, and the flow is diverted into either the bottom port or the right-hand port. Note that the inlet flow tends to open either port. The hydraulic forces created are opposed by either the spring force or diaphragm air pressure which eliminates excessive slamming when the valve is fully stroked.

Rotary valves such as the ball or plug cock valve have also been modified to three-way designs that are suitable for use with blending, on-off, and diversion applications.

Valve Type Selection

Another important consideration in valve selection is the type of valve to be employed. A globe valve, for instance, is a very rugged, multipurpose valve, but considerable savings could be brought about if a less expensive, lighter duty valve such as a butterfly could be used. Figure 18-26 shows the pressure

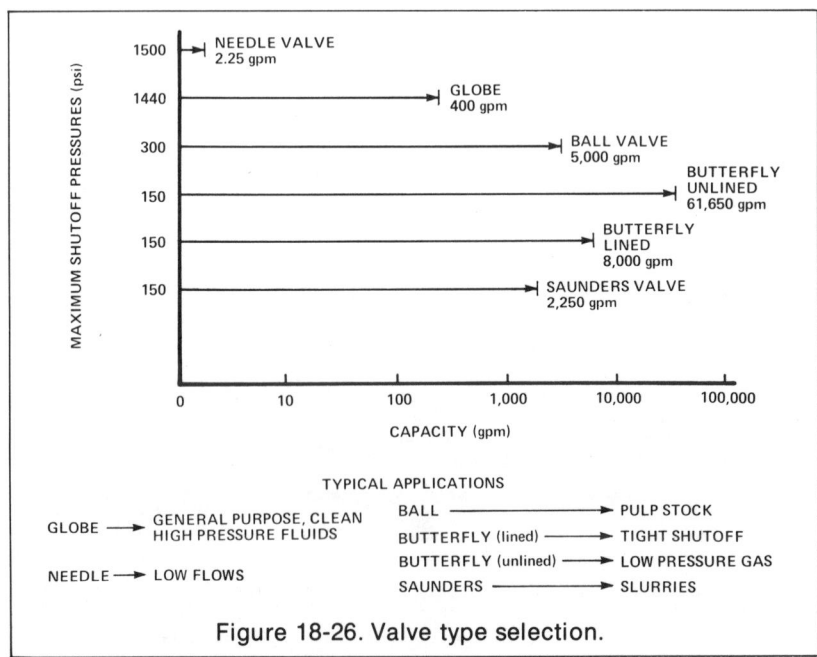

Figure 18-26. Valve type selection.

limits and maximum capacities of the various types found in pulp and paper mills. Also shown are common applications for different valves.

Corrosion Considerations

Process fluid corrosive effect, which is another consideration in valve selection, depends heavily on its concentration, temperature, and the presence of other chemicals either as impurities or as part of the composition. Table 18-B is a general guide to typical materials of construction often found adequate on some stream conditions that it is possible to encounter in the paper industry.

Since only the user is in a position to know the exact chemical content of the stream, and to know and control such important aspects as temperature of the fluid, it is necessary that the final responsibility for choice of suitable materials remain with the person who will use them.

VALVE POSITIONERS

In certain control valve applications, factors such as high stem friction, valve plug and stem imbalance, and diaphragm and spring hysteresis cause the valve plug to assume a position not strictly proportional to the instrument signal. Also, there are valve installations which involve long transmission distances and extreme or variable line pressures. When these conditions exist, an auxiliary device operated by the instrument but with a separate air supply is used to provide the necessary pressure to the actuator for accurate positioning in accordance with the demands of the controller. Some of the conditions under which valve positioners have been specified are:

1. Split range operation, for example, where one valve moves open-to-close with a control air pressure change between 3 psi and 9 psi and another between 9 psi and 15 psi, both valves being connected to the same control air pressure.
2. Valves handling viscous fluids, slurries, or solids in suspension.
3. Single-seated valves handling high pressure drops at high flow velocities requiring large stem thrusts (includes Saunders patent type valves).
4. Three-way control valves.
5. Operation where more than normal tightening of the packing gland is required.
6. Operation where increased speed is desired.
7. Positioning of springless cylinders or diaphragm actuators.
8. Operation of very large valves requiring large air volumes.

Many valve positioners are available but, in principle, all are essentially the same; they are mechanically connected to the valve stem so that actual position can be compared to desired position. Positioners are available in two basic mounting configurations: as an integral part of the actuator or yoke-mounted, allowing removal from the valve without disturbing normal operation. A

TABLE 18-B

VALVE MATERIAL CONSIDERATION

Fluid	Body	Trim	Diaphragms (Saunders) or Liners (Butterfly)
Air	Cast iron	17-4PH	Butyl
Alcohol, methyl	Cast iron	17-4PH	TFE
Aluminum sulfate	316 S.S.	17-4PH	EPR
Ammonia	Carbon steel	17-4PH	TFE
Ammonium hydroxide	Alloy 20	Alloy 20	TFE
Ammonium sulfate	Hastelloy C	Hastelloy C	Neoprene
Ammonium sulfide	316 S.S.	17-4PH	TFE
Black liquor	Cast iron	17-4PH	EPR
Brine	Bronze	316 S.S.	TFE
Calcium chloride	Alloy 20	Alloy Trim	Penton
Calcium hydroxide	Cast iron	17-4PH	EPR
Calcium hypochlorite	316 S.S.	17-4PH	TFE
Calcium sulfate	316 S.S.	17-4PH	Neoprene
Carbon dioxide	Cast iron	17-4PH	EPR
Carbon monoxide	Cast iron	17-4PH	TFE
Chemical pulp	316 S.S.	17-4PH	TFE
Chlorine (dry)	316 S.S.	17-4PH	TFE
Chlorine (wet)	Hastelloy C	Hastelloy C	TFE
Glue	Cast iron	17-4PH	TFE
Hydrochloric acid	Hastelloy C	Hastelloy C	TFE
Hydrogen peroxide	316 S.S.	17-4PH	TFE
Hydrogen sulfide	316 S.S.	17-4PH	TFE
Magnesium hydroxide	316 S.S.	17-4PH	EPR
Magnesium sulfate	316 S.S.	17-4PH	EPR
Natural gas	Cast iron	17-4PH	TFE
Nitric acid	Alloy 20	Alloy 20	TFE
Oxygen	Cast iron	17-4PH	TFE
Peroxide bleach	316 S.S.	17-4PH	TFE
Size	316 S.S.	17-4PH	EPR
Sodium chloride	Hastelloy B	Hastelloy B	TFE
Sodium hydroxide	Cast iron	17-4PH	EPR
Starch solutions	Cast iron	17-4PH	TFE
Sulfate liquor	316 S.S.	17-4PH	TFE
Sulfite liquor	Alloy 20	Alloy 20	TFE
Sulfur dioxide gas	316 S.S.	17-4PH	TFE
Sulfuric acid	Monel	Monel	TFE
Sulfurous acid	Alloy 20	Alloy 20	TFE
Water	Cast iron	17-4PH	TFE
White liquor	Cast iron	17-4PH	EPR

yoke-mounted positioner is shown in Figure 18-27. A positioner is a kind of proportional controller in itself. It contains a complete feedback loop which senses valve stem position and its deviation from the expected value. As with controllers, the basic designs of the positioners are either force balance or position balance.

Force-Balance Positioner. The basic force-balance positioner is shown in Figure 18-28. This consists of a bellows that receives the instrument signal, a beam fixed to the bellows at one end and linked to the valve stem at the other, a pilot valve connected to the beam, a fixed pivot point, and a spring that produces a force proportional to the valve stem position. The pilot is at equilibrium whenever the force produced by the bellows is balanced by the force produced by the spring. For every force produced by the bellows there is a corresponding force from the spring that returns the pilot to equilibrium.

Figure 18-29 is a schematic illustration of a typical force-balance positioner. In operation, the controller or instrument air acting on the bellows is balanced by a spring in compression. The spring force is changed by valve stem movement which is transmitted to the spring through a compound lever system. An increased instrument pressure moves the bellows assembly to the left to open the supply valve allowing the operating medium to flow to the control valve diaphragm. The diaphragm pressure increases until stem movement creates enough spring force to return the bellows to its original position, closing the supply valve. The unit is again at equilibrium but at a higher value of instrument pressure and at a new stem position. On decreasing instrument pressure, the exhaust valve in the bellows moves to the right, relieving diaphragm pressure to atmosphere. The valve stem moves up, decreasing the spring force

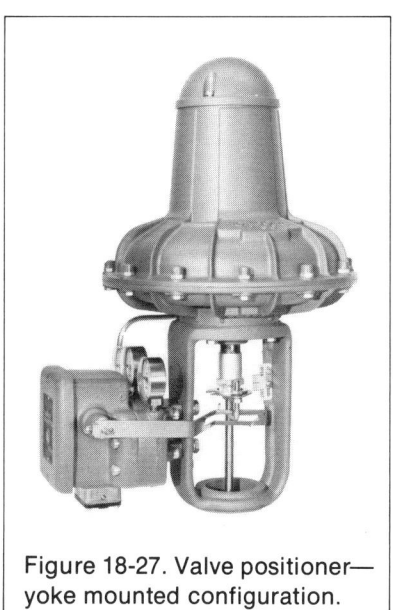

Figure 18-27. Valve positioner—yoke mounted configuration.

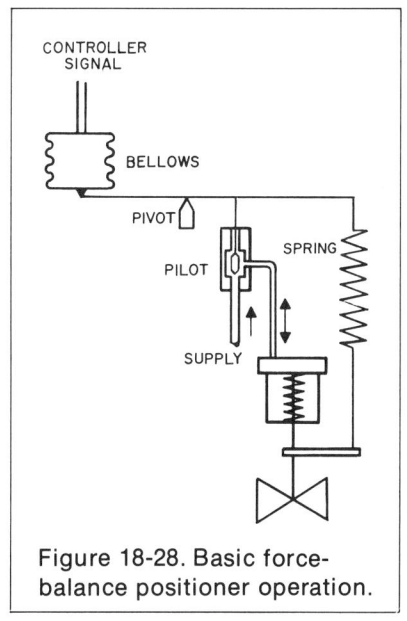

Figure 18-28. Basic force-balance positioner operation.

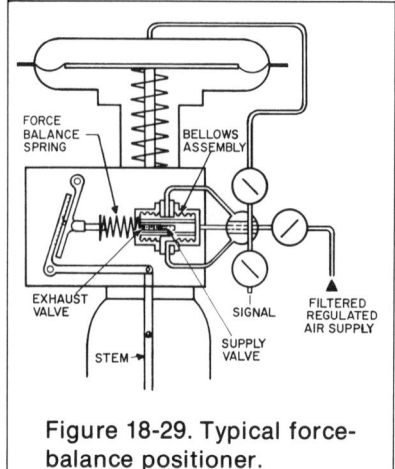

Figure 18-29. Typical force-balance positioner.

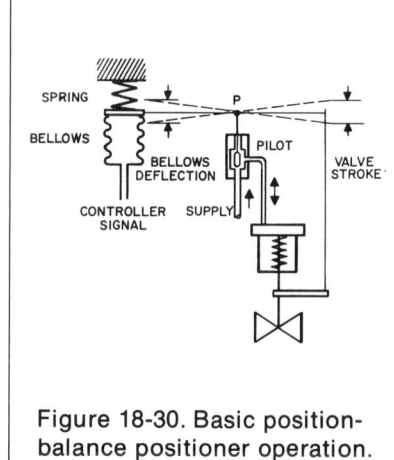

Figure 18-30. Basic position-balance positioner operation.

on the bellows until spring compression is again equal to the bellows force, closing the exhaust valve.

Position-Balance Positioner. The position-balance principle is shown in Figure 18-30. The components are similar to the basic force-balance positioner except there is no fixed pivot point or balancing spring, and point P (where the pilot valve is connected to the beam) is movable. Whenever the bellows moves in response to an instrument signal, the pilot moves, either admitting air to or bleeding air from the diaphragm until the valve stem position corresponds to the instrument air signal. At this point the pilot is again at its neutral or equilibrium position. There is one valve position for every corresponding bellows position when the pilot is at equilibrium. Although point P moves when the valve is not at equilibrium, it returns to within a few thousandths of an inch of the point shown when the valve is at equilibrium. This is essentially a floating pivot point, and this action causes the motion of the valve end of the beam to be exactly proportional to the motion of the bellows end of the beam.

A position-balance valve positioner is shown in Figure 18-31A. Here, a flapper-nozzle relationship is produced by mechanically positioning a flexure assembly which is affected by valve stem movements on one hand and instrument output on the other. The 3-15 psig controller air signal pressure in bellows A opposes flexure assembly B when the system is in balance. An increase in controller output pressure causes bellows A to expand, moving the ball tip of the flexure assembly away from flapper E which, in turn, covers nozzle F. As pressure on the diaphragm of control relay G builds up, the pressure (20 psig) then passes to the positioner output which is applied to the diaphragm actuator, stroking the valve stem and positioning the valve plug. By a direct mechanical linkage, stem motion is translated to rotary motion at shaft C which moves the flexure assembly so that the ball tip allows equilibrium between the flapper and nozzle. The position of the ball tip in the plane indicated by 1 is directly proportional to the signal pressure in the bellows. The position

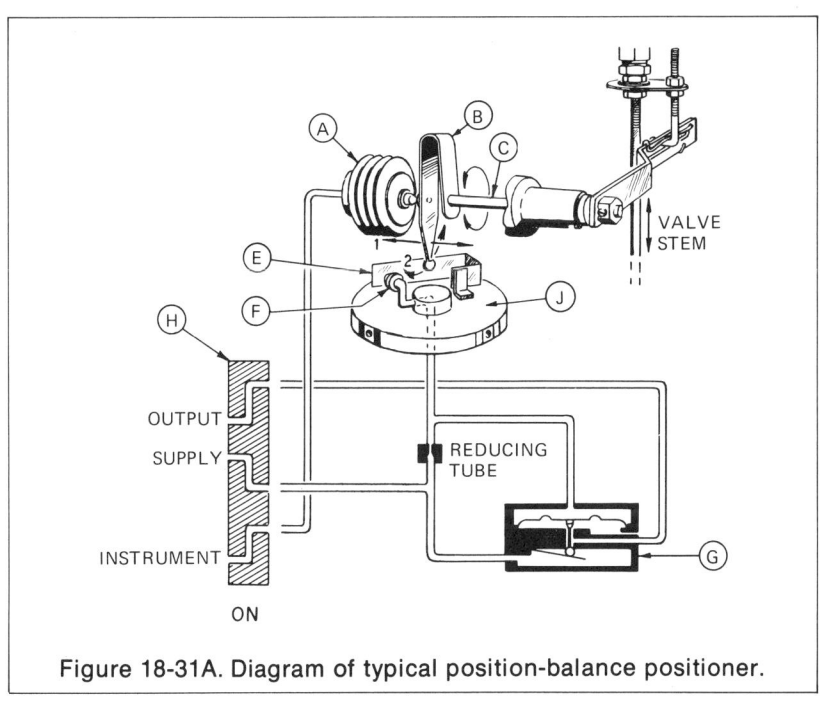

VALVE
STEM

OUTPUT

SUPPLY

REDUCING
TUBE

INSTRUMENT

ON

Figure 18-31A. Diagram of typical position-balance positioner.

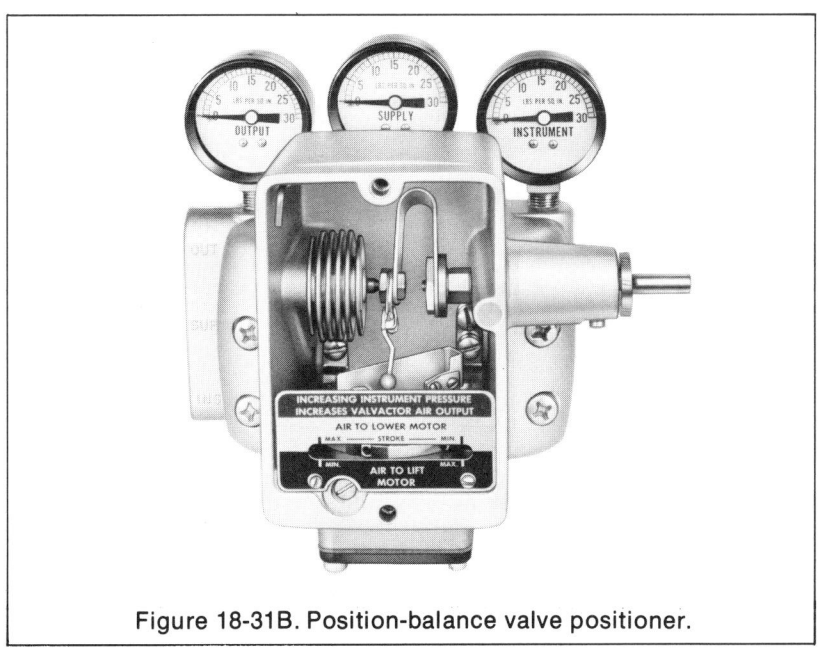

Figure 18-31B. Position-balance valve positioner.

of the ball tip in the plane indicated by 2 is exactly proportional to the amount of the imbalance in plane 1. This means that the position of the valve stem is directly proportional to the signal air pressure in the bellows. Rotating color-coded disc J accomplishes reversal of the positioner control action (for air-to-lower or air-to-raise actuators) by changing the ball tip-flapper relationship. The white sector is visible when an increase in pressure lowers the valve stem; the black sector is visible when a pressure increase raises the valve stem. An actual photograph of a position-balance valve positioner with cover off is shown in Figure 18-31B.

Electronic Valve Positioner. This kind of unit is essentially a current-to-air converter with a mechanical valve stem position feedback (Figure 18-32). In operation, a milliampere signal is imposed on coil H, and the field of the permanent magnet repels the coil. This pushes rod J against assembly K and changes the moment about pivot A, causing rotation about the pivot and causing the flapper to cover nozzle G. This increases the back pressure on the relay, increasing pressure output to the actuator. This lifts the valve stem to

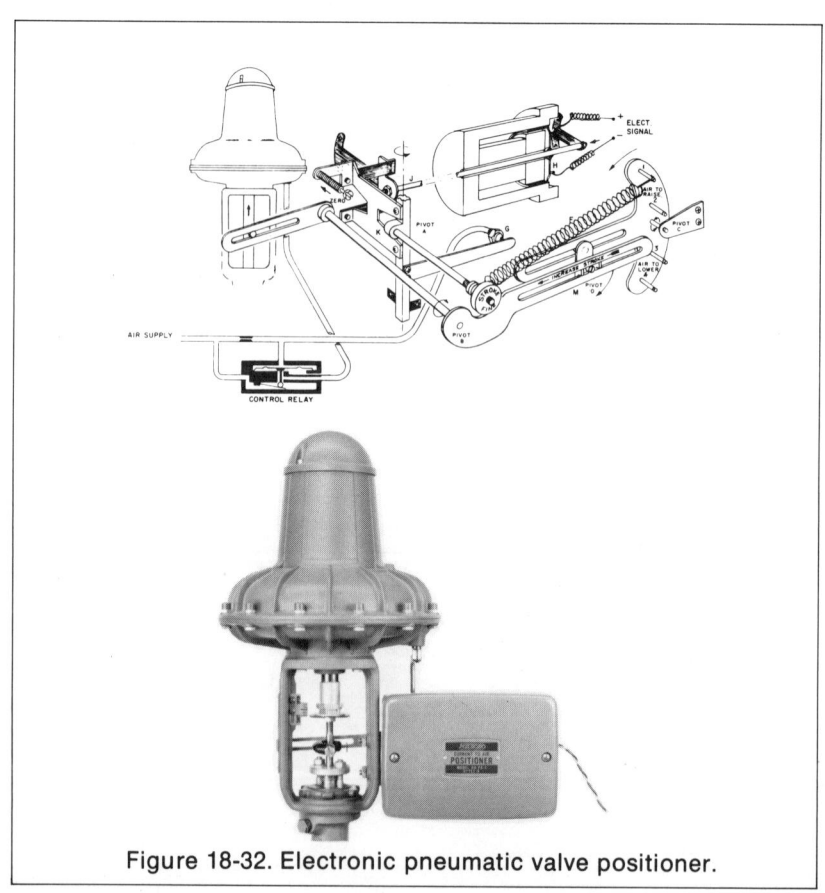

Figure 18-32. Electronic pneumatic valve positioner.

which the feedback lever is attached. Rotation of the feedback lever causes rotation of one lever in assembly M to rotate in the opposite direction about pivot C. The ratio of rotation between the two levers is adjustable by means of sliding pivot D in assembly M. The rotation of the second lever relaxes the tension on spring F, depressing the moment force on assembly K about pivot A and allowing it to move the flapper away from the nozzle. When the actuator has stroked the required distance for the spring force to be equal and opposite to coil force, the flapper-nozzle will throttle and the actuator will stop stroking.

CONTROL VALVE SIZING

For satisfactory control and adequate valve life, the selection of proper valve size is as important as the design and mechanical features of the valve. Pipeline size is rarely used anymore to determine valve size. The line size in a given piping system is generally selected to give a fluid velocity in accordance with good engineering practice, whereas valve size is based on the quantity of fluid flowing and the pressure drop condition which must be met.

Too large a valve causes rapid wear to inner valve parts because the valve plug necessarily operates close to the seat, causing wire drawing or erosion. Also, control is unsatisfactory because only a small portion of the total valve stroke is required to handle the flow range from minimum to maximum demand, necessitating more accurate valve positioning than is really needed.

Therefore, selection of suitable flow data for sizing calculations should proceed with great care, and realistic values of minimum and maximum flow conditions are essential. Unreliable, inconsistent, and inaccurate assumed conditions are the most frequent reasons for incorrect sizing. Since the control valve is an adjustable orifice, its area is automatically changed by the controlling instrument to produce a desired rate of flow under varying pressure drop conditions. Therefore, available operating range is sacrificed when pressure drops are specified too low and when safety factors are unnecessarily high.

The amount of operational pressure drop required for good control is a function of pressure differential across the valve with respect to the drop through the entire system. The operating pressure drop will normally be one-third to one-half of the total system pressure drop if all of the hydraulic components have been properly selected. Several methods have been used to size valves, the universal one being the determination of a C_v rating from available flow data by the use of calculations, slide rules, nomographs, or charts.

C_v Rating Calculations

The use of the valve coefficient C_v has materially simplified the problem of control valve sizing to a simple method applicable to a wide variety of valve constructions, valve sizes, and field services. The capacity rating C_v is an expression of a valve's capacity for handling flow, that is, the number of U.S. gallons per minute of water discharged through a wide open valve subjected to a pressure drop of 1 psi. The capacity of valves for nonaqueous liquids, gas, or steam can also be related to this quantity. When capacity rating has been determined, it is then a simple matter to select the applicable size of valve from the list of C_v values published by all valve manufacturers for the valve under consideration.

A C_v rating can be derived from one of three basic equations, one each for liquids, gases, and vapors, all of which are based on Bernouilli's theorem on fluid flow, $V^2 = 2gh$. The maximum required rating is obtained by using anticipated flow rate with minimum pressure drop.

Liquids. The liquid flow (gpm) is measured at flowing temperature and referred to water $(G = 1.0)$.

$$C_v = V\sqrt{\frac{G}{\Delta p}}$$

where V = flow (U.S. gpm)
        G = specific gravity
        $\Delta p = P_1 - P_2$ = pressure drop through valve in psi
        $(P_1$ = upstream pressure; P_2 = downstream pressure)

Gases. Gas flow and specific gravity are referred to standard conditions, 60°F air, and must be corrected for flowing temperature thus:

$$C_v = \frac{Q}{1360}\sqrt{\frac{T_f G}{\Delta p\,(P_2)}}$$

where Q = flow (scfh)
        G = specific gravity
        P_1 = upstream pressure in psia
        P_2 = downstream pressure in psia
        $\Delta_p = P_1 - P_2$ = pressure drop through valve

T_f = flowing temperature, absolute scale (°F + 460); when T_f *is between 0 and 120°F*, this equation may be simplified to:

$$C_v = \frac{Q}{60}\sqrt{\frac{G}{\Delta p\,(P_2)}}$$

Inlet and outlet line pressure must be expressed in psia. If the pressure drop exceeds one-half of the inlet pressure absolute, the flow is in the critical flow range and a maximum value of its $\frac{P_1}{2}$ is used for $\triangle p$.

Vapors. This includes steam and others. When the pressure drop exceeds one-half of the inlet pressure absolute, P_1 must be used for $\triangle p$.

Using outlet absolute pressure rating but no less than $\frac{P_1}{2}$, density in pounds per cubic foot can be obtained from saturated steam tables found in practically all engineering handbooks. Where superheated steam is a process factor, the downstream density is obtained from a Mollier chart, also in the handbooks.

The equation then becomes:

$$C_v = \frac{W}{63.3 \sqrt{\Delta p\,(w)}}$$

where W = steam or other vapor flow in pounds per hour
$\Delta p = P_1$ abs $- P_2$ abs = pressure drop across valve
w = density of the steam in pounds per cubic foot at P_2 abs, or $\frac{P_1}{2}$ abs, whichever condition applies.

Pulp Stock. C_v ratings of valves to be used on pulp stock slurries are sometimes derived by using the same basic equation for liquid flow and applying a correction factor which is based on consistency in accordance with Table 18-C.

$$C_v = \frac{V\sqrt{\dfrac{G}{\Delta p}}}{Fc}$$

where V = flow (U.S. gpm)
G = specific gravity (relative to water, standard temperature)
$\Delta p = P_1 - P_2$ = pressure drop through valve in psi
P_1 = upstream pressure; P_2 = downstream pressure
Fc = pulp stock correction factor (see Table 18-C)

Slide Rules

Specially calibrated slide rules (Figure 18-33) provide a simple and convenient means for making the preceding calculations; these devices are available from various manufacturers.

TABLE 18-C

PULP STOCK CORRECTION FACTOR (Fc)

Percent Consistency	(Fc) Chemical Stock	(Fc) Mechanical (Groundwood) Stock
Below 2%	1.0	1.0
2% to 3%	0.97	0.99
3% to 4%	0.90	0.95
4% to 5%	0.84	0.92
Above 5%	0.80	0.90

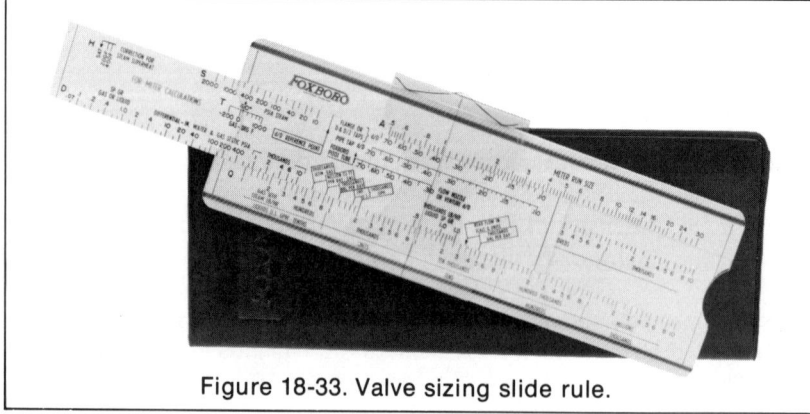

Figure 18-33. Valve sizing slide rule.

OTHER FINAL CONTROL ELEMENTS

Control valves are by far the most extensively used final control elements in the paper industry. On a more limited basis, however, other mechanisms are used. These include louvers, dampers, pumps, and motors.

Louvers and Dampers

A louver is composed of a number of parallel rectangular vanes, as shown in Figure 18-34. It is used for the control of the flow of air and other gases around lime kilns, power and recovery boilers, and the like.

Dampers are used for similar service and consist of a circular or rectangular slide in a circular casing as shown in Figure 18-35. A slide damper is generally installed slightly off vertical in order to obtain a tight seating by using the weight of the damper against the slide.

Pumps

Both variable-speed centrifugal pumps and so-called controlled volume or positive displacement pumps have been used as final control elements. Centrifugal pumps have been controlled through their electrical motor drives.

A positive displacement pump delivers quantities of fluid and capacity depends on its speed, length of piston stroke, and cylinder dimensions. Usually, either speed or piston stroke, or both, are used for control applications to regulate the pump output in response to deviations of a controlled variable. A typical design is shown in Figure 18-36.

Another type uses a combination of a piston for displacement and a diaphragm to eliminate the possibility of leakage which could occur from the stuffing box of a piston-only pump. Its operation is illustrated in Figure 18-37.

Motors

Another final control element is a speed-regulated electric motor. It is an inherent characteristic of dc motors that their speed can be regulated quite

Figure 18-34. Louver type final control element.

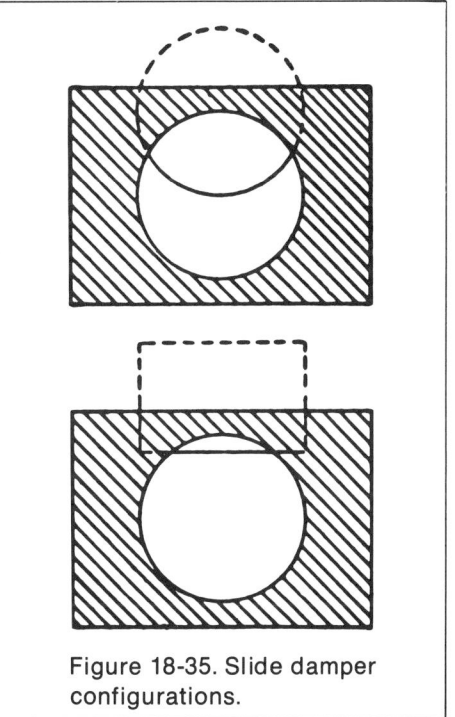

Figure 18-35. Slide damper configurations.

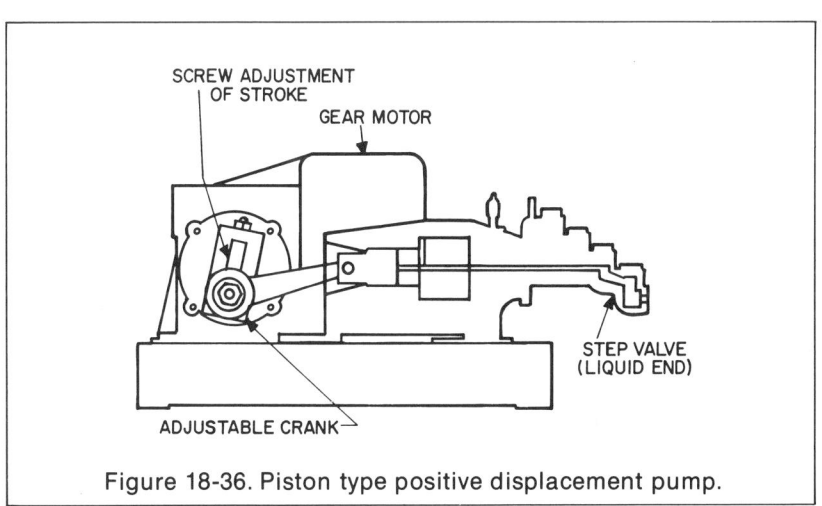

Figure 18-36. Piston type positive displacement pump.

easily. Where dc power is available, automatic speed control can readily be obtained by either a motor or pneumatically driven rheostat, or some other suitable electric arrangement. In case of ac power supply, it is necessary to convert to dc. A motor controller which uses electronic tubes to convert ac to

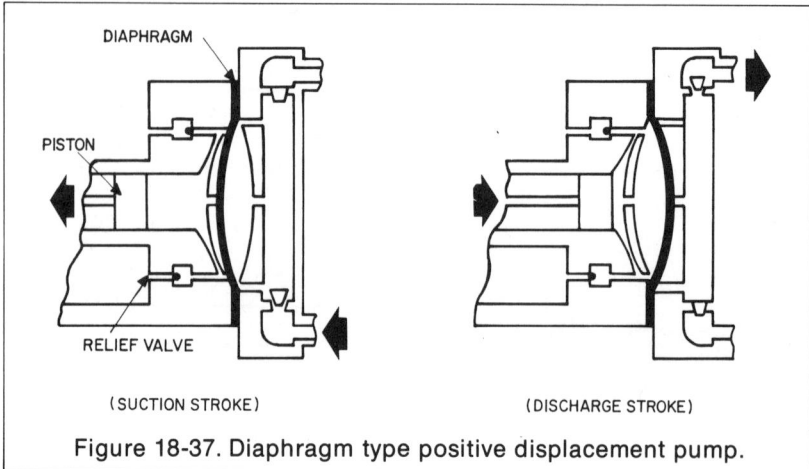

DIAPHRAGM

PISTON

RELIEF VALVE

(SUCTION STROKE)

(DISCHARGE STROKE)

Figure 18-37. Diaphragm type positive displacement pump.

controlled dc has been used for operating dc motors. Suitable rectifier tubes provide voltage for the field and armature of the motor. A group of small control tubes regulates the output of these rectifiers in such a manner as to provide for starting and accelerating the motor, as well as for controlling its speed over a wide range.

BIBLIOGRAPHY

Ball Control Valves. Foxboro, MA: The Foxboro Company, Bulletin J-14, January 1970.

Beard, C. S. *Final Control Elements–Valves and Actuators.* Philadelphia: Rimbach Publications Division, Chilton Company, 1969.

Considine, D. M. *Process Instruments and Controls Handbook.* New York: McGraw-Hill Book Company, 1957.

Control Valves. Foxboro, MA: The Foxboro Company, Bulletin J-12, January 1971.

Eckman, D. P. *Principles of Industrial Process Control.* New York: John Wiley & Sons, 1965.

Fisher Governor Company. *Control Valve Handbook.* 1st ed. St. Louis: John S. Swift Company, 1965.

Hutchison, J. W. *ISA Handbook of Control Valves.* Pittsburgh: Instrument Society of America, 1974.

Model V9000 Ball Valve. Foxboro, MA: The Foxboro Company, TI 31-70a, December 1969.

Process Control Instrumentation. Foxboro, MA: The Foxboro Company, Pub. 105A, 1970.

Saunders Type Dual Range Control Valve. Foxboro, MA: The Foxboro Company, TI 31-15a, May 1970.

Three-Way Control Valves. Foxboro, MA: The Foxboro Company, TI 31-12b, April 1967.

STUDY QUESTIONS

Indicate whether following statements are True or False by inserting T or F in parentheses:

1. In an air-to-lower diaphragm actuator, the controller signal is admitted to the lower side of the diaphragm. (   )

2. For the throttling applications, the springless valve actuator requires the use of a positioner. (   )

3. Cylinder actuators can handle greater positioning loads than diaphragm actuators. (   )

4. Trim consists of those parts of a valve that do not directly contact the process fluid. (   )

5. The needle valve can be used to provide precise control at low flows. (   )

6. In the balanced trim cage style valve, the characteristics depend on the shape of the cage parts. (   )

7. The major advantage of the Saunders type valve is that it provides good throttling characteristics. (   )

8. Ball and butterfly valves can be used on pulpy or other difficult fluids. (   )

9. The total capacity of a three-way valve is the difference between the two discharge or inlet ports, depending on the direction of flow. (   )

10. The design of a three-way valve for blending allows for smooth action by a pneumatic actuator.

11. Positioners have been used where increased speed of operation is desired. (   )

12. A positioner is a kind of derivative controller in itself. (   )

13. The operation of a motion balance positioner involves a floating pivot point. (   )

14. To reverse motion balance positioner control action, the flapper-nozzle relationship must be changed. (   )

15. Valve sizing and line sizing are based on the same requirements. (   )

16. Erosion is caused by using too small a valve. (   )

17. The rating C_v is an expression of a valve's capacity for handling flow. (   )

18. Pressure drop and flow are needed to calculate C_v. (   )

19. To calculate C_v for gas flow, specific gravity need not be corrected to flowing temperature. (   )

20. Speed and/or piston stroke are usually controlled to regulate pump output. (   )

(See Appendix for answers)

19. Instrument Applications

The previous chapters have been concerned with the basic principles of operation and design of all of the instruments used in the measurement and control systems for the paper industry. This instrumentation can be classed into three groups:

1. Primary elements—measurement
2. Secondary elements
   a) Indicators
   b) Recorders
   c) Controllers
3. Final control elements
   a) Valves
   b) Louvers and dampers
   c) Pumps
   d) Motors

Although all systems consist of combinations of these elements, system requirements will vary from mill to mill due to the type of process, the material being processed, the products manufactured, and equipment variations.

Depending on the type of process employed for manufacturing pulp, mills are referred to as sulfate (kraft), sulfite, soda, groundwood, semichemical, or chemigroundwood mills. The predominate process used in the making of pulp and paper is the kraft process shown in block form in Figure 19-1. A list of typical instrument applications found in this process is shown in Table 19-A. The applications are indeed many, and this chapter will attempt to present some of these applications in an orderly fashion.

INSTRUMENTATION SYMBOLS

Many flow diagrams will be presented. It will be helpful to review at this point some aspects of these diagrams. Instrumentation diagrams depict the use of measurement and control components and their connecting lines in applications. These diagrams use a system of simplified letter and pictorial symbols in place of the actual instruments. A summary of pictorial symbols used in flow diagrams is shown in Figure 19-2. A combination of letters is used to identify each function as shown in Table 19-B.

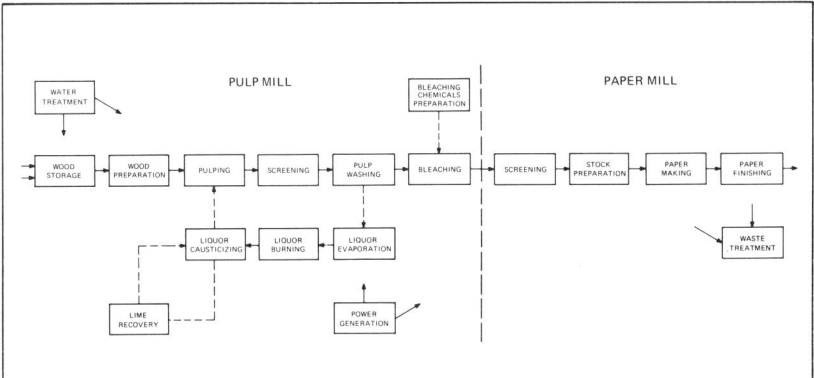

Figure 19-1. Kraft pulp and paper manufacturing block diagram.

TABLE 19-A

KRAFT MILL INSTRUMENT APPLICATION

A. Digester room
  1. Liquor drawdown and refill
  2. Total steam flow
  3. Heater condensate level
  4. Heater condensate conductivity
  5. Steaming
  6. Temperature-pressure
  7. Gas-off
  8. Blow tank level
  9. Turpentine condenser temperature
  10. Liquor circulation flow
B. Brown stock washer
  1. Blow tank level
  2. Blow tank motor load dilution
  3. Stock to washer flow
  4. Knotter dilution
  5. Shower
  6. Filtrate tank level
  7. Repulper dilution
  8. Strong liquor Baumé
  9. Strong liquor filter
  10. Liquor to evaporator—flow
  11. Strong liquor storage level
  12. Washed stock storage level
  13. H.D. chest dilution
C. Screen room
  1. Washed stock storage level
  2. Consistency
  3. Stock flow
  4. Dilution
  5. Screened stock storage level

D. Bleach plant
  1. Screened stock storage level
  2. Consistency
  3. Chemical flow
  4. Tower levels
  5. Tower temperatures
  6. Dilution
  7. Mixer temperatures
  8. Shower flows
  9. Total steam flow
  10. Total water flow
  11. Fresh water makeup flow
  12. Filtrate overflow to sewer
  13. Bleach stock storage level
  14. H.D. storage dilution
  15. Chemical storage level
  16. Effluent to sewer—flow
  17. Hot water temperature
E. Grinder
  1. Temperature
  2. Consistency
  3. pH
F. Stock preparation and blending
  1. Level
  2. Consistency
  3. Vortrap, Jordan, etc.—pressure
  4. Ratio control of furnish
  5. Parshall flume
G. Paper machine (Fourdrinier)
  1. Wet end
    a) Consistency
    b) Headbox level

(continued on next page)

TABLE 19-A (Continued)

c) Headbox temperature
d) Flat box vacuum
e) Couch vacuum
f) Press vacuums
g) Flat box seal box level
h) Couch pit level
i) Pneumatic loading
(press rolls)
j) Shower water pressures
k) White water tank level
l) Saveall level
m) effluent weir
2. Dry end
a) Steam turbine back
pressure
b) Header pressure
c) Diff. pressure
(between sections)
d) Felt dryer temperatures
e) Steam flow
f) Machine speed
g) Sheet tension
h) Pneumatic loading for
calender stack and reel
i) Sheet moisture
j) Basis weight
k) Hood ventilation
H. Blow steam heat recovery
1. Condenser temperature
2. Condenser water makeup
temperature
3. Heat exchanger temperature
4. Clean hot water tank level
5. Clean hot water tank
temperature
I. Black liquor evaporator
1. Strong liquor storage level
2. Strong liquor Baumé
3. Flow ratio—liquor to 5
and 6 effects
4. Temperatures
5. Pressure and vacuum
6. Soap tank level
7. Steam flow and pressure
8. Boiling point rise
9. Condensate level
10. Thick liquor flash tank level
11. Condensate diversion
12. Thick liquor storage level
J. Recovery
1. Thick liquor storage level
2. Thick liquor flow
3. Cascade evaporator level
4. Precipitator wet bottom level

5. Multiple indicating draft gauge
6. Temperature in and out—
cascade
7. Liquor-salt cake ratio
8. Primary heater temperature
9. Secondary heater temperature
10. Liquor to nozzles pressure
11. Feed water control system
12. Oxygen
13. Steam temperature and
pressure
14. Air flow
15. Dissolving tank density
16. Dissolving tank level
K. Causticizing
1. Liquor storage level
2. Raw green liquor flow
3. Green liquor temperature
4. Hot water temperature
5. Clarifier rake-torque
6. Water to lime mud filter
7. Lime mud density
8. Lime mud filter shower flow
9. Filter cake thickness
10. Lime mud filter vat—level
11. White liquor flow
L. Kiln
1. Oil temperature
2. Oil pressure
3. Oil storage level
4. Hot end temperature
5. Exit gas temperature
6. Draft control and gauge
7. Atomizing steam pressure
8. Kiln speed
9. Drive motor load
10. Scrubber level
M. Chemical makeup and handling
1. Storage levels
2. Chlorine gas pressure
3. Chlorine vaporizer temperature
4. Caustic strength of solution
5. Strong caustic temperature
6. Hypo mixing
7. Chemical flows
8. Lime feeder
N. Chlorine dioxide plant
1. Chemical storage levels
2. Chemical mixing
3. Temperature
4. Cold water makeup
5. Spent liquor pH
6. Chlorine dioxide flow
7. High-low alarms

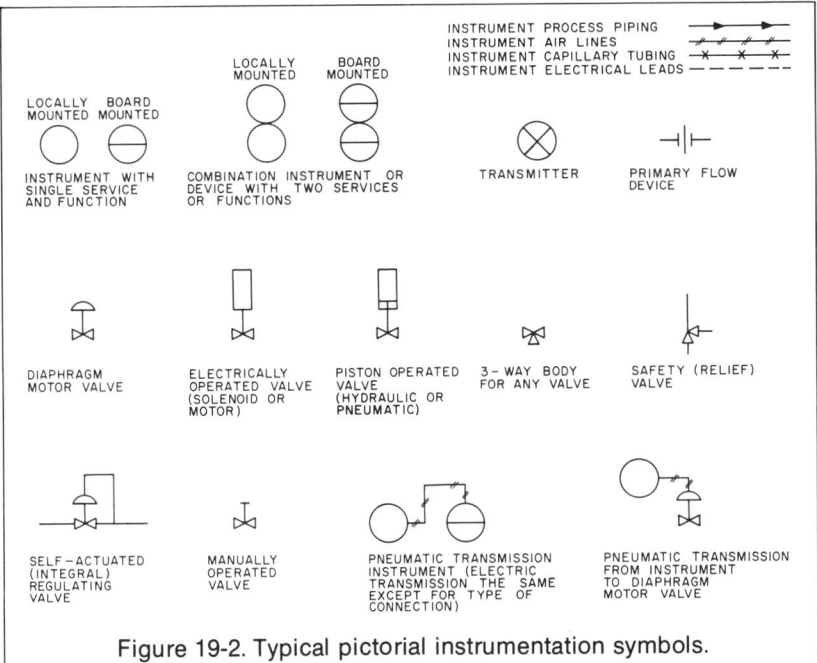

Figure 19-2. Typical pictorial instrumentation symbols.

A number is assigned to each instrument or loop depending on user practice. Combining the symbols and the numbers defines a particular circuit or device as illustrated by the basic loops shown in Figure 19-3. Although this method of depicting instrumentation on flow diagrams is generally accepted in the industry, modifications are made by some users to satisfy special circumstances. When a digital computer is involved, for example, some symbols must be changed to include this function.

For the purposes of this discussion, the instrument applications for pulp and paper processes will be shown pneumatically on the diagrams. Electronic diagrams, however, would be similar; only changes in the connecting lines and insertion of proper signal converters would be necessary.

DIGESTERS

After the initial preparation of the primary raw material, which is chiefly wood, the beginning phase in the manufacture of paper is pulp preparation. The first step is pulping, defined as the reduction of a fibrous material into its fiber components for easier handling in subsequent treatment. Pulping is accomplished chemically by dissolving the lignin, that material which binds the fibers in the natural state. This is done through cooking the raw material, usually chips, in suitable chemicals in a digester under controlled conditions of temperature, pressure, time, and chemical (liquor) concentration. Two general digester types, batch and continuous, are used in the cooking process.

Instrument Applications   359

TABLE 19-B

TYPICAL LETTER INSTRUMENTATION SYMBOLS

Process Variable	First letter	Measuring Indicating -I	Measuring Recording -R	Measuring Totalizing or Integrating[4] -Q	Measuring Glass Device for Observation Only -G	Transmitting Blind -T	Transmitting Indicating -IT	Transmitting Recording -RT	Controlling Blind -C	Controlling Indicating -IC	Controlling Recording -RC	Controlling Self-actuated Valve -CV	Controlling Safety Valve -SV	Controlling Control Valve -V	Controlling Powered Operator -P	Alarm[5] Blind -A	Alarm[5] Indicating -IA	Alarm[5] Recording -RA	Primary Elements -E
Analysis[1]	A-	AI	AR	AQ		AT	AIT	ART	AC	AIC	ARC			AV	AP	AA	AIA	ARA	AE
Burner Flame	B-	BI				BT	BIT		BC	BIC				BV	BP	BA	BIA		BE
Consistency	C-	CI	CR			CT	CIT	CRT	CC	CIC	CRC			CV	CP	CA	CIA	CRA	CE
Density	D-	DI	DR			DT	DIT	DRT	DC	DIC	DRC			DV	DP	DA	DIA	DRA	DE
Electric[2]	E-	EI	ER	EQ		ET	EIT	ERT	EC	EIC	ERC	ECV		EV	EP	EA	EIA	ERA	EE
Flow[3]	F-	FI	FR	FQ	FG	FT	FIT	FRT	FC	FIC	FRC	FCV	FSV	FV	FP	FA	FIA	FRA	FE
Hand	H-	HI	HR						HC					HV	HP				
Level	L-	LI	LR		LG	LT	LIT	LRT	LC	LIC	LRC	LCV		LV	LP	LA	LIA	LRA	LE
Moisture	M-	MI	MR			MT	MIT	MRT	MC	MIC	MRC			MV	MP	MA	MIA	MRA	ME
Pressure[3]	P-	PI	PR			PT		PRT	PC	PIC	PRC	PCV	PSV	PV	PP	PA	PIA	PRA	PE
Speed	S-	SI	SR	SQ		ST	SIT	SRT	SC	SIC	SRC	SCV	SSV	SV	SP	SA	SIA	SRA	SE
Temperature[3]	T-	TI	TR			TT	TIT	TRT	TC	TIC	TRC	TCV	TSV	TV	TP	TA	TIA	TRA	TE
Viscosity	V-	VI	VR		VG	VT	VIT	VRT	VC	VIC	VRC			VV	VP	VA	VIA	VRA	VE
Weight	W-	WI	WR	WQ		WT	WIT	WRT	WC	WIC	WRC	WCV		WV	WP	WA	WIA	WRA	WE

Notes:

1. Readily recognized self-defining symbols such as CO_2, O_2, pH, and ORP may be used in place of A.
2. Letter subscripts c (current), v (voltage), or w (power) may be used after E to designate type of measurement.
3. Lower case letters r, d, and t may be inserted to distinguish ratio, difference, and time respectively.
4. Q may also be used with letters I, R, and C where indicating, recording, or control functions also exist.
5. Lower case letter subscripts l and h may be used to denote low or high alarm functions.

If required note (1) could be expanded to cover any first letter, not in conflict with those listed, as long as a description for such a designation is part of the flow diagram.

Batch Digester

The batch digester is the traditional method used in the kraft process. In operation, the digester, typically a one-piece pressure vessel of welded mild steel construction, is filled with the chips plus the cooking chemicals. The cook itself consists of: (1) a heating period with steam (2) a length of time at pressure or cooking temperature to relieve the noncondensible gases formed during reaction, and (3) the "blowing" or emptying of the digester. An instrument system for such a digester would, then, have to control steaming rate for maximum steam efficiency and good circulation, record digester temperature and pressure, control relief, minimize liquor pullover, and control plugging of the relief screens. The overall objective is to produce a uniform pulp with minimum rejects with lowest possible steam consumption in each cook. Typical instrumentation for a direct-heated kraft digester is shown in Figure 19-4.

Steaming. At the start of a cook, steam is injected into the vessel at a uniform rate, regulated for optimum operation. Steam flow is measured from an orifice plate and flow transmitter (FT-1). This signal is transmitted to a steam flow recorder-controller (FRC-1). The output of the flow controller goes through a low selector relay (LSR-1) to the steaming control valve (FV-1). A pressure transmitter (PT-2) sends the vessel pressure signal to a time schedule cam pressure controller (CPC-2). The output from this controller also feeds into the low selector relay. As the output pressure from the flow controller

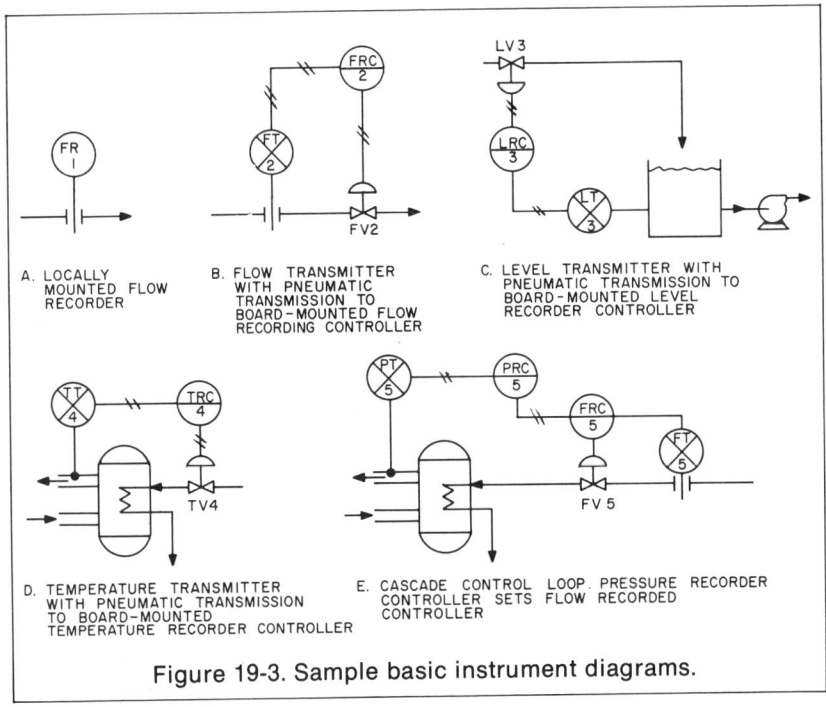

A. LOCALLY MOUNTED FLOW RECORDER

B. FLOW TRANSMITTER WITH PNEUMATIC TRANSMISSION TO BOARD-MOUNTED FLOW RECORDING CONTROLLER

C. LEVEL TRANSMITTER WITH PNEUMATIC TRANSMISSION TO BOARD-MOUNTED LEVEL RECORDER CONTROLLER

D. TEMPERATURE TRANSMITTER WITH PNEUMATIC TRANSMISSION TO BOARD-MOUNTED TEMPERATURE RECORDER CONTROLLER

E. CASCADE CONTROL LOOP. PRESSURE RECORDER CONTROLLER SETS FLOW RECORDED CONTROLLER

Figure 19-3. Sample basic instrument diagrams.

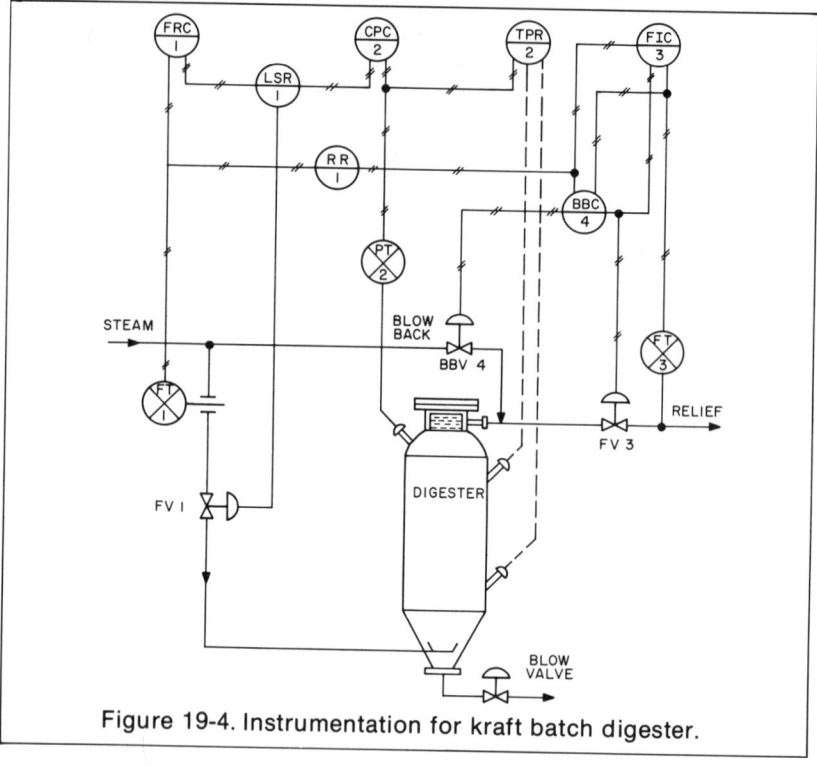

Figure 19-4. Instrumentation for kraft batch digester.

rises above the cam controller output, the relay smoothly transfers control of the steaming valve to the cam pressure controller. The pressure at which the transfer occurs is determined by the controlled rate of steam flow and the shape of the cam which is cut for optimum cooking conditions.

Temperature and Pressure Recording. The digester temperature pressure recorder (TPR-2) records the digester's top pressure and top and bottom temperature. The chart scales are set up in accordance with saturated steam temperature-pressure relationship. The temperature measurement is automatically switched periodically between top and bottom bulbs. Circulation within the digester can be determined by comparison of these measurements. With good circulation, top and bottom temperatures should be close together at top pressure. Top temperature will exceed the equivalent saturated steam pressure by several degrees because the cooking liquor, which is greater in specific gravity than water, boils at a higher temperature than water. When noncondensable gases are not adequately relieved during the cooking process the pressure will exceed the equivalent temperature.

Relief. The optimum rate of flow of relief gases, which is dependent on the rate of steam flow to the digester, will prevent excessive liquor pullover. A high steam flow requires a low relief flow, and a low steam flow will allow a relatively high relief flow. The relief flow transmitter (FT-3) measures the flow of relief gases and transmits a signal to the relief flow indicating controller (FIC-3). The output from the steam flow transmitter (FT-1) is inversed by reversing relay RR-1. This signal is used to continuously set the control point of the relief flow controller, the output of which is used to throttle the relief flow valve (EV-3). Digester circulation depends largely on the steaming rate and relief rate, and is measured by the temperature difference between the top and bottom of the digester. The schedule for rate of relief gases on each digester is determined by experience, and the record of the top digester temperature and pressure difference is essential for establishing the relief flow operation.

Blowback. The plugging of the screen located in the neck of the digester is controlled by the blowback controller (BBC-4). Blowback consists of opening the blowback control valve (BBV-4) and closing the relief flow control valve (FV-3) for a fixed period of time. The blowback controller measures the difference between the set point signal and measurement signal to the relief flow controller. When the measurement signal is sufficiently less than the set point signal, the relief valve is wide open and the screen is plugging up. The blowback control will close the relief valve and open the blowback valve energizing a timer which reverses the operation after a set time has elapsed. If measurement and set point are still not satisfactory for proper relief, the operation will be repeated until the proper relationship is obtained.

Continuous Digester

In recent years, much progress has been made in the continuous cooking of wood chips. There has been a large increase in the number of continuous systems used, with many mills converting from the batch method. The various configuration designs that have evolved during the development of continuous digesters can be generally classified in one of three categories:

1. Vertical—types with either downflow, upflow, or countercurrent flow of wood chips and cooking liquors.
2. Horizontal—types with forced movement of chips and cooking liquor.
3. Inclined—types with forced movement of chips and cooking liquors.

Pulping systems have been built consisting of one single unit or a multiple of units in series, depending on the type of wood being cooked and the final type of product desired.

Vertical Type

Although there have been a number of vertical type digesters used in the paper industry, one of the more widely used systems utilizes the chips and cooking liquor downflow method of operation. Figure 19-5 is a schematic diagram depicting such a system with the instrumentation normally used.

Process. In operation, wood chips enter from a surge bin discharging via a chip hopper into a chip meter whose speed determines the production rate of the system. From the meter the chips pass through a low-pressure feeder into a

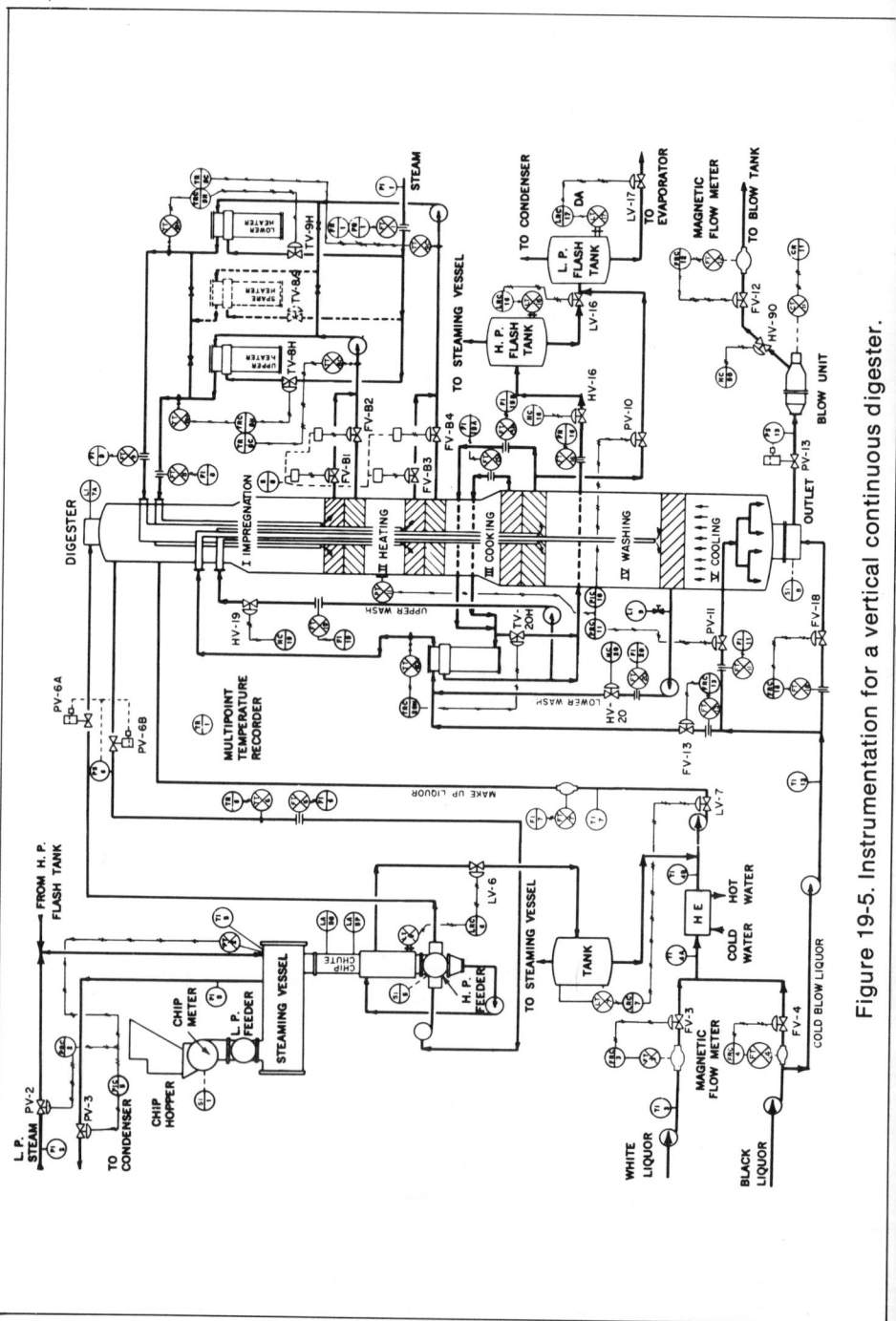

Figure 19-5. Instrumentation for a vertical continuous digester.

steaming vessel. After presteaming, the chips enter a high-pressure feeder whose rotor inlet port is totally submerged, and liquor level is controlled in a preceding chip chute. The chips are flushed into a feeder.

As the feeder rotor turns, the chips enter the digester's high-pressure system and are conveyed to the top of the digester by a circulation system. The liquor required for circulation is extracted through a strainer which is cleaned by a screw equipped with a special torsion type level indicator that measures the chip level in the top of the digester. The volume of cooking liquor is controlled by use of a magnetic flowmeter control system. As chips and cooking liquor are moved down through the digester they pass through five consecutive process zones: impregnation, heating, cooking, washing, and cooling. Cold blowback liquor is put into the bottom of the digester to dilute the charge for easy removal and to lower temperature before discharge. The cooked pulp is discharged to the blow tank through an outlet device, and blow is controlled by a magnetic flowmeter control system.

Instrumentation. The most import control loops on this type of continuous digester system (see Figure 19-5) are those recording and controlling the flow of white liquor (FT-3, FRC-3, FV-3); upper and lower cooking temperature (TT-8H, TRC-8H, TV-8H, and TT-9H, TRC-9H, TV-9H); digester pressure (PT-11, PRC-11, PV-11 and PIC-10, PV-10); and pulp flow to the blow tank (FT-12, FRC-12, FV-12). These instruments, along with the chip speed indicator (SI-1), determine the input feed rate, retention time, and cooking conditions in the digester, as well as operation of the washing portion of the digester. Steam for the vessel is supplied from the high-pressure flash tank and a pressure recording control loop (PT-2, PRC-2, PV-2) controls any necessary makeup. The venting of noncondensables from the steaming vessel is accomplished by an indicating control loop (PV-2, PIT-2, PV-2).

The level in the chip chute is controlled by overflowing the excess liquor to a tank whose level is controlled (LT-7 and LRC-7) by throttling valve LV-7 in the digester liquor makeup pipeline. Since the amount of white liquor is controlled with recording control flow loop FT-3, FRC-3, FV-3 and black liquor, with FT-4, FRC-4, FV-4 and both flow through valve LV-7, the level tank acts as a surge vessel whose level actually is the determining factor to the quantity of fresh cooking liquor going to the top of the digester. This flow is indicated by loop FT-7, FI-7 and circulation, by loop FT-6, FI-6. Temperature of the circulation flow is recorded by loop TT-6, TR-6.

The upper and lower cooking zone temperatures are controlled by control loops at the outlet liquor lines from the upper and lower heaters. Inlet heater liquor temperatures are recorded by loops TT-8C, TR-8C and TT-9H, TR-9H, and their flows are indicated by FT-8, FI-8 and FT-9, TI-9. There is a spare heater whose valve (TV-8/9) may be operated from either of the two existing temperature controllers. Valves FV-B1, FV-B2, FV-B3, and FV-B4 are butterfly valves on the upper and lower cooking zone screen outlets which are alternately opened and closed by a sequence timer (S-8) thereby maintaining screens in a clean condition, providing free flow through them.

The liquor flow from the upper washing zone to the heat exchanger is indicated by FT-15A, FI-15A and FT-15B. Washing zone recirculated liquor is indicated by loop FT-19, FI-19 and controlled manually by HC-19, HV-19. The extracted flow to a high-pressure flash tank is recorded by FT-16, FR-16. The temperature of the recirculated upper washing liquor is controlled by TT-20H, TRC-20H, TV-20H, by locating the valve at the heater bypass. Recirculated liquor flow in the lower washer zone is measured by FT-20, FI-20, set manually by HC-20, HV-20, and mixed with cold blow liquor whose flow is controlled by loop FT-18, FRC-13, FV-13. Digester pressure control loop PT-11, PRC-11, PV-11 regulates dilution liquor to the bottom of the digest-er which is indicated by FT-11, FI-11. Flow control loop FT-18, FRC-18, FV-18 controls the counterwash liquor flow to the outlet device. A second digester indicating pressure controller (PIC-10) operating from the same pressure trans-mitter output, is set to act as a safety valve with PV-10 venting excess liquor to a low-pressure flash tank. Levels on both the high- and low-pressure flash tanks are controlled and, by loops LT-16, LRC-16, LV-16 and LT-17, LRC-17, LV-17, recorded with the control valves located in the tank outlets.

A tachometer generator (SI-8) is used to measure and indicate the speed of the outlet device on the discharge system. Consistency of the pulp is recorded by a current system on the blow unit (CT-11, CR-11). Consistency variations are corrected by adjusting the counterwash flow control system. The blow-down valve (HV-90) is manually operated by HC-90, with the pulp flow to the blow tank being measured, recorded, and controlled by FT-12, FRC-12, FV-12.

On temporary shutdowns, cylinder-operated valves PV-6A, PV-6B, and PV-13 are used to "shut-in" the digester. Pressure switches PS-6 and PS-13 prevent opening of these valves until pressure in the top circulating line and blow unit is high enough to provide necessary system pressure balance. Level indicator LI-9 is used primarily during initial filling of the digester.

Several local temperature, pressure, and level gauges are used, some of which are shown on the flow diagram. TR-1 is a multipoint temperature recorder used to record selected temperatures throughout the system. Orifice plates with differential pressure transmitters are used quite extensively for liquor flow measurement, and flanged flush diaphragm type level transmit-ters are recommended for liquor and flash tanks. Magnetic flowmeters are normally used on white and black liquor and pulp flow measurements.

Horizontal and Inclined Types

The horizontal and inclined tube type continuous digesters were first used for the manufacture of semichemical pulps and were primarily suitable for cooking sawdust and other sawmill residues. They were later adapted and are being used for the production of full chemical pulps from a full range and combination of wood chips. The inclined digester is another type which has become readily accepted in the paper industry for this purpose.

Process. The digester is essentially a cylindrical pressure vessel usually mounted at a 45 degree angle. Figure 19-6 is a schematic drawing showing a typical basic unit used to cook wood chips. As shown, wood chips from the preparation and storage areas are fed in metered amounts through a feeder presteamer into a specially designed inlet rotary valve.

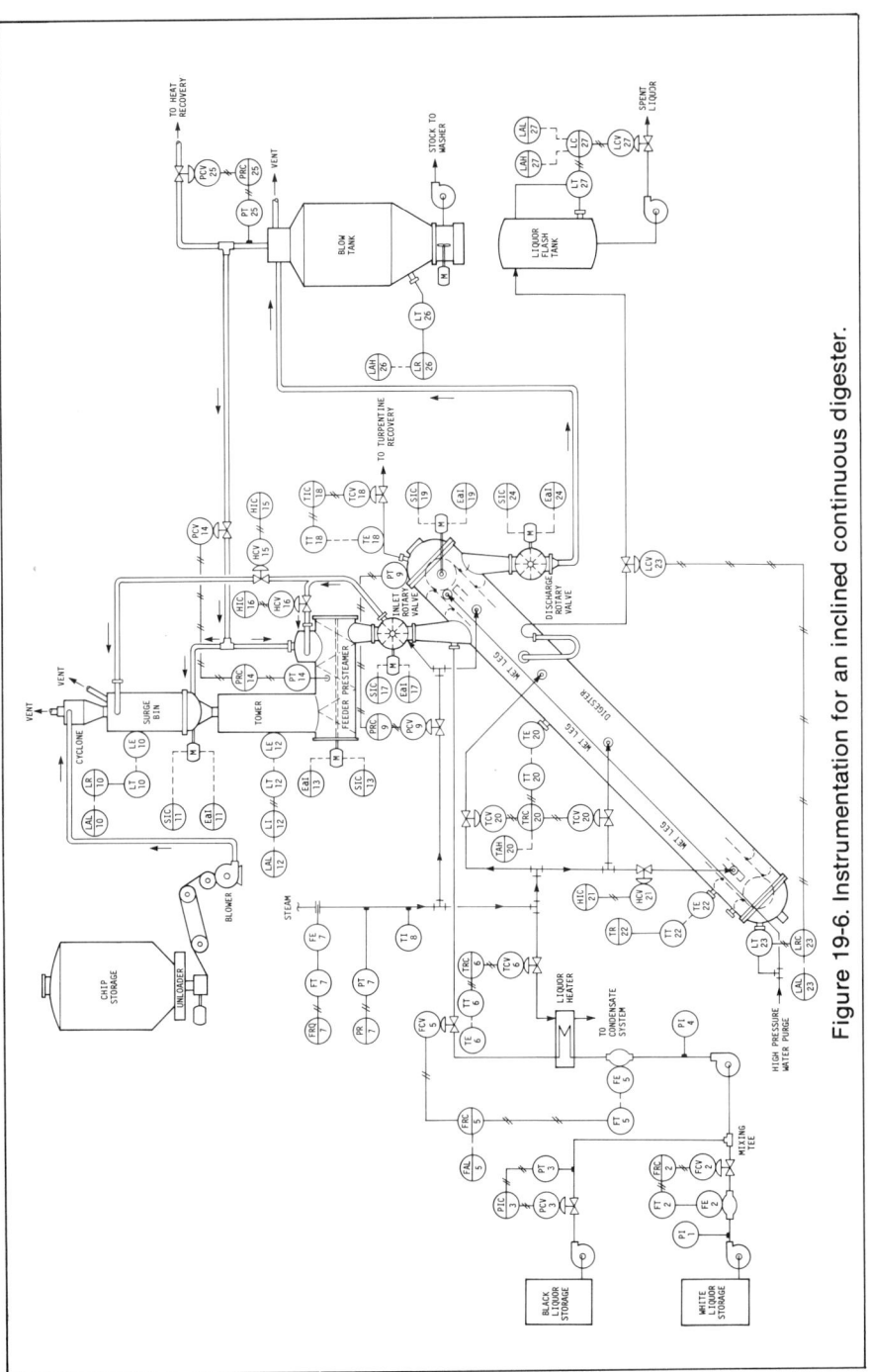

Figure 19-6. Instrumentation for an inclined continuous digester.

The inlet rotary valve is provided with a special exhaust vent to relieve vapor pressure in the rotor pocket after it has discharged its chips and before it reaches the low-pressure inlet side of the valve. When a closed vent chip charging system is used (shown in Figure 19-6 as the feeder presteamer and tower), the exhaust from this vent together with blow steam from the blow tank is used to preheat the incoming chips in a feeder presteamer. The discharge rotary valve does not have such a vent as the pressure is needed to blow the pulp into the blow tank and is relieved accordingly.

The wood chips from the inlet rotary valve drop into a series of traveling compartments in the digester. These compartments, formed by conveyor flights, are in series and are chain driven. A variable-speed reducer controls the conveyor movement. The chip-laden flights travel downward on the top side of an internal dividing plate which separates the tube in halves, and returns on the opposite or bottom side. Heated cooking liquor is added just below the inlet rotary valve. The lower section of the entire inclined tube contains liquor which can be maintained at or varied from any predetermined level. As the flights move downward through the liquor the chips are steamed. Impregnation takes place while the chips are dragged by the conveyor through the section of the inclined tube containing liquor. Vapor phase cooking takes place while the chips travel from the liquor to the specially designed discharge rotary valve.

Time for each phase of the process is controlled by: (1) the level of the liquor, (2) location of inlet and discharge nozzles, and (3) the speed control of the digester's conveyor drive system. All chips proceed through the digester at a uniform rate to achieve uniform cooking.

Several cooking zones can be established in the digester as desired, such as steaming, impregnating, and cooking (either submerged or vapor phase). The type of cook, yield, conditions, and tonnage governs the specifications for a particular unit. Also, the single unit can be coupled with other complete tubes, each having a specific kind of zone varying the temperature, pressure, or liquor as desired.

The cooking features include such things as the use of a variety of cooking methods utilizing any type of liquor with temperature maintained through direct or indirect heating. Liquor is constantly agitated by movement of chips through the unit and, if desired, liquor can be circulated and heated at various points for required temperatures at various stages in the liquid cooking phase.

Installation features include the location of both inlet and discharge on the same operating floor level. Building requirements and installation costs are lessened by locating 90 percent of the equipment outside the building and the angle of the tube can be varied to meet structural restrictions.

Instrumentation

The drawing in Figure 19-6 shows the white and black liquor being mixed in a mixing tee to make up the cooking liquor. The white liquor is added to the tee under flow control. The flow measurement (FE-2) is usually made with a magnetic type flowmeter. This control loop is set to maintain proper chemical-to-wood ratio based on wood chip conditions. The black liquor flow to the tee is under pressure control with the pressure transmitter (PT-3) having a suitable seal to prevent plugging of the measurement connection. This loop automatically adds black liquor to maintain the proper liquor-to-wood ratio. If a mixing tank were used instead of a tee to prepare the cooking liquor, the black liquor pressure control loop would be replaced with a level control loop on the tank controlling the flow of black liquor to it.

Total cooking liquor to the digester is added under flow control, also using a magnetic flowmeter (FE-5). Cooking liquor is heated in a surface type heat exchanger with the flow of steam regulated by a temperature control loop using a resistance type temperature detecting element (TE-6). Cooking liquor is sometimes heated by direct injection of steam. The steam supply flow is measured by an orifice plate and a differential pressure transmitter which is recorded and totalized. Steam supply pressure is also measured and recorded.

The pressure in the digester is maintained at a desired value with a pressure control loop regulating the steam supply. A sealed connection at the point of measurement on the dome of the digester is also required on this transmitter (PT-9). Temperatures along the digester zones are controlled by temperature control loops regulating steam flow to optional steam inlets along the tube. Resistance type temperature detecting elements are used here also (TE-20, TE-22). Liquor level in the digester is adjusted and controlled regulating the flow of excess liquor to the liquor flash with a level control loop utilizing a diaphragm-sealed flange-mounted differential type level measurement (LT-23) which is provided with a wet leg seal on the low-pressure side to obtain proper liquor level measurement.

Chip level measurements LE-10, LE-12 on the surge bin and presteamer tower are usually made with radioactive detecting devices. Pressure in the feeder presteamer is controlled by a pressure controller regulating the flow of exhaust steam from the blow tank. The speed control on the feeder presteamer determines the rate at which chips are fed to the inlet rotary valve and digester unit. Speed controllers are also used to adjust and regulate operation of the rotary valves and internal conveyor in relation to speed of the feeder presteamer.

Level measurements on the blow tank (LT-26) and liquor flash tank (LT-27) are diaphragm-sealed flange-mounted differential types with wet leg seals on the low pressure side also.

BLOW STEAM CONDENSER

The function of this system is to recover maximum heat that is released in the form of hot steam and vapors when the digester is discharged or "blown" after the cooking of chips is completed. A typical system with its associated instrumentation is shown in Figure 19-7.

Process. After the cooking of chips is completed, the contents of the digester are discharged under pressure into a cyclone which is located above a blow tank. The hot steam and vapors are separated from the pulp and liquor by the cyclone. The pulp and liquor drop down into the blow tank while the hot steam and vapors are sent to a jet condenser where heat is removed by water

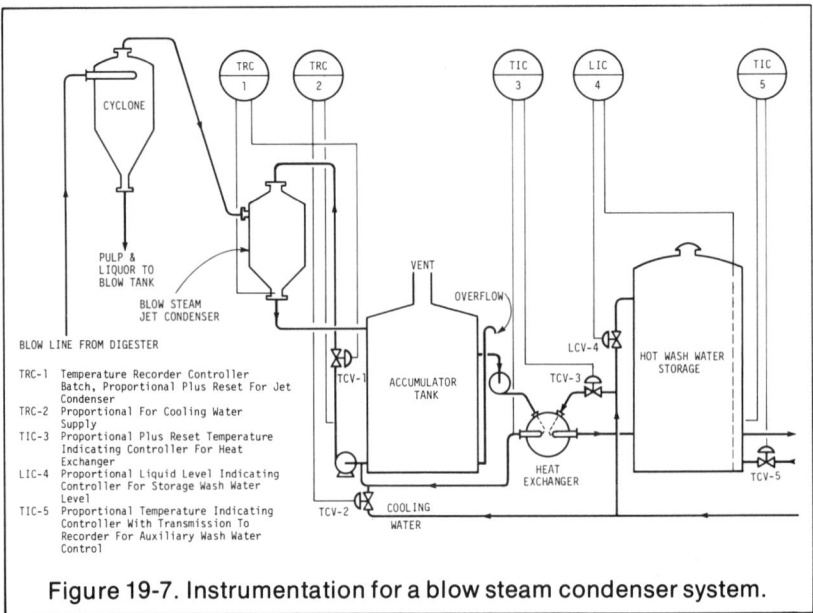

Figure 19-7. Instrumentation for a blow steam condenser system.

TRC-1 Temperature Recorder Controller Batch, Proportional Plus Reset For Jet Condenser
TRC-2 Proportional For Cooling Water Supply
TIC-3 Proportional Plus Reset Temperature Indicating Controller For Heat Exchanger
LIC-4 Proportional Liquid Level Indicating Controller For Storage Wash Water Level
TIC-5 Proportional Temperature Indicating Controller With Transmission To Recorder For Auxiliary Wash Water Control

which is stored in an accumulator tank. This hot water is used to heat the hot washer in a heat exchanger. The heated water is then used in the mill for washing, dilution, and other process purposes.

Instrumentation. The temperature-sensing bulb of TRC-1 is located in the output of the jet condenser. When this bulb senses heat from the blow steam, valve TCV-1 in the condenser water line is opened rapidly. This controller is furnished with the batch function to eliminate overshooting the control point on startup, effecting improved steam savings.

The water pump supplying water from the accumulator tank to the jet condenser runs continuously. If the water from the accumulator is too hot (as measured by the temperature-sensing bulb of controller TRC-2), valve TCV-2 in the makeup water line to the pump inlet is opened.

Temperature controller TIC-3 measures the temperature of the accumulator tank water outlet from the heat exchanger and regulates the flow of clean, cold makeup water through valve TCV-3 for maximum heat recovery.

The liquid level controller (LIC-4) maintains the storage level at a prescribed setting by adding makeup water through valve LCV-4 regardless of the wash water demand by the mill. Temperature controller TIC-5 provides auxiliary heating for wash water if the demand exceeds the capacity of the blow steam system.

SCREENING

Screening is essentially the removal of oversized material such as uncooked knots, large slivers, extraneous dirt, and debris from the cooked pulp by the use of mechanical separation from acceptable pulp fibers. This is normally accomplished by coarse or fine screens which would be located between the blow tank and the pulp washing operation. There are three general types of coarse screens: basket, inclined plate, and cylinder. The fine screens are available as the flat perforated type or as the rotary type.

Hot Stock Screening

The more favorable economics of screening at higher consistencies and relatively higher temperature has led to the development and use of pressure type rotary screens. The process is commonly known today as hot stock screening.

Process. Figure 19-8 depicts how typical hot stock screens operate and Figure 19-9 shows their use in a system. Cooked pulp is drawn from a blow tank and passed through the screens, which in this case operate on a pressure differential between inlet and outlet sides of the screen. This is accomplished by developing pulsations across the screen by a foil sweeping over the screen face at close tolerances. The unscreened stock is introduced in a cylindrical basket type screen in a tangential direction and the screening action, as well as the movement of the rejects, is carried out by the rotating foils. The number, arrangement, clearance, and form of the foils are matters of design which varies from manufacturer to manufacturer. Reject removal and the addition and control of dilution and purge water are usually designed to suit the

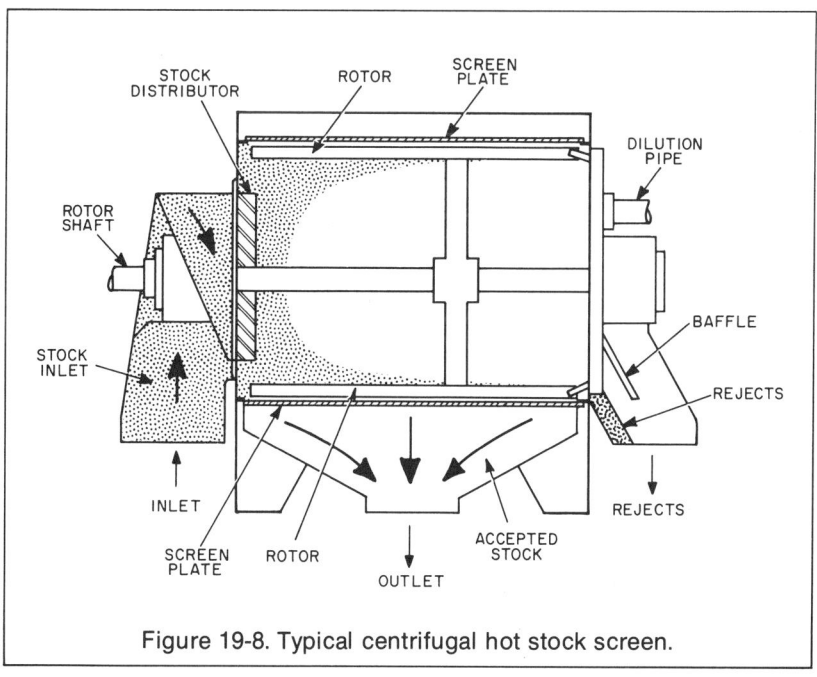

Figure 19-8. Typical centrifugal hot stock screen.

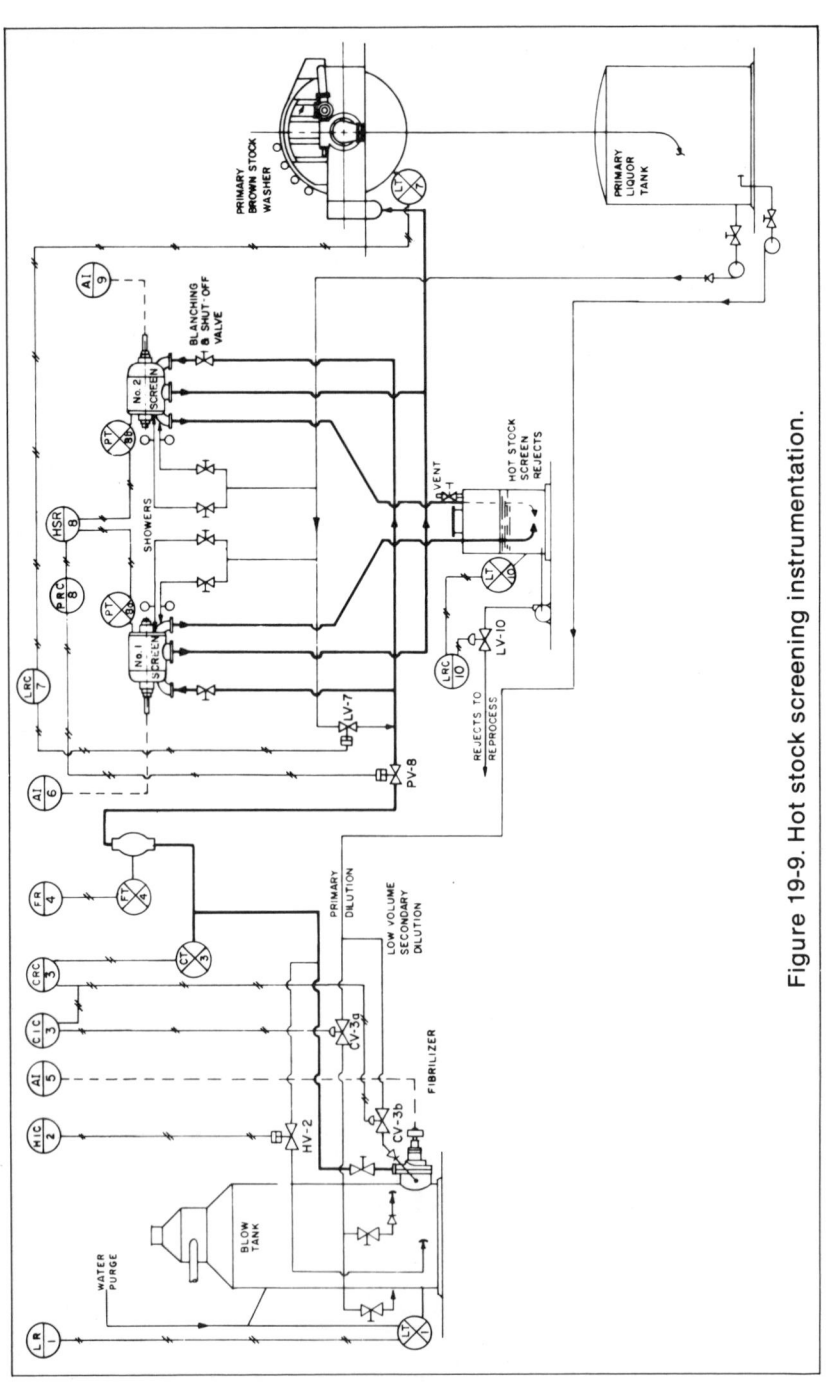

Figure 19-9. Hot stock screening instrumentation.

characteristics of the specific screen unit. Numerous arrangements are possible and exist in practice. Such combinations and the recycling of accepted pulp from the various stages are dependent on individual mill situations and cleanliness requirements.

Instrumentation. The major objectives of the instrumentation involved in a hot stock screening operation are to control:

1. The consistency of unscreened stock to the screens.
2. The flow of unscreened stock to the screens.
3. The flow of dilution water to the screens.
4. The level of the rejects storage tank.

Instruments LT-1, LR-1 measure and record level in the blow tank. The recommended level transmitter used here is a flush diaphragm flanged type. An indicating hand control station (HIC-2) is used to regulate the recirculation of unscreened pulp from the fibrillizer back to the blow tank by adjusting control valve HV-2. A pressurized in-line type consistency transmitter (CT-3) detects variations in consistency of unscreened pulp to the screens and transmits a signal representing this change to consistency recording controller CRC-3. The output from the controller is used to operate secondary dilution valve CV-3b. This output is also recorded as representing the valve position. The same signal is also used as the measurement to consistency-indicating controller CIC-3, which operates primary dilution control valve CV-3a in relation to secondary dilution valve CV-3b. Total flow of unscreened pulp is measured with an electromagnetic flow transmitter (FT-4) and is recorded on flow recorder FR-4. Load conditions on the fibrillizer and screen motors are monitored by current indicators AI-5, AI-6, AI-9. A flush diaphragm flanged type level transmitter (LT-7) measures the level in the primary brown stock washer and transmits the signal to level recording controller LRC-7, which controls the level in the washer vat by throttling level control valve LV-7 in the black liquor dilution line from the brown stock washer vat to the screens. The vacuums in No. 1 and No. 2 screens are measured by flush diaphragm flanged type vacuum transmitters. A high selector relay automatically selects and transmits the higher of the two output signals from the vacuum transmitter to vacuum recorder-controller PRC-8. The output from this control adjusts control valve PV-8 in the supply line to the screens in order to maintain optimum vacuum and screening conditions in the screens. Level transmitter LT-10 measures the level in the screen rejects tank and sends a signal to level recorder controller LRC-10, which controls tank level by modulating a control valve in the discharge side of the rejects pump.

PULP WASHING

Pulp washing is the removal of washable spent cooking liquor and intercellular matter dissolved by cooking liquors during the cooking process. This must be accomplished without diluting the wash liquors any more than is absolutely necessary, as all of these liquors have to be evaporated later in the process of chemical recovery. Incomplete washing renders subsequent bleaching of the pulp difficult and impairs the value of the pulp. Excessive washing renders recovery of chemicals too costly. Obviously, the maximum amount of wash liquors must be conserved and none should be allowed to escape to waste.

Process. Present-day washing is accomplished by the use of rotating vacuum washers as schematically shown in Figure 19-10.

In this system, pulp from the blow tank is discharged at regulated consistency to a pump which delivers pulp to a series of vacuum washers. These vacuum washers consist of wire screen-covered cylinders suspended in vats and provided with a means of applying vacuum to the inside. The slurry in the vats is drawn against the wire screen by siphon action of filtrate falling through the down legs, causing the pulp fiber to form a mat on the surface. Three stages of washing are shown, but systems can consist of a number of stages. Hot water is used in the last stage of the wash, with the effluent or filtrate from this stage used as wash water on the preceding stage, and so on to the first stage. Three definite mechanical actions take place during this operation:

1. Dilution of the pulp slurry with wash liquors.
2. Filtering of the pulp fiber from the slurry over the wire-covered drums.
3. Displacing of cooking liquors from the pulp mass by progressively weaker black liquor and eventually hot water.

The filtrate from the first stage contains approximately 14 to 20 percent solids, and is pumped to a storage tank prior to subsequent processing for recovery of chemicals.

Instrumentation. The major objectives of the instrumentation involved in pulp washing operation are to control:

1. The flow rate of unwashed pulp per process requirements.
2. The flow rate of recycled filtrate in fixed ratio to stock flow.
3. The flow rate of wash water input in accordance with washing demand.
4. The stage-to-stage balancing of filtrate counterflow.

Blow tank level is measured and recorded by loop LT-1, LR-1 and provides the operator with a guide to the system's ability to keep up with digester production. Consistency of the stock being pumped from the blow tank is detected by an agitator motor load measurement (CT-11) using a thermal watt converter. It is controlled by consistency recording controller CRC-11, modulating control valve CV-11 in the black liquor dilution line from the first-stage filtrate tank. Inflow rate is measured by magnetic flow transmitter FT-2 and is controlled by flow indicating controller FIC-2 and flow control valve FV-2 in the pulp line to the washers. The pulp flow signal, together with the recirculated filtrate flow measured by FT-2a, is also transmitted to flow ratio recording controller FrRC-2, which maintains proper inflow consistency by operating a flow control valve, FV-2a, on the recirculated filtrate flow line.

Level control for each washer is accomplished by level indicating controllers LIC-3, LIC-4, LIC-5 from signals transmitted by diaphragm flanged type level transmitters LT-3, LT-4, LT-5 and by operating level control valves LV-3, LV-4, LV-5 in filtrate recirculation lines. These controls, combined

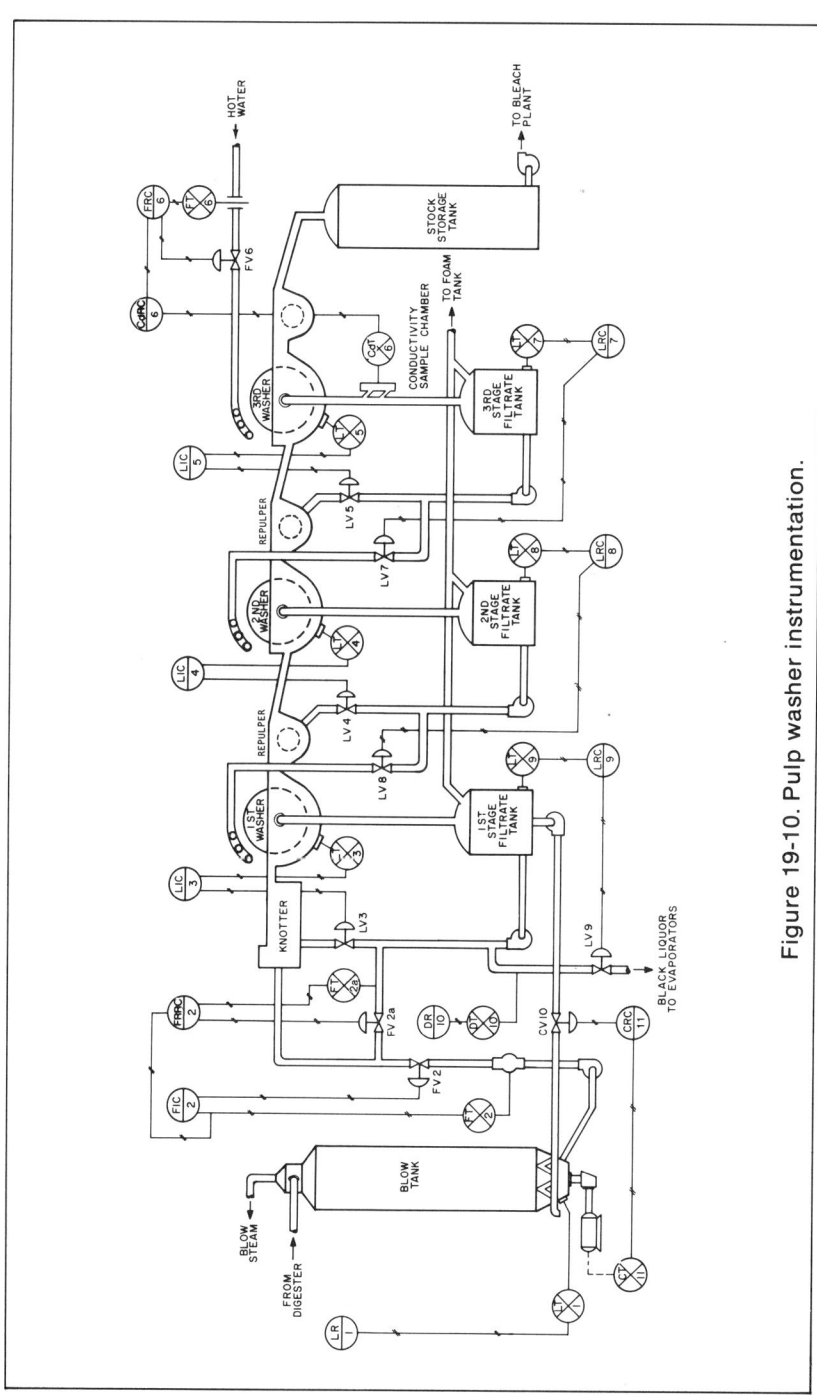

Figure 19-10. Pulp washer instrumentation.

Instrument Applications   375

with filtrate tank level controllers LRC-7, LRC-8, LRC-9, throttling control valves LV-7, LV-8, LV-9, using signals from the same type transmitters (LT-7, LT-8, LT-9) on the filtrate tanks, maintain balanced stage-to-stage counterflow operation. Conductivity of the filtrate from the third stage washer is measured by conductivity transmitter CdT-6 as an indication of its concentration. This signal is transmitted to conductivity recorder controller CdRC-6, which is the primary instrument of a cascade loop, by setting the control point of hot wash water secondary flow recorder controller FRC-6. A differential type flow transmitter (FT-6) is used to measure this flow, and flow control valve FV-6 in the hot water line is throttled to maintain final stage filtrate at constant concentration.

Density recorder DR-10 provides a continuous record of recovered liquor density (normally in terms of Baumé) as an indication of reclaimable solids from a signal received from density transmitter DT-10 in the liquor line from the system. This density transmitter is usually of the nuclear radiation or refractive index type.

BLEACHING

After cooking, screening, and washing, the next step in the process of manufacturing pulp intended for use in white paper and paperboard is bleaching. Pulp bleaching is essentially the removal of coloring impurities in the fibers by: (1) destruction and dissolving of coloring matter and (2) removal of residual lignin material. In this sense, pulp bleaching can be regarded as a continuation of the stepwise isolation and purification of wood fibers begun in the digester. Lignin materials and colored compounds are converted into a water-soluble form, while an attack on the desired fiber is held at a minimum.

Process. All bleaching systems are not exactly similar in design and operation but result from research and planning to fulfill the objective of a given mill. However, the basic principles are the same and can be found in a typical four-stage bleach plant diagrammed in Figure 19-11.

As shown, bleaching of pulp is accomplished by a series of chemical reactions. These reactions must be closely controlled in order not to destroy the desirable fiber component and affect its strength. Various bleach chemicals are mixed with the wet pulp, heated to the desired temperature with live steam, and retained for a period of time as it passes through a tower. On leaving the tower, the chemicals are washed out by dilution and dewatering by suction on interstage washers similar to those used in the pulp washing operation. The washed pulp then goes to the next stage for another treatment.

The first stage of modern multiple-stage pulp bleaching operations is the low-density chlorination where chlorine mixed with the pulp slurry is allowed to react with residual lignins. After washing, the pulp enters the second caustic extraction stage where chlorinated fiber residues and other alkali-soluble constituents are dissolved in sodium hydroxide. The pulp is washed again and enters the third hypochlorite stage. This is the whitening stage where coloring matter is destroyed by a hypochlorite solution, usually sodium hypochlorite, but sometimes calcium hypochlorite. After another washing the pulp is further whitened by the use of

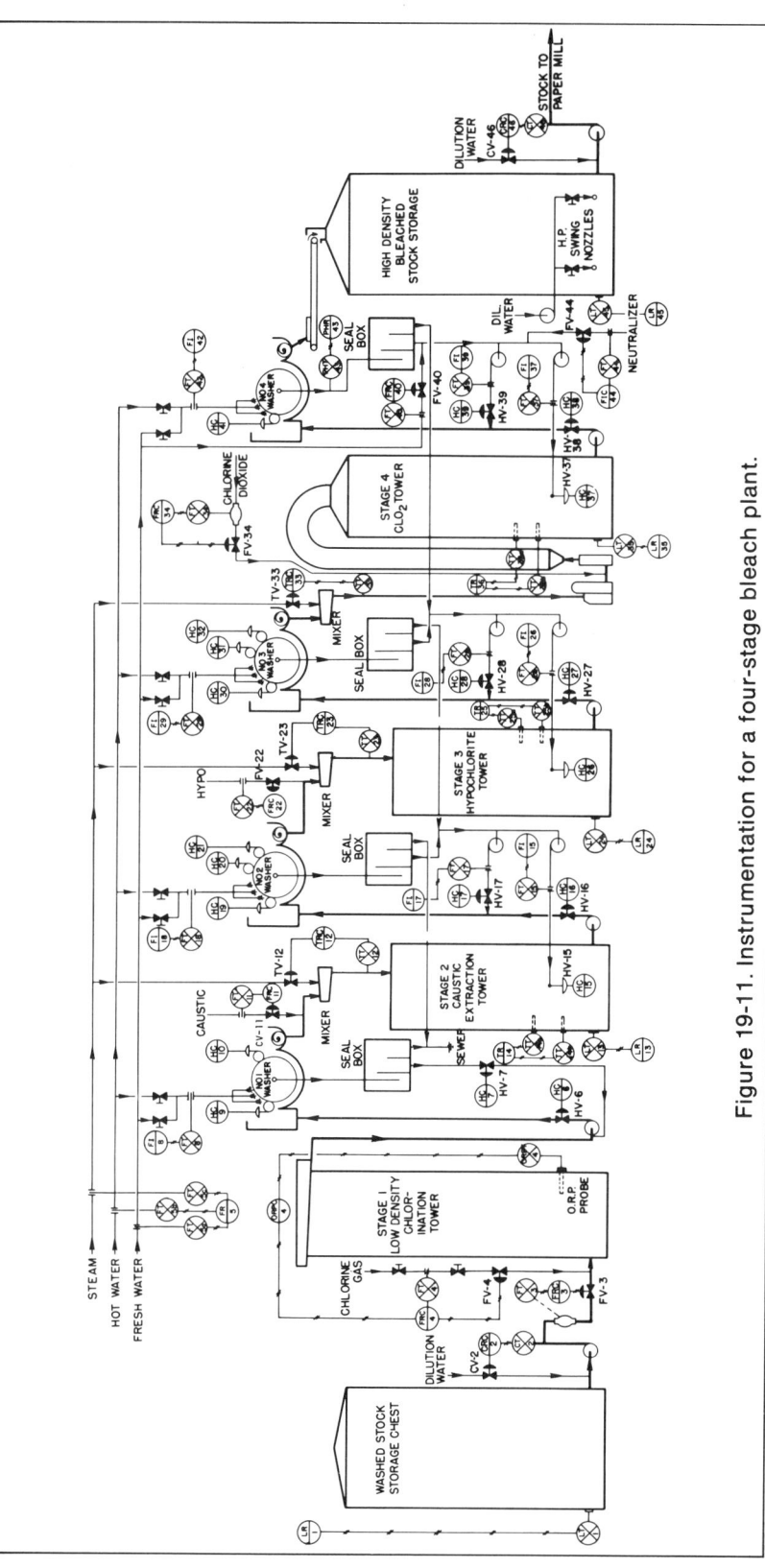

Figure 19-11. Instrumentation for a four-stage bleach plant.

chlorine dioxide (ClO_2) in the fourth stage followed by a final washing. It is then stored in storage towers prior to being sent to the paper mill to be prepared for papermaking.

Instrumentation. Level recorder loops LT-1, LR-1 and LT-45, LR-45 on washed and bleached stock storage tanks are used as operator guides to stock supply and paper mill demand. The desired pulp throughput, on which chemical additions and retention time in reaction towers depend, is controlled by flow control loop FT-3, FRC-3, FV-3 and consistency control loop CT-2, CRC-2, CV-2. Chlorine dosage is based on the residual chlorine (active chlorine left at the end of chlorination). An oxidation-reduction potential (ORP) measurement made at the proper location in the chlorination tower is related to the residual chlorine. Therefore, chlorine addition can be controlled by an oxidation-reduction potential controller (ORPC-4) cascading its output to the set point of chlorine flow recorder controller FRC-4, which regulates chlorine control valve FV-4 in the chlorine supply line.

Level recorders LR-13, LR-24, and LR-25 are used to maintain proper retention time in each tower. The chlorination tower operates under full condition at all times, hence no level recorder is required. Flow loops FRC-11, FI-11, FV-11 and FT-22, FRC-22, FV-22 and FT-34, FRC-23, FV-34 measure and control exact amounts of caustic hypochlorite solution, and chlorine dioxide solution going to the pulp. Settings are made by the operator periodically, depending on the efficiency of the chlorination stage and degree of pulp whiteness (referred to in the industry as *brightness*) which may be desired in the finished product.

Temperature controllers TRC-12, TRC-23, and TRC-33 control the amount of steam used for heating the pulp to the temperature necessary for optimum action of the chemical on the impurities.

Temperature recorders TR-14, TR-25, TR-36 are used to detect approximate interface level between the heavy stock and diluted stock in the bottom of the towers. Modern bleach plants operate the caustic extraction, hypochlorite, and chlorine dioxide stages at high consistencies. However, to pump the pulp to the washer it must be diluted. Manual dilution is achieved by the operation of valves HV-7, HV-15, HV-17, HV-26, HV-28, HV-37, and HV-39 by hand control stations HC-7, HC-15, HC-17, HC-26, HC-28, HC-37, and HC-39. Under normal conditions the upper bulb will sense the warm pulp and the lower bulb will sense cooler diluted stock. Dilution water is added to maintain interface level between the two bulbs.

FR-5 records steam and water flows to the bleach plant and FRC-40 regulates the amount of fresh water added to the system. Recorder pHR-43 supplies a record of the acidity or alkalinity of the ClO_2 washer effluent which is used as a guide in determining the set point of the neutralizer flow controller FRC-44.

The electromagnetic flow transmitters for pulp slurry and chemical flow measurements are selected to meet specific plant requirements. Tower temperature measurements use special heavy-duty thermal wells to withstand stress and strain produced by heavy moving pulp. All instrument materials are selected to meet the severe corrosive services that are to be found in bleach plants used in industry operations.

BLEACHING CHEMICALS PREPARATION

Most of the chemicals used in pulp bleaching are manufactured in-plant for both economical and technical reasons. Chlorine, the most important of the bleaching chemicals, is used both in the elemental and combined form as chlorine gas in the chlorinated stage, as sodium or calcium hypochlorite in the hypochlorite stage, and as chlorine dioxide solution in the chlorine dioxide stage. At one time, practically all pulp mills manufactured their own chlorine by electrolysis of saturated brine solution. One of the by-products was sodium hydroxide which was also used in the manufacture of sodium hypochlorite and in bleaching. More recently, however, it has become economical for most pulp mills to purchase liquid chlorine from large chemical manufacturers and vaporize it at the mill site for bleaching and preparing hypochlorite solutions. This also applies to sodium hydroxide which is purchased in concentrated solutions and diluted to proper concentration for use in pulp bleaching and in the preparation of hypochlorite solutions.

Chlorine dioxide is an explosive gas that cannot presently be commercially prepared safely in undiluted form, nor can it be liquified for transport and storage. Chlorine dioxide solutions made from it are also very unstable. The gas must be generated at the bleach plant site and solutions made from it must be used in bleaching with the shortest practical storage time.

Hypochlorite Bleach Liquor Making

Process. A typical flow diagram of the sodium hypochlorite manufacturing process is shown in Figure 19-12. Chlorine is received at the mill site in liquid form in single-unit tank cars. The tank cars are connected to the mill supply and emptied by introducing compressed air into the tank car. The liquid chlorine is forced through a pipeline provided with appropriate air-filled expansion chambers and in-line pressure regulators to an evaporator which converts it into a gas through vaporization by heat. It is then piped to the bleach plant for pulp bleaching and to the sodium hypochlorite bleach liquor makeup.

Caustic soda solution arrives at the pulp mill in tank car lots at a concentration of 50 percent. Desired concentration of 5 percent dilute caustic solution is continuously produced by mixing with the proper amounts of water in two steps. The first step dilutes from 50 percent to 25 percent and the second, from 25 percent to 5 percent. The dilute caustic solution and gaseous chlorine are reacted in proper proportions to produce the finished sodium hypochlorite bleach liquor for use in the plant. Some of the 25 percent and 5 percent solutions are also used directly in the bleaching process.

Instrumentation. The primary control requirements for the instrumentation in manufacturing sodium hypochlorite bleach liquor are to maintain proper:

1. Pressure of air padding to the chlorine cars.
2. Operation of the chlorine vaporizer by controlling temperatures of chlorine gas discharge.
3. Pressure of the chlorine gas to bleach plant and hypochlorite bleach liquor makeup.

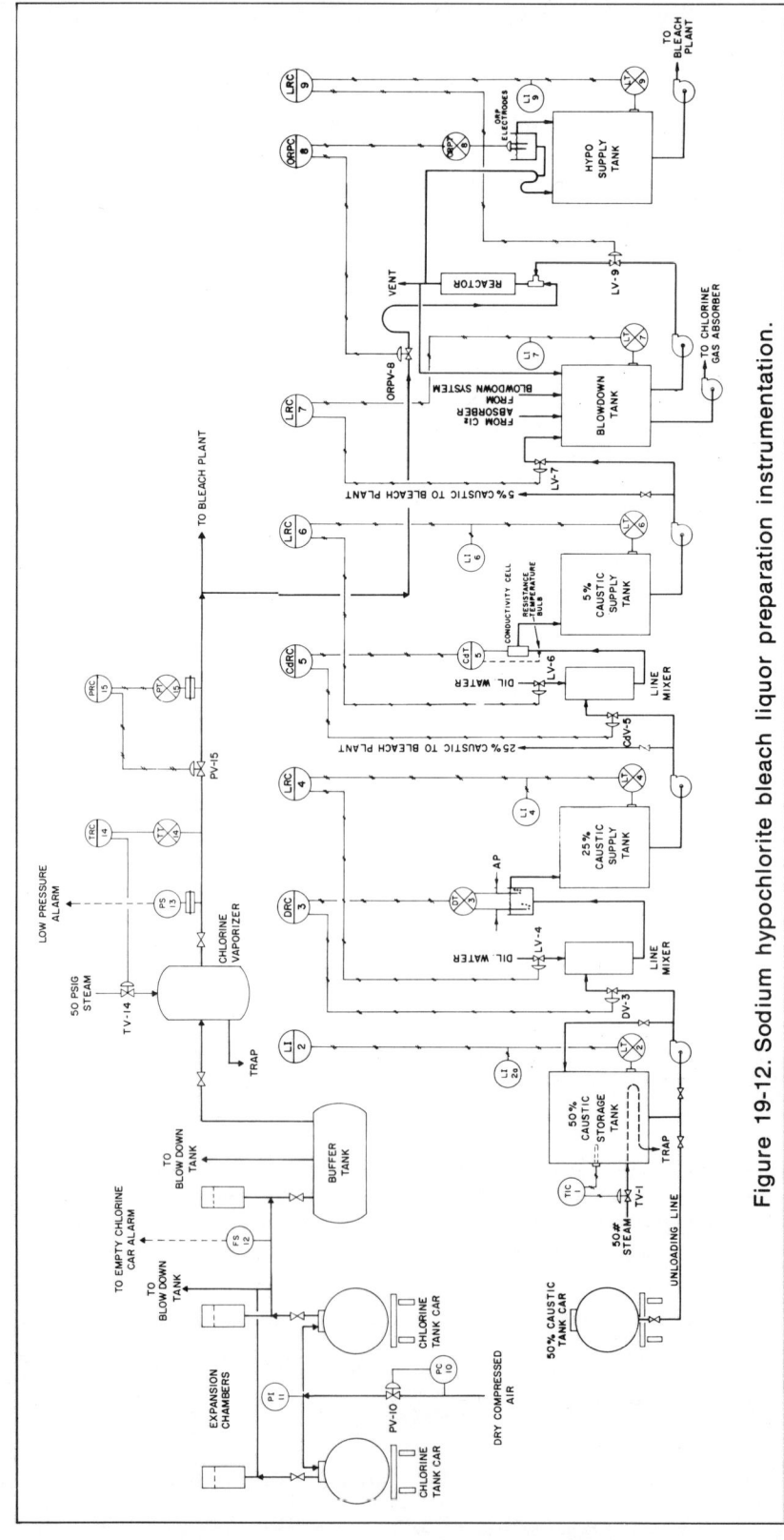

Figure 19-12. Sodium hypochlorite bleach liquor preparation instrumentation.

4. Dilution of the caustic solutions.
5. Mixing of the chlorine gas with caustic solution to produce desired strength hypochlorite bleaching chemical.

Pressure of the padding air to the chlorine tank cars is controlled by in-line adjustable type regulators PC-10, PV-10 and is indicated locally by a direct-connected indicating gauge (PI-11). Flow switch FS-12 actuates an alarm on low flow, signaling an empty chlorine car, and pressure switch PS-13 actuates an alarm on low pressure in the chlorine gas line to the bleach plant and hypochlorite makeup. Control loop TT-14, TRC-14 controls the temperature of the chlorine gas by regulating the steam flow valve (TV-14) to the chlorine vaporizer, and control loop PT-15, PRC-15, PV-15 controls the chlorine gas pressure. All pressure measurements must be protected from the corrosive effect of the chlorine by chemical seals of proper materials of construction.

The caustic is received at the mill in tank car units of 50 percent concentration solution and pumped to a storage supply tank. To prevent the 50 percent caustic solution from crystallizing in the storage tank, it is kept warm by steam controlled by a locally mounted temperature indicating controller and valve (TIC-1, TV-1). The level is measured and indicated both locally and remotely by LT-2, LI-2a. It is then diluted down to 25 percent concentration by mixing with water in a line mixer. A low differential pressure type transmitter (DT-3) measures the line mixer discharge density. This signal is used by density recorder controller DRC-3 to regulate the flow of 50 percent caustic to the line mixer to maintain proper density level equivalent to 25 percent caustic concentration. The flow of dilution water to the line mixer is regulated by level loop LT-4, LRC-4, LV-4, maintaining desired level in the 25 percent caustic supply tank. The same method is used to dilute 25 percent caustic solution to 5 percent concentration except that a conductivity measurement (CdT-5) is made after mixing with water, and a conductivity recorder controller maintains a conductivity value equivalent to 5 percent caustic concentration by throttling the 25 percent caustic flow to the line mixer. The dilution water to the line mixer is regulated on the basis of the level in the 5 percent caustic supply tank by instruments LT-6, LRC-6, LV-6.

Due to the type of chemical reaction which occurs when chlorine is mixed with a caustic solution, a relationship exists between the oxidation-reduction potential generated and the strength of the hypochlorite produced. This measurement (made by ORPT-8) is used by controller ORPC-8 to regulate the amount of chlorine being reacted with the 5 percent caustic solution by adjusting ORPV-8 to maintain the corresponding hypochlorite strength being made. Level in the hypochlorite supply tank is maintained by loop LT-9, LRC-9, LV-9, regulating the flow of 5 percent caustic to the reactor. This level is also indicated locally by level indicator LI-9.

Chlorine Dioxide Manufacturing

Chlorine dioxide is a highly toxic gas with an odor resembling chlorine and possesses over twice the oxidizing power. It oxidizes residual lignins and extraneous coloring matter to water-soluble, colorless materials with minimum detrimental effect on the desirable cellulose fiber component of pulp. As a result, its use as a bleaching agent by the pulp and paper industry has increased over recent years. A number of processes have been developed for the production of chlorine dioxide. Although they differ in design and method, most of them are similar in that the basic chemical reaction is the reduction of sodium chlorate. Sulfur dioxide gas, chromic sulfate, hydrochloric acid, methyl alcohol, and sodium chloride are listed among the chemicals used as reducing agents.

Process. Figure 19-13 illustrates a typical chlorine dioxide bleach liquor manufacturing process used on the reduction of sodium chlorate by sodium chloride. Sulfuric acid and sodium chlorate solutions are delivered to the mill site by tank car and pumped into storage tanks. The sodium chloride (salt), brought by truck or rail car, is dissolved in a lixator and the sodium chloride solution (brine) is mixed with the sodium chlorate solution to form a sodium chloride/chlorate solution containing approximately equimolar concentrations of sodium chloride and sodium chlorate. This solution and sulfuric acid are fed to the generator where, under acid conditions, the sodium chloride reacts with the sodium chlorate, reducing it to form chlorine dioxide and chlorine in about a 2:1 ratio. The gases formed are stripped out with air and diluted to a safe concentration. The gas mixture first passes through an absorption tower where the chlorine dioxide and about 25 percent of the chlorine are absorbed in chilled water. The gas leaving the absorber, which still contains valuable chlorine, is then passed through a smaller packed tower irrigated with dilute caustic soda or through a scrubber-eductor fed with dilute caustic solution to recover the chlorine as hypochlorite.

Instrumentation. Control requirements of the instrumentation system:

1. The control of acid flow to maintain proper acid conditions in the chlorine dioxide generator.
2. The maintaining of necessary flow of brine and levels in sodium chloride/chlorate storage tanks to assure proper ratio of sodium chloride to sodium chlorate.
3. The regulation of flows of sodium chloride/chlorate solutions and air to keep a balanced stripping of chlorine dioxide and chlorine gas produced.
4. The monitoring and control of temperatures in order to initiate emergency safety measures on the development of hazardous conditions causing sudden temperature increase in the process.

The level device on the brine well is a simple one-point capacity probe on-off controller (LC-1) which admits fresh water to the lixator to maintain level below a maximum in the well. Indicating flow controller FIC-2A receives a flow signal from a positive displacement type measuring element in the fresh water brine and operates an on-off valve (FV-2A) to batch fixed quantities into the mixing tank while controller FIC-2B performs the same function by starting and stopping a pump on the brine line. Temperatures of the mixing and storage tanks are controlled by instrument loops TT-3, TIC-3, TV-3 and TT-4, TIC-4, TV-4 by regulating steam to coil heaters located in the tanks. Their levels are recorded by loops LT-4, LR-4 and LT-5, LR-5. LT-7 and LR-7 are

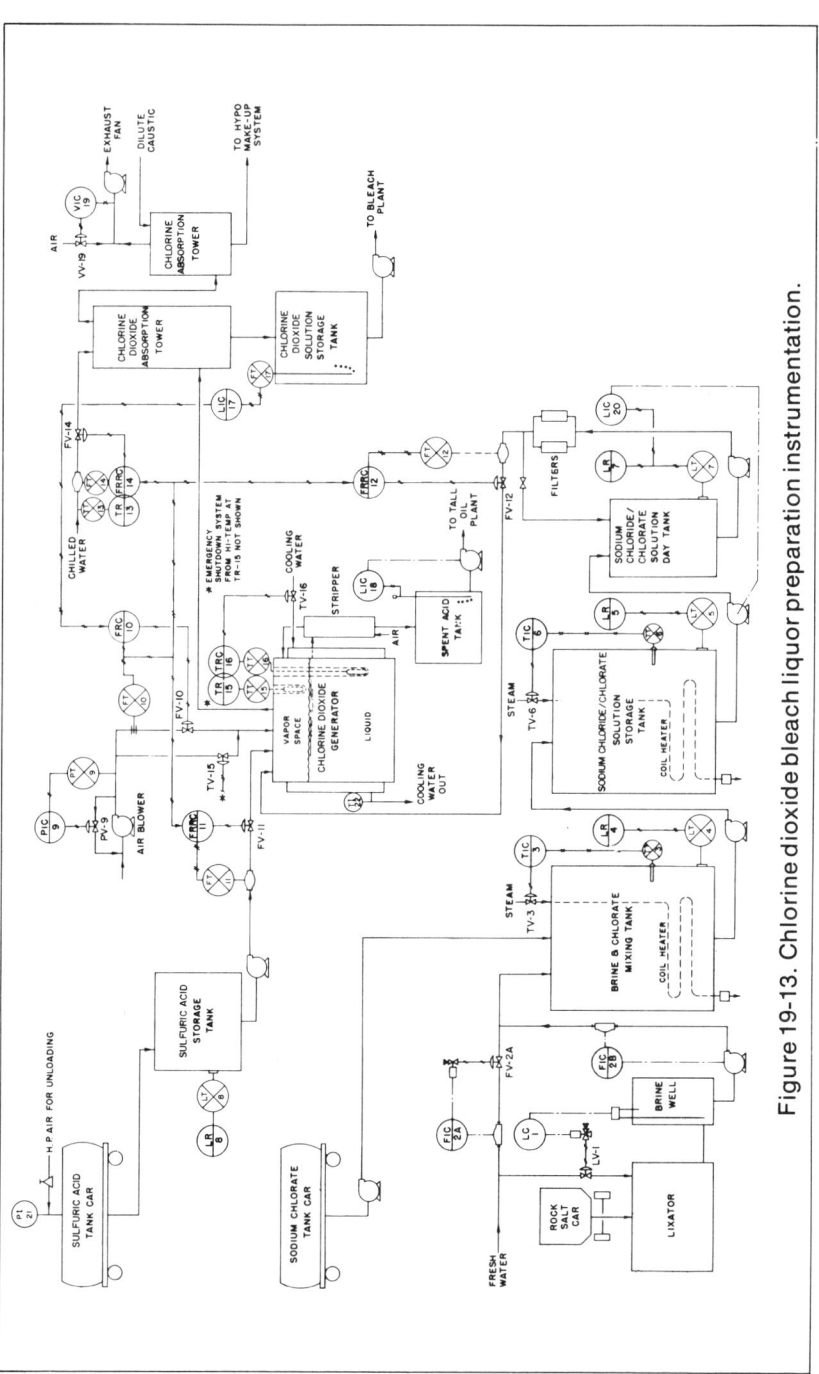

Figure 19-13. Chlorine dioxide bleach liquor preparation instrumentation.

used to measure and record level in the sodium chloride/chlorate solution day tank. It is controlled by indicating controller LIC-20, starting and stopping the pump between storage and day tank.

Pressure gauge PI-21 is used to locally monitor the pressure in the sulfuric acid car as a guide to when it is empty. LT-8, LR-8 are used to measure and record level in the sulfuric acid storage tank. The pressure of the stripping air to the generator is controlled by throttling a bypass valve (PV-9) around the air blower with an indicating pressure controller (PIC-9) from a pressure measurement being made with transmitter PT-9. The air flow is measured by an orifice plate and flow transmitter (FT-10) to operate valve FV-10, to control air flow to the generator. This same signal is transmitted as the primary signal to flow ratio controllers FRRC-11, FRRC-12, and FRRC-14. A controlled amount of sulfuric acid and sodium chloride/chlorate solution passes to the generator, and chilled water goes to the chlorine dioxide absorption tower in relation to the stripping air flow. With this arrangement, all flows would automatically shut down if there were a failure of air flow to the generator, preventing the development of a dangerous accumulation of unpurged gases in the generator. Any malfunction causing an excessively hazardous temperature in the generator is detected by temperature recorder TR-15 which actuates an emergency shutdown system, stopping all flows to the generator and flooding it with water. Temperature recorder controller TRC-16 operates control valve TV-6 on cooling water to the generator jacket to maintain temperature at optimum operating and safety values. Level indicating controller LIC-17 on the chlorine dioxide solution storage tank remotely sets the control point of flow controller TRC-10 on the air. This adjusts flow of air and all other ratioed flows to the generator and absorption tower as the level tends to vary in the storage tank (caused by the amount of chlorine dioxide bleach solution being used by the bleach plant). The temperature of the generator cooling water outlet is measured by a locally mounted temperature indicator (TI-22) while a locally mounted bubble type measurement level indicating controller (LIC-18) maintains a constant level in the spent acid tank. Another local controller (VIC-19) regulates the vacuum of the evacuating system on the chlorine absorption tower.

Electromagnetic flowmeters are used to make the chemical and water flow measurements to the chlorine dioxide generator.

STOCK PREPARATION

Stock preparation may be defined as that part of the papermaking process in which pulp is mechanically treated or mixed with other pulps, chemicals, additives, and dyes in order to prepare it for sheet formation on the paper machine. Pulp as it comes from the pulp mill does not exhibit all the satisfactory mechanical and physical characteristics required for the formulation of the many different final products. Desired mechanical characteristics are imparted to the pulp by beating and/or refining. Other desired physical properties are achieved by mixing with other pulps already containing these qualities or are implanted by the addition of chemicals, additives, and dyes. This operation is commonly known as stock proportioning or stock blending, stock being the fibrous slurry resulting from the mechanical treatment and mixing of pulp.

Mechanical Treatment

Process. In the early days of papermaking, the mechanical treatment of pulp to improve its matting or felting properties was accomplished manually. One of the first machines used was known as the *beater* or Hollander. Essentially, it was a batch operation. Although there are many batch type beaters still in use today, advancing technology toward continuous processing of pulp led to the development of continuous refining equipment. These refiners may be of conical or disc design. The basic elements are a stationary housing and movable rotating members, and the instrumentation for these is similar.

Instrumentation. The primary objective of the control scheme is to regulate the amount of mechanical work done on the pulp by positioning of the rotating component (referred to as a *plug* or *disc*). Useful measurements here include plug pressure, paper machine couch vacuum, stock freeness refiner drive motor power, temperature difference across the refiner, or a combination of these factors.

Figure 19-14 illustrates a typical system using a combination of drive motor power and temperature difference across the refiner to control plug position. Temperature detectors are located at the inlet and outlet of the refiner. These temperatures are transmitted by TT-2a and TT-2b to temperature difference converter TdCon-2 which subtracts the inlet temperature from the outlet temperature

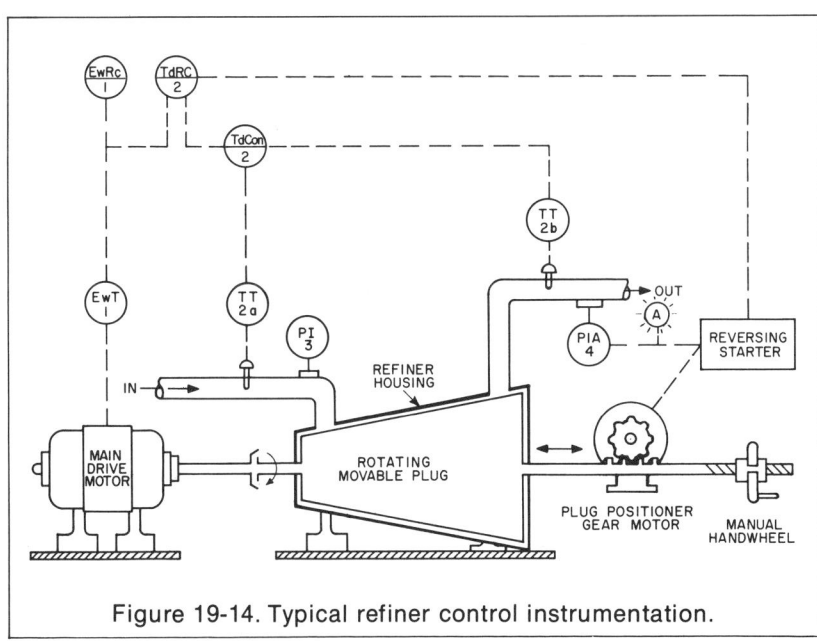

Figure 19-14. Typical refiner control instrumentation.

and transmits a signal proportional to temperature rise across the refiner to temperature difference recorder controller TdRC-2. This is a pulse duration type controller which emits a pulse type signal to energize the appropriate relay in the reversing starter of the plug positioning motor. This action moves the plug in or out (or varies the distance between discs on a disc type refiner) as required, to maintain proper level of mechanical action (work) on the pulp.

Main drive motor load power is measured and transmitted by transmitter EwT-1 to recorder EwRC-1. This same signal is fed to the temperature difference recorder controller (TdRC-2) and used to supplement its control action. When an upset occurs, a time lag exists until its effect is detected by the differential temperature measurement and the plug is repositioned. The more sensitive motor load power measurement detects upsets more rapidly and sends a corresponding signal to the temperature difference controller which uses it to make a compensating adjustment to its control action. A manual handwheel is also provided.

Pressure indicator PI-3 is used as a guide to possible plugging or loss of flow to the refiner. Pressure indicating alarm PIA-4 is also used for this purpose but, in addition, it will automatically withdraw the rotating plug on low flow by overriding the control signal from the temperature difference controller to the reversing starter. At the same time, it energizes a visible and/or audible alarm to call the operator's attention to this condition.

Stock Proportioning

Process. An early method used to mix pulps with varying characteristics was to batch blend predetermined amounts in mixing tanks or machine chests prior to running the mixed pulps on the paper machine. Chemical additives and dyes were also added to the mixing tanks in the same manner in order to impart certain interior and exterior properties to the final sheet of paper. Subsequent continuous mechanical proportioning systems, many of which are still in use today, consisted of individual positive displacement metering compartments, one for each pulp, dye, and additive as required. The compartment rotors are driven by individual PIVs which, in turn, are connected to the variable-speed shaft of the master PIV. Thus, the speed of each can be adjusted to give the required proportions. The master PIV is controlled by the mixing chest level; the speed recorders on the compartment rotors are used to record the proportions. More recently, the continuous flow metering method of stock proportioning has been developed. With the magnetic flowmeter now being used as the flow measuring device, the continuous flow method has become the predominant one used for blending pulps, additives, and dyes in the paper industry today.

Instrumentation. Figure 19-15 shows that a typical continuous flow type stock proportioning system blends stocks, dyes, and additives by using a magnetic flowmeter, a flow rate controller, and flow valve. A level controller on the mixing chest automatically adjusts the ratio flow controller to regulate amounts of blend components to the mixing chest. Level recorder controller LRC-9 maintains level in the machine chest by adjusting valve LV-9 in the

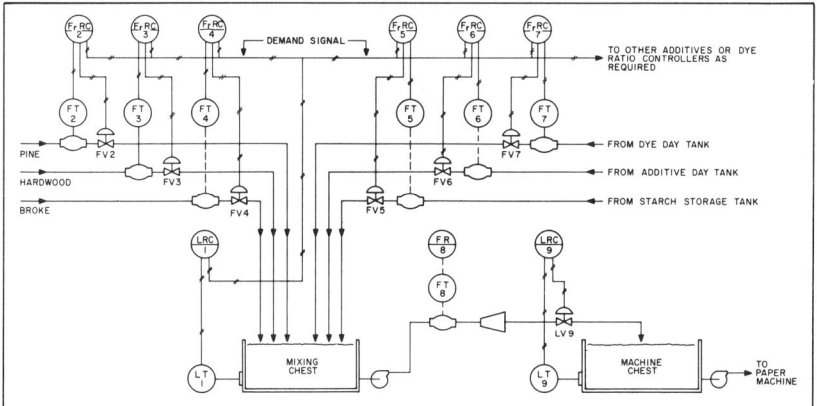

Figure 19-15. Representative scheme for stock proportioning system instrumentation.

stock line from the mixing chest to make up for changes in paper machine demands. The quantity is recorded by FR-8 from a magnetic flow transmitter (FT-8) in the flow line. Level recorder controller LRC-1 maintains level in the mixing chest in response to stock drawn from it by the action of level controller LRC-9. Therefore, the mixing chest level will, in turn, reflect demand changes by the paper machine. As the mixing chest tries to vary in response to paper machine demand changes, level controller LRC-1 varies its output to ratio mechanisms located in the flow ratio recording controllers on pine (FrRC-2), hardwood, (FrRC-3), and broke (FrRC-4) pulp flows. This signal is also transmitted to similar ratio mechanisms associated with dye, additive, and starch flows (FrRC-5, FrRC-6, and FrRC-7). The ratio devices adjust the set points of each controller in accordance with the desired amount of each ingredient relative to total demand flow, which is previously determined and set on the ratio devices. Flows of pulp components are measured by magnetic flowmeters FT-2, FT-3, and FT-4, and dye, additive, and starch flows are measured by magnetic flow transmitters FT-5, FT-6, and FT-7. These flow signals are transmitted to their respective flow ratio controllers. Level transmitters on mixing chest LT-1 and machine chest LT-9 are usually of the flanged diaphragm type, although the bubble type can also be used.

PAPERMAKING

The formation of a sheet of paper from stock is accomplished on papermaking machines located in an area usually called *the machine room.*

Paper Machines

The types of paper machines used to manufacture the many grades of paper can be classified into two general categories: (1) the fourdrinier machine and (2) the cylinder machine.

Instrument Applications   387

Additionally, a combination of the two has been used to manufacture a particular product and, more recently, some new types have been designed and introduced. The innovations have been primarily around the sheet formation or wet end section of the machine. Notwithstanding the type, the basic general procedure used by all is essentially the same. At the wet end, the web of paper is first formed from an aqueous suspension of fibers on a traveling wire screen in the case of the fourdrinier, or on a cylinder mold in the cylinder machine. As it approaches the dry end, this wet sheet is further dewatered by the press section of the machine, dried by the dryers, and the surface is finished or smoothed by the calender.

The fourdrinier is by far the most extensively used type of machine in the paper industry. Older, smaller machines were designed to run at speeds up to 300 fpm, producing a sheet of paper between 72 and 100 inches in width. Today, machines are being designed to run at speeds approaching 5000 fpm, making sheets of over 380 inches in width.

Wet End Instrumentation. The principal components of a wet end of a fourdrinier paper machine are shown in Figure 19-16, and include the machine storage chest, refiner, stuff regulating box, fan pump, cleaners, screens, headbox, wire screen, white water tank, wire pit, couch pit, presses, saveall, broke pulper, and all associated equipment.

Beginning with the machine chest, instruments LT-1, LRC-1, and LV-1 measure and maintain normal operating stock level in response to machine demand or production. Consistency loop CT-2, CRC-2, and CV-2 ensures uniform stock consistency to the refiner by automatically adding water to the suction side of the pump. The degree to which the stock is refined is controlled by refiner motor load controller EwRC-11A, operating the refiner plug positioning motor in response to couch roll vacuum transmitter VT-11A, which remotely sets its control point accordingly. Temperature difference and freeness measurements have also been used for this control. Depending on the grades of paper being made, a paper machine may or may not have a stuff or regulating box whose level is measured and controlled by instruments LT-3, LRC-3, and LV-3. The rate of stock delivery to the machine is automatically controlled by having a basis weight controller at the dry end to cascade set the control point of magnetic type stock flowmetering control system FT-4, FRC-4.

Many grades of paper are sized for resistance to water penetration by adding rosin to stock before it goes to the machine. Alum is generally used to "set" the size on the individual fibers. This is best accomplished at a definite pH level. A pH recording control loop (pHT-5, pHRC-5, pHV-5) regulates the addition of alum to the paper stock. To attain a uniform flow through the slice onto the wire and to ensure good sheet formation, both total head (pressure) and level must be controlled in the headbox. Level controller LRC-6 controls headbox level by adjusting valve LV-6 in the bypass stock line around the fan pump. The total head is controlled by instrument PRC-7, operating a vent valve (PV-7) in a vacuum pressure exhaust system attached to the headbox. Efficient and safe water removal on the wire is further enhanced by controlling the vacuum in the individual flat boxes with vacuum instrument loops VIC-8,

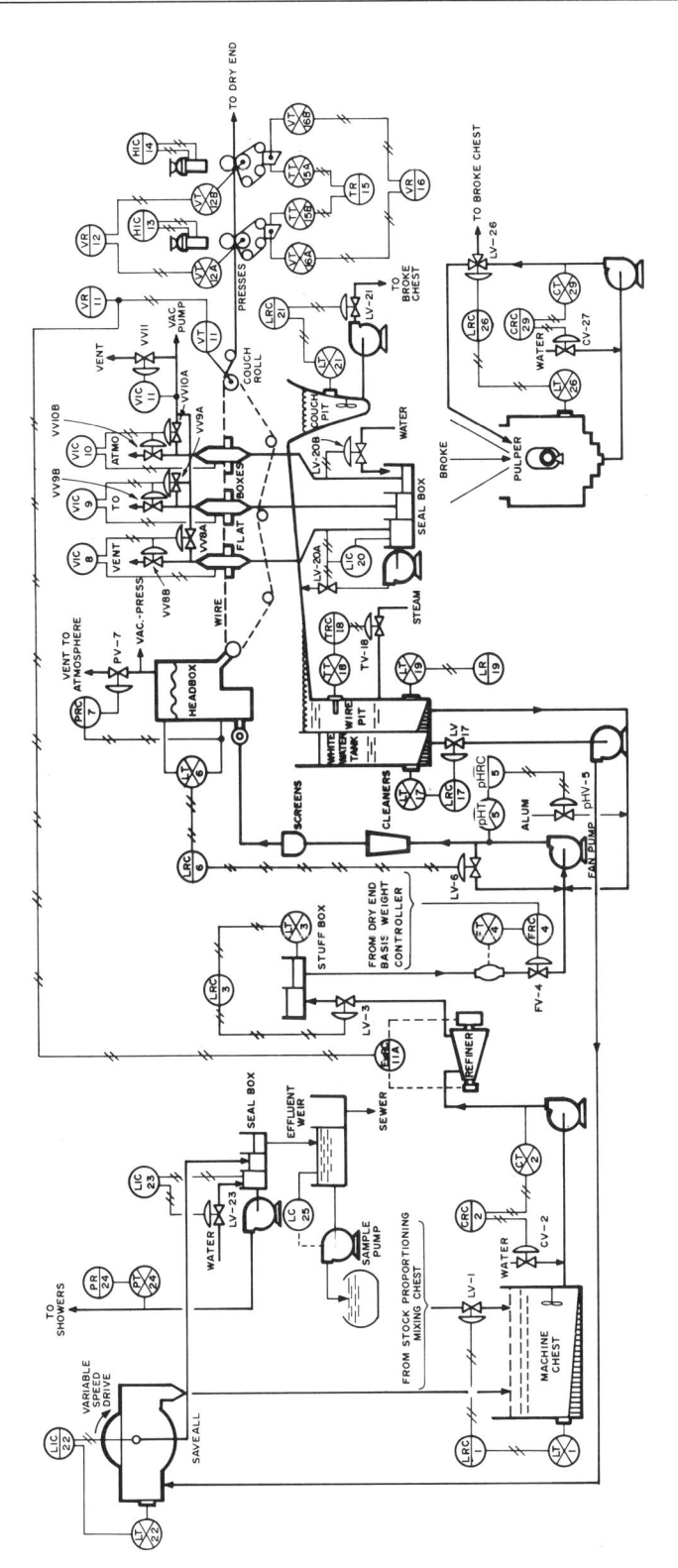

Figure 19-16. Paper machine wet end controls.

VV-8A, VV-8B, VIC-9, VV-9A, VV-9B and VIC-10, VV-10A, VV-10B, with loop VIC-11, VV-11 controlling the vacuum in the header to the vacuum pump.

The couch roll drives the wire and is evacuated to create a vacuum under the wire. This measurement is recorded on VR-11 and also transmitted to EwRC-11A to cascade set the load on the refiners. Vacuums on the presses are recorded on a two-pen vacuum recorder (VR-12) while remote loading of press rolls is accomplished by hand-indicating control stations HIC-13 and 14. Instrument loops VT-16A, VT-16B, VR-16 and TT-15A, TT-15B, and TR-15 and record vacuums and temperatures of felt conditioners as a guide to proper operation.

The vacuum system seal box level is maintained at a uniform level by indicating level controller LIC-20, operating level control valve LV-20A, which is split-ranged with control valve LV-20B on the fresh water line, to ensure a minimum water level in the seal box to protect the pump suction. Instrument loop LT-21, LRC-21, and LV-21 controls level in the couch pit, while instrument LIC-22 controls level in the saveall vat by adjusting a variable speed drive on the drum. Shower pump suction on the saveall seal box is protected by a locally mounted level indicating controller (LIC-23) and the shower water pressure is recorded on PR-24. A locally mounted level controller on the mill effluent weir is used to operate a sample pump in order to provide means of checking fiber losses in the paper mill effluent.

Basic level and consistency control loops used on broke pulpers in paper mills are illustrated by instruments LT-26, LRC-26, LV-26 and CT-27, CRC-27, CV-27.

Dry End Instrumentation. A typical dry end section of a paper machine includes dryers, a size press, calenders, and reels. The dryers consist of a series of steam-heated cylinders. The two sides of the sheet are brought alternately into contact with the heated cylinders. Water vaporized from the paper is carried away by exhaust fans through a paper machine ventilating hood. The size press is a method of applying size to the surface of the paper, while the calender conditions the surface of the paper.

Some typical instrumentation used to control the operation of the dry end of a paper machine is shown in Figure 19-17. To obtain efficient extraction from input steam, the machine is divided into sections and the differential pressure between sections is regulated. With this arrangement, the condensate will flash from each successive separator to the following section inlet header.

Steam used in the paper drying operation is measured, recorded, and totalized by instruments FT-1 and FR-1. Control loops dPRC-2, dPRC-3, dPRC-5, and dPRC-9 maintain the necessary differential pressure across each section of dryers in order to ensure good condensate and air removal from the dryers for efficient drying. Auto-selector type pressure controllers PsC-3, PsC-5, and PsC-9 establish the necessary differential gradient between sections. These controllers are provided with auxiliary mechanisms which will automatically reduce pressure in the subsequent sections if pressures approach values too close for good condensate removal. The cascade moisture control system MT-7, MRC-7, and PRC-7 controls the sheet moisture prior to the size application in the size press. The moisture measurement is made by a

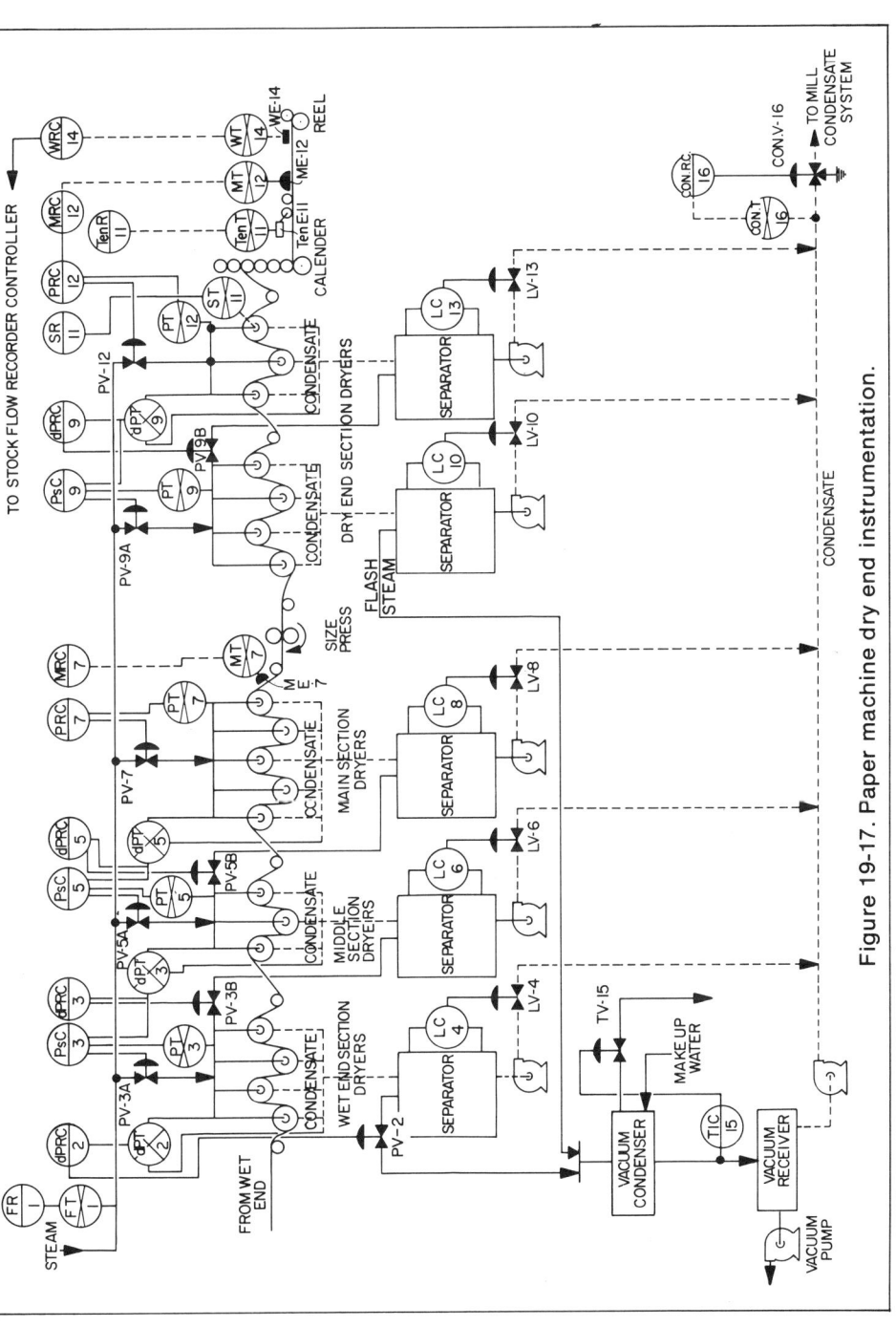

Figure 19-17. Paper machine dry end instrumentation.

variety of methods as described in the chapter on moisture measurements. The output of the moisture controller automatically adjusts the pressure controller set point in a cascade fashion in order to control at the desired steam pressure level to maintain a constant moisture value. In the same manner, the moisture control system (MT-12, MRC-12, PT-12, and PRC-12) maintains uniform sheet moisture going to the reel.

Also by cascade control, the basis weight of the sheet going to the reel is maintained at predetermined levels by instrument system WT-14 and WRC-14 automatically adjusting the set point of the flow controller on the stock to the fan pump of the paper machine. Different methods of basis weight measurement have previously been described in the chapter on basis weight measurements. Sheet tension to the reel is measured by TenT-13 and recorded on TenR-12. Speed of the machine is measured by speed transmitter ST-11 which is usually a tachometer generator, and a record is made on recorder SR-11.

Level control loops LC-4, LC-6, LC-8, LC-10, and LC-13 ensure proper level in order to achieve satisfactory condensate flashing in the separator tank and maintain a positive head of condensate to protect the suction of the pump.

Temperature loop TIC-15, TV-15 controls the outlet temperature of the vacuum condenser, while condensate recorder controller ConRC-16 diverts condensate to the sewer when it reaches a dangerously dirty condition, allowing only clean condensate to pass back to the power boiler feedwater system.

CHEMICAL RECOVERY

Liquor Evaporation

Black liquor is pumped from the pulp washers to a thin liquor storage tank (see Figure 19-18). From there it passes through a set of multiple-effect evaporators where it is initially concentrated in order to achieve a density sufficiently high to support combustion in the liquor-burning stage of the chemical recovery cycle. A countercurrent backward-feed system is used, with the weak liquor entering the sixth effect at 15 to 20 percent solids and leaving the first effect at 50 to 55 percent solids.

Evaporator Instrumentation. Sixty-pound steam at 300°F is admitted to the first effect under pressure control by PRC-1, whose set point is automatically set in a cascade mode from the output of boiling point rise recording controller BPRC-2. This controller uses resistance type thermal elements to obtain a temperature difference between the boiling liquor and saturated steam at the same pressure. A known relationship exists between boiling point elevation and density of specific types of liquors. Where this relationship is not known, it can be easily established by each mill. Bulb X is so located as to ensure continual submergence in the liquor by level control loop LT-12, LC-12, LV-12, while bulb Y is installed in a condensing chamber tied into the vapor line which serves to desuperheat the vapor by radiation to atmosphere, thus providing a saturated steam temperature at evaporator pressure.

A variation in liquor density to the first effect will cause a pressure change. This will be partially corrected at the pressure controller as it modifies the

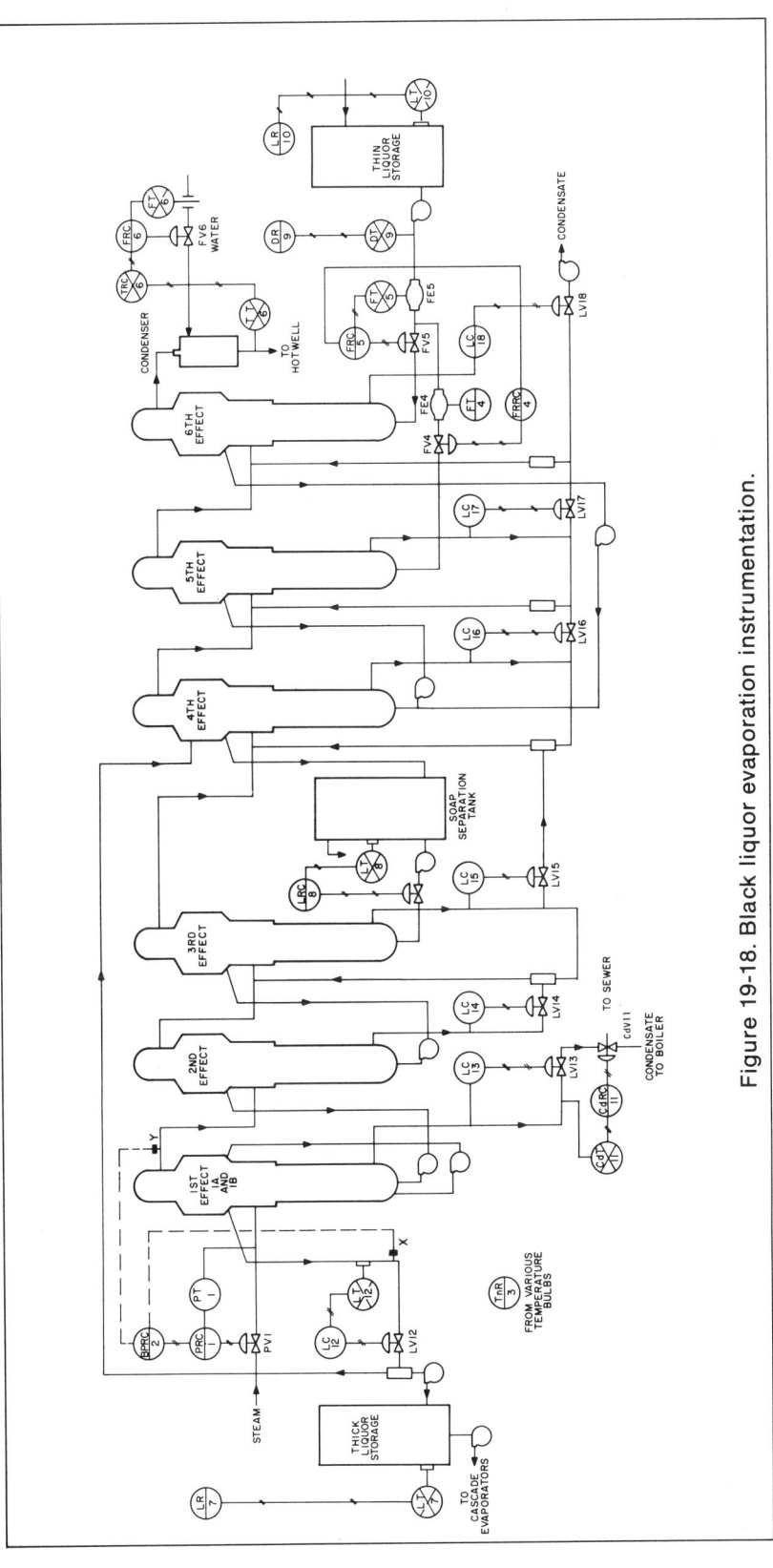

Figure 19-18. Black liquor evaporation instrumentation.

steam flow rate. The boiling point rise measurement will sense the change in temperature difference and gradually alter the set point of the pressure controller to a new value. This will cause a further change in steam flow rate to complete the compensation and reestablish the boiling point rise to the original and desired value.

Condensate from the first effect is returned to the boiler house and is controlled by level controller LC-13 to maintain level of condensate below the evaporator tubes and prevent steam blow through the condensate line. If contamination with black liquor occurs, conductivity recording controller CdRC-11 immediately operates the three-way diversion valve (CdV-11) to divert the condensate flow to the sewer.

Thin liquor feed to the evaporator is divided between the fifth and sixth effects because, in most cases, the sixth effect alone could not handle the total feed rate without flooding. Instruments FRC-5 and FRRC-4 control the feed of this thin black liquor. The feed is usually split as shown with flows automatically ratioed in desired proportions. The magentic type primary flow devices are commonly used here because they do not obstruct entrained solids.

DR-9 is a continuous Baumé density recorder on the liquor feed which receives its measurement signal from transmitter DT-9. It gives a continuous trend of weak liquor gravity changes to the operator so he may periodically reset the total volume feed rate flow recording controller (FRC-5) to maintain a constant total solids throughput. The density measurement transmitter (DT-9) is usually of the bubble tube type; however, radioactive and refractive index density measuring elements have also been used here.

Liquor leaving the fourth effect has been concentrated to the point where resin soaps formed during the cooking operation will no longer stay dissolved but will tend to separate out. If allowed to continue into the following effects, the soaps would cause excessive foaming, scaling, and reduced capacity. The soap separates in the separation tank and is skimmed off the top of the black liquor. Proper liquor level is maintained by level recording controller LRC-8, regulating feed rate to the third effect with valve LV-8. Level transmitter LT-8 can be a bubble tube or flanged flush-mounted diaphragm force-balance type.

Level controllers LC-14, LC-15, LC-16, LC-17, and LC-18 hold condensate seals on each effect and allow flash condensate from the control valves (LV-14, LV-15, LV-16, and LV-17) to add to the overhead vapor of each effect. Flow recorder controller FRC-6 regulates the water rate to the barometric condenser in accordance with its set point as cascade set from discharge temperature.

Liquor Burning

Process. Black liquor for the evaporators at about 45 to 50 percent solids is further evaporated and then burned in a recovery boiler which is part of a closed cycle in the recovery process. This process consists of washing the digested pulp, evaporization of the weak black liquor, liquor burning (to produce green liquor), and causticizing the green liquor to produce white cooking liquor for pulping.

The three basic reasons for liquor burning are:

1. To recover inorganic chemicals for reprocessing.
2. To recover energy in the form of steam generation through incineration of the organic constituents of black liquor.
3. To eliminate or reduce both air and water pollution.

A cascade or cyclone type evaporator is used, depending on the manufacturer of the recovery furnace system used by the mill. Figure 19-19 shows a typical installation in which the 50 percent solids liquor is exposed to a hot gas stream from the recovery boiler in a cascade type evaporator for further removal of water. The liquor serves as a wetting agent to remove dust from furnace gas. Thick liquor feed passes through the cascade evaporators over rotating drums or discs where water is removed by contact with the hot flue gases. Liquor leaving the cascade evaporators at a concentration of 65 to 70 percent solids is combined with salt cake and sulfur to replace sodium salts and sulfur which are lost during the pulping and recovery cycle. After makeup, the fortified liquor is passed through the liquor heater which raises the temperature to 220-240°F. Liquor is introduced into the furnace through an oscillating spray nozzle which projects the liquor onto the furnace walls where the remaining water content is removed. The dehydrated liquor falls to the smelt bed at the base of the furnace. An alternate firing system introduces the black liquor into the hot gas stream where it is dehydrated as it falls to the hearth. Air for combustion is supplied by a forced draft fan and then passed through an air heater. It is then introduced into the furnace through primary air ports near the bottom of the furnace and through secondary air ports above the bed. Primary air is used in the combustion of carbon compounds in the black liquor and also maintains a reducing atmosphere for the reduction of the sodium sulfate to sodium sulfide in the smelt. The secondary air completes the combustion of volatile gases.

The hot gases, at about 1800-2000°F through the boiler tubes, enter the direct-contact evaporator at 600-700°F. Approximately 95-98 percent of the gas-borne dust is removed by the cascade evaporator and precipitator.

The molten smelt, made up of soluble salts and some impurities, flows from the bed of the furnace through water cooled smelt spouts into an agitated smelt bed dissolving tank where it is mixed with weak green liquor from the recausticizing plant and pumped to the recausticizing plant as green liquor.

Instrumentation. Some of the more important instrumentation used on a typical modern high-capacity recovery furnace is shown in Figure 19-19.

Level controller LIC-1 maintains level in the back of the precipitator by regulating recirculation flow of black liquor from the salt cake mixing tank. The level of the thick liquor storage tank is recorded on LR-2 and the black liquor pumping pressure to the cascade evaporator is monitored by pressure indicator PI-3. Level controller LRC-2A maintains uniform pickup by the drums or discs of the evaporator while density recording controller DRC-8 maintains a constant liquor density going to the burners by sensing the load on the drum drive motors.

Salt cake makeup to the liquor leaving the cascade is added by flow ratio

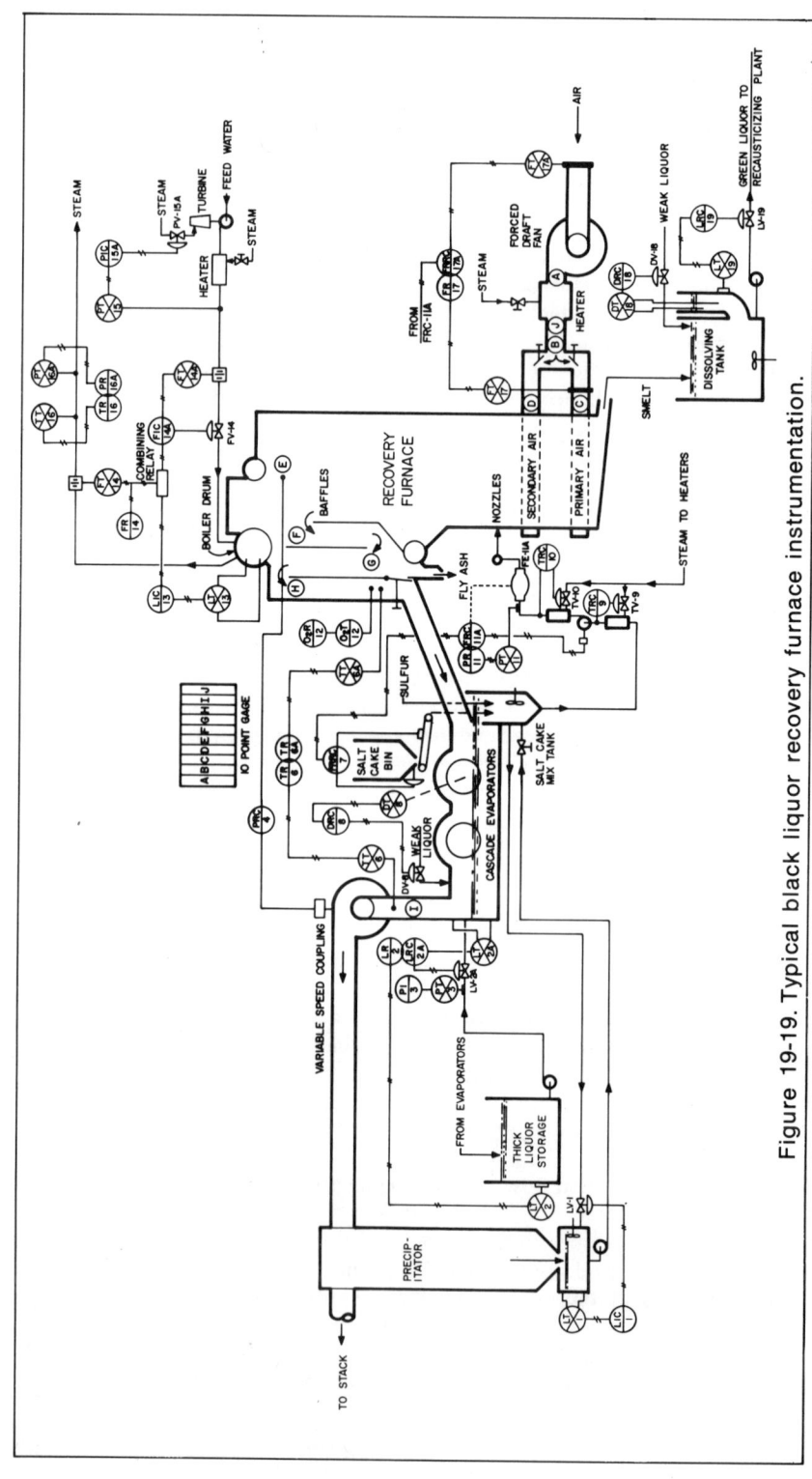

Figure 19-19. Typical black liquor recovery furnace instrumentation.

recording controller FRRC-7 in accordance with the rate of flow of black liquor to the furnace as measured by an electromagnetic flow element (FE-11A). Reduction of salt cake in the furnace yields sodium sulfide which replaces the sulfide lost in the cooking and washing operations. The liquor is further heated in the primary and secondary heaters whose temperatures are controlled by temperature recording controllers TRC-9 and TRC-10. The liquor flow rate is measured by an electromagnetic flow element (FE-11A) and is controlled by FRC-11, which regulates a variable-speed pump. The liquor is then sprayed through burner nozzles into the furnace where it ignites and burns.

To ensure proper liquor-to-air mixture, the total air flow is ratioed to liquor flow and controlled by FRRC-17A.

Boiler drum level measurement is made with a differential type level transmitter (LT-13) and is fed to a level indicating controller (LIC-13) whose output biases the boiler steam flow signal from transmitter FT-14 through a combining relay. The output of the combining relay regulates the set point of the feedwater flow controller (FIC-14A) to maintain feedwater flow based on boiler outlet steam flow and trimmed by the drum level controller.

Oxygen recorder O_2R-12 in the rear pass of the boiler records combustion efficiency so that proper fuel-air ratio settings can be maintained. Draft controller PRC-4 regulates the induced draft fan speed to ensure proper negative pressure in the furnace. Various temperatures, pressures, and drafts are recorded or indicated, depending on the type of furnace. This gives the operator all the necessary information for complete furnace control.

The smelt passes from the furnace floor into a dissolving tank where it becomes green liquor. Tank level is maintained by level recording controller LRC-19. Density recording controller DRC-18 maintains constant density of the green liquor in order to ensure uniform gravity necessary for efficient operation of the next process, liquor causticizing.

Liquor Causticizing

Process. The purpose of the liquor causticizing process is to convert the sodium carbonate in green liquor to an active chemical (sodium bydioxide), thereby producing white liquor which is recycled back to the digesters to be reused as cooking liquor. In order to accomplish this, raw green liquor from the smelt dissolving tank in the recovery boiler area is pumped to the causticizing area where lime is added. The resultant white liquor is separated from the sludge and pumped to the digester house. The sludge, which is primarily calcium carbonate, is pumped to the lime which is used to recausticize more green liquor (see Figure 19-20).

The liquor causticizing process can be divided into a number of separate functions:

1. Green liquor clarification where foreign material (dregs) is removed.
2. The washing of green liquor to remove entrained sodium salts.
3. Mixing lime and green liquor in the shaker.
4. Allowing the lime and green liquor to react in agitated tanks called causticizers.

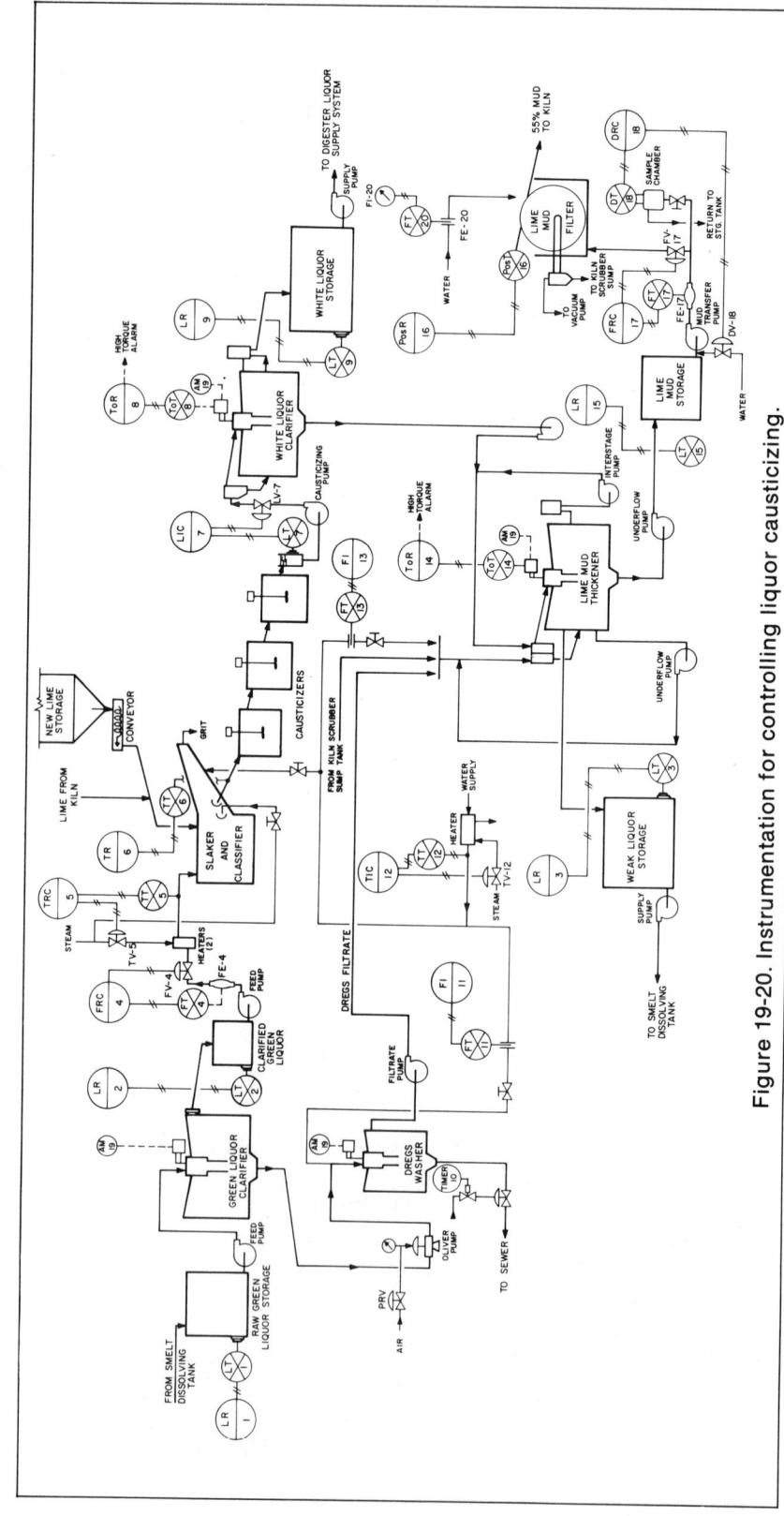

Figure 19-20. Instrumentation for controlling liquor causticizing.

5. The settling out and separation of sludge (lime mud) from the white liquor in a clarifier.
6. Thickening of lime mud.
7. Removing excess moisture from lime mud before burning in the lime kiln.

Instrumentation. Typical instrumentation required for control of the liquor causticizing process for alkaline pulp cooking liquor is shown in Figure 19-20. Raw green liquor flow to the slaker and classifier is regulated by recording flow controller FRC-4. The primary measuring system (FE-4 and FT-4) is an electromagnetic flow transmitter; its lined straight-through metering tube makes it suitable for handling the suspended solids and corrosive liquor. Flow control of uniform-density green liquor supply stabilizes the system input to secure optimum clarification and slaking efficiency.

Lime slaking is best done within fairly close temperature limits. At optimum temperature, the reaction proceeds more completely, unreacted lime settles more rapidly in the white liquor clarifier, and there is less of it to be recirculated in the kiln. Recording temperature controller TRC-5 controls the system input to the green liquor heater to maintain the optimum temperature of the green liquor entering the slaker regardless of load variation. Temperature recorder TR-6 shows the temperature of the slurry leaving the slaker. The record will show if the final desired temperature has been reached, indicating completion of reaction. A drop in temperature will reveal any interruptions in the lime feed.

Wash water temperature is important. Optimum temperature in the washer accelerates the settling of the lime mud and the dregs, ultimately resulting in clearer white liquor. It reduces soda losses occurring through incompletely washed lime mud or green liquor dregs. Indicating temperature controller TIC-12 maintains the desired wash water temperature regardless of load fluctuations by controlling the steam input to the heater.

At times, the lime mud may build up in the white liquor clarifier or lime mud thickener, overloading the rakes and resulting in excessive maintenance costs and production losses. Torque recorders TOR-8 and TOR-14, with high-load alarms, automatically warn the operator of damaging overloads on these rakes. The lime mud filter operates best if supplied with constant-density slurry. Density recording controller DRC-18 maintains this desired density, thus stabilizing filter operation and improving calcination in the lime kiln.

The lime mud filters also require a stabilized vat level for best operation. Otherwise, a filter cake of varying moisture will be produced which constantly upsets kiln operation. Flow recording controller FRC-17 maintains the slurry feed at a constant flow rate to the filter by use of an electromagnetic flow transmission system (FE-17 and FT-17) thereby regulating the level in the vat. Filter cake thickness recorder PosR-16 records the variations in cake thickness as transmitted by position transmitter PosT-16. The operator, guided by this record, can maintain desired thickness.

Showers are used to give the lime slurry a final washing on the mud filter drum. If enough hot water is used, the soda chemical will largely be recovered before the sludge enters the kiln, with reduction in formation of the undesirable "mud rings" in the kiln. On the other hand, excessive water usage may produce a "sloppy" filter cake which will not be readily calcined. This water flow is indicated on FI-20 with orifice plate FE-20 and differential pressure transmitter FT-20 used as the measurement system.

Chemicals used in the causticizing operation are expensive and any loss from overflow of storage tanks should be avoided. For this reason, recording level instruments LR-1, LR-2, LR-3, LR-9, and LR-15 are used for the various storage tanks in the process. The flange-mounted level transmitter with flush diaphragm element is normally used for the primary measurement (in these cases, LT-1, LT-2, LT-3, LT-9, and LT-15). The level recording instruments can be equipped with appropriate high and low alarms, or independent alarm systems can be used with level sensing electrodes or level switches installed on the tanks.

Lime Recovery

Process. The purpose of the lime recovery process is to reclaim the lime used in the causticizing process. The lime mud from the filter in the causticizing plant, with solids content of about 50 percent, is fed into the upper end of a rotary lime kiln. The kiln is essentially a long inclined horizontal cylinder mounted on rollers and rotated by a variable-speed electric motor drive through a girth gear assembly around the kiln, as shown in Figure 19-21. As the lime mud works down the kiln, flowing countercurrent to the heat, moisture is evaporated. The carbon dioxide is driven off converting the mud, which is primarily calcium carbonate, into burned lime which is calcium oxide. The necessary heat is obtained by the combustion of natural gas or fuel oil in burners at the lower end of the kiln. The diagram shows an installation set up to use either fuel. Some heat is also generated by the burning of the carbonate itself. The lime mud goes through several definite phases during its travel through the kiln as it is converted to burned lime: (1) the initial drying stage, (2) the stage in which the mud temperature is increased during which it passes a plastic state, and (3) the stage in which the carbon dioxide is evolved from the calcium carbonate to yield burned lime or calcium oxide in the high-temperature zone.

Instrumentation. In order to produce good-quality lime at minimum fuel consumption, the instrumentation shown in Figure 19-21 holds the burning temperature as constant as possible by controlling the air necessary to support combustion in order to completely consume the fuel, but at the same time avoiding any excess air. The control system also maintains moisture content of the entering fuel as constant as possible; it maintains constant optimum retention time in the kiln.

Temperature recorder TR-3 keeps a record of temperatures of the entire system from measurements made by thermocouple-sensing devices located in strategic locations such as the feed end of the kiln, the firing end of the kiln, gas stream, and other critical areas. These temperatures can be used in an alarm and safety system to automatically shut off the fuel supply when undesirable or dangerous temperature conditions develop. A continuous oxygen recording controller (O_2RC-2) is used at the feed end of the kiln to establish and maintain

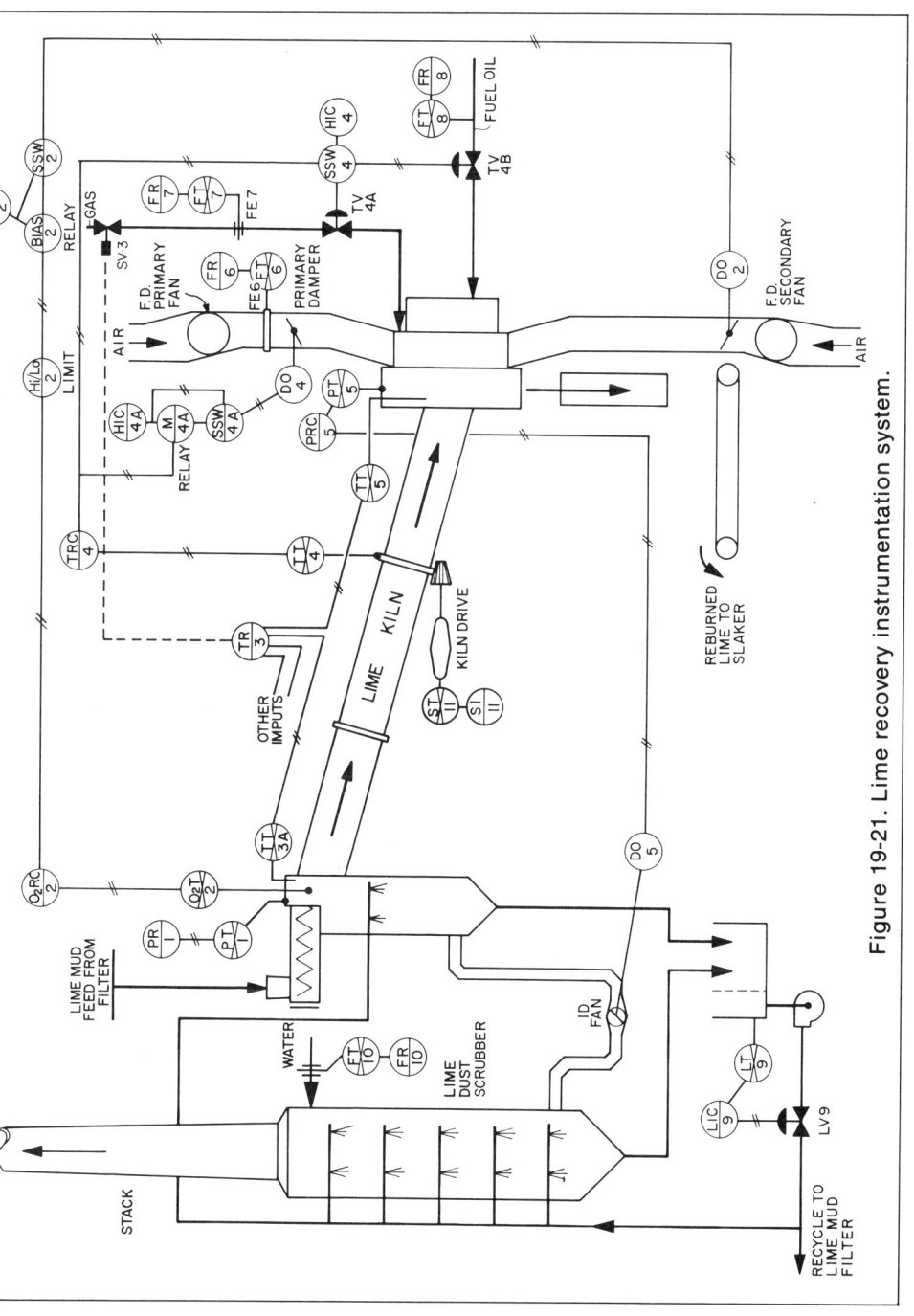

Figure 19-21. Lime recovery instrumentation system.

optimum oxygen levels by adjustment of secondary air damper DO-2 at the firing end of the kiln. Burning zone temperature is controlled by a recording temperature controller (TRC-4) by adjusting a flow control valve on the gas or fuel oil supply as selected by selector switch SSW-4. Although a thermocouple-sensing device is shown, this measurement can also be made with a radiation type pyrometer "looking" into the firing end of the kiln.

The draft at the firing end of the kiln is recorded and controlled by PRC-5 through the regulation of the induced draft fan speed by the adjustable operator (DO-5). The draft on the feed end is recorded on PR-1.

The kiln speed is monitored by an indicator (SI-4) which receives its signal from magnetogenerator type speed transmitter ST-11, mounted on the shaft of the kiln drive. Flows of gas and fuel oil are recorded and totalized by instruments FR-7 and FR-8. The primary flow device for gas (FE-7) is usually an orifice plate, while for fuel oil (FE-8) it is usually a target type or positive displacement type flowmeter. The flow of water to the lime dust scrubber is recorded on FR-10 with an orifice plate flow device (FE-10) and differential type transmitter FT-10.

WATER TREATMENT

Process. The paper and pulp industry uses large quantities of water. Although some water can be used directly from its source in such areas as turbine condensers and log flumes, much of the water must be conditioned in some way before use.

Water treatment processes and equipment vary widely from mill to mill, depending not only on the quality of the source but also on the quality of the water required by the type of product being manufactured. Nearly all surface waters and some ground water supplies are turbid and colored by finely divided suspended matter. These impurities are usually of such nature that even after extended periods of time they will not settle out completely. The most common method of removing these impurities is by coagulation, followed by sedimentation and filtration.

Coagulation is the addition of chemicals, usually alum, under controlled pH conditions to remove finely divided or colloidal suspended matter. The introduction of the coagulant chemical affects the surface charge of the particles causing them to come together into larger, denser masses called *floc*, which will settle out more readily. Coagulation is followed by sedimentation where, under quiescent conditions, most of the floc settles out, the remainder to be removed by filtration. Filtration removes the remaining flocculent material— silt, clay, and algae. The filtered water then flows to a clear well to be stored until demanded by the process.

Filtration consists of passing the water through a bed of sand and gravel, filtering any remaining particles; as filtration proceeds, material accumulates on the filter causing increased friction as the water passes through the filter and is monitored by measuring the head dissipated across the filter. This loss-of-head value is allowed to increase to some predetermined maximum value indicating that the filter must be cleaned by backwashing.

Instrumentation. Figure 19-22 shows the instrumentation found in a typical pulp and paper water treatment plant. The control systems required can be broken into three general categories: (1) plant throughput control, (2) backwash control, and (3) chemical addition control.

There are several basic control systems used to govern plant throughput. One of the most common systems allows the level in the clear well to determine the setting of the filter effluent rate controller on each filter with raw water influent flow rate being controlled by filter level. If a change in demand occurs, the change in level in the clear well causes an increase in the filtration rate, and the lowering of the level over the filters will admit more water to the plant. Plant output is thus matched against the process demand. A variation of this system allows the level of the clear well to determine the plant influent flow rate with the filter level setting the filtration rate. This system will not control the level of the clear well as closely but changes in the filtration rate will be slower, giving a more stable, uniform process.

Control during backwashing is dependent on the proper sequencing of valves in preparing the filter. Completely automatic control consists of sequencing valves on a time basis with initiation either from high loss-of-head or by the operator.

Chemical addition is normally done by ratioing flows to the influent raw water flow rate. A pH trim function can be added although the pH tends to be fairly stable for most sources.

WASTE TREATMENT

Process. During the early days of the pulp and paper industry, the discharge of untreated effluents into receiving waterways was tolerated because of the relatively small volumes involved and the ability of these waterways to easily assimilate these flows. But times have changed. In recent years, with rapid expansion of the industry and increasing competition for the use of our rivers and streams, this practice has become unacceptable. Encouraged by the public and enlightened management, the paper industry has increased its emphasis on the treatment of its effluents.

The selection of a satisfactory waste treatment system for a mill has become just as important to its economic health as other primary factors such as adequate wood, transportation, good water, and qualified personnel.

Waste treatment is usually accomplished by some combination of the following processes:

1. Recovery.
2. Primary treatment by coagulation and sedimentation to remove suspended matter.
3. Secondary treatment by one of several biological processes such as activated sludge and trickling filter, to remove oxygen-demanding matter.
4. Lagooning for final settling and polishing.

Of the systems mentioned, most are fairly simple and require a mimimum of instrumentation with the possible exception of the activated sludge process

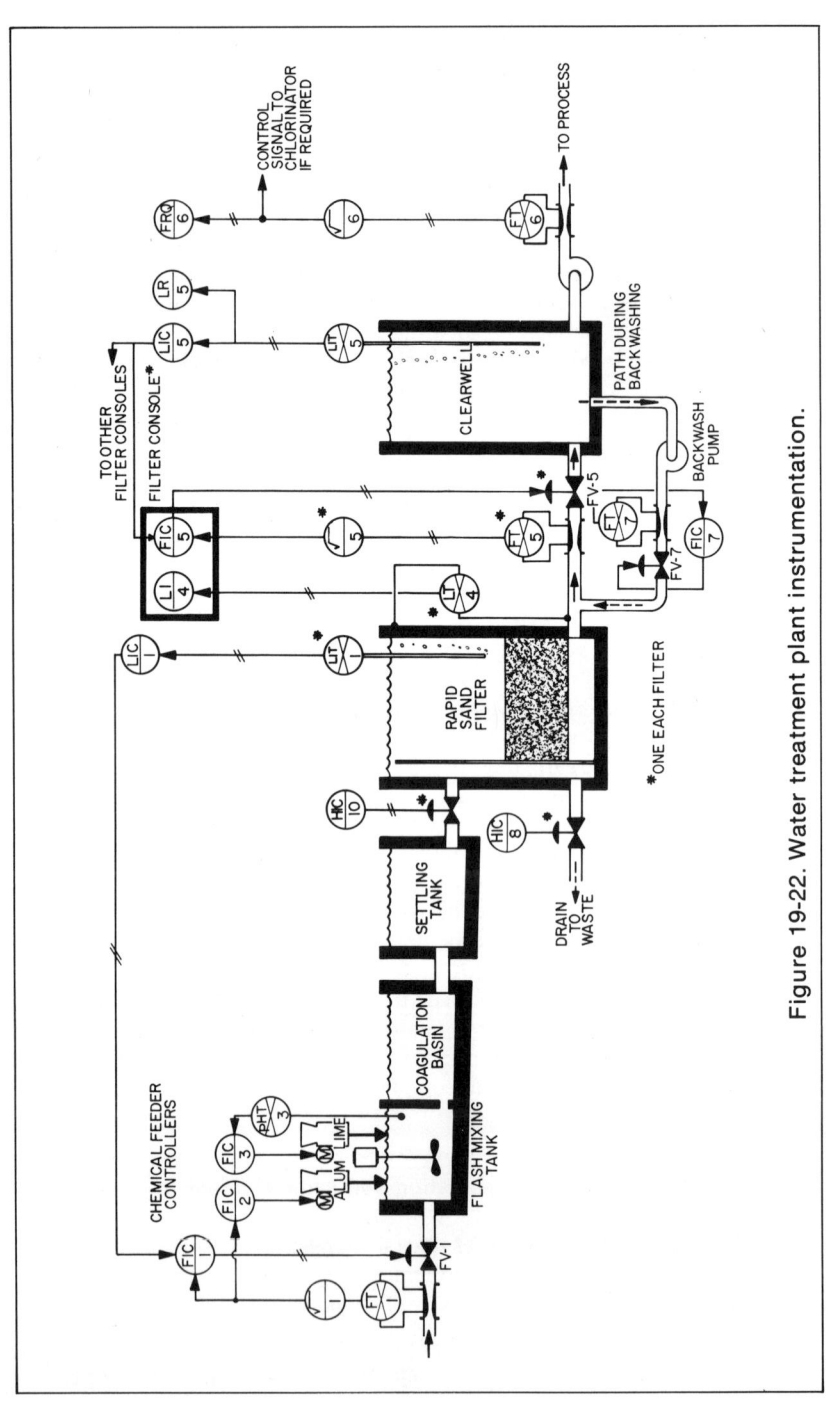

Figure 19-22. Water treatment plant instrumentation.

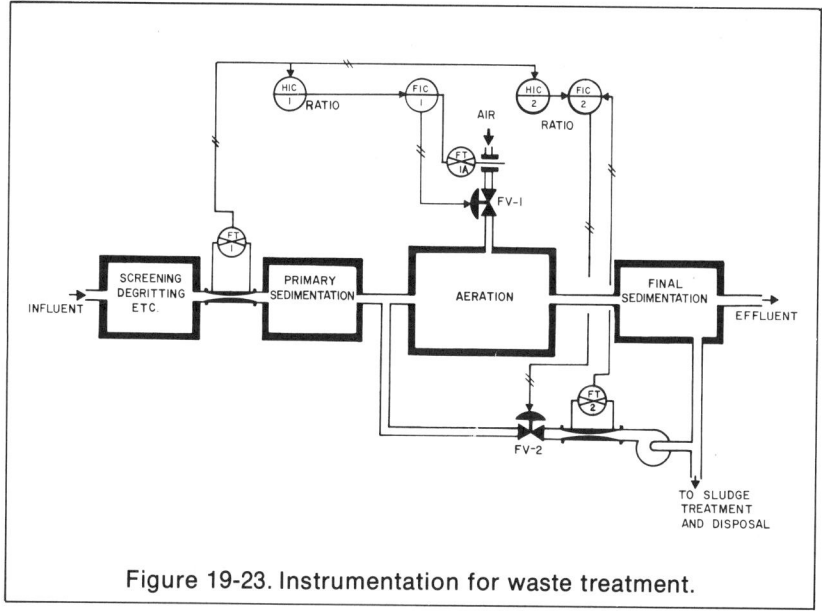

Figure 19-23. Instrumentation for waste treatment.

and the various recovery methods covered earlier in this chapter. The activated sludge process is a biological process in which nonsettling substances occurring in dissolved and colloidal form are converted by microorganisms to a settling form. The settled material that develops is called activated sludge. A typical flow diagram is shown in Figure 19-23. Settled waste from the primary tanks is mixed with a portion of activated sludge previously developed and enters the aeration tank. Aeration is necessary as the bacteria and other microorganisms which make up the activated sludge must have oxygen for their life processes. The aerated mixture then passes to the final settling tank in which the activated sludge is separated, leaving a clear liquid to be discharged to the receiving water course.

Instrumentation. The control of the activated sludge process revolves primarily around control of the return activated sludge flow and control of the oxygen content in the aeration tank. The most common technique involves the simple ratioing of both return activated sludge flow and air to influent flow. A primary flow device and transmitter along with the associated controllers and valves are utilized on the return activated sludge line. A transmitted signal from the raw waste flow transmitter goes through a ratio relay which then sets the index of a flow controller on the return activated line. The ratio of raw flow to return flow can now be adjusted by supervisory personnel, based on plant laboratory tests. Control of air flow can be accomplished in a similar manner with a low limit relay to ensure that the air flow does not fall below a desired minimum. With the advent of reliable continuous analyzers for dissolved oxygen it is now becoming practical to control air flow directly based on a measurement of dissolved oxygen in the aeration tank.

Instrument Applications   405

BIBLIOGRAPHY

Britt, K. W. *Handbook of Pulp and Paper Technology.* New York: Reinhold Publishing Corporation, 1964.

Calkin, J. B. *Modern Pulp and Paper Making.* New York: Reinhold Publishing Corporation, 1957.

Cole, E. J., and Todd, M. *Pulp and Paper Mill Instrumentation.* New York: Lockwood Trade Journal Company, 1967.

Considine, D. M., and Ross, S. D. *Handbook of Applied Instrumentation.* New York: McGraw-Hill Book Company, 1964.

Evans, J. C. W. *Pulp and Paper Mill Process Instrumentation.* New York: Lockwood Publishing Company, 1969.

Lavigne, J. R. *Instrumentation Applications for the Pulp and Paper Industry.* Foxboro, MA: The Foxboro Company, 1975.

MacDonald, R. G., and Franklin, J. N. *Pulp and Paper Manufacture.* 2nd ed., vol. 1, 2, and 3. New York: McGraw-Hill Book Company, 1969.

STUDY QUESTIONS

Select the answer or answers which correctly apply to the following statements:

1. The predominant process used in the paper industry is the
   a) semichemical process.
   b) sulfite process.
   c) sulfate process.

2. The letters LRC in an instrument diagram usually indicate
   a) a load recording controller.
   b) an indicating level controller.
   c) a level controller with recording capability.

3. In an instrument diagram, a single circle with a horizontal line through the center indicates
   a) an instrument mounted on a panelboard.
   b) instruments located in the field.
   c) a transmitter.

4. Chemical pulping of wood is primarily the dissolving of its
   a) fiber components.
   b) lignin.
   c) fiber binding material.

5. When a low selector relay device is used on the steaming system of a batch digester, the steam flow is controlled by
   a) flow recording controller.
   b) cam pressure controller.
   c) local recording controller.

6. In a continuous digester the steam flow is controlled by
   a) flow recording controller.
   b) pressure recording controller.
   c) level recording controller.

7. A major objective of the hot stock screening instrumentation is to
   a) control temperature in the rejects tank.
   b) control dilution water level in the screens.
   c) control flow and consistency of unscreened stock.

8. Instrumentation used in the pulp washing area is designed to control the flow rates of
   a) wash water input.
   b) washed pulp.
   c) recycled filtrate.

9. Pulp bleaching is essentially the removal of coloring impurities in the fibers by
   a) dissolving the coloring matter.
   b) filtering out the coloring matter.
   c) settling out the coloring matter.

10. In the manufacture of hypochlorite bleach liquor, the amount of chlorine reacting with caustic solution is controlled by a
    a) temperature controller.
    b) pressure controller.
    c) oxidation-reduction potential controller.

Indicate whether the following statements are True or False by inserting T or F in the parentheses:

11. During the stock preparation phase of papermaking, desired mechanical characteristics are impacted to the pulp by the addition of selective dyes. (   )

12. Both temperature difference and motor load power measurements are utilized in the automatic control of pulp refiner operation. (   )

13. The basic primary purpose of a paper machine is to form a sheet of paper from a stock slurry. (   )

14. The rate of stock delivery to the paper machine is automatically controlled by controlling the level in the machine chest. (   )

15. In the automatic control of broke pulper operation, it is most important to control level and consistency. (   )

16. The principal function of the dry end of the paper machine is to coat the surface of the paper. (   )

17. The most important control loop in a multiple-effect black liquor evaporator is level in the condensate system. (   )

18. Uniform specific gravity of the green liquor is maintained by controlling the level in the smelt tank. (   )

19. Temperature control is important in order to maintain optimum conditions during the lime slaking operation. (   )

20. Discharge of untreated effluents was tolerated in the early days of the pulp and paper industry, because they contained no obnoxious chemicals at that time. (   )

(See Appendix for answers)

20. Computers in the Paper Industry

There are two principal classes of computers used in the pulp and paper industry—analog and digital. The basic distinction is that the analog computers work with continually varying quantities or inputs such as amplitudes of voltage, current, or pressure, while the digital computer works with numbers in the form of pulses, representing small increments of a variable. Basically, the analog controller, described previously in the chapter on automatic control, is an analog computer. It computes equations which relate various controller settings of proportional band, integral, and derivative. It accepts a continually varying measurement signal, compares it with a set point, and continually changes controller output to correct for any differences between the two. Aside from controllers, there are analog computers composed of components designed specifically to perform certain arithmetic operations.

Since digital computers are more widely used in the pulp and paper industry, the discussions in this chapter will restrict themselves to digital computers and how they are used in the industry's instrumentation systems.

THE DIGITAL COMPUTER

The digital computer is a device that works with numbers. Input numbers are fed into the computer and arithmetic operations are performed on these numbers. Simple logical decisions are made according to the instructions (called a *program*) to produce output numbers. The input numbers might be numerical numbers associated with a basic arithmetic problem or those numbers representing process measurements or other information. The arithmetic operations can be addition, subtraction, multiplication, division, or any combination thereof. A typical function in logical decision might be to compare two numbers to determine which is larger. The output numbers represent the results after the input numbers are processed by the computer, and there is usually some external device which translates the output numbers into a form for further use.

Basic Functional Units

All digital computers consist of five major functional units (Figure 20-1):

1. Input section
2. Control section
3. Memory section

4. Arithmetic section
5. Output section

Physically, the memory, control, and arithmetic sections can be considered as a single piece of equipment usually referred to as the *central processing unit*, or CPU. The input and output sections are usually combined into a single unit called the *input/output processor*; this is the main channel of communications between the computer and the "outside world."

Input Section. In the operation and utilization of the computer, various types of information are entered into the computer by means of the input section, which is essentially a communications device through which problem numbers, instructions, and operational data are received and fed into the machine. Every digital computer has its own "language." Therefore, the problem numbers and instructions must be translated into a form the computer can understand. Punched cards, perforated tape, and magnetic tapes are used by interpreters, the programmers, to "talk" to the computer. The input section reads the information they contain and passes it on to the appropriate section.

Any process operational data upon which the computer is to act is received from measuring devices in the form of either a series of numerical signals (digital) or continuously varying signals (analog). Equipment outside the basic computer system converts these signals into the type of electrical pulses that can be utilized by the digital computer. These pulses are accumulated into significant groups by the input unit and then forwarded to the computer control section. The external equipment (Figure 20-2) used to converse with the computer through the input unit is collectively referred to as *peripheral* devices and may include:

1. Pulse counters
2. Analog-to-digital converters
3. Card read-punch units
4. Printers
5. Typers
6. Bulk memories
7. Operator's console

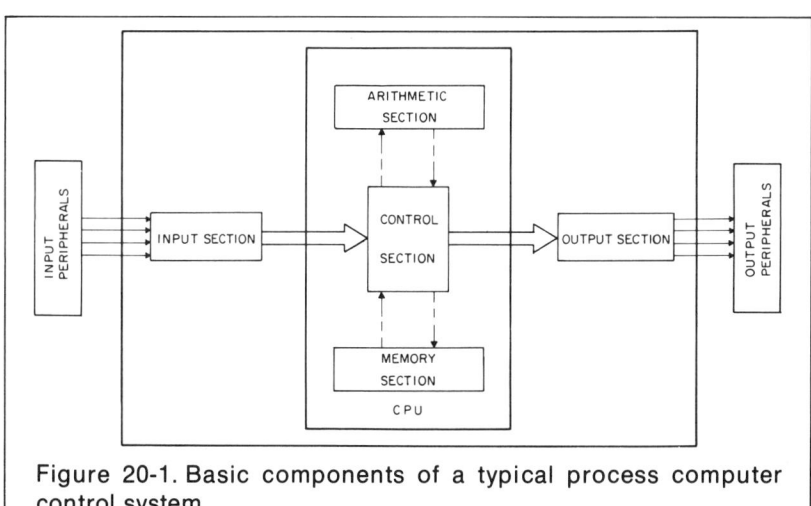

Figure 20-1. Basic components of a typical process computer control system.

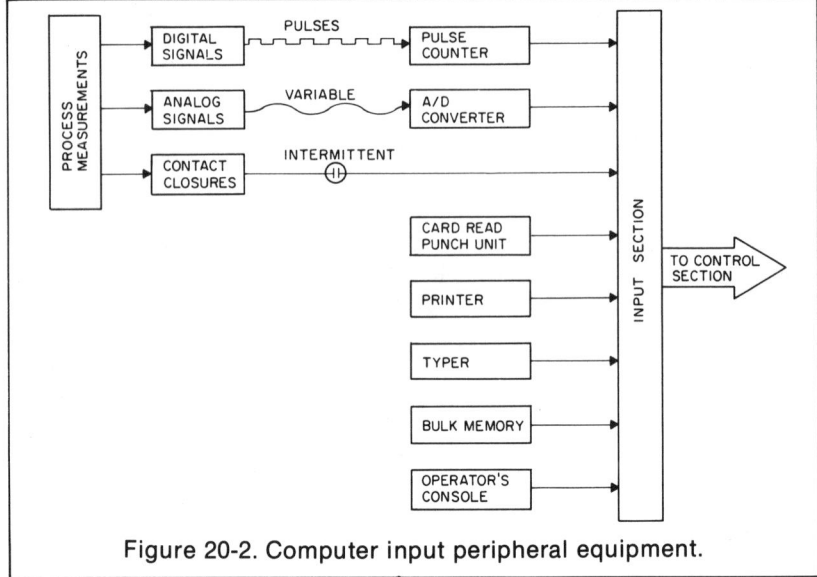

Figure 20-2. Computer input peripheral equipment.

Control Section. It is the function of the computer's control section to execute the instructions of the computer program. In effect, it executes the programmer's orders transmitted through the input section by decoding the instructions and generating electrical signals which tell the other sections what to do. The control section is considered the heart of every digital computer because it synchronizes all computer operations. The unit interprets instructions; controls input and output units, transfer of data, and all calculations.

Memory Section. All information then proceeds from the control section to the memory section, also called *storage* because it provides extensive storage and rapid retrieval facilities for programs and other data. Stored data include problems to be solved, input data for problems, constants for values for problems, programs, intermediate results, and final results. Memories are either of the magnetic or mechanical type and can be divided into two categories—bulk and working or core. The working or core memory contains programs and data actively being used by the computer and is of the magnetic type. Shown in Figure 20-3, magnetic core memories are units made up of tiny ferrite rings called *cores* strung on a complex arrangement of wires which can sense the polarity of magnetism in the cores and can also change the polarity.

Bulk memories can be of either the magnetic or mechanical type. The magnetic types are typically drums, discs, or tapes. Magnetic drum, disc, and tape memories utilize a layer of ferrous oxide coating on the surface of drums, discs, and Mylar for the same purpose of detecting polarity and storing electrical charges. Mechanical memories employ coded punched holes on cards and tapes so photoelectric cells or wire contacts can sense and decode data.

The total memory is located partly inside the computer or CPU and partly outside the computer. That portion that is located in the mainframe of the computer is referred to as *core memory*. That portion located outside is known as *bulk storage* and is part of the peripheral equipment. The reason for this division is that core memory receives and delivers information at speeds much higher than the bulk storage; and, although it would be ideal if the whole memory of a computer could be of this type, as the amount of data increases it becomes more economical to store in backup bulk memories because of the high cost of core memories. Accordingly, only programs and data currently essential to the functioning of the control processor are held in core memory; the rest are held in bulk storage and transferred to the computer as required.

Arithmetic Section. All computations and logical operations of the computer are performed by the arithmetic section, sometimes called the *logic* section. This section accepts data from the memory section, makes necessary calculations in accordance with stored programs of instructions, and transfers the answer back to the memory section, all under the direction of the control section. The most common work performed in the arithmetic section includes adding, subtracting, multiplying, dividing, and performing logical functions such as comparisons.

Output Section. Under the direction of the control section, the output section accepts data after processing by the memory and arithmetic sections, where the data are represented by electrical pulses. These pulses are accumulated into significant groups and, on signal from the control section, the output section transmits them to peripheral equipment for use as digital or analog signals. These signals are then transmitted to process control devices or to other peripheral equipment such as typewriters, printers, digital displays, magnetic tape, punched tape, and cathode ray tubes (CRTs). See Figure 20-4.

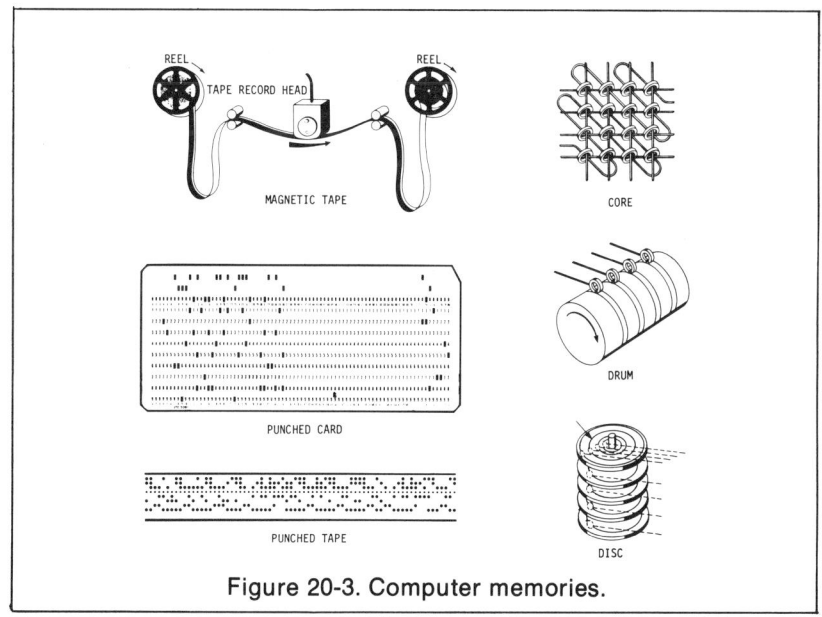

Figure 20-3. Computer memories.

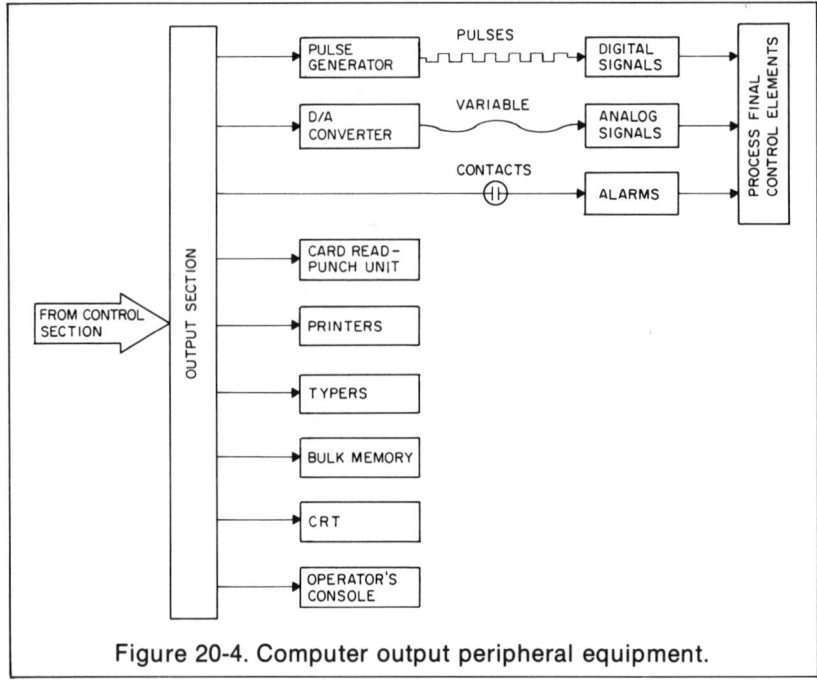

Figure 20-4. Computer output peripheral equipment.

Operation

The basic steps in the functional operation of a typical computer begin with the problem informational and instructional data being broken down into a form called a *program*, which can be understood and processed by the computer. This program, including problem numbers together with appropriately processed information from a process and other sources, is fed into the input section. This information is placed in appropriate memory section locations under direction of the control section. According to instructions in the program, the control section takes over and processes the stored data numbers by regulating the operation of and the flow of information between the memory and arithmetic sections. The result of this data manipulation is sent to the output section and displayed in a variety of understandable forms on peripheral equipment or transmitted to converters where it is conditioned for use by process control devices.

The operation of all digital computers depends on electronic circuits and magnetic elements which are stable in either of two conditions. The control, arithmetic, and memory sections are composed of a combination of electronic circuits, while the magnetic elements are employed in devices used in the memory section to store the numbers and programs of instructions. Due to the fact that computer elements have two stable states which can be produced and duplicated by pulses from electrical switches, relays, and other electronic and magnetic devices, these on-off characteristics are represented by two basic numbers, 1 and 0. These are organized into a code to represent numbers, letters, and symbols necessary to perform communication and computation functions; this code is known as the Binary Number System. Binary numbers are, however, awkward to read and relate to the more generally used decimal number system. Computers, therefore, generally convert the inputs from decimal to binary and the outputs from binary to decimal. Another approach used by computers is to express each decimal digit by a combination of binary numbers known as binary coded decimals (BCD). In this system, numbers greater than one are represented by a combination of the two basic numbers as shown in Table 20-A.

Each 1 and 0 used in a binary number is called a *bit*, a contraction for binary digit. A computer usually deals with information consisting of a fixed number of bits which make up a *word*. In a computer which uses a 24-bit word, the number 6 would appear as follows:

000000   000000   000000   000110

In the same manner, the binary system can be used to represent alphabet characters such as shown in Table 20-B.

Using a 24-bit word, four letter A's would appear as follows:

010001   010001   010001   010001

References made to the "size" of a computer do not apply to its physical dimensions but rather to the number of words it can store in its memory section, expressed in units rounded down to the nearest thousand. For instance, a computer with a memory storage capacity of 32,768 words may be referred to as a *32K* machine. Most process computers have at least 4000 words of memory and some have as many as 132,000 words.

TABLE 20-A
DECIMAL AND BINARY NUMBER EQUIVALENTS

Decimal Number	Binary Number
0	000000
1	000001
2	000010
3	000011
4	000100
5	000101
6	000110
7	000111
8	001000
9	001001
10	001010
11	001011
12	001100
13	001101
14	001110
15	001111
etc.	

TABLE 20-B
ALPHABETIC CHARACTER AND BINARY NUMBER EQUIVALENTS

Alphabetic Character	Binary Number
A	010001
B	010010
C	010011
D	010100
etc.	

Languages

The set of binary numbers representing the various computer operations in a coded instruction is known as *machine language*. This set of instructions is used by the control section to cause the arithmetic to perform the required operations. Since it is tedious and confusing to write programs as lengthy combinations of 1 and 0, intermediate languages oriented to the application of the computer have been developed. These are much easier for the programmer to write; the computer then translates these into machine language in accordance with a previously stored program. Languages which resemble written or mathematical English are referred to as *compiler languages*; those which do not, are referred to as *assembler languages*.

A typical widely used, technically oriented compiler language developed for the convenient use of computers by engineers and scientists is FORTRAN, which is a contraction of Formula Translator. It allows an intermediate program to be written by formulas with which these users are already familiar. Other languages are also available and completely written programs for many of the standard calculations are available from both computer manufacturers and users.

DIGITAL COMPUTER USES

Business and Scientific Computers

As in other industries, the first uses for the digital computer in the paper industry were in business-oriented operations. Consequently, operations such as making inventory reports, maintenance schedules, payrolls, production schedules, processing orders and invoices, sales forecasting, determining costs, profits, etc., using the digital computer have become commonplace in most large pulp and paper organizations. Eventually the digital computer was pressed into service by engineering personnel to perform more technical functions. In order to make complex mathematical computations involving simultaneous equations, linear programming, differential equations, and other interrelated mathematical functions, scientists have also made extensive use of the digital computer to solve problems by simulation, in which a mathematical model of the operation to be explored is constructed and a random selection of probabilities is used in the model to determine their effect and establish those that will produce the optimum solution to the problem. Computers used in this way are generally called *electronic data processing* (EDP) *computers*.

Process Computers

Naturally, the next step in the ever expanding use of the digital computer in industry has evolved in its application to the process itself. When the computer is used in this manner it is considered to be a *process computer*, and it usually operates in *real time* as contrasted to the business and scientific EDP computers which do not. The term *real time* refers to the actual time during which the process is going on, with input information and process variables such as pressure, temperature, etc., being continuously monitored and fed to the computer. The EDP computer mode of operation is principally based on input data being held in the form of punched tape and cards with subsequent handling at the computer's convenience. This mode of computer operation is also known as *off-line*. Process computers may also operate off-line when there is no physical contact between the process and the computer. Communication between the two is provided by the operator who selects process variables from operating logs, feeds them into the computer, and takes action based on its readout. On the other hand, when input data enter the computer directly from the process it is considered to be operating *on-line*.

Priority interrupt, in which variables being monitored are given an order of importance with the most significant acted on first, therefore becomes a characteristic feature with process computers operating on-line and in real time. If two variables change at the same time, the one with the higher priority is attended to first. In cases of a particular urgency, the computer immediately interrupts its operations already in progress to act on the process upset detected with high priority assigned to it.

The process computer can be used in several ways: (1) as a data processing device which receives signals from process instruments and presents computed values for use by an operator; (2) as a sequence controller which obtains most of its data from an operator, then exercises direct control over the process; or (3) as a full-fledged process controller which notes process conditions and takes action without operator intervention. These uses are usually associated with the analog instrumentation generally employed for process regulation. The block diagram in Figure 20-5 shows this instrumentation arranged in the basic loop configuration, similarly shown in the chapter on automatic control. The three process computer schemes just mentioned involve this loop or variations of it.

Data Logging. The first attempts using the digital computer for process-related regulation purposes were as data loggers to automatically collect process operation data. In essence, data logging has been practiced in the paper industry for many years, using conventional graphical circular chart and strip chart recorders and/or indicating instruments. However, digital data loggers convert analog process measurements into more concisely recorded digital data and present this information in the form of a list (or log) printed out on a typewriter at regular intervals. A typical log would contain the time, the point

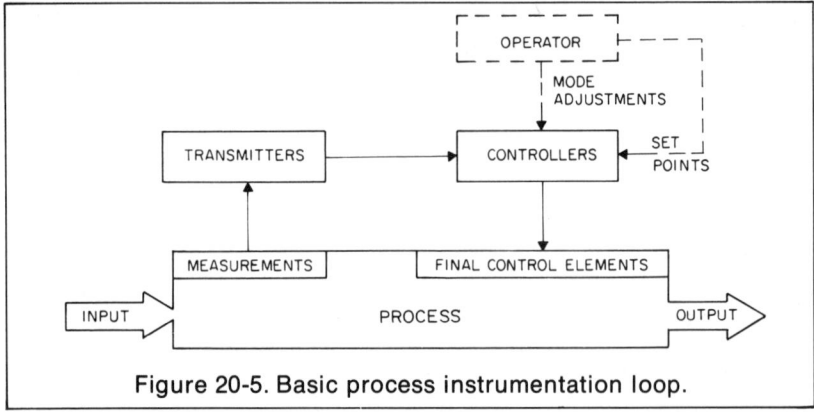

Figure 20-5. Basic process instrumentation loop.

and location, and the value expressed as a decimal number in engineering units of the particular measurements being periodically scanned by the computer. A log, consisting of a single point or a complete list, can be obtained on demand at any time at the plant operator's request. Limited after-the-fact logs can also be made to study history of process operation data by calling up information previously stored in memory. Data trends and average logs can also be produced automatically or on demand. In addition, data loggers can perform other functions such as flow integration, alarm scanning, and simple calculations such as mass flow.

Although initially data loggers were primarily used for the acquisition of data for management information purposes, they constituted an important step toward the use of digital computers for direct process control in the paper industry by introducing data gathering and handling techniques to the field of process control. These techniques eventually resulted in the use of this information by plant operators as a guide. In this case, the computer not only gathers information but also performs the mathematics and logic necessary to determine what changes should be made in plant operation so as to advise the operator of what should be done. The operator then makes the necessary changes to process equipment and/or set point and mode adjustments to the controller in process control loops as instructed by the computer through messages automatically written out on a typewriter. A horn or light is usually provided to advise the operator that there is a message for him on the typewriter. This is sometimes referred to as using the process computer in an *open loop* control fashion, which implies that a man or operator is required to complete the cycle from the process to the computer and back to the process (see Figure 20-6).

Sequence Controlling. When the computer is considered as a sequence controller, the operator feeds into the computer information which he obtains by observing the process operation, raw material characteristics, product specifications, and equipment conditions. Then the computer performs any mathematics or logic required in the data and adjusts process conditions by direct manipulation of analog control instrumentation or process equipment in accordance with the results obtained from this information on a preestablished programmed basis. Figure 20-7 illustrates this concept with the operator being interfaced between the process and computer. In this respect, the concept can also be considered as open loop control. This approach can be used to start up a machine or system on a precalculated schedule. It is a similar function to cam controllers such as conventionally used for batch digester cooking control, but has the means provided by which information on wood species, chip-to-liquor ratios, time-temperature-pressure program, and the desired operation of the steam flow and relief valves for optimum loading conditions can be communicated through the computer.

Supervisory Process Computer Control. An extension of the sequence computer control concept in which direct communication between the computer and process is established without direct operator intervention is commonly known as *supervisory process computer control.* As illustrated by the block diagram in Figure 20-8, the operator is not included in the operation of the control cycle so this concept can be classified as a *closed loop* type operation. In this kind of system, the computer receives process measurement instrument signals directly, calculates best operating conditions, and automatically adjusts the analog controller set point. In essence, the computer performs the same function as the operator in setting the set points of the analog controllers in a conventional process control loop. However, the controllers' mode settings are still made by the operator. Although it is possible to design a system to permit the computer to automatically adjust the controller modes as well (called *adaptive control action*), this is not practical with this type of computer system. Other provisions (typically typewriters) are made available for direct operator-to-computer communications. This configuration of process computer control has also been designated as *set point control* and *digital-directed analog control.* In case of computer outages due to failures and maintenance, backup process control operation is provided by designing the system so that it will operate on local conventional control loops with set point adjustments being made by the operator. Automatic/manual stations are provided in conjunction with, and without, the analog controller to allow back up control of the process when the computer is out of service.

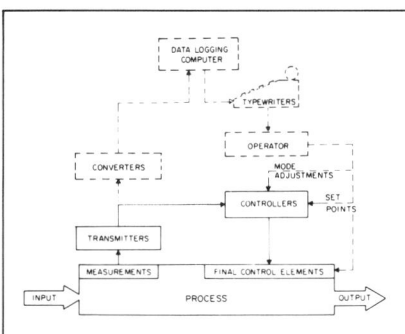

Figure 20-6. Data logger used as operator's guide.

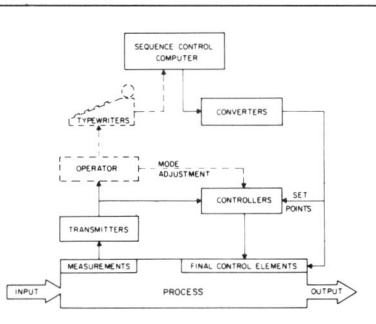

Figure 20-7. Process computer used as a sequence controller.

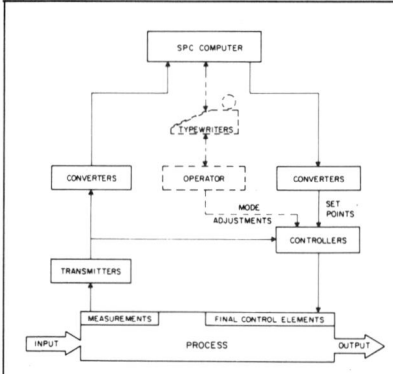

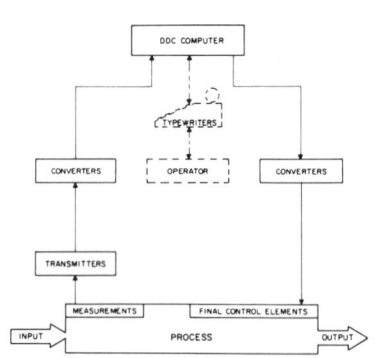

Figure 20-8. Process computer used for supervisory control.

Figure 20-9. Process computer used for direct digital control.

Direct Digital Control. As previously mentioned, the controller in a conventional control loop is basically an analog computer. It actually computes equations which relate the various controller settings of proportional band, integral, and derivative. It accepts a measurement signal, compares it with a set point, and changes controller output to correct for any differences. Therefore, the controller's function is merely calculation. A digital computer is ideally suited for making calculations and this led to the next logical consideration of further extending the computer's use for process control. That is, not only can it replace the operator's function in the conventional control loop, but it also can replace and perform the function of the analog controller—and send its output directly to the valve and/or other desired final control elements—as well as perform other needed or desired functions. This then becomes another type of closed loop computer control concept called *direct digital control* (DDC) as depicted in Figure 20-9. Although this illustration shows only one complete loop in actual operation, the computer is time-shared among many control loops. It samples a measurement, performs the necessary calculations based on the set point and other control loop data stored in the memory, and makes necessary changes to the output signal to the corresponding final control element. It then samples the next measurement and performs the calculations for that loop, and so on, repeating the cycle for the remaining loops. The computer performs this operation at such extremely fast speeds that it appears as if continuous analog control is being accomplished on each loop. With this scheme, the proportional band, integral, and derivative responses of the control loop are part of the computer program, making automatic adjustments of these modes by the computer a practical approach to achieving automatic tuning and adaptive control functions.

DDC saves a good part of the cost of individual controllers; however, the process control is more completely dependent upon the computer. Therefore, provisions for backup process control become a very important consideration when the computer is out of service for any reason at all. There are several methods being used today to ensure continuity of process operation during computer outage. The first is to pass all computer output signals through manual stations so that the process may be operated by manual control until the computer is brought back on line. The second method is to pass all vital output signals through backup analog controllers so that the process may be run by conventional automatic control when the computer is down. A third method is to employ two digital computers, each with the facility for operating the process should a failure occur in the other. The backup system must always be ready to take over control; therefore, it must contain self-checking procedures which ensure that the system is available for backup. Provisions must also be made for continuously updating the backup system with the latest process information, such as set points and tuning parameters, in order to prevent upsets during the transfer of control from one computer to another. The other method of providing backup protection consists of any and all combinations of the first three methods.

Special-Purpose Computers. Most of the computer techniques mentioned so far fall into the class of general-purpose computers. Another approach is offered by a number of so-called special-purpose computers, which are designed on the same lines and employ much of the technique of the general-purpose computers but are generally smaller and dedicated to a specific purpose. Due to smaller size, they have sometimes been referred to as *minicomputers*. Several such special-purpose computers would appear to be a much more economical proposition than a full-sized general-purpose computer. However, if the number of such computers on any one application continues to increase, a stage will soon be reached when a large-sized general-purpose computer would be technically and economically more desirable. This concept has gained increasingly wide acceptance on and around the paper machine with some units being used in other areas of the pulpmaking and papermaking operations.

COMPUTER APPLICATIONS IN THE PAPER INDUSTRY

Among the primary incentives for installing a digital process computer control system are manufacturing cost reduction, improved production efficiency, improved product quality, process control enforcement, and safer process operating conditions. Numerous applications exist for the installation in the paper industry of process control computers which can be classified under three general categories: (1) those found on chemical type processes primarily in the pulp mill; (2) those concerning the physical operations of stock preparation and sheet manufacturing in the paper mill; and (3) those operations involving utilities and by-product processing such as water and waste treatment, power boilers, and tall oil plants.

These processes—which require such inputs as steam, chemicals, air, water, electricity, and wood—need interrelated close control in order to achieve the most efficient utilization in their processing and final conversion to paper. Figure 20-10 illustrates a feasible approach to a control system to accomplish this task. The entire system is depicted as consisting of a number of

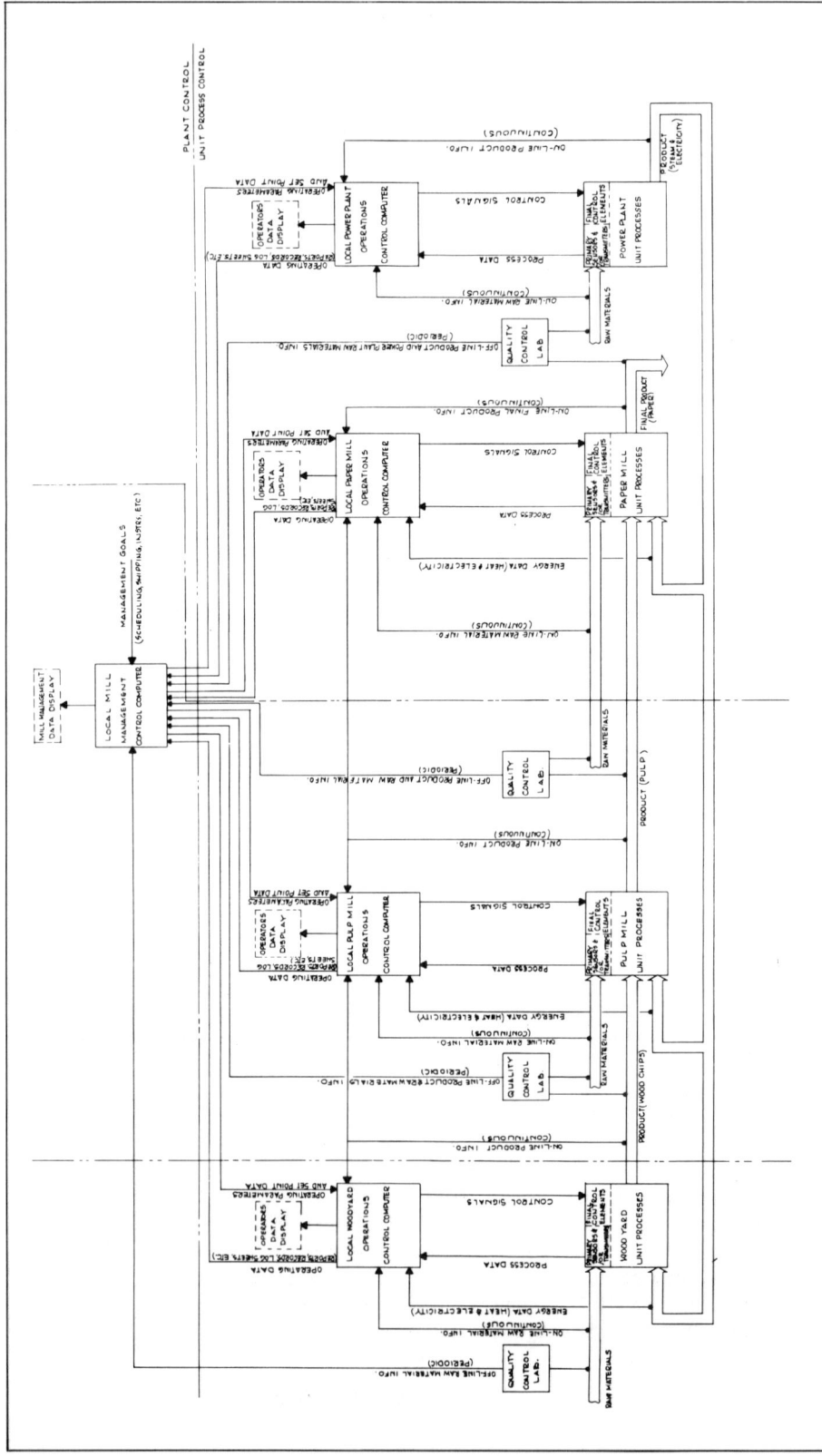

computers organized like a corporation with individual units assigned to do specific operations in the local mill areas under the direction of a centralized management control type computer. The following discussion will concern itself with some of the specific things that can be and are being done by the process control computers in the various pulp and paper mill areas.

Pulp Mill

Applications in the pulp mill, although not as numerous as those found in the paper mill, also possess the potential of realizing improvement of process operation by the use of computer control techniques. Figure 20-11 shows the unit processes in the pulp mill to which a computer control system can be applied to perform such major functions as: (1) scheduling of various units to provide efficient utilization of existing equipment; (2) prediction of process loading to avoid upset conditions; (3) tracing various types and grades of pulp through several units to provide profiles on processing conditions and pulp quality; and (4) providing "operator guide" instructions which will result in more uniform operations.

Batch Digesters. In chemical pulping operations using batch type digesters, computers are capable of production scheduling so that digester discharge blow tanks will not run dry or overflow and will still maintain the required number of batches per day in the most efficient manner through the operator guide concept. Computers can be used to automatically calculate the chip weight required for maximum digester loading based on such input information as size of digester, type of wood, wood moisture, and density of chip packing. In the same manner, the computer can calculate correct digester liquor charge based on type of wood, chip moisture, final weight of chip load, type of cook, size of digester, white liquor strength, total liquor volume requirements, and corrections required due to K or Kappa number test on pulp from the previous cook. The computer can also provide for liquor fill, steaming, and control of functions. Additionally, the relative cooking rate could be calculated and integrated against time in order to obtain a factor to determine when the cook is finished, sometimes referred to as the H factor. The computer could have control of the digester from the time the cook capped the digester until it alerted the operator that it was time for the digester to be blown. Monitoring of the cooking cycle and calculation of the proper additions of white and black cooking liquors would provide a more uniform pulp with less screenings and greater yields, more constant K numbers, and increased turpentine yields.

Several basic concepts have been used in the application of computer control systems to batch type digesters. A common one is to have the computer calculate the cooking temperature and time required to reach a given K or Kappa number by using an H factor correction in conjunction with a predetermined predictive model of the cooking process. Figure 20-12 is a block diagram of such a system. The structure is shown as consisting of four general sections:

1. Basic control functions
2. Coordinated control
3. Optimization
4. Information system

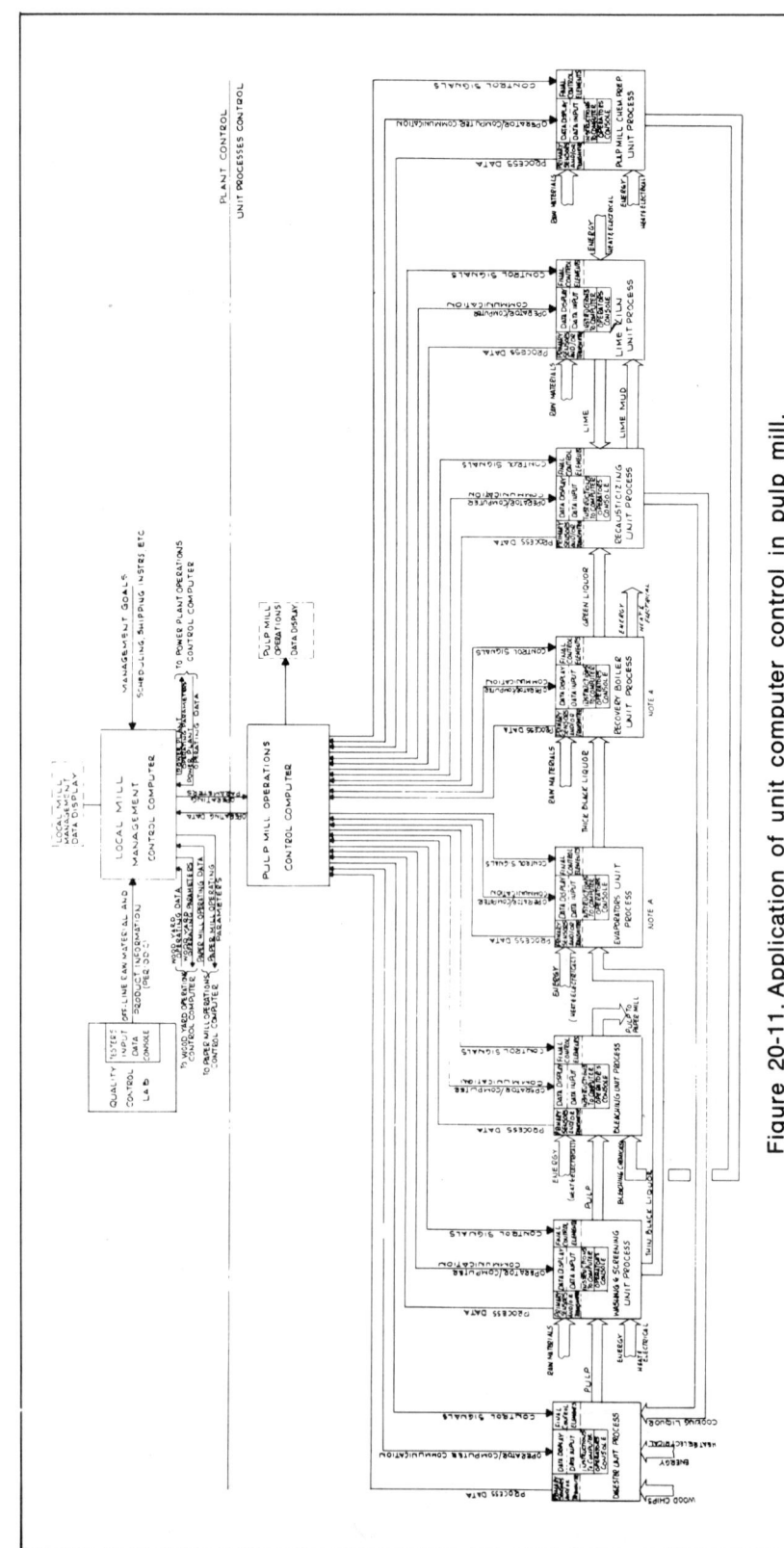

Figure 20-11. Application of unit computer control in pulp mill.

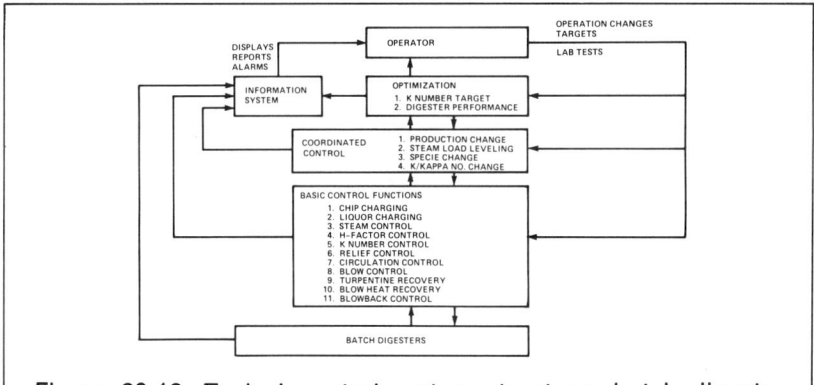

Figure 20-12. Typical control system structure—batch digester computer control system.

This system concept uses specified target set points, measurements, and laboratory test results to compute the appropriate amount of cooking liquors (chemicals), heat, and then sets the blow times to achieve the required quality of pulp at the desired production rate. Production changes and K or Kappa number changes are coordinated so they can be accomplished with minimum upset to the process. Steam load leveling is used to coordinate the digester and minimize steam load upsets. Target optimization is done on the K number variations. The operator interacts with the system via an operator's console using interactive display. These displays provide for initiating production changes and steam load leveling, for changing target set points, and for entering laboratory test results. A demand digester graphic display with periodically updated variable values is also provided for each digester. The system normally includes the provision for producing hard copy management reports. Typical basic control functions performed by such a batch digester computer control system as shown in Figure 20-13 are:

1. Chip charging
2. Liquor charging
3. Steam control using H factor correction
4. K or Kappa number control
5. Relief control
6. Circulation control
7. Blow control
8. Automatic blowback
9. Production change coordination
10. Blow heat recovery
11. Steam load leveling

Another concept of a batch digester computer control system is designed to provide real time calculations which are specific for each cook, with the computer selecting the best set of cooking conditions to reach the targeted K

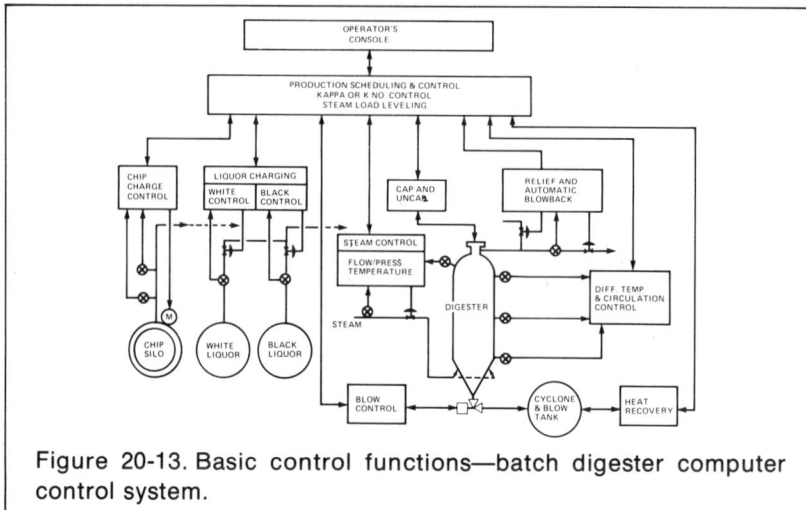

Figure 20-13. Basic control functions—batch digester computer control system.

number. A simplified schematic of such a system is shown in Figure 20-14. Control over an individual cook is achieved by measuring the effective alkali concentration during an early stage of the cook. The rate of delignification (pulping) is primarily controlled by three cooking factors—time, temperature, and alkali concentration. Based on the determination of the effective alkali concentration, the time-temperature cycle of the remaining part of the cook is adjusted to achieve the H factor required to reach the targeted K number for that specific cook. Effective alkali can be determined by a conductimetric titration with standardized reagent on intermittent samples taken from the digester liquor circulating lines. Another method to determine the effective alkali along with sulfidity, which is another important cooking variable, is by the use of a multiple ion-selective electrode measuring sensor on a continuous sample stream from the digester during the cooking process.

Continuous Digesters. When chemical pulping is done in a continuous type digester, computers can be used for many of the same functions. Heat and material balances are important for successful control of this type digester. The mathematics of calculating these balances are performed on-line by the computer. From input information on chip feed, black and white liquor flows, steam flows, and temperatures it is possible for the computer to regulate the alkali-to-wood ratio, K number of the pulp, and chip load in the digester. To bring about the correct alkali-to-wood ratio, the operator enters the desired target into the computer which makes the necessary corrections in white liquor flow to achieve the target band on species production (chip flow) and white liquor (cooking chemical) strength. The desired K or Kappa number (degree of cooking) of the pulp is maintained through the regulation of temperatures. Samples of pulp are taken periodically, tested in the laboratory for K number, and the results of the test are entered into the computer. The computer then makes the necessary corrections to cooking temperatures to bring the K or

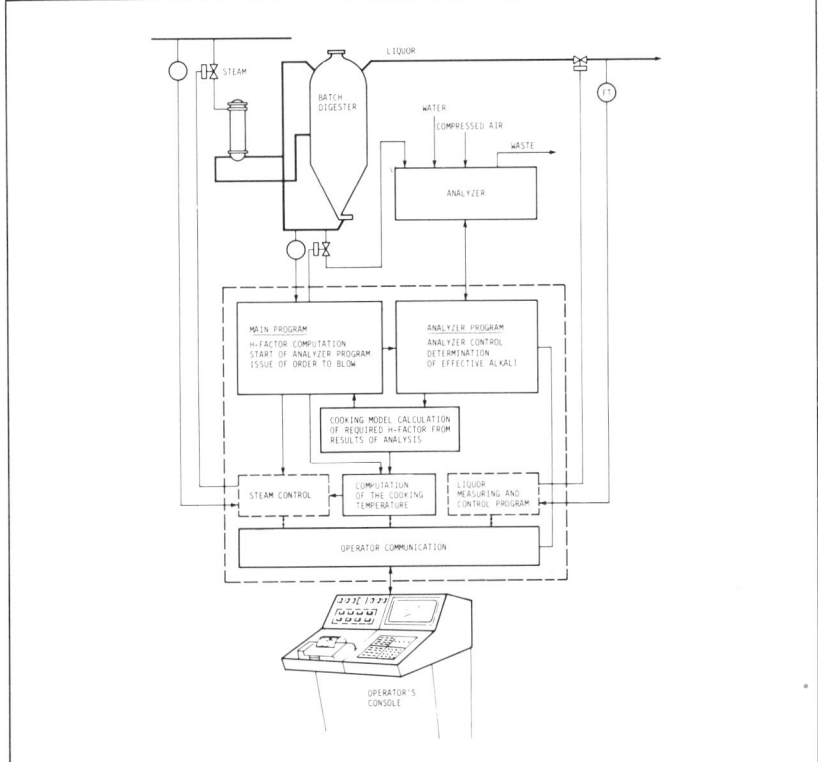

Figure 20-14. Schematic of typical batch digester computer control system using effective alkali analysis.

Kappa numbers into the target range in accordance with a program for that particular species. Difficult measurements to make, such as cooking zone and chip moisture, are determined by having the computer make heat and mass balances automatically. The computer can also be programmed to take the time delay through the digester (residence time) into account so that it can anticipate changes in material input to the digester. Experience has shown that major savings from computer control of a continuous digester have come from: (1) higher yields, (2) reduced chemical usage, (3) increased throughput from existing facilities, and (4) a more uniform product.

All of these are directly related to closer, faster, more accurate control, preventing up and down excursions in quality.

The structure for a typical continuous digester computer control system is shown in Figure 20-15. It can be divided into four general sections, similar to the basic digester computer control system.

1. Basic control functions
2. Coordinated control
3. Optimization
4. Information system

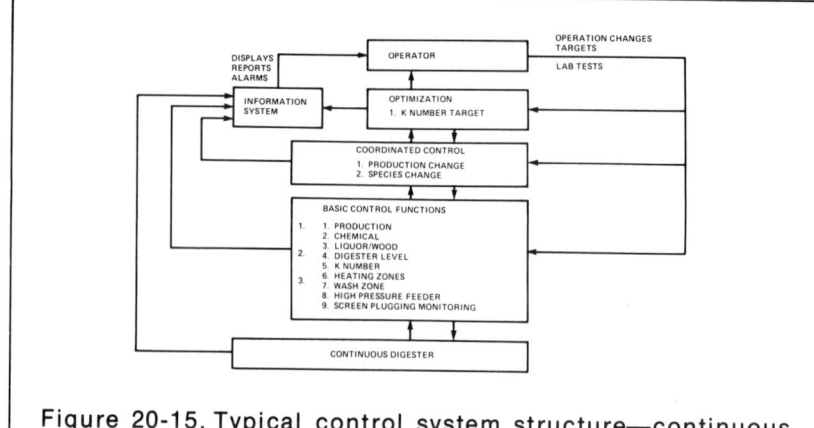

Figure 20-15. Typical control system structure—continuous digester computer control system.

Figure 20-16 is an overview of the basic control functions found in such a system, listed as follows:

1. Production control
2. Chemical control
3. Liquor/wood control
4. Digester level control
5. K or Kappa number control
6. Upper and lower heater temperature control
7. Wash zone control
8. High-pressure feeder control

This control strategy provides for production control in order to maintain the balance for wood-in and pulp-out, and to control production at the desired set point. The level control is used to trim the production control for any variations in the balance of wood-in and pulp-out that cannot be eliminated by production control alone.

The appropriate amount of chemicals is provided by the chemical control which maintains a desired alkali-to-wood ratio by manipulating the white liquor flow to the digester. Liquor-to-wood control provides for maintaining a desired liquor-to-wood ratio. Wash zone control achieves a liquor balance in the wash zone and maintains desired wash upflow target set point.

The control of the temperature of the upper and lower heaters provides for maintaining the temperature at the top of the cook zone at a desired set point. K or Kappa number control varies the temperature set point based on current alkali-to-wood ratio, residence time, and temperature in order to minimize the K number variation and control the K number at the desired target set point.

Product change coordination is accomplished by making appropriate changes to temperature and chemical targets at appropriate times and rates, as determined by the computer, to minimize upsets during species changes, for example, softwood to hardwood and vice versa. This is done by changing the appropriate target set points from those for the old species to those of the new species at various points in time as the tracked interface of the two species moves through the digester zones.

K number optimization is used to shift the K number target based on K number laboratory results which provide for higher K numbers based on smaller deviations under computer control.

Bleaching. Like the continuous digesters, multistage bleach plants, with their long time delays for pulp to pass through the system, are good candidates for the application of computer control. Since typical bleaching operations consist of a combination of various stages arranged in sequence—for example, chlorination, caustic extraction, hypochlorite, and chlorine dioxide—the primary reasons for using computer control are to:

1. Save chemicals by ensuring that only sufficient amounts are used for each stage of bleaching to obtain the correct amount of liquor removal or oxidation required.
2. Obtain higher yields with the computer system maintaining operating conditions in each stage to minimize degrading the pulp by decreased yield and strength due to action of chemicals on cellulose.
3. Improve quality with computer control maintaining output brightness of pulp at the required level under both steady state and dynamic operating conditions.
4. Increase throughput.
5. Decrease steam usage.
6. Minimize total volume of effluent from the plant.

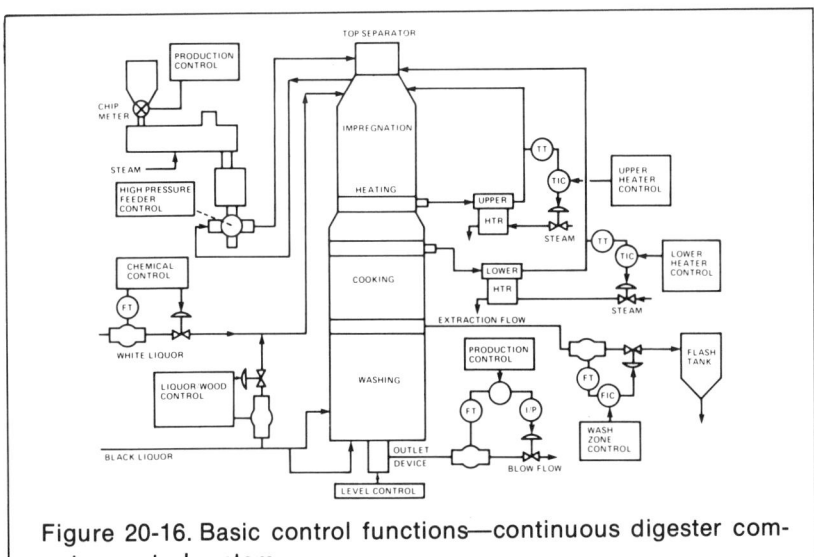

Figure 20-16. Basic control functions—continuous digester computer control system.

The computer system achieves these objectives by employing a control scheme that optimizes each stage, using quality indicators as input information. This information can be used as feedback corrections of previous stages and feedforward corrections in succeeding stages. The logical strategy on which these corrections are based can be stored in the computer. Such indicators as brightness, viscosity, and K number measurements could be used as inputs with the computer calculating and making the required readjustments in chemical flows, retention times, temperatures, pH, etc.

A possible computer control scheme is to have the computer predict the degree of bleachability (K number) of the incoming stock on a fixed rate scan. The operator then has immediate access to the instantaneous K number entering the bleaching system. When the tonnage rate changes the operator enters new target tonnages in the computer. The computer immediately gives him the correct pulp flow to the bleach plant. Then the desired chlorine flow is calculated on the basis of predicted K number, the new target tonnage, temperature, retention time, or other variables which might affect it. The chlorine flow is then set and the computer returns to its normal cycle, allowing time for the chlorine to react and the ORP measurement to reflect the change. After a precalculated time, the computer returns to the ORP measurement, reads the value, and establishes this control set point until another tonnage rate change. Thus, chlorination control and caustic addition based on the degree of chlorination can be calculated by the computer. A constant desired brightness can be obtained with the computer controlling the hypochlorite stage by predicting the chemical addition value (based on the K number obtained after the extraction stage), retention time, and temperature at a constant pH. The same computer control strategy can be used on the chlorine dioxide stage with brightness being a function of hypochlorite brightness, temperature, tonnage rate, and retention time.

As in the digester computer control systems, the structure of a typical bleach plant computer control system, shown in Figure 20-17, can be considered as consisting of four similar units, namely:

1. Basic control functions
2. Coordinated control
3. Optimization
4. Information system

The basic control functions performed by a typical four-stage bleach plant computer control system are shown in Figure 20-18 and consist of:

1. Chlorination stage brown stock and chlorine flow control.
2. Extraction stage caustic flow and temperature control.
3. Bleaching stages hypochlorite flow, chlorine dioxide flow, buffer flow, and temperature control.

A bleach plant computer control system can be designed for any number and combination of bleaching sequences of chlorination, extraction, hypochlorite, or chlorine dioxide stages. In the chlorination stage, a target chemical residual or extracted K number is maintained. Extraction stages are controlled

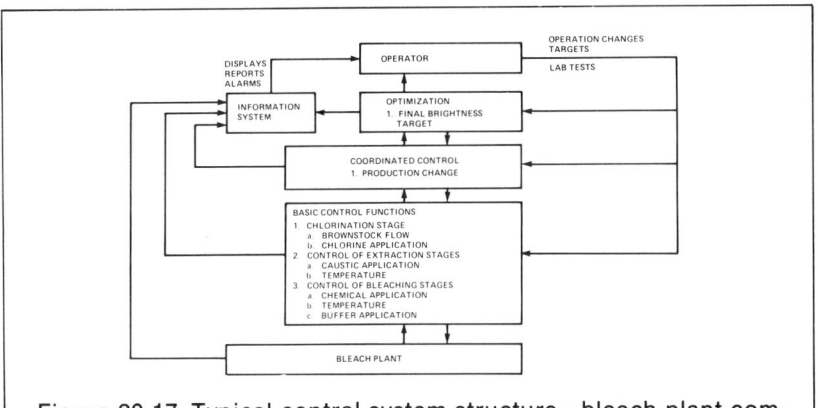

Figure 20-17. Typical control system structure—bleach plant computer control system.

to a target pH and temperature. Bleaching stages such as hypochlorite and chlorine dioxide are controlled to a target brightness, a target residual, and a target temperature. The control objective is usually to maintain pulp quality while minimizing the usage of chemicals and energy.

Other Pulp Mill Applications. Although the application of computer control techniques has not been as extensive in other pulp mill areas as in the digesters and bleach plants, some work has been done or contemplated to: (1) maintain proper stock consistencies and soda carryover from the washers and evaluate evaporator heat transfer characteristics to define optimum cleaning schedules; (2) aid recovery furnace and associated steam and electric generating units to meet surges in demand due to startup and shutdown of process equipment, as well as possible reduction of auxiliary fuel requirements; (3) control the density of green liquor going to the recausticizing plant; (4) improve the throughput of the tall oil plant by computing and specifying such variables as correct reflux ratio to use to obtain a satisfactory overhead fraction in refining crude tall oil by successive distillation.

Paper Mill

The control problems caused by complex interaction of variables traditionally subjected to independent control action throughout a paper mill have led to the use of digital computer process control systems designed to take such interaction into account. Figure 20-19 illustrates a possible configuration of a computer scheme that is arranged to accomplish this. The drawing shows the four major unit process areas. However, the majority of paper mill computers have been concerned with the stock preparation and paper machine areas.

Stock Preparation. Uniform paper manufacturing is easier using stock whose composition and physical properties such as freeness are not subject to rapid variation. Computer control in the stock preparation during the refining and blending operation minimizes these variations in the stock.

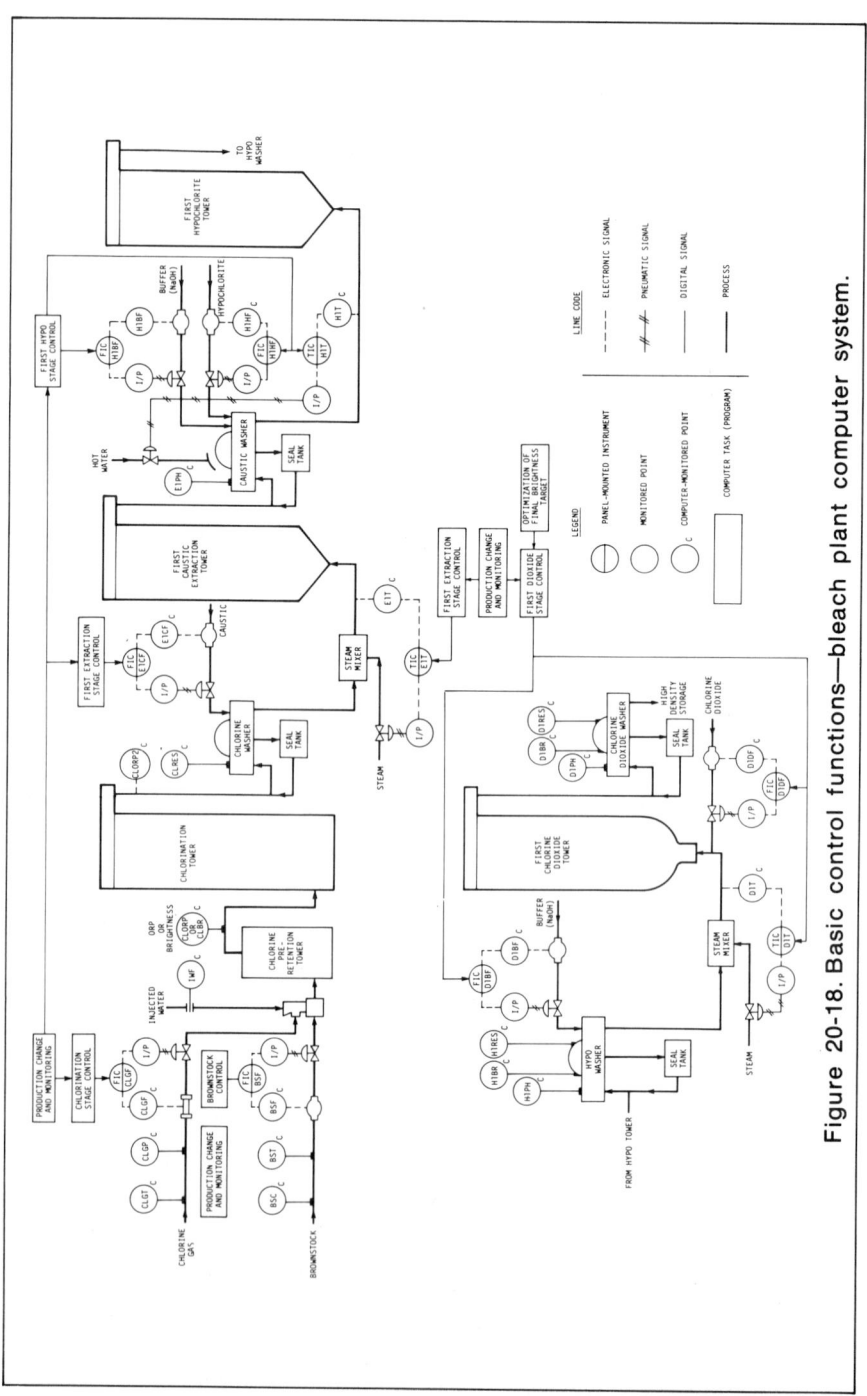

Figure 20-18. Basic control functions—bleach plant computer system.

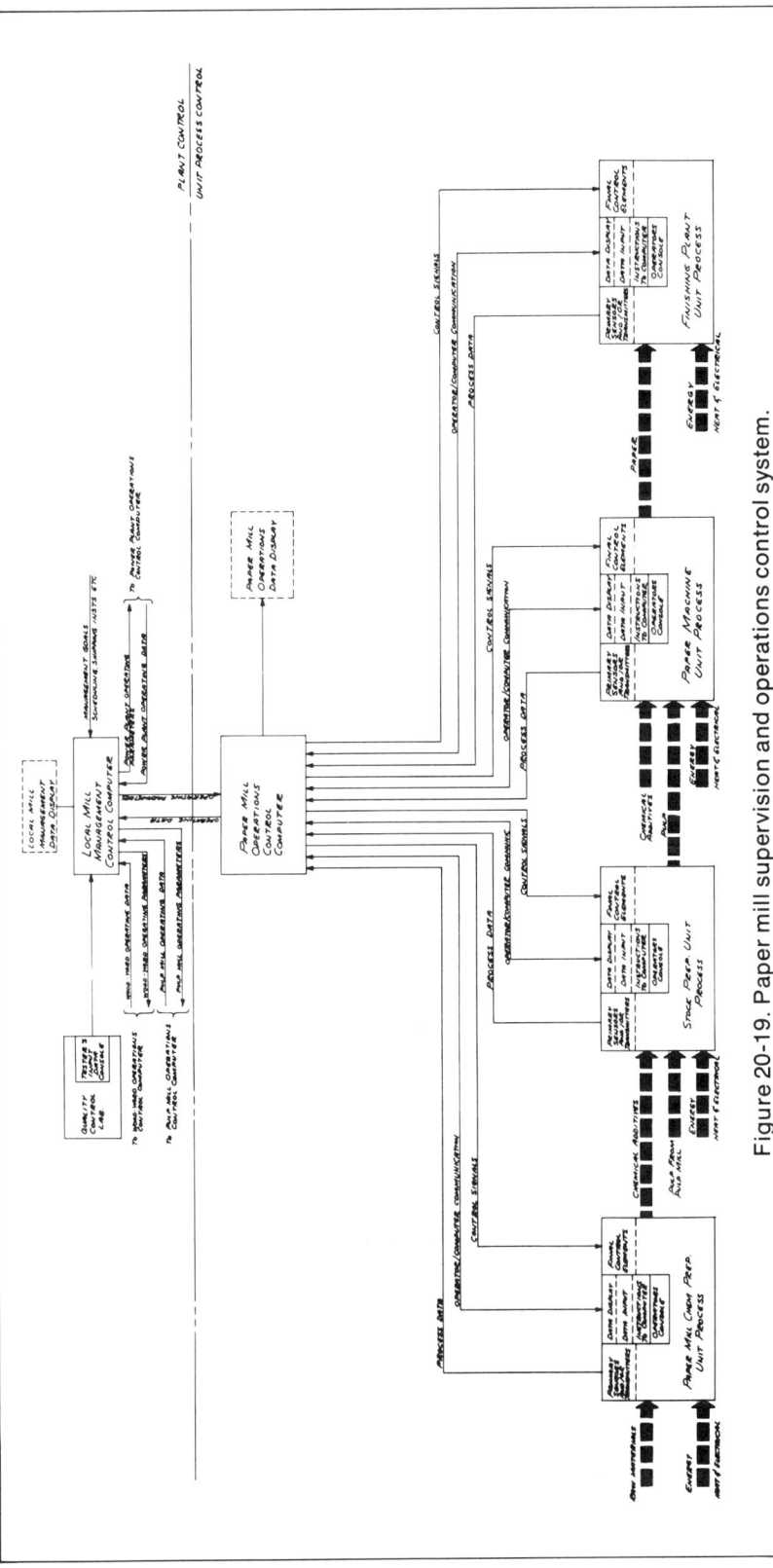

Figure 20-19. Paper mill supervision and operations control system.

A computer scheme that can be used for refiner control to maintain constant freeness is shown in Figure 20-20, which employs both feedback and feedforward control concepts. Continuous drainage measurements and periodic freeness laboratory tests provide feedback information on actual freeness being obtained. While the actual values deviate from the desired value, the computer calculates and executes a change in the computed intermediate variable—which can be horsepower days per ton, kilowatts per ton, or temperature rise of the refiner—that will correct the error. This system will also provide feedforward control over disturbances resulting from changes in stock consistency, composition, and flow. The periodic refined stock freeness tests are used by the computer to update the relationship between hp days/ton, kw/ton, or ΔT to freeness change. As the varying characteristics of the unrefined fibers are reflected in the refined stock freeness, the intermediate variable of the refiner is adjusted by the computer to return the actual freeness to the desired value.

Stock Blending. A computer scheme to proportion dry weights of pulps, chemicals, and additives is shown in Figure 20-21. The weight of these materials is calculated from on-line measurements of all flows, consistency measurements, and periodic off-line measurements of the solid content of additives. The system functions like a flow-ratio controller whose set points are adjusted by the computer to compensate for variations in percent solids and consistencies of materials being metered. The computer can also provide the blend of all components on a bone-dry basis. Consistency and flows of all stock components are measured, and consistency measurements can be compensated if the influence of other variables such as degree of refining, temperature, and flow are known. Using compensated consistency and flow, the bone-dry tons/hour flow of each component can be determined and controlled to produce the desired bone-dry fraction being blended in accordance with mill demand requirements.

Paper Machine. The first applications for computer control in the pulp and paper industry have been on the paper machine because initially the wet end of the fourdrinier machine was considered as the last point in the manufacture of

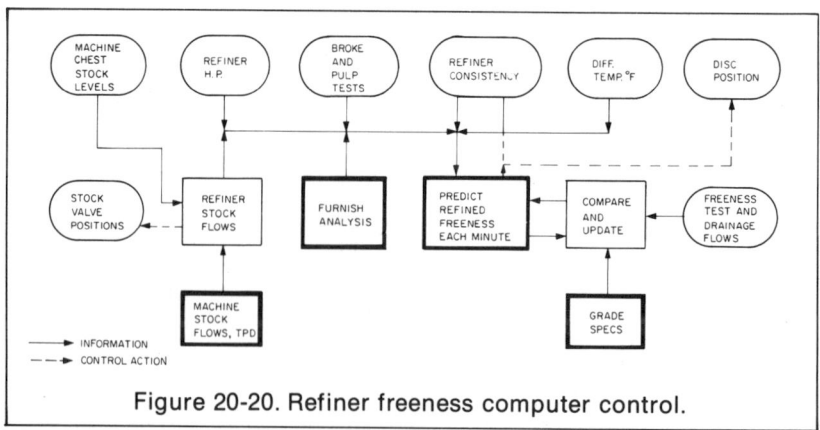

Figure 20-20. Refiner freeness computer control.

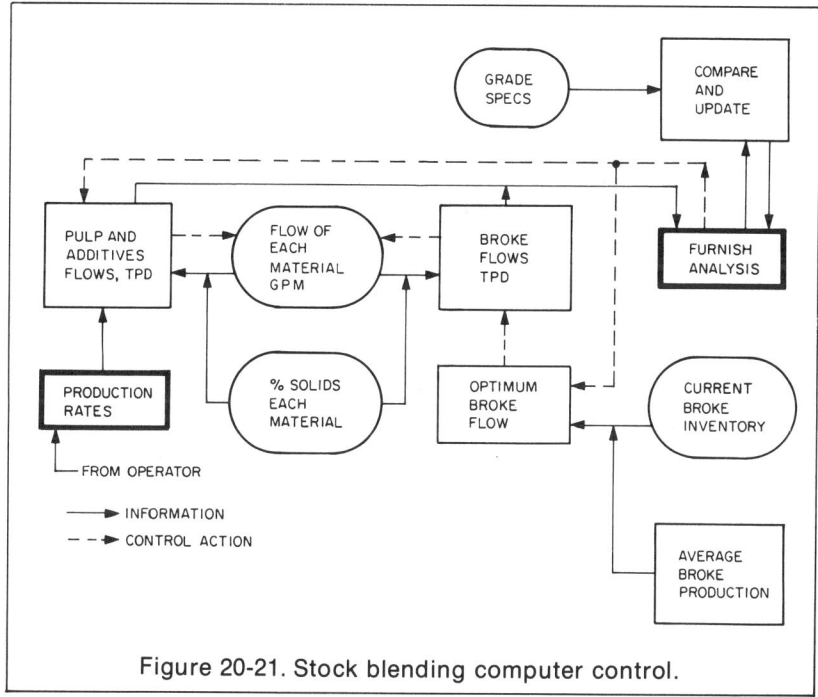

Figure 20-21. Stock blending computer control.

paper where the greatest influence could be applied to affect the properties of the final paper product. Logically, this concept extended itself to include the dry end of the paper machine. Paper machine configurations vary from mill to mill and ranges of grades manufactured also differ, requiring different operating economics. The computer concept utilization can only be determined for a specific machine under consideration. However, there are some general control functions which can be applied to practically all paper machines.

Around the wet end of the paper machine a computer can be used to compute headbox consistency for which there is presently no practical measurement device. It does this by continually computing material balance equations using measurements of thick stock flow to the paper machine, thin stock flow, additive flow, and thick stock consistency. Freeness measurements can also be made by using the same material balance strategy around the wet end.

By adjusting the ratio between wire speed and the speed at which the thin stock is discharged from the slice (called *rush/drag* or *efflux ratio*) and headbox consistency, the computer can change sheet formation. This is done by having the computer compute the correct headbox pressure value from the wire speed and control it in order to achieve a ratio required to maintain the proper rush/drag relationship. If a change in consistency is also required, the computer is instructed accordingly and it adjusts the various flows, slice, and headbox pressure in correct time relationship without disturbing the basis weight.

The computer can handle grade changes by referring to stored standard running conditions for all probable grades and then setting up the desired operating conditions automatically.

Basis weight of the paper can be controlled by the computer by first setting up standard running conditions for the specific grade being made and then trimming these by feedback from an on-line basis weight measuring gauge. If an error is observed by the computer, the measurement is used to predict what the basis weight would be at a given suitable future moment. A control adjustment is made from this to compensate for the error by taking into account the process dynamics. Values of constants used in the calculations are continually updated. Paper moisture can be controlled by a similar method and interactive effects can be compensated for.

Changes in speed of the machine can also be made in a calculated, coordinated manner by a digital computer. When a speed change is required it is ramped from one level to another, with simultaneous ramping and controlling of all other interrelated variables, and the computer will adjust the master speed accordingly. The computer can also be used to monitor the various machine dryer section speeds and calculate the percentage draws along the machine with digital display to the operating personnel.

In order to ensure consistent operation of the dry end of the paper machine with respect to other process variables, the steam and dryer drainage system can also be brought under computer control. This will ensure that the dryer pressures are compatible with speed and basis weight under steady state and transient conditions. To accomplish this, the computer control system must measure the dryer pressures and differentials, measure and calculate bone-dry heavy stock flow, and control the dryer section pressure and/or differential. By the use of a feedforward control scheme, an increase in dryer steam pressures is caused by an increase in heavy stock flow at a rate consistent with the transport between the heavy stock valve and dryer sections. Differentials are automatically set in response to changes in machine speeds ensuring adequate differentials to evacuate dryers and at the same time limit excessive pressure drops.

Some work has also been done in the area of using the computer in color control of the paper sheet. The measurement of color has been standardized so a color can be specified by three factors. Thus, a computer program can be written in terms of the effects of specific dyes on different grades. It is possible to feed the characteristics of available dyes, the furnish to be colored, and the final shade required into the computer and obtain a recipe for the exact match or the nearest match with the available dyes together with the indication of the shade error. The computer can then be programmed to make the necessary adjustments to dye flows automatically, or print this information out to be used as an operator's guide.

A computer-based system can be used for trimming paper machines. When applied for this purpose, it combines sophisticated and reiterative mathematical methods, and the computer's speed of calculation, to match a paper company's order on hand with the machine on which the paper is to be made. In this way it determines the best finishing roll sizes so as to keep the machine winder trim loss at a minimum, consistent with finishing process equipment utilization.

The coordinated computer control concept applied to overall paper machine operation allows large and small production changes to be made in a smooth and rapid manner with stable operation at the product target values. This concept is illustrated in Figure 20-22. It offers certain advantages over conventional manual and set point control techniques. Changes in production are conventionally done by adjusting set points or manual controls slowly to avoid upsets; otherwise the adjustments are liable to be made too rapidly and production is lost until the upsets settle. Coordinated control provides an improved way to make grade and speed changes and avoids some of the undesirable features of conventional methods. It also offers improvement for control of basis weight and for wet end adjustments during normal production, such as adjustments in the wet end consistency (water in the sheet).

Production changes on a paper machine are typically involved operations consisting of many separate adjustments. The calculation of the correct control actions is greatly aided by the use of a process model. The success of the coordinated control scheme, both in terms of technical achievements and in terms of reasonable implementation cost, depends on the approach taken toward process modeling. One practical approach takes advantage of the ''building block'' or modular organization of the system.

An on-line simulation model is usually used in coordinated paper machine computer control. The simulation model is based on physical material balance equations, mixing, and transport delay. The basic building blocks in the model are made up of sets of algorithms. The parameters of the simulation model can be changed from an operator's panel and subsequent correction of the model as machine speeds and wet-end flows change can be done automatically. Some of the model parameters (such as retention values) and some of the model inputs (such as dye concentrations) are set to approximate values. This, together with other approximations in the model, gives an overall approximation to the actual process. This results in an overall control system which

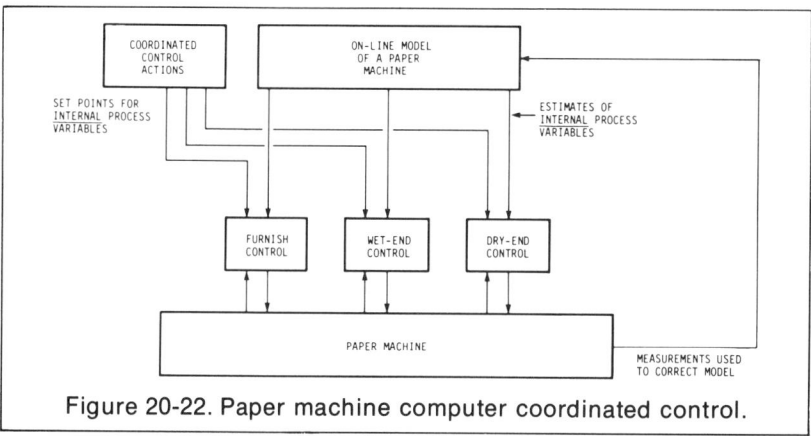

Figure 20-22. Paper machine computer coordinated control.

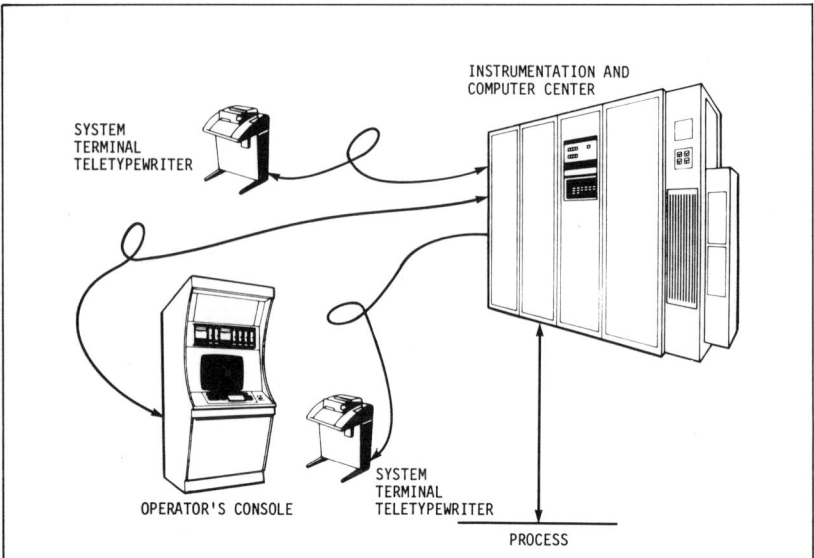

Figure 20-23. Computer control system configuration—single unit process.

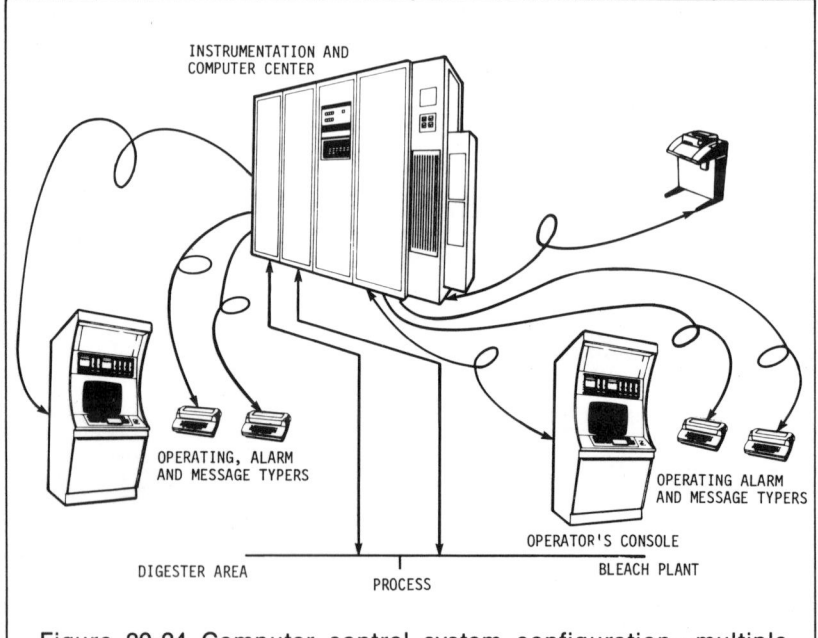

Figure 20-24. Computer control system configuration—multiple unit process.

operates to minimize disturbances. The quality of the control will depend on the accuracy of the model. The computer provides coordination of the process inputs by providing control based on the estimated values of internal process variables. The major part of the control action is based on the on-line simulation model of the process. Feedback control is used to correct the residual errors due to model inaccuracy and unmeasured disturbances. The paper machine operators also provide some of the "feedback" in terms of adjusting the process inputs.

Computer Systems Configuration. A typical configuration of a computer system used to control a single-unit process such as a digester, bleach plant, or paper machine is shown in Figure 20-23. Computer control systems have also been designed and are being used to control a number of similar unit processes or a combination of different unit processes. For instance, one system can be used to control a number of digesters or bleach plants. It can also be used to control a single digester and bleach plant. This configuration is shown in Figure 20-24.

BIBLIOGRAPHY

Benray, R. *Understanding Digital Computers.* New York: John F. Rider Publisher, 1965.

Black, W. W., Gordon, J., and Breach, A. *An Introduction to On-Line Computers.* New York: Science Publishers, Inc., 1972.

Collins, C. E. "Direct Digital Control Applied to Paper Mill Processes." *Paper Trade Journal,* August 7, 1967.

Grant, R. L. *Introduction to the Terminology, Equipment and Concepts of Computers.* New York: Lockwood Publishing Company, 1971.

Heller, S. *Digital Computers Made Simple.* Boston: Cahners Publishing Company, Inc., 1970.

Joint Textbooks Committee of the Paper Industry.

Pulp and Paper Manufacture, vol. 2. 2nd ed. New York: McGraw-Hill Book Company, 1970.

Lowe, K.E., ed. *Practical Computer Applications for the Pulp and Paper Industry.* San Francisco: Miller Freeman Publications, Inc., 1975.

Murphy, B. H. "Understanding Digital Computer Process Control." *Automation,* January 1965.

Savas, E.S. *Computer Control of Industrial Processes.* New York: McGraw-Hill Book Company, 1965.

Stout, T.M. "Computer Control in the Pulp and Paper Industry." *Tappi,* vol. 43. no. 4 (April 1960).

STUDY QUESTIONS

Indicate whether following statements are True or False by inserting T or F in parentheses:

1. Signals from the input peripherals are fed directly to the control section of the digital computer. ( )

2. The control section is the heart of every digital computer because it synchronizes all operations in the computer. ( )

3. Core memory devices can transmit information at much higher speeds than bulk memory devices. ( )

4. Given a set of instructions in FORTRAN, the computer will translate these into machine language and then perform the required operations. ( )

5. Process control using computer data logging does not require participation by the operator. ( )

6. When the computer is used as a sequence controller, the computer automatically collects all information required for control, and the operator manipulates the control devices or process equipment. ( )

7. In supervisory process computer control, the computer performs set point changes, while the operator makes the controller mode adjustments. ( )

8. With direct digital control, the only way to provide backup control is to use another computer. ( )

9. Safer process operating conditions are the only incentive for installing a digital process computer control system. ( )

10. The majority of the computers installed in paper mills are concerned with the stock preparation and paper machine areas. ( )

(See Appendix for answers)

APPENDIX A

Abbreviations Used in this Book

abs	absolute	mm	millimeter
ac	alternating current	mv	millivolt
a.d.	air dry	NPT	normal pressure and tem-
API	American Paper Institute		perature
BCD	binary coded decimals	OD	outside diameter
b.d.	bone dry	ORP	oxidation-reduction
Btu	British thermal unit		potential
cf	cubic foot	pH	hydrogen-ion concentration
cfs	cubic feet per second	PIV	peak inverse voltage
c.g.s.	centimeter-gram-second	psi	pounds per square inch
CPU	central processing unit	psig	pounds per square inch
CRT	cathode-ray tube		gauge
CSF	Canadian Standard freeness	rev	revolutions
C_v	valve-coefficient	rpm	revolutions per minute
dc	direct current	RTD	resistance temperature
DDC	direct digital control		detector
Diff	differential	SAMA	Scientific Apparatus
EDP	electronic data processing		Makers Association
emf	electromotive force	sec	second
F	Fahrenheit	scfh	standard cubic foot/hr
fpm	feet per minute	sq ft	square foot
ft	foot	SR	Schopper-Riegler
gpm	gallons per minute	SSF	Saybolt ferrol second
H.D.	high density	SSU	Saybolt universal second
hp	horsepower	TAPPI	Technical Association of
hr	hour		the Pulp and Paper
Hz	hertz		Industry
ID	inside diameter	temp	temperature
K	Kelvin	U.S.	United States
kw	kilowatt	v	volts
max	maximum	vs	versus

APPENDIX B

Glossary

The following is not intended to be either a complete or an official list of analog and digital process control terms and definitions of any one society or organization. Although these terms have not all been used in the text, the purpose is to give the reader knowledge of the ones commonly being used in the field of process control today.

ABSOLUTE ADDRESS. (1) An address that is permanently assigned by the machine designer to a storage location. (2) A pattern of characters that identifies a unique storage location without further modification.

ABSOLUTE ALARM. An alarm caused by the detection of a variable which has exceeded a set of prescribed high or low limit conditions.

ABSOLUTE CODING. Coding written in language acceptable to a computer without further modification. Synonymous with Machine Language.

ACCESS. Pertaining to the ability to place information into, or retrieve information from, a storage device.

ACCESS TIME. (1) The time interval between the instant at which data are called for from a storage device and the instant delivery is completed, i.e., the read time. (2) The time interval between the instant at which data are requested to be stored and the instant at which storage is completed, i.e., the write time.

ACCUMULATOR. A register in which the result of an arithmetic or logic operation is formed.

ACCURACY. (1) The degree of freedom from error, that is, the degree of conformity to truth or to a rule. Accuracy is contrasted with precision, e.g., four-place numerals; nevertheless, a properly computed four-place numeral might be more accurate than an improperly computed six-place numeral. (2) A number or quantity which defines the limit of error under reference operating conditions.

NOTE 1. Unless otherwise specified, accuracy is defined as that in effect under reference operating conditions.

NOTE 2. Accuracy includes the combined conformity, hysteresis, and repeatability errors. The units being used are to be stated explicitly. It is preferred that a $+$ and/or $-$ sign precede the number or quantity. The absence of a sign infers a $+$ and $-$ sign. It can be expressed in a number of forms. The following five examples are typical:

a) Accuracy expressed in output units. Typical expression: the accuracy is ± 1 F.

b) Accuracy expressed in percent of output span. Typical expression: the accuracy is $\pm 1/2\%$ output span. (This percentage is calculated using scale units such as F, psi, etc.)

c) Accuracy expressed in percent of the upper range-value. Typical expression: the accuracy is $\pm 1/2\%$ of upper range-value. (This percentage is calculated using scale units such as F, psi, etc.)

d) Accuracy expressed in percent scale length. Typical expression: the accuracy is ±1/2% of scale length.

e) Accuracy expressed in percent of actual output reading. Typical expression: the accuracy is ±1% of actual output reading. (This percentage is calculated using scale units such as F, psi, etc.)

ACTUATING ERROR SIGNAL. The reference input signal minus the feedback signal.

ADAPTIVE CONTROL ACTION. Control action whereby automatic means are used to change the type and/or influence of control parameters in such a way as to improve the performance of the control system.

ADDER. A device whose output is a representation of the sum of the quantities represented by its inputs.

ADDRESS. The identification of a location in memory.

AIR CONSUMPTION. The maximum rate at which air is consumed by a device within its operating range during steady state signal conditions.

ALARM. An audible or visible signal that indicates an abnormal or out-of-limits condition in the plant or control system.

ALPHANUMERIC. *See* Character.

ALLOCATION. The assignment of blocks of data to specified blocks of storage.

AMPLIFICATION FACTOR. The dimensionless ratio of output/input.

AMPLIFIER. A device whose output is, by design, an enlarged reproduction of the input signal and which is energized from a source other than the signal.

A/M STATION. (Automatic/Manual Station.) A device that permits a system to be run manually by the process operator.

AMBIENT PRESSURE. The pressure of the medium surrounding a device.

AMBIENT TEMPERATURE. The temperature of the medium surrounding a device.

NOTE 1. For devices which do not generate heat this temperature is the same as the temperature of the medium at the point of device location when the device is not present.

NOTE 2. For devices which do generate heat this temperature is the temperature of the medium surrounding the device when it is present and dissipating heat.

NOTE 3. Allowable ambient temperature limits are based on the assumption that the device in question is not exposed to significant radiant energy sources.

ANALOG. The representation of quantities by means of continuously variable physical quantities such as voltage, current, resistance, rotation, etc. Contrast with Digital.

ANALOG BACKUP. An alternative method of process control by conventional analog instrumentation in the event of a failure in the computer system.

ANALOG COMPUTER. A computer which manipulates numerical quantities represented as electrical and physical variables for solutions to mathematical problems. Contrast with Digital Computer.

ASYNCHRONOUS COMPUTER. A computer in which each event or the performance of each operation starts as a result of a signal generated by the completion of the previous event or operation, or by the availability of the parts of the computer required for the next event or operation.

ASSEMBLER. A program which converts symbolic language to machine language by substitution of absolute operation codes for symbolic operation codes and absolute or relocatable addresses for symbolic addresses.

ATTENUATION. (1) A decrease in signal magnitude between two points, or between two frequencies. (2) The reciprocal of gain, when the gain is less than one.

AUCTIONEERING DEVICE. A device which automatically selects either the highest or the lowest input signal from among two or more input signals.

NOTE. This is frequently referred to as a high or low signal selector.

AUTOMATION. The act or method of making a processing or manufacturing system perform without the necessity of operator intervention or supervision. The common word designating the state of being automatic.

AUTOMATIC CONTROLLER. A device, or combination of devices, which measures the value of a variable, load, or condition and operates so as to correct or limit deviation of the controlled variable from a selected reference.

AUTOMATIC CONTROL SYSTEM. An operable arrangement of one or more automatic controllers along with their associated equipment connected in loops with one or more processes.

AUTOMATIC PROGRAMMING. The process of using a computer to perform some stages of the work involved in preparing a program.

AVAILABILITY. The total amount of time that a computer is properly operating.

BACKUP. Provision of alternative means of operation in case of a failure of the primary means of operation. *See* Analog Backup, Digital Backup, and Manual Backup.

BINARY. (1) Pertaining to a characteristic or property involving a selection, choice, or condition in which there are two possibilities. (2) Pertaining to the numeration system with a radix of two.

BINARY CODED DECIMAL (BCD). Pertaining to a decimal notation in which the individual decimal digits are each represented by a group of binary digits. In the 8-4-2-1 binary coded decimal notation, the number twenty-three is represented as 0010 0011, whereas in binary notation twenty-three is represented as 10111.

BINARY DIGIT. A character used to represent one of the two digits in the binary number system.

BIT. Contracted form of "binary digit." *See* Binary Digit.

BLOCK DIAGRAM. A diagram of a system, instrument, computer, or program in which selected portions are represented by annotated boxes and interconnecting lines.

BOOTSTRAP. A technique or device designed to bring itself into a desired state by means of its own action.

BUFFER. (1) A storage device used to compensate for a difference in rate of flow of data, or time of occurrence of events, when transmitting data from one device to another. (2) An isolating circuit used to prevent a driven circuit from influencing the driving circuit.

BUG. An error or malfunction in a program or hardware.

BULK MEMORY. An auxiliary memory device with storage capacity greatly in excess of working (core) memory; e.g., disc file, drum.

BUS. One or more conductors used for transmitting signals or power.

BYTE. A sequence of adjacent binary digits operated upon as a unit and usually shorter than a word.

CALIBRATE. (1) To ascertain, usually by comparison with a standard, the locations at which scale/chart graduations should be placed to correspond to a series of values of the quantity which the instrument is to measure, receive, or transmit. (2) To adjust the output of a device, to bring it to a desired value, within a specified tolerance, for a particular value of the input. (3) To ascertain the error in the output of a device by checking it against a standard.

CAPACITY. In computer terminology, the quantity of information that can be contained in a storage device defined in terms of the basic information size such as words or characters.

CASCADE CONTROL ACTION. Control action where the output of one controller is the set point for another controller.

CATHODE RAY TUBE (CRT). A TV tube upon whose screen data are displayed.

CENTRAL PROCESSOR UNIT (CPU). That portion of any computer system that performs the actual computation. It usually consists of the arithmetic and control units and working memory.

CHARACTER. One of a set of elementary symbols which express information. The set may be alphanumeric, including the decimal digits 0 through 9, the letters A through Z, and special symbols used to denote functions or numeric including only digit 0 through 9.

CLEAR. To erase the information in a storage device by replacing its contents with zeros.

CLOSED LOOP. A signal path which includes a forward path, a feedback path and a summing point, and forms a closed circuit.

COMMON MODE INTERFERENCE. A form of interference which appears between any measuring circuit terminals and ground.

COMMON MODE REJECTION. The ability of a circuit to discriminate against common mode voltage; usually expressed as a ratio or in decibels.

COMMON MODE VOLTAGE. A signal of the same polarity on both sides of a differential input.

COMPILER. A program which translates a problem-oriented language to a machine-oriented language. Example: FORTRAN, ALGOL. A compiler, as contrasted with an assembler, can substitute subroutines as well as single machine instructions for certain symbolic inputs. A compiler which translates directly from source language to machine language is known as a single-pass compiler. A compiler which generates an interim object language which requires further translation or modification is known as a multiple-pass compiler.

COMPUTER. A device capable of accepting information, performing prescribed operation on the information and providing the results of these operations. Its major elements usually include memory, control, arithmetic, logical, and input and output facilities.

CONTROL ACTION. Of a controller or a controlling system, the nature of the change of the output effected by the input.

NOTE. The output may be a signal or the value of a manipulated variable. The input may be the control loop feedback signal when the set point is constant, an actuating error signal, or the output of another controller.

CONTROL ALGORITHM. A mathematical representation of the control action to be performed.

CONTROL MODE. A specific type of control action such as proportional, integral, or derivative.

CONTROL SYSTEM. A system in which deliberate guidance or manipulation is used to achieve a prescribed value of a variable.

NOTE. It is subdivided into a controlling system and a controlled system or process.

CONTROL UNIT. In a digital computer, those parts that effect the retrieval of instructions in proper sequence, the interpretation of each instruction, and the application of the proper signals to the arithmetic unit and other parts, in accordance with this interpretation.

CONTROLLED SYSTEM. *See* Process.

CONTROLLER. A device which operates automatically to regulate a controlled variable.

CONTROLLING SYSTEM. (1) Of a feedback control system, that portion which compares functions of a directly controlled variable and a set point, and adjusts a manipulated variable as a function of the difference. It includes the reference input elements; summing point: forward and final controlling elements; and feedback elements (including sensing element). (2) Of an automatic control system without feedback, that portion of the control system which manipulates the controlled system.

CONVERTER. A transducer which converts a measured signal into a standard transmission signal.

CORE MEMORY. A high-speed random-access storage device utilizing matrix arrays of ferrite cores usually used as the computer's working memory.

CORE RESIDENT. A term pertaining to certain pivotal programs permanently stored in core memory for frequent execution.

CORRECTION. The difference between the true value and the indication of the measured quantity.

NOTE. A positive correction denotes that the indication is less than the true value.

Correction = true − indication

COUNTER. A device or memory location whose contents can be successively incremented or decremented.

CYCLING. A periodic change in the factor under control; often sinusoidal with equal excursions above and below the control point.

CYCLE TIME. The basic unit of computer speed, usually the time required for a read and a write operation in core memory.

DAMPING. The action by which the output settles to a steady state value after a change in the value of measured signal.

NOTE. When the time response to an abrupt stimulus is as fast as possible without overshoot, the response is said to be "critically damped"; "underdamped" when overshoot occurs; "overdamped" when response is slower than critical.

DATA. The general term used to denote any information which can be processed or produced by a computer or control system.

DATA BREAK. An automatic input-output channel which provides external equipment with direct access to core memory.

DEADBAND. The range through which an input can be varied without initiating response.

NOTE. Deadband is usually expressed in percent of span. Resolution sensitivity and ultimate sensitivity have been defined as one-half deadband. However, such usage conflicts with accepted standard definitions of these terms and their use is depreciated in the sense of deadband.

DEADTIME. The interval of time between initiation of an input change or stimulus and the start of the resulting response.

DEAD ZONE. A zone in which no value of the output exists.

DEBUG. To detect, locate, and remove mistakes from a program or malfunctions from a computer. Synonymous with Trouble Shoot, when applied to a computer system.

DERIVATIVE ACTION GAIN (RATE GAIN). The ratio of maximum gain resulting from proportional-plus-derivative control action to the gain due to proportional action alone.

DERIVATIVE ACTION TIME. In proportional-plus-derivative control action, for a unit ramp signal input, the advance in time of the output signal (after transients have subsided) caused by derivative control action as compared to the output signal due to proportional control action only.

DESIRED VALUE. The value of the controlled variable wanted or chosen.

NOTE. The desired value equals the ideal value in an idealized system.

DEVIATION. Any departure from a desired or expected value or pattern.

DEVIATION ALARM. An alarm caused by a variable departing from its desired value by a specified amount.

DEVICE. An apparatus for performing a prescribed function.

DIAGNOSTIC ROUTINE. A program designed to locate malfunctions in computer hardware or software.

DIFFERENTIAL GAP. Applies to two-position controller action, is the smallest range of values through which the controlled variable must pass in order to move the final control element from one to the other of its fixed positions.

DIGITAL. Pertaining to quantized data which may be expressed in the form of digits. Contrast with Analog.

DIGITAL BACKUP. An alternative method of digital process control initiated by use of special-purpose digital logic in the event of a failure in the computer system.

DIGITAL COMPUTER. A computer that operates on discrete data by performing arithmetic and logic processes on these data. Contrast with Analog Computer.

DIRECT ACTING. Operation of a final control element directly proportional to the control output.

DIRECT ACTING CONTROLLER. A controller in which the absolute value of the output signal increases as the absolute value of the input (measured variable) increases.

DIRECT DIGITAL CONTROL ACTION (DDC). Control action in which control is performed by a digital device (usually a digital computer) which establishes the signal to the final controlling element.

NOTE. Examples of possible digital (D) and analog (A) combinations for this definition are:

	Feedback Elements	Controller	Final Controlling Element
1.	D	D	D
2.	A	D	D
3.	A	D	A
4.	D	D	A

DIRECTLY CONTROLLED SYSTEM. The body, process, or machine directly guided or restrained by final controlling element to achieve a prescribed value of the directly controlled variable.

DIRECTLY CONTROLLED VARIABLE. In a control loop, that variable whose value is sensed to originate a feedback signal.

DISC. A flat circular plate with a magnetic surface on which data can be stored by selective magnetization of portions of the flat surface.

DISCRETE COMPONENT CIRCUIT. Circuit implemented by use of individual transistors, resistors, diodes, capacitors, etc. Contrast with Integrated Circuit.

DISTORTION. An undesired change in waveshape.

DISTURBANCE. An undesired change in a variable applied to a system which tends to affect adversely the value of a controlled variable.

DOUBLE PRECISION. Pertaining to the use of two computer words to represent a number.

DOWNTIME. The time interval during which a device is malfunctioning or nonfunctioning.

DRIFT. An undesired change in output over a period of time, which change is unrelated to input, operating conditions, or load.

NOTE. Drift is usually expressed as the change in output over a specified time with fixed input and operating conditions.

DRUM. A circular cylinder with a magnetic surface on which data can be stored by selective magnetization of portions of the curved surface.

DYNAMIC GAIN. The magnitude ratio of the steady state amplitude of the output signal from an element or system to the amplitude of the input signal to that element or system, for a sinusoidal signal.

NOTE. It may be expressed as a ratio, or in decibels as 20 times the log of that ratio for a specified frequency.

DYNAMIC RESPONSE. The behavior of the output of a device as a function of the input, both with respect to time.

ELEMENT. A component of a device or system.

ENGINEERING UNITS. Units of measure as applied to a process variable. Example: psi, degrees F, etc.

EQUILIBRIUM. When all inputs and outputs (supply and demand) have settled down and are in balance.

ERROR. The difference between the indication and the true value of the measured signal.

NOTE. A positive error denotes that the indication of the instrument is greater than the true value.
ERROR = indication − true

ERROR SIGNAL. In a closed loop, the signal resulting from subtracting a particular return signal from its corresponding input signal. *See* Actuating Error Signal.

EXECUTIVE PROGRAM. A program which controls the execution of all other programs in the computer based on established hardware and software priorities and real time or demand requirements.

FEEDBACK CONTROL ACTION. Control action in which a measured variable is compared to its desired value to produce an actuating error signal which is acted upon in such a way as to reduce the magnitude of the error.

FEEDBACK ELEMENTS. Those elements in the controlling system which change the feedback signal in response to the directly controlled variable.

FEEDBACK SIGNAL. That return signal which results from a measurement of the directly controlled variable.

FEEDFORWARD CONTROL ACTION. Control action in which information concerning one or more conditions that can disturb the controlled variable is converted into corrective action to minimize deviations of the controlled variable.

FINAL CONTROL ELEMENT. That forward controlling element which directly changes the value of the manipulated variable.

FIXED HEADS. Pertaining to the use of stationary, rigidly mounted, reading and writing heads on bulk memory devices.

FIXED POINT. Pertaining to a numeration system in which the position of the point is fixed with respect to one end of the numerals according to some convention.

FLAG. An indication, by presence or absence of a signal, of a hardware condition or program status.

FLIP-FLOP. A circuit or device containing active elements, capable of assuming either one of two stable states at a given time.

FLOATING CONTROLLER. A controller in which the rate of change of the output is a continuous (or at least a piecewise continuous) function of the actuating error signal. The output of the controller can remain at any value in its operating range when the actuating error signal is zero and constant. Hence the output is said to float. When the controller has integral action only, the mode of control has been called "proportional speed floating." The use of the term "integral control action" is recommended as a replacement for "proportional speed floating control." *See* Single-speed Floating Controller, Multiple-speed Floating Controller, Integral (Reset) Controller.

FLOW CHART. A graphical representation for the definition, analysis, or solution of a problem, in which symbols are used to represent operations, data, flow, and equipment.

FORTRAN (FORmula TRANslating system.) A procedure-oriented language designed for solution of arithmetic and logical programs.

FORWARD CONTROLLING ELEMENTS. Those elements in the controlling system which change a variable in response to the actuating error signal.

FREQUENCY. Occurrence of a periodic function (with time as the independent variable), generally specified in a certain number of cycles per unit time.

GAIN (MAGNITUDE RATIO). The ratio of change in output divided by the change in input which caused it. Both output and input must be expressed in the same units making gain a pure (dimensionless) number.

HARDWARE. Physical equipment, e.g., mechanical, magnetic, electrical, or electronic devices. Contrast with Software.

HEAD. A device that reads, records, or erases data on a storage medium, e.g., a small electromagnet used to read, write, or erase data on a magnetic drum or tape, or the set of perforating reading, or making devices used for punching, reading, or printing on paper tape.

HIGH LIMITING CONTROL ACTION. Control action in which the output never exceeds a predetermined high limit value.

HUNTING. Oscillation or cycling that may be of appreciable amplitude caused by the system in spite of zealous effort to achieve a prescribed level of control.

IDEAL VALUE. The value of the indication, output, or ultimately controlled variable of an idealized device or system.

NOTE. It is assumed that an ideal value always exists even though it may be impossible to determine.

IDEALIZED SYSTEM. An imaginary system whose ultimately controlled variable has a stipulated relationship to specified set points.

NOTE. It is a basis for performance standards.

IMMERSION LENGTH. The length from the free end of the bulb or well to the point of immersion in the medium, the temperature of which is being measured.

INDICATING INSTRUMENT. A measuring instrument in which the value of the measured quantity is visually indicated.

INDICATOR TRAVEL. The length of the path described by the indicating means or the tip of the pointer in moving from one end of the scale to the other.

NOTE 1. The path may be an arc or a straight line.

NOTE 2. In the case of knife-edge pointers and others extending beyond the scale division marks, the pointer shall be considered as ending at the outer end of the shortest scale division marks.

INDIRECTLY CONTROLLED SYSTEM. That portion of the controlled system in which the indirectly controlled variable is changed in response to changes in the controlled variable.

INDIRECTLY CONTROLLED VARIABLE. A variable which does not originate a feedback signal, but which is related to, and influenced by, the directly controlled variable.

INITIALIZE. To set counters, switches, and addresses to zero or other starting values at the beginning of, or at prescribed points in, a computer routine.

INPUT. (1) The data to be processed. (2) The state or sequence of states occurring on a specified input channel. (3) The device or collective set of devices used for bringing data into another device. (4) A channel for impressing a state on a device or logic element. (5) The process of transferring data from an external storage to internal storage.

INPUT/OUTPUT CHANNELS. Central processor communication paths, consisting in Foxboro systems of a high-speed data bus, a high-speed input/output bus for computer peripheral equipment, and a low-speed bus buffer channel for slower process equipment.

INPUT/OUTPUT EQUIPMENT. Devices for putting information into and receiving information from the computer system.

INPUT IMPEDANCE. The impedance presented by a device to the source.

INPUT SIGNAL. A signal applied to a device, element, or system.

INSTRUCTION. A statement that specifies an operation and the values or location of its operands.

INTEGRAL ACTION RATE (Reset Rate). (1) Of proportional-plus-integral or proportional-plus-integral-plus-derivative control action devices; for a step input, the ratio of the initial rate of change of output due to integral control action to the change in steady state output due to proportional control action. (2) Of integral control action devices; for a step input, the ratio of the initial rate of change of output to the input change.

NOTE. Integral action rate is often expressed as the number of repeats per minute because it is equal to the number of times per minute that the proportional response to a step input is repeated by the initial integral response.

INTEGRAL (Reset) CONTROL ACTION. Control action in which the output is proportional to the time integral of the input, i.e., the rate of change of output is proportional to the input.

INTEGRAL (Reset) CONTROLLER. A controller which produces integral control action only.

INTEGRATED CIRCUIT. A circuit element incorporating transistor, diode, and resistor elements in the same semiconductor chip. Contrast with Discrete Component Circuit.

INTEGRATOR. A device often employed with a flow meter to totalize the area under the flow rate record, (e.g., gals/min X min = total gals) which produces a numerical readout of total flow.

INTERACTING CONTROL. Control action which is produced by an algorithm whose various terms are interdependent.

INTERFACE LOGIC. Logic necessary to provide electrical and communication compatibility between two devices.

INTERFERENCE (Electrical). Any spurious voltage or current arising from external sources and appearing in the circuits of a device. *See* Noise.

INTERMEDIATE ZONE. Any range of input values not bounded by a range limit.

LAG. A delay in output with respect to a change in input.

LIBRARY. A collection of standard programs with which problems or parts of problems may be solved.

LINEARITY. The closeness to which a curve approximates a straight line.

NOTE. It is usually expressed as a nonlinearity, e.g., a maximum deviation between an average curve and a straight line. The average curve is determined after making two or more full range traverses in opposite directions. The value of nonlinearity is referred to the output unless otherwise stated.

LOAD. Change in level of material, force, torque, energy, power, or other variables applied or removed from a process or other component in the system.

LOG. A periodic hard copy summary of process operation data.

LOGIC. (1) The science dealing with the interim or formal principles of reasoning and thought. (2) The systematic scheme which defines the interaction of signals in the design of an automatic data processing system.

LOOP GAIN (Open-Loop Gain). The ratio of the change in the return signal to the change in its corresponding error signal at a specified frequency.

LOOP GAIN CHARACTERISTIC. Of a closed loop, the characteristic curve of the ratio of the change in the return signal to the change in its corresponding error signal for all real frequencies.

LOW LIMITING CONTROL ACTION. Control action in which the output is never less than a predetermined low limit value.

MAGNETIC FIELD INTERFERENCE. A form of interference induced in the circuits of a device due to the presence of a magnetic field.

NOTE. It may appear as common mode or normal mode interference in the measuring circuits.

MANIPULATED VARIABLE. A quantity or condition which is varied as a function of the actuating signal so as to change the value of the directly controlled variable.

NOTE. In any practical control system, there may be more than one manipulated variable. Accordingly, when using the term, it is necessary to state which manipulated variable is being discussed. In process control work, the one immediately preceding the directly controlled system is usually intended.

MANUAL BACKUP. An alternative method of process control by means of manual adjustment of final control elements in the event of a failure in the computer system.

MATHEMATICAL MODEL. A mathematical representation of a process, device, or concept.

MAXIMUM HYSTERESIS. The maximum difference for the same input between the upscale and downscale indication of the measured signal during full range traverses.

NOTE 1. The difference is expressed as a percent of span.

NOTE 2. Statement used in this definition is a common usage definition and includes hysteretic error and deadband.

MEASURING ELEMENT. The element which converts into a form or language that the controller can understand.

MEASURED SIGNAL. The electrical, mechanical, pneumatic, or other variable applied to the input of a device. It is the analog of the measured variable produced by a transducer (when such is used). *See* Measured Variable.

NOTE. For example: In a thermocouple thermometer system, the measured signal is an emf which is the electrical analog of the temperature applied to the thermocouple.

In a flow meter, the measured signal may be a differential pressure which is the analog of the rate of flow through the orifice.

In an electric tachometer system, the measured signal may be a voltage which is the electrical analog of the speed of rotation of the part coupled to the tachometer generator.

MEASURED VARIABLE. The physical quantity, property, or condition which is to be measured. *See* Measured Signal.

NOTE 1. It is sometimes referred to as the measurand.

NOTE 2. Common measured variables are temperature, pressure, rate of flow, thickness, speed, etc.

MEASURING INSTRUMENT. A device for ascertaining the magnitude of a physical quantity or condition presented to it.

MECHANICAL SHOCK. The momentary application of an acceleration force to a device.

NOTE. It is usually expressed in multiples of the acceleration due to gravity.

MEMORY. Equipment for the storage of information.

MEMORY PROTECT. A technique of protecting the contents of sections of memory from alteration by inhibiting the execution of any memory modification instruction upon detection of the presence of a guard bit associated with the accessed memory location. Memory modification instructions accessing protected memory are usually executed as a no-operation and a memory protect violation program interrupt is generated.

MICROSECOND. One millionth of a second (10^{-6} sec.)

MILLISECOND. One thousandth of a second (10^{-3} sec.)

MNEMONIC. An alphanumeric designation, easy to remember and commonly used to designate a member location or computer operation; e.g., START might represent the location of the first instruction in a routine.

MODULATION. The process, or result of the process, whereby some characteristic of one wave is varied in accordance with some characteristic of another wave.

MOUNTING POSITION. The position of a device relative to physical surroundings.

MOVABLE HEADS. Pertaining to the use of movable reading and writing heads on bulk memory devices.

MULTI-ELEMENT CONTROL SYSTEM. A control system utilizing input signals derived from two or more process variables for the purpose of jointly affecting the action of the control system.

NOTE. Examples are input signals from pressure and temperature or from speed and flow, etc.

MULTIPLE-SPEED FLOATING CONTROLLER. A floating controller in which the output may change at two or more rates, each corresponding in a definite range of values of the actuating error signal.

MULTIPLEX. To interleave or simultaneously transmit two or more messages on a single channel.

MULTI-POSITION CONTROLLER. A controller having two or more discrete values of output.

NANOSECOND. One billionth of a second (10^{-9} sec.)

NEUTRAL ZONE. A zone in which the previously existing output value is not changed.

NOISE. An unwanted component of a signal or variable which obscures its information content. *See* Interference (Electrical).

NOTE. It may be expressed in units of the output or in percent of output span.

NONINTERACTING CONTROL SYSTEM. A multi-element control system designed to avoid disturbances to other controlled variables due to the process input adjustments which are made for the purpose of controlling a particular process variable.

NONLINEAR SYSTEM. Any system whose operation cannot be represented by a first-order mathematical equation.

NORMAL MODE INTERFERENCE. A form of interference which appears between measuring circuit terminals.

NORMAL MODE REJECTION. The ability of a circuit to discriminate against normal mode voltage; usually expressed as a ratio or in decibels.

NORMAL MODE VOLTAGE. An extraneous voltage induced across the circuit path (transverse mode voltage).

NORMAL OPERATING CONDITIONS. The range of operating conditions within which a device is designed to operate and within which operating influences are usually stated.

NORMALIZE. To shift the representation of a quantity so that the representation lies in a prescribed range.

OFFLINE. (1) Pertaining to equipment or programs not under the direct control of the central processor. (2) Pertaining to a computer that is not actively monitoring or controlling a process or operation, or pertaining to a computer operation performed while the computer is not monitoring or controlling a process or operation.

OFFSET. The steady state deviation of the controlled variable when the set point is fixed. *See* Steady State Deviation.

NOTE. The offset resulting from a no-load to a full-load change (or other specified limits) is often called "droop" or "load regulation."

ON LINE. (1) Pertaining to equipment or programs under direct control of a central processor. (2) Pertaining to a computer that is actively monitoring or controlling a process or operation, or pertaining to a computer operation performed while the computer is monitoring or controlling a process or operation.

ON-OFF CONTROLLER. A multiposition controller having two discrete values of output, fully on, or fully off.

OPEN LOOP. A signal path without feedback.

OPERATING CONDITIONS. Conditions, such as ambient temperature, ambient pressure, vibration, etc., to which a device is subjected, but not including the variable measured by the device.

OPERATING INFLUENCE. The change in a designated performance characteristic caused solely by a prescribed change in a specified operating variable from reference operating condition to another specified operating condition, all other operating variables being held within the limits of reference operating conditions.

NOTE. The specified operating conditions are usually the limits of the normal operating conditions.

OPERATIVE LIMITS. The range of operating conditions to which a device may be subjected without permanent impairment of operating characteristics.

NOTE 1. In general, performance characteristics will not be stated in the region between the limits of normal operating conditions and operative limits.

NOTE 2. Upon returning within the limits of normal operating conditions, a device may require adjustments to restore normal performance.

OPERATOR. (1) The person who initiates and monitors the operation of a computer. (2) The person who initiates and monitors the operation of a process.

OPTIMIZE. To establish control parameters so as to make control as effective as possible.

OPTIMIZING CONTROL ACTION. Control action that automatically seeks and maintains the most advantageous value of a specified variable, rather than maintaining it at one set value.

OPTIMUM. The highest obtainable proficiency of control, e.g., supply equals demand and offset has been reduced to a minimum.

OUTPUT SIGNAL. A signal delivered by a device, element, or system.

OVERSHOOT. The amount that a process output (controlled variable) exceeds its desired value after a change of input.

PARAMETER. A controllable or variable characteristic of a system or device, temporarily regarded as a constant, the respective values of which serve to distinguish the various specific states of a (the) system or device.

PEN TRAVEL. The length of the path described by the pen in moving from one end of the chart scale to the other. The path may be an arc or a straight line.

PERIPHERAL. Equipment used for entering data into, or receiving data from, the computer.

PHASE. The relationship with respect to time that exists between a periodic function and a reference (generally another periodic function), e.g., the time relationship between two sine waves.

PHASE DIFFERENCE. The time, usually expressed in electrical degrees, by which one wave leads or lags another.

PNEUMATIC EXHAUST CAPACITY. The maximum rate at which a device can exhaust air from a load at a 1 psi drop in pressure, at a specified pressure level.

NOTE. The pressure drop is to be measured across the device only.

PNEUMATIC SUPPLY CAPACITY. The maximum rate at which a device can supply air to a load at a 1 psi drop in pressure, at a specified pressure level.

NOTE. The pressure drop is to be measured across the device only.

POSITION. Of a multi-position controller, a discrete value of the output signal.

POWER CONSUMPTION. The maximum wattage used by a device within its operating range during steady state signal condition.

NOTE. For a power factor other than unity, power consumption shall also be stated as maximum volt-amperes used under above stated conditions.

PRECISION. The degree of discrimination with which a quantity is stated, e.g., a three-digit numeral discriminates among 1000 possibilities.

PRIMARY ELEMENT (DETECTOR). The first system element that responds quantitatively to the measured variable and performs the initial measurement operation.

NOTE 1. A primary element performs the initial conversion of measurement energy.

NOTE 2. For transmitters not used with external primary elements, the sensing portion is the primary element.

PRIORITY. Level of importance of a program or device.

PRIORITY INTERRUPT. The temporary suspension of a program currently being executed in order to execute a program of higher priority. Priority interrupt functions usually include distinguishing the highest priority interrupt active, remembering lower priority interrupts which are active, selectively enabling or disabling priority interrupts, executing a jump instruction to a specific memory location(s), and storing the program counter register in a specific location(s).

PROCESS. The collective functions performed in and by the equipment in which a variable or variables is or are to be controlled.

NOTE. Equipment as embodied in this definition should be understood not to include any automatic control equipment. The process may also be referred to as the controlled system.

PROCESS CONTROL LOOP. A system of control devices linked together to control one phase of a process.

PROCESS PRESSURE. The pressure of the process medium at the sensing element.

PROCESS TEMPERATURE. The temperature of the process medium at the sensing element.

PROGRAM. A series of routines which logically solve a given problem.

PROGRAMMER. A person who prepares computer operation procedures by means of flow charts and coding.

PROPORTIONAL BAND. The change in input required to produce a full range change in output, due to proportional control action.

NOTE 1. It is reciprocally related to proportional gain.

NOTE 2. It may be stated in input units or as a percent of the input span (usually the indicated or recorded input span). The preferred term is "Proportional Gain."

PROPORTIONAL CONTROL ACTION. Control action in which there is a continuous linear relation between the output and the input.

NOTE. This condition applies when both the output and input are within their normal operating ranges and when operation is at a frequency below a limiting value. *See note* under Control Action.

PROPORTIONAL CONTROLLER (P). A controller which produces proportional control action only.

PROPORTIONAL GAIN. The ratio of the change in output due to proportional control action to the change in input.

Illustration: $Y = \pm PX$

where P = proportional gain
X = input transform
Y = output transform

PROPORTIONAL-PLUS-DERIVATIVE (Rate) CONTROL ACTION. Control action in which the output is proportional to a linear combination of the input and the time rate-of-change of input.

NOTE. In the practical embodiment of proportional-plus-derivative control action, the relationship between output and input, neglecting high frequency terms, is:

$$\frac{Y}{X} = \pm P \frac{1 + sD}{1 + sD/a} \quad a > 1$$

where a = derivative action gain
D = derivative action time constant
P = proportional gain
s = complex variable
X = input transform
Y = output transform

See note under Control Action

PROPORTIONAL-PLUS-DERIVATIVE (Rate) CONTROLLER (PD). A controller which produces proportional-plus-derivative (rate) control action.

PROPORTIONAL-PLUS-INTEGRAL (Reset) CONTROL ACTION. Control action in which the output is proportional to a linear combination of the input and the time integral of the input.

NOTE. In the practical embodiment of proportional-plus-integral action, the relation between output and input, neglecting high frequency terms, is:

$$\frac{Y}{X} = \pm P \frac{I/s + 1}{bI/s + 1} \quad 0 \leqslant b \ll 1$$

where b = proportional gain/static gain
I = integral action rate
P = proportional gain
s = complex variable
X = input transform
Y = output transform

See note under Control Action.

PROPORTIONAL-PLUS-INTEGRAL (Reset) CONTROLLER. A controller which produces proportional-plus-integral (reset) control action.

PROPORTIONAL-PLUS-INTEGRAL (Reset)-PLUS-DERIVATIVE (Rate) CONTROL ACTION. Control action in which the output is proportional to a linear combination of the input, the time integral of input and the time rate-of-change of input.

NOTE. In the practical embodiment of proportional-plus-integral-plus-derivative control action, the relationship of output to input, neglecting high frequency terms, is:

$$\frac{Y}{X} = \pm P \frac{I/s + 1 + Ds}{bI/s + 1 + Ds/a} \qquad a > 1; \; 0 \leqslant b < 1$$

where a = derivative action gain
b = proportional gain/static gain
D = derivative action time constant
I = integral action rate
P = proportional gain
s = complex variable
X = input transform
Y = output transform

See note under Control Action.

PROPORTIONAL-PLUS-INTEGRAL (Reset)-PLUS-DERIVATIVE (Rate) CONTROLLER (PID). A controller which produces proportional-plus-integral (reset)-plus derivative-(rate) control action.

PULSE RATE. Repetition rate of a pulse signal.

PULSE SIGNAL. A change of signal level away from and back to a base level, the amplitude, duration, or repetition rate of which conveys intelligence.

RAMP ACTION. An action in which the set point is changed at a steady rate.

RAMP RESPONSE. The total (transient plus steady state) time response resulting from a sudden increase in the rate of change in the input from zero to some finite value.

RAMP RESPONSE TIME. The time interval by which an output lags an input, when both are varying at a constant rate.

RANDOM ACCESS. (1) Pertaining to the process of obtaining data from, or placing data into, storage where the time required for such access is independent of the location of the data most recently obtained or placed in storage. (2) Pertaining to a storage device in which the access time is effectively independent of the location of the data.

RATIO CONTROL. Control of a secondary variable by making its magnitude follow a primary variable at a set ratio.

RATIO CONTROLLER. A controller that maintains a predetermined ratio between two or more variables.

REAL TIME PROGRAM. A program which operates concurrently with an external process which it is monitoring or controlling, meeting the needs of that process with respect to time.

RECORDING INSTRUMENT. A measuring instrument in which the values of the measured quantity are recorded.

REFERENCE-INPUT ELEMENTS. That portion of the controlling system which changes the reference input signal in response to the set point.

REFERENCE-INPUT SIGNAL. One external to a control loop which serves as the standard of comparison for the directly controlled variable.

REFERENCE OPERATING CONDITIONS. The range of operating conditions of a device, within which operating influences are negligible.

NOTE 1. The range is usually narrow.

NOTE 2. They are the conditions under which reference performance is stated and the base from which the values of operating influences are determined.

REFERENCE PERFORMANCE. Performance attained under reference operating conditions.

NOTE. Performance includes such things as accuracy, deadband, repeatability, hysteresis, linearity, etc.

REGISTER. A device composed of flip-flops that temporarily holds words while they are being processed or waiting to be passed on to another component.

REPEATABILITY. The closeness of agreement among a number of consecutive measurements of the output for the same value of the measured signal under the same operating conditions, approaching from the same direction, for full range traverses.

NOTE. It is usually expressed as a maximum nonrepeatability in percent of span. It does not include hysteresis.

REPRODUCIBILITY. The closeness of agreement among repeated measurements of the output for the same value of input made under the same operating conditions over a period of time, approaching from either direction.

NOTE. It is expressed as a maximum nonreproducibility in percent of span for a specified time. Normally, this implies a long period of time, but under certain conditions the period may be a short time during which drift may not be included.

Reproducibility includes hysteresis, drift, repeatability, and deadband.

Operating conditions and input may vary between normal operating limits between measurements.

RESET RATE. *See* Integral Action Rate.

RESET TIME. The calibrated time on the controller reset dial which represents the time that will elapse while the open loop controller repeats proportional action.

RESOLUTION. The least interval between two adjacent discrete details which can be distinguished one from the other.

NOTE. Resolution is termed fine or coarse as the interval is small or large. For continuous-reading devices, S.A.M.A. depreciates use of the term in the sense of "resolution sensitivity" because "specified conditions" are so seldom given.

RESONANCE. Of system or element, a condition evidenced by large oscillatory amplitude, which results when a small amplitude of a periodic input has a frequency approaching one of the natural frequencies of the driven system.

RESPONSE. Reaction to a forcing function applied to the input, e.g., the variation in measured variables which occur as the result of step, sinusoidal, ramp, or other known type inputs.

RETURN SIGNAL. In a closed loop, the signal resulting from a particular input signal, and transmitted by the loop and to be subtracted from the input signal.

REVERSE ACTING CONTROLLER. A controller in which the absolute value of the output signal decreases as the absolute value of the input (measured variable) increases.

RISE TIME. The time required for the output of a system (other than first order) to make change from a small specified percentage (often 5 or 10) of the steady state increment to a large specified percentage (often 90 or 95), either before overshoot or in the absence of overshoot.

NOTE. If the term is unqualified, response to a unit step stimulus is understood, otherwise the pattern and magnitude of the stimulus should be specified.

ROUTINE. A series of computer instructions which performs a specific task.

SAMPLING CONTROLLER. A controller using intermittently observed values of a signal, such as the set point signal, the actuating error signal, or the signal representing the controlled variable, to effect control action.

SAMPLING PERIOD. The time interval between observations in a periodic sampling control system.

SCALE. To change a quantity by a factor in order to bring its range within prescribed limits.

SCALE FACTOR. A number used as a multiplier, so chosen that it will cause a set of quantities to fall within a given range of values. To scale the values 856, 432, −95, and −182 between −1 +1, a scale factor of 1/1000 would be suitable.

SCAN. (1) Collection of data from process sensors by a computer for use in calculations, usually obtained through a multiplexer. (2) Sequential interrogation of device or lists of information under program control.

SECONDARY STORAGE. Same as Bulk Memory.

SELF-OPERATED CONTROLLER. A control device in which all the energy to operate the final controlling element is derived from the controlled system through the sensing element.

SELF-TUNING. The technique of automatic modification of control algorithm constants based upon process conditions.

SENSE. To detect the presence of a condition.

SENSING ELEMENT. The portion of a device directly responsive to the value of the measured quantity.

NOTE. It may include the case protecting the sensitive portion.

SENSING ELEMENT ELEVATION. The difference in elevation between the sensing element and the case.

NOTE. The elevation is considered positive when the sensing element is above the case.

SENSITIVITY. The ratio of a change in output to the change of input which causes it, after steady state has been reached.

NOTE. It is expressed as a numerical ratio with the units of measurement of two quantities stated.

SERVOMECHANISM. (1) A feedback control system in which at least one of the system signals represents mechanical motion. (2) Any feedback control system.

SET POINT (COMMAND). Any input variable which sets the desired value of the controlled variable.

NOTE 1. The input variable may be manually set, automatically set, or programmed.

NOTE 2. It is expressed in the same units as the controlled variable.

SET POINT CONTROL. A control technique in which the computer supplies a calculated set point to a conventional analog instrumentation control loop.

SHARED TIME CONTROL ACTION. Control action where one controller divides its computation of control time among several control loops rather than acting on all loops simultaneously.

SIGN BIT. A single bit, usually the most significant bit in a word, which is used to designate the algebraic sign of the information contained in the remainder of the word.

SIGNAL TO NOISE RATIO. Ratio of signal amplitude to noise amplitude.

NOTE. For sinusoidal signals amplitude may be peak or rms. For nonsinusoidal signals, peak values should be used.

SIGNAL TRANSDUCER. A transducer which converts one standardized transmission signal to another.

SIMULATION. Using an analog or digital computer, the representation of physical systems (such as a chemical process) in which information provided to the computer represents process variables, the processing done by the computer represents the process itself, and information produced by the computer represents the results of the process.

SIMULATOR. A device or computer system that performs simulation.

SINGLE-SPEED FLOATING CONTROLLER. A controller in which the output changes at a fixed rate, increasing or decreasing depending on the sign of the actuating error signal.

NOTE. A neutral zone of values of the actuating error signal in which no action occurs may be used.

SMOOTH. To apply procedures that decrease or eliminate rapid fluctuations in data.

SOFTWARE. (1) The collection of programs and routines associated with a computer, e.g., compilers; library routines. (2) Also the documents associated with a computer, e.g., manuals, circuit diagrams. Contrast with Hardware.

SOURCE LANGUAGE. A program language that is used as an input to a translation program such as an assembler or compiler.

SPAN. The algebraic difference between the upper and lower range values.

NOTE 1. For example:

a) range 0 to 150 F, span 150 F
b) range −20 to 200 F, span 220 F
c) range 20 to 150 F, span 130 F

NOTE 2. The following compound terms are used with suitable modifications in the units: measured variable span, measured signal span, etc.

NOTE 3. For multirange devices, this definition applies to the particular range that the device is set to measure.

SPAN ERROR. The difference between the actual span and the ideal span.

NOTE. It is usually expressed as a percent of ideal span.

STATIC GAIN. The ratio of an output change to an input change after steady state has been reached.

NOTE. Sometimes referred to as zero frequency gain. It may be expressed as a ratio, or in decibels as 20 times the log of that ratio.

STEADY STATE. A characteristic of a condition, such as value, rate, periodicity, or amplitude, exhibiting only negligible change over an arbitrary long period of time.

NOTE. It may describe a condition in which some characteristics are static, others dynamic.

STEADY STATE DEVIATION. The system deviation after transients have expired. *See* Offset.

STEP RESPONSE. The time response of an instrument when subjected to an instantaneous change in input from one steady state value to another.

STEP RESPONSE TIME. Of a system or an element, the time required for an output to make the change from an initial value to a large specified percentage of a steady state either before or in the absence of overshoot, as a result of a step change to the input.

NOTE. Usually states for 90, 95, 99 percent change. *See* Time Constant for use of 63% value.

STORAGE. (1) Pertaining to a device into which data can be entered, in which it can be held, and from which it can be retrieved at a later time. (2) Loosely, any device that can store data. Synonymous with Memory.

SUPERVISORY. (1) A process computer application wherein the computer performs higher level process calculations but does not actuate the final element, e.g., a valve. Contrast with Direct Digital Control Action. For example, the computer may handle mathematical models of the process, or may perform process calculations and relay the results to controllers for valve actuation. (2) That computer in the Foxboro PCP 88 configuration which handles higher level functions such as optimizing and feedforward computation, production scheduling, inventory control, data logging, and alarming.

SUPERVISORY CONTROL ACTION. Control action in which the control loops operate independently subject to intermittent corrective action, e.g., set point changes from an external source.

SUPPLY PRESSURE. The pressure at the supply port of the device.

SWITCHING POINT. A point in the input span of a multi-position controller at which the output signal changes from one position to another.

SYNCHRONOUS COMPUTER. A computer in which each event, or the performance of each operation, starts as a result of a signal generated by a clock.

SYSTEM. A collection of hardware and software organized in such a way as to achieve an operational objective.

SYSTEMS ANALYSIS. The definition of a control problem and the development of a solution to the control problem.

SYSTEMS ENGINEERING. The implementation of a hardware and software system resulting from analysis of a control problem.

TERMINATION RACK. An equipment rack containing field wiring terminals and associated signal conditioning equipment. It provides the termination interface between a computer control system and field-mounted instrumentation.

THERMAL SHOCK. An abrupt temperature change applied to a device.

THREE POSITION CONTROLLER. A multi-position controller having three discrete values of output.

TIME CONSTANT. For a first order system, the time required for the output to complete 63.2% of the total rise or decay as a result of a step change of the input.

TIME PROPORTIONING CONTROLLER. A controller whose output consists of periodic pulses whose duration is varied to relate, in some prescribed manner, the time average of the output to the actuating error signal.

TIME SCHEDULE CONTROLLER. A controller in which the set point (or reference input signal) automatically adheres to a predetermined time schedule.

TIME SHARING. Pertaining to the interleaved use of the time of a device.

TOGGLE SWITCH. A manually operated electrical switch which may be placed in either of two or three positions by a projecting arm or knob.

TRACK. The portion of a moving storage medium, such as a drum, tape, or disc, that is accessible to a given reading head position.

TRANSDUCER. An element or device which receives information in the form of one physical quantity and converts it to information in the form of the same or other quantity.

NOTE. This is a general definition and applies to specific classes of devices such as primary element, signal transducer, and transmitter. *See* Converter, Primary Element, Signal Transducer, and Transmitter.

TRANSMITTER. A transducer which responds to a measured variable by means of a sensing element, and converts it to a standardized transmission signal which is a function only of the measurement.

TROUBLE SHOOT. Search of entire system, including the process, to locate source of difficulty. Called Debug when the search is restricted to a computer or its programs.

TUNING. The adjustment of control constants in algorithms or analog controllers to produce the desired control effect.

TWO-POSITION CONTROLLER. A multi-position controller having two discrete values of output.

VALVE CONTROL AMPLIFIER (VCA). An integrating amplifier which accepts analog signals from a time-shared valve output module, provides memory and conditioning for the signal, and generates the control output signal.

VALVE OUTPUT MODULE (VOM). A device which translates the computer's output data into analog signals which are suitable to position and control valves or other devices.

VARIABLE. A level, quantity, or other condition which is subject to change; this may be regulated, e.g., the controlled variable, or be simply a measured variable.

VELOCITY LIMIT. A limit which the rate of change of a specified variable cannot exceed.

VELOCITY LIMITING CONTROL ACTION. Control action in which the rate of change of a specified variable will not exceed a predetermined limit.

WARM-UP PERIOD. The time required after energizing a device before its rated performance characteristics apply.

WORD. A sequence of bits or characters treated as a unit and capable of being stored in one computer location.

WORD-TIME. The data transfer rate (words per second) between a device and the computer.

WRITE. To deliver data to a medium such as storage.

ZERO-BASED CONFORMITY. The value of nonconformity determined after any translation and/or rotation of the actual curve is made to make it coincide with zero on the specified curve and that minimizes the maximum deviation.

ZERO-BASED LINEARITY. The value of nonlinearity determined after any translation and/or rotation of the actual curve is made to make it coincide with a straight line through zero that minimizes the maximum deviation.

ZERO ERROR. The error of a device operating under the specified conditions of use when the input is at the lower range value.

ZERO SHIFT. Any parallel shift of the input-output curve.

ZONE. On a multi-position controller, the range of input values between selected switching points or any switching point and range limit.

APPENDIX C

Answers to Study Questions

Chapter 1
1. c
2. b
3. c
4. b
5. b
6. a
7. b
8. b
9. a
10. c
11. F
12. T
13. F
14. T
15. F
16. F
17. T
18. T
19. F
20. F
21. c
22. a
23. c
24. a
25. c
26. b
27. a
28. c
29. a
30. b

Chapter 2
1. T
2. F
3. T
4. T
5. F
6. F
7. T
8. F
9. T
10. T

Chapter 3
1. c
2. a
3. c
4. a
5. c
6. b
7. a
8. b
9. c
10. a

Chapter 4
1. F
2. F
3. T
4. F
5. T
6. T
7. T
8. F
9. T
10. F
11. b
12. c
13. c
14. b
15. b
16. a & b
17. c
18. a
19. a & c
20. a & b

Chapter 5
1. F
2. T
3. T
4. F
5. F
6. T
7. F

8. T
9. F
10. T
11. b & c
12. c
13. b
14. a
15. b
16. a & b
17. c
18. b
19. a
20. c
21. F
22. T
23. T
24. T
25. F
26. T
27. T
28. T
29. T
30. F

Chapter 6
1. a & b
2. b
3. a
4. a & c
5. c
6. c
7. b
8. a
9. a
10. b

Chapter 7
1. T
2. F
3. F
4. T
5. F

6. T
7. F
8. T
9. T
10. F

Chapter 8
1. a & c
2. c
3. b & c
4. b
5. a
6. b & c
7. b
8. c
9. a & c
10. c

Chapter 9
1. T
2. F
3. F
4. T
5. T
6. T
7. F
8. T
9. F
10. F

Chapter 10
1. b
2. a
3. c
4. a & b
5. a & c
6. a
7. c
8. b
9. a
10. a & b

Chapter 11
1. T
2. T
3. F
4. T
5. F

Chapter 12
1. T
2. T
3. F
4. T
5. F

Chapter 13
1. c
2. a
3. c
4. a
5. b
6. a & b & c
7. b

Chapter 14
1. b & c
2. a & b
3. b & c
4. a & b
5. b & c

Chapter 15
1. F
2. T
3. F
4. F
5. T
6. F
7. F
8. T
9. T
10. T
11. F
12. T
13. F
14. T
15. F

Chapter 16
1. F
2. T
3. T
4. F
5. F
6. F
7. T
8. T
9. F
10. T

Chapter 17
1. b
2. a

3. b & c
4. c
5. a & b
6. a
7. c
8. b
9. b
10. a & b

Chapter 18
1. F
2. T
3. T
4. F
5. T
6. T
7. F
8. T
9. F
10. T
11. T
12. F
13. T
14. F
15. F
16. F
17. T
18. T
19. F
20. T

Chapter 19
1. c

2. c
3. a
4. b & c
5. a & b
6. b
7. c
8. a & c
9. a
10. c
11. F
12. T
13. T
14. F
15. T
16. F
17. F
18. F
19. T
20. F

Chapter 20
1. F
2. T
3. T
4. T
5. F
6. F
7. T
8. F
9. F
10. T

APPENDIX D

Units and Conversion Tables

These conversion factors are among those found most useful to pulp and paper industry personnel involved with instrumentation. English units of measurement are still in common use in pulp and paper mills in the United States, particularly for nonelectrical quantities. However, the movement toward use of the International System of Units, commonly referred to as SI units (for Système International d'Unités) or metric system, is becoming rapid. Hence, these tables have been arranged to facilitate conversions into SI units by placing these conversions first, followed by the other useful conversions.

ATMOSPHERES — atm (Standard at sea-level pressure)
× 101.325 = Kilopascals (kPa) absolute
× 14.696 = Pounds-force per square inch absolute (psia)
× 76.00 = Centimetres of mercury (cmHg) at 0°C
× 29.92 = Inches of mercury (inHg) at 0°C
× 33.96 = Feet of water (ftH$_2$O) at 68°F
× 1.01325 = Bars (bar) absolute
× 1.0332 = Kilograms-force per square centimetre (kg/cm^2) absolute
× 1.0581 = Tons-force per square foot (tonf/ft^2) absolute
× 760 = Torr (torr) (= mmHg at 0°C)

BARRELS, LIQUID, U.S. — bbl
× 0.11924 = Cubic metres (m^3)
× 31.5 = U.S. gallons (U.S. gal) liquid

BARRELS, PETROLEUM — bbl
× 0.15899 = Cubic metres (m^3)
× 42 = U.S. gallons (U.S. gal) oil

BARS — bar
× 100 = Kilopascals (kPa)
× 14.504 = Pounds-force per square inch (psi)
× 33.52 = Feet of water (ftH$_2$O) at 68°F
× 29.53 = Inches of mercury (inHg) at 0°C
× 1.0197 = Kilograms-force per square centimetre (kg/cm^2)
× 0.98692 = Atmospheres (atm) sea-level standard
× 1.0443 = Tons-force per square foot (tonf/ft^2)
× 750.06 = Torr (torr) (= mmHg at 0°C)

BRITISH THERMAL UNITS — Btu (See note)
× 1055 = Joules (J)
× 778 = Foot-pounds-force (ft · lbf)
× 0.252 = Kilocalories (kcal)
× 107.6 = Kilogram-force-metres (kgf · m)
× 2.93 × 10^{-4} = Kilowatt-hours (kW · h)
× 3.93 × 10^{-4} = Horsepower-hours (hp · h)

BRITISH THERMAL UNITS PER MINUTE — Btu/min (See note)
× 17.58 = Watts (W)
× 12.97 = Foot-pounds-force per second (ft · lbf/s)
× 0.02358 = Horsepower (hp)

CENTARES
× 1 = Square metres (m^2)

CENTIMETRES — cm
× 0.3937 = Inches (in)

CENTIMETRES OF MERCURY — cmHg, at 0°C
× 1.3332 = Kilopascals (kPa)
× 0.013332 = Bars (bar)
× 0.4468 = Feet of water (ftH$_2$O) at 68°F
× 5.362 = Inches of water (inH$_2$O) at 68°F
× 0.013595 = Kilograms-force per square centimetre (kg/cm^2)
× 27.85 = Pounds-force per square foot (lbf/ft^2)
× 0.19337 = Pounds-force per square inch (psi)
× 0.013158 = Atmospheres (atm) standard
× 10 = Torr (torr) (= mmHg at 0°C)

CENTIMETRES PER SECOND — cm/s
× 1.9685 = Feet per minute (ft/min)
× 0.03281 = Feet per second (ft/s)
× 0.03600 = Kilometres per hour (km/h)
× 0.6000 = Metres per minute (m/min)
× 0.02237 = Miles per hour (mph)

CUBIC CENTIMETRES — cm³
× 3.5315 × 10⁻⁵ = Cubic feet (ft³)
× 6.1024 × 10⁻² = Cubic inches (in³)
× 1.308 × 10⁻⁶ = Cubic yards (yd³)
× 2.642 × 10⁻⁴ = U.S. gallons (U.S. gal)
× 2.200 × 10⁻⁴ = Imperial gallons (imp gal)
× 1.000 × 10⁻³ = Litres (l)

CUBIC FEET — ft³
× 0.02832 = Cubic metres (m³)
× 2.832 × 10⁴ = Cubic centimetres (cm³)
× 1728 = Cubic inches (in³)
× 0.03704 = Cubic yards (yd³)
× 7.481 = U.S. gallons (U.S. gal)
× 6.229 = Imperial gallons (imp gal)
× 28.32 = Litres (l)

CUBIC FEET PER MINUTE — cfm
× 472.0 = Cubic centimetres per second (cm³/s)
× 1.699 = Cubic metres per hour (m³/h)
× 0.4720 = Litres per second (l/s)
× 0.1247 = U.S. gallons per second (U.S. gps)
× 62.30 = Pounds of water per minute (lbH₂O/min) at 68°F

CUBIC FEET PER SECOND — cfs
× 0.02832 = Cubic metres per second (m³/s)
× 1.699 = Cubic metres per minute (m³/min)
× 448.8 = U.S. gallons per minute (U.S. gpm)
× 0.6463 = Million U.S. gallons per day (U.S. gpd)

CUBIC INCHES — in³
× 1.6387 × 10⁻⁵ = Cubic metres (m³)
× 16.387 = Cubic centimetres (cm³)
× 0.016387 = Litres (l)
× 5.787 × 10⁻⁴ = Cubic feet (ft³)
× 2.143 × 10⁻⁵ = Cubic yards (yd³)
× 4.329 × 10⁻³ = U.S. gallons (U.S. gal)
× 3.605 × 10⁻³ = Imperial gallons (imp gal)

CUBIC METRES — m³
× 1000 = Litres (l)
× 35.315 = Cubic feet (ft³)
× 61.024 × 10³ = Cubic inches (in³)
× 1.3080 = Cubic yards (yd³)
× 264.2 = U.S. gallons (U.S. gal)
× 220.0 = Imperial gallons (imp gal)

CUBIC METRES PER HOUR — m³/h
× 0.2778 = Litres per second (l/s)
× 2.778 × 10⁻⁴ = Cubic metres per second (m³/s)
× 4.403 = U.S. gallons per minute (U.S. gpm)

CUBIC METRES PER SECOND — m³/s
× 3600 = Cubic metres per hour (m³/h)
× 15.85 × 10³ = U.S. gallons per minute (U.S. gpm)

CUBIC YARDS — yd³
× 0.7646 = Cubic metres (m³)
× 764.6 = Litres (l)
× 7.646 × 10⁵ = Cubic centimetres (cm³)
× 27 = Cubic feet (ft³)
× 46,656 = Cubic inches (in³)
× 201.97 = U.S. gallons (U.S. gal)
× 168.17 = Imperial gallons (imp gal)

DEGREES, ANGULAR (°)
× 0.017453 = Radians (rad)
× 60 = Minutes (')
× 3600 = Seconds (")
× 1.111 = Grade (gon)

DEGREES PER SECOND, ANGULAR (°/s)
x 0.017453 = Radians per second (rad/s)
x 0.16667 = Revolutions per minute (r/min)
x 2.7778 x 10⁻³ = Revolutions per second (r/s)

DRAMS (dr)
x 1.7718 = Grams (g)
x 27.344 = Grains (gr)
x 0.0625 = Ounces (oz)

FATHOMS
x 1.8288 = Metres (m)
x 6 = Feet (ft)

FEET — ft
x 0.3048 = Metres (m)
x 30.480 = Centimetres (cm)
x 12 = Inches (in)
x 0.3333 = Yards (yd)

FEET OF WATER — ftH₂O, at 68°F
x 2.984 = Kilopascals (kPa)
x 0.02984 = Bars (bar)
x 0.8811 = Inches of mercury (inHg) at 0°C
x 0.03042 = Kilograms-force per square centimetre (kg/cm²)
x 62.32 = Pounds-force per square foot (lbf/ft²)
x 0.4328 = Pounds-force per square inch (psi)
x 0.02945 = Standard atmospheres

FEET PER MINUTE — ft/min
x 0.5080 = Centimetres per second (cm/s)
x 0.01829 = Kilometres per hour (km/h)
x 0.3048 = Metres per minute (m/min)
x 0.016667 = Feet per second (ft/s)
x 0.01136 = Miles per hour (mph)

FEET PER SECOND PER SECOND — ft/s²
x 0.3048 = Metres per second per second (m/s²)
x 30.48 = Centimetres per second per second (cm/s²)

FOOT-POUNDS-FORCE — ft • lbf
x 1.356 = Joules (J)
x 1.285 x 10⁻³ = British thermal units (Btu) (see note)
x 3.239 x 10⁻⁴ = Kilocalories (kcal)
x 0.13825 = Kilogram-force-metres (kgf • m)
x 5.050 x 10⁻⁷ = Horsepower-hours (hp • h)
x 3.766 x 10⁻⁷ = Kilowatt-hours (kW • h)

GALLONS, U.S. — U.S. gal
x 3785.4 = Cubic centimetres (cm³)
x 3.7854 = Litres (l)
x 3.7854 x 10⁻³ = Cubic metres (m³)
x 231 = Cubic inches (in³)
x 0.13368 = Cubic feet (ft³)
x 4.951 x 10⁻³ = Cubic yards (yd³)
x 8 = Pints (pt) liquid
x 4 = Quarts (qt) liquid
x 0.8327 = Imperial gallons (imp gal)
x 8.328 = Pounds of water at 60°F in air
x 8.337 = Pounds of water at 60°F in vacuo

GALLONS, IMPERIAL — imp gal
x 4546 = Cubic centimetres (cm³)
x 4.546 = Litres (l)
x 4.546 x 10⁻³ = Cubic metres (m³)
x 0.16054 = Cubic feet (ft³)
x 5.946 x 10⁻³ = Cubic yards (yd³)
x 1.20094 = U.S. gallons (U.S. gal)
x 10.000 = Pounds of water at 62°F in air

GALLONS, PER MINUTE, U.S. — U.S. gpm
x 0.22715 = Cubic metres per hour (m³/h)
x 0.06309 = Litres per second (l/s)
x 8.021 = Cubic feet per hour (cfh)
x 2.228 x 10⁻³ = Cubic feet per second (cfs)

GRAINS — gr av. or troy
x 0.0648 = Grams (g)

GRAINS PER U.S. GALLON — gr/U.S. gal at 60°F
- x 17.12 = Grams per cubic metre (g/m³)
- x 17.15 = Parts per million by weight in water
- x 142.9 = Pounds per million gallons

GRAINS PER IMPERIAL GALLON — gr/imp gal at 62°F
- x 14.25 = Grams per cubic metre (g/m³)
- x 14.29 = Parts per million by weight in water

GRAMS — g
- x 15.432 = Grains (gr)
- x 0.035274 = Ounces (oz) av.
- x 0.032151 = Ounces (oz) troy
- x 2.2046 x 10⁻³ = Pounds (lb)

GRAMS-FORCE — gf
- x 9.807 x 10⁻³ = Newtons (N)

GRAMS-FORCE PER CENTIMETRE — gf/cm
- x 98.07 = Newtons per metre (N/m)
- x 5.600 x 10⁻³ = Pounds-force per inch (lbf/in)

GRAMS PER CUBIC CENTIMETRE — g/cm³
- x 62.43 = Pounds per cubic foot (lb/ft³)
- x 0.03613 = Pounds per cubic inch (lb/in³)

GRAMS PER LITRE — g/l
- x 58.42 = Grains per U.S. gallon (gr/U.S. gal)
- x 8.345 = Pounds per 1000 U.S. gallons
- x 0.06243 = Pounds per cubic foot (lb/ft³)
- x 1002 = Parts per million by mass (weight) in water at 60°F

HECTARES — ha
- x 1.000 x 10⁴ = Square metres (m²)
- x 1.0764 x 10⁵ = Square feet (ft²)

HORSEPOWER — hp
- x 745.7 = Watts (W)
- x 0.7457 = Kilowatts (kW)
- x 33,000 = Foot-pounds-force per minute (ft · lbf/min)
- x 550 = Foot-pounds-force per second (ft · lbf/s)
- x 42.43 = British thermal units per minute (Btu/min) (see note)
- x 10.69 = Kilocalories per minute (kcal/min)
- x 1.0139 = Horsepower (metric)

HORSEPOWER — hp boiler
- x 33,480 = British thermal units per hour (Btu/h) (see note)
- x 9.809 = Kilowatts (kW)

HORSEPOWER-HOURS — hp · h
- x 0.7457 = Kilowatt-hours (kW · h)
- x 1.976 x 10⁶ = Foot-pounds-force (ft · lbf)
- x 2545 = British thermal units (Btu) (see note)
- x 641.5 = Kilocalories (kcal)
- x 2.732 x 10⁵ = Kilogram-force-metres (kgf · m)

INCHES — in
- x 2.540 = Centimetres (cm)

INCHES OF MERCURY — inHg at 0°C
- x 3.3864 = Kilopascals (kPa)
- x 0.03386 = Bars (bar)
- x 1.135 = Feet of water (ftH₂O) at 68°F
- x 13.62 = Inches of water (inH₂O) at 68°F
- x 0.03453 = Kilograms-force per square centimetre (kg/cm²)
- x 70.73 = Pounds-force per square foot (lbf/ft²)
- x 0.4912 = Pounds-force per square inch (psi)
- x 0.03342 = Standard atmospheres

INCHES OF WATER — inH₂O at 68°F
- x 0.2487 = Kilopascals (kPa)
- x 2.487 x 10⁻³ = Bars (bar)
- x 0.07342 = Inches of mercury (inHg) at 0°C
- x 2.535 x 10⁻³ = Kilograms-force per square centimetre (kg/cm²)
- x 0.5770 = Ounces-force per square inch (ozf/in²)

x 5.193 = Pounds-force per square foot (lbf/ft²)
x 0.03606 = Pounds-force per square inch (psi)
x 2.454 x 10⁻³ = Standard atmospheres

JOULES — J
x 0.9484 x 10⁻³ = British thermal units (Btu) (see note)
x 0.2390 = Calories (cal) thermochemical
x 0.7376 = Foot-pounds-force (ft • lbf)
x 2.778 x 10⁻⁴ = Watt-hours (W • h)

KILOGRAMS — kg
x 2.2046 = Pounds (lb)
x 1.102 x 10⁻³ = Tons (ton) short

KILOGRAMS-FORCE — kgf
x 9.807 = Newtons (N)
x 2.205 = Pounds-force (lbf)

KILOGRAMS-FORCE PER METRE — kgf/m
x 9.807 = Newtons per metre (N/m)
x 0.6721 = Pounds-force per foot (lbf/ft)

KILOGRAMS-FORCE PER SQUARE CENTIMETRE — kg/cm²
x 98.07 = Kilopascals (kPa)
x 0.9807 = Bars (bar)
x 32.87 = Feet of water (ftH₂O) at 68°F
x 28.96 = Inches of mercury (inHg) at 0°C
x 2048 = Pounds-force per square foot (lbf/ft²)
x 14.223 = Pounds-force per square inch (psi)
x 0.9678 = Standard atmospheres

KILOGRAMS-FORCE PER SQUARE MILLIMETRE — kgf/mm²
x 9.807 = Megapascals (MPa)
x 1.000 x 10⁶ = Kilograms-force per square metre (kgf/m²)

KILOMETRES PER HOUR — km/h
x 27.78 = Centimetres per second (cm/s)
x 0.9113 = Feet per second (ft/s)
x 54.68 = Feet per minute (ft/min)
x 16.667 = Metres per minute (m/min)

x 0.53996 = International knots (kn)
x 0.6214 = Miles per hour (mph)

KILOMETRES PER HOUR PER SECOND — km • h⁻¹ • s⁻¹
x 0.2778 = Metres per second per second (m/s²)
x 27.78 = Centimetres per second per second (cm/s²)
x 0.9113 = Feet per second per second (ft/s²)

KILOMETRES PER SECOND — km/s
x 37.28 = Miles per minute (mi/min)

KILOPASCALS — kPa
x 10³ = Pascals (Pa) or newtons per square metre (N/m²)
x 0.1450 = Pounds-force per square inch (psi)
x 0.010197 = Kilograms-force per square centimetre (kg/cm²)
x 0.2953 = Inches of mercury (inHg) at 32°F
x 0.3351 = Feet of water (ftH₂O) at 68°F
x 4.021 = Inches of water (inH₂O) at 68°F

KILOWATTS — kW
x 4.425 x 10⁴ = Foot-pounds-force per minute (ft • lbf/min)
x 737.6 = Foot-pounds-force per second (ft • lbf/s)
x 56.90 = British thermal units per minute (Btu/min) (see note)
x 14.33 = Kilocalories per minute (kcal/min)
x 1.3410 = Horsepower (hp)

KILOWATT-HOURS — kW • h
x 3.6 x 10⁶ = Joules (J)
x 2.655 x 10⁶ = Foot-pounds-force (ft • lbf)
x 3413 = British thermal units (Btu) (see note)
x 860 = Kilocalories (kcal)
x 3.671 x 10⁵ = Kilogram-force metres (kgf • m)
x 1.3410 = Horsepower-hours (hp • h)

KNOTS — kn (International)
x 0.5144 = Metres per second (m/s)
x 1.151 = Miles per hour (mph)

LITRES — l
× 1000 = Cubic centimetres (cm³)
× 0.035315 = Cubic feet (ft³)
× 61.024 = Cubic inches (in³)
× 1.308 × 10⁻³ = Cubic yards (yd³)
× 0.2642 = U.S. gallons (U.S. gal)
× 0.2200 = Imperial gallons (imp gal)

LITRES PER MINUTE — l/min
× 0.01667 = Litres per second (l/s)
× 5.885 × 10⁻⁴ = Cubic feet per second (cfs)
× 4.403 × 10⁻³ = U.S. gallons per second (U.S. gal/s)
× 3.666 × 10⁻³ = Imperial gallons per second (imp gal/s)

LITRES PER SECOND — l/s
× 10⁻³ = Cubic metres per second (m³/s)
× 3.600 = Cubic metres per hour (m³/h)
× 60 = Litres per minute (l/min)
× 15.85 = U.S. gallons per minute (U.S. gpm)
× 13.20 = Imperial gallons per minute (imp gpm)

MEGAPASCALS — MPa
× 10⁶ = Pascals (Pa) or newtons per square metre (N/m²)
× 10³ = Kilopascals (kPa)
× 145.0 = Pounds-force per square inch (psi)
× 0.1020 = Kilograms-force per square millimetre (kgf/mm²)

METRES — m
× 3.281 = Feet (ft)
× 39.37 = Inches (in)
× 1.0936 = Yards (yd)

METRES PER MINUTE — m/min
× 1.6667 = Centimetres per second (cm/s)
× 0.0600 = Kilometres per hour (km/h)
× 3.281 = Feet per minute (ft/min)
× 0.05468 = Feet per second (ft/s)
× 0.03728 = Miles per hour (mph)

METRES PER SECOND — m/s
× 3.600 = Kilometres per hour (km/h)
× 0.0600 = Kilometres per minute (km/min)
× 196.8 = Feet per minute (ft/min)
× 3.281 = Feet per second (ft/s)
× 2.237 = Miles per hour (mph)
× 0.03728 = Miles per minute (mi/min)

MICROMETRES — μm formerly micron
× 10⁻⁶ = Metres (m)

MILES — mi
× 1.6093 × 10³ = Metres (m)
× 1.6093 = Kilometres (km)
× 5280 = Feet (ft)
× 1760 = Yards (yd)

MILES PER HOUR — mph
× 44.70 = Centimetres per second (cm/s)
× 1.6093 = Kilometres per hour (km/h)
× 26.82 = Metres per minute (m/min)
× 88 = Feet per minute (ft/min)
× 1.4667 = Feet per second (ft/s)
× 0.8690 = International knots (kn)

MILES PER MINUTE — mi/min
× 1.6093 = Kilometres per minute (km/min)
× 2682 = Centimetres per second (cm/s)
× 88 = Feet per second (ft/s)
× 60 = Miles per hour (mph)

MINUTES, ANGULAR — (')
× 2.909 × 10⁻⁴ = Radians (rad)

NEWTONS — N
× 0.10197 = Kilograms-force (kgf)
× 0.2248 = Pounds-force (lbf)
× 7.233 = Poundals
× 10⁵ = Dynes

OUNCES — oz av.
x 28.35 = Grams (g)
x 2.835 x 10^{-5} = Tonnes (t) metric ton
x 16 = Drams (dr) av.
x 437.5 = Grains (gr)
x 0.06250 = Pounds (lb) av.
x 0.9115 = Ounces (oz) troy
x 2.790 x 10^{-5} = Tons (ton) long

OUNCES — oz troy
x 31.103 = Grams (g)
x 480 = Grains (gr)
x 20 = Pennyweights (dwt) troy
x 0.08333 = Pounds (lb) troy
x 0.06857 = Pounds (lb) av.
x 1.0971 = Ounces (oz) av.

OUNCES — oz U.S. fluid
x 0.02957 = Litres (l)
x 1.8046 = Cubic inches (in)

OUNCES-FORCE PER SQUARE INCH — ozf/in²
x 43.1 = Pascals (Pa)
x 0.06250 = Pounds-force per square inch (psi)
x 4.395 = Grams-force per square centimetre (gf/cm²)

PARTS PER MILLION BY MASS — mass (weight) in water
x 0.9991 = Grams per cubic metre (g/m³) at 15°C
x 0.0583 = Grains per U.S. gallon (gr/U.S. gal) at 60°F
x 0.0700 = Grains per imperial gallon (gr/imp gal) at 62°F
x 8.328 = Pounds per million U.S. gallons at 60°F

PASCALS — Pa
x 1 = Newtons per square metre (N/m²)
x 1.450 x 10^{-4} = Pounds-force per square inch (psi)
x 1.0197 x 10^{-5} = Kilograms-force per square centimetre (kg/cm²)
x 10^{-3} = Kilopascals (kPa)

PENNYWEIGHTS — dwt troy
x 1.5552 = Grams (g)
x 24 = Grains (gr)

POISES — P
x 0.1000 = Newton-seconds per square metre (N · s/m²)
x 100 = Centipoises (cP)
x 2.0886 x 10^{-3} = Pound-force-seconds per square foot (lbf · s/ft²)
x 0.06721 = Pounds per foot second (lb/ft · s)

POUNDS-FORCE — lbf av.
x 4.448 = Newtons(N)
x 0.4536 = Kilograms-force (kgf)

POUNDS — lb av.
x 453.6 = Grams (g)
x 16 = Ounces (oz) av.
x 256 = Drams (dr) av.
x 7000 = Grains (gr)
x 5 x 10^{-4} = Tons (ton) short
x 1.2153 = Pounds (lb) troy

POUNDS — lb troy
x 373.2 = Grams (g)
x 12 = Ounces (oz) troy
x 240 = Pennyweights (dwt) troy
x 5760 = Grains (gr)
x 0.8229 = Pounds (lb) av.
x 13.166 = Ounces (oz) av.
x 3.6735 x 10^{-4} = Tons (ton) long
x 4.1143 x 10^{-4} = Tons (ton) short
x 3.7324 x 10^{-4} = Tonnes (t) metric tons

POUNDS-MASS OF WATER AT 60°F
x 453.98 = Grams (g)
x 0.45398 = Litres (l)
x 0.01603 = Cubic feet (ft³)
x 27.70 = Cubic inches (in³)
x 0.1199 = U.S. gallons (U.S. gal)

POUNDS OF WATER PER MINUTE AT 60°F
x 7.576 = Cubic centimetres per second (cm³/s)
x 2.675 x 10⁻⁴ = Cubic feet per second (cfs)

POUNDS PER CUBIC FOOT — lb/ft³
x 16.018 = Kilograms per cubic metre (kg/m³)
x 0.016018 = Grams per cubic centimetre (g/cm³)
x 5.787 x 10⁻⁴ = Pounds per cubic inch (lb/in³)

POUNDS PER CUBIC INCH — lb/in³
x 2.768 x 10⁴ = Kilograms per cubic metre (kg/m³)
x 27.68 = Grams per cubic centimetre (g/cm³)
x 1728 = Pounds per cubic foot (lb/ft³)

POUNDS-FORCE PER FOOT — lbf/ft
x 14.59 = Newtons per metre (N/m)
x 1.488 = Kilograms-force per metre (kgf/m)
x 14.88 = Grams-force per centimetre (gf/cm)

POUNDS-FORCE PER SQUARE FOOT — lbf/ft²
x 47.88 = Pascals (Pa)
x 0.01605 = Feet of water (ftH₂O) at 68°F
x 4.882 x 10⁻⁴ = Kilograms-force per square centimetre (kg/cm²)
x 6.944 x 10⁻³ = Pounds-force per square inch (psi)

POUNDS-FORCE PER SQUARE INCH — psi
x 6.895 = Kilopascals (kPa)
x 0.06805 = Standard atmospheres
x 2.311 = Feet of water (ftH₂O) at 68°F
x 27.73 = Inches of water (inH₂O) at 68°F
x 2.036 = Inches of mercury (inHg) at 0°C
x 0.07031 = Kilograms-force per square centimetre (kg/cm²)

QUARTS — qt dry
x 1101 = Cubic centimetres (cm³)
x 67.20 = Cubic inches (in³)

QUARTS — qt liquid
x 946.4 = Cubic centimetres (cm³)
x 57.75 = Cubic inches (in³)

QUINTALS — obsolete metric mass term
x 100 = Kilograms (kg)
x 220.46 = Pounds (lb) U.S. av.
x 101.28 = Pounds (lb) Argentina
x 129.54 = Pounds (lb) Brazil
x 101.41 = Pounds (lb) Chile
x 101.47 = Pounds (lb) Mexico
x 101.43 = Pounds (lb) Peru

RADIANS — rad
x 57.30 = Degrees (°) angular

RADIANS PER SECOND — rad/s
x 57.30 = Degrees per second (°/s) angular

STANDARD CUBIC FEET PER MINUTE — scfm (at 14.696 psia and 60°F)
x 0.4474 = Litres per second (l/s) at standard conditions (760 mmHg and 0°C)
x 1.608 = Cubic metres per hour (m³/h) at standard conditions (760 mmHg and 0°C)

STOKES — St
x 10⁻⁴ = Square metres per second (m²/s)
x 1.076 x 10⁻³ = Square feet per second (ft²/s)

TONS-MASS — tonm long
x 1016 = Kilograms (kg)
x 2240 = Pounds (lb) av.
x 1.1200 = Tons (ton) short

TONNES — t metric ton, millier
x 1000 = Kilograms (kg)
x 2204.6 = Pounds (lb)

TONNES-FORCE — tf metric ton-force
× 980.7 = Newtons (N)

TONS — ton short
× 907.2 = Kilograms (kg)
× 0.9072 = Tonnes (t)
× 2000 = Pounds (lb) av.
× 32000 = Ounces (oz) av.
× 2430.6 = Pounds (lb) troy
× 0.8929 = Tons (ton) long

TONS OF WATER PER 24 HOURS AT 60°F
× 0.03789 = Cubic metres per hour (m³/h)
× 83.33 = Pounds of water per hour (lb/r H_2O) at 60°F
× 0.1668 = U.S. gallons per minute (U.S. gpm)
× 1.338 = Cubic feet per hour (cfh)

WATTS — W
× 0.05590 = British thermal units per minute (Btu/min) (see note)
× 44.25 = Foot-pounds-force per minute (ft · lbf/min)
× 0.7376 = Foot-pounds-force per second (ft · lbf/s)
× 1.341 × 10⁻³ = Horsepower (hp)
× 0.01433 = Kilocalories per minute (kcal/min)

WATT-HOURS — W · h
× 3600 = Joules (J)
× 3.413 = British thermal units (Btu) (see note)
× 2655 = Foot-pounds-force (ft · lbf)
× 1.341 × 10⁻³ = Horsepower-hours (hp · h)
× 0.860 = Kilocalories (kcal)
× 367.1 = Kilogram-force-metres (kgf · m)

NOTE: SIGNIFICANT FIGURES The precision to which a given conversion factor is known, and its application, determine the number of significant figures which should be used. While many handbooks and standards give factors contained in this table to six or more significant figures, the fact that different sources disagree, in many cases, in the fifth or further figure indicates that four or five significant figures represent the precision for these factors fairly. At present the accuracy of process instrumentation, analog or digital, is in the tenth percent region at best, thus needing only three significant figures. Hence this table is confined to four or five significant figures. The advent of the pocket calculator (and the use of digital computers in process instrumentation) tends to lead to use of as many figures as the calculator will handle. However, when this exceeds the precision of the data, or the accuracy of the application, such a practice is misleading and timewasting.

NOTE: BRITISH THERMAL UNIT When making calculations involving Btu it must be remembered that there are several definitions of the Btu. The first three significant figures of the conversion factors given in this table are common to most definitions of the Btu. However, if four or more significant figures are needed in the calculation, the appropriate handbooks and standards should be consulted to be sure the proper definition and factor are being used.

TEMPERATURE CONVERSION TABLES

Fahrenheit and Celsius (Centigrade)

C	*	F	C	*	F	C	*	F	C	*	F	C	*	F
-273.15	-459.67		-17.2	1	33.8	10.6	51	123.8	43	110	230	266	510	950
-268	-450	-454	-16.7	2	35.6	11.1	52	125.6	49	120	248	271	520	968
-262	-440	-436	-16.1	3	37.4	11.7	53	127.4	54	130	266	277	530	986
-257	-430	-418	-15.6	4	39.2	12.2	54	129.2	60	140	284	282	540	1004
-251	-420	-400	-15.0	5	41.0	12.8	55	131.0	66	150	302	288	550	1022
-246	-410		-14.4	6	42.8	13.3	56	132.8	71	160	320	293	560	1040
-240	-400		-13.9	7	44.6	13.9	57	134.6	77	170	338	299	570	1058
-234	-390		-13.3	8	46.4	14.4	58	136.4	82	180	356	304	580	1076
-229	-380		-12.8	9	48.2	15.0	59	138.2	88	190	374	310	590	1094
-223	-370		-12.2	10	50.0	15.6	60	140.0	93	200	392	316	600	1112
-218	-360		-11.7	11	51.8	16.1	61	141.8	99	210	410	321	610	1130
-212	-350		-11.1	12	53.6	16.7	62	143.6				327	620	1148
-207	-340		-10.6	13	55.4	17.2	63	145.4				332	630	1166
-201	-330		-10.0	14	57.2	17.8	64	147.2				338	640	1184
-196	-320		-9.4	15	59.0	18.3	65	149.0				343	650	1202
-190	-310		-8.9	16	60.8	18.9	66	150.8	100	212	413	349	660	1220
-184	-300		-8.3	17	62.6	19.4	67	152.6				354	670	1238
-179	-290		-7.8	18	64.4	20.0	68	154.4				360	680	1256
-173	-280		-7.2	19	66.2	20.6	69	156.2				366	690	1274
-169	-273	-459.4	-6.7	20	68.0	21.1	70	158.0				371	700	1292
-168	-270	-454	-6.1	21	69.8	21.7	71	159.8				377	710	1310
-162	-260	-436	-5.6	22	71.6	22.2	72	161.6	104	220	428	382	720	1328
-157	-250	-418	-5.0	23	73.4	22.8	73	163.4	110	230	446	388	730	1346
-151	-240	-400	-4.4	24	75.2	23.3	74	165.2	116	240	464	393	740	1364
-146	-230	-382	-3.9	25	77.0	23.9	75	167.0	121	250	482	399	750	1382

C	*	F
−140	−220	−364
−134	−210	−346
−129	−200	−328
−123	−190	−310
−118	−180	−292
−112	−170	−274
−107	−160	−256
−101	−150	−238
−95.6	−140	−220
−90.0	−130	−202
−84.4	−120	−184
−78.9	−110	−166
−73.3	−100	−148
−67.8	−90	−130
−62.2	−80	−112
−56.7	−70	−94
−51.1	−60	−76
−45.6	−50	−58
−40.0	−40	−40
−34.4	−30	−22
−28.9	−20	−4
−23.3	−10	14
−17.8	0	32
−3.3	26	78.8
−2.8	27	80.6
−2.2	28	82.4
−1.7	29	84.2
−1.1	30	86.0
−0.6	31	87.8
0	32	89.6
0.6	33	91.4
1.1	34	93.2
1.7	35	95.0
2.2	36	96.8
2.8	37	98.6
3.3	38	100.4
3.9	39	102.2
4.4	40	104.0
5.0	41	105.8
5.6	42	107.6
6.1	43	109.4
6.7	44	111.2
7.2	45	113.0
7.8	46	114.8
8.3	47	116.6
8.9	48	118.4
9.4	49	120.2
10.0	50	122.0
24.4	76	168.8
25.0	77	170.6
25.6	78	172.4
26.1	79	174.2
26.7	80	176.0
27.2	81	177.8
27.8	82	179.6
28.3	83	181.4
28.9	84	183.2
29.4	85	185.0
30.0	86	186.8
30.6	87	188.6
31.1	88	190.4
31.7	89	192.2
32.2	90	194.0
32.8	91	195.8
33.3	92	197.6
33.9	93	199.4
34.4	94	201.2
35.0	95	203.0
35.6	96	204.8
36.1	97	206.6
36.7	98	208.4
37.2	99	210.2
37.8	100	212.0
127	260	500
132	270	518
138	280	536
143	290	554
149	300	572
154	310	590
160	320	608
166	330	626
171	340	644
177	350	662
182	360	680
188	370	698
193	380	716
199	390	734
204	400	752
210	410	770
216	420	788
221	430	806
227	440	824
232	450	842
238	460	860
243	470	878
249	480	896
254	490	914
260	500	932
404	760	1400
410	770	1418
416	780	1436
421	790	1454
427	800	1472
432	810	1490
438	820	1508
443	830	1526
449	840	1544
454	850	1562
460	860	1580
466	870	1598
471	880	1616
477	890	1634
482	900	1652
488	910	1670
493	920	1688
499	930	1706
504	940	1724
510	950	1742
516	960	1760
521	970	1778
527	980	1796
532	990	1814
538	1000	1832

INTERPOLATION VALUES

C	*	F	C	*	F
0.56	1	1.8	3.33	6	10.8
1.11	2	3.6	3.89	7	12.6
1.67	3	5.4	4.44	8	14.4
2.22	4	7.2	5.00	9	16.2
2.78	5	9.0	5.56	10	18.0

*In the center column, find the temperature to be converted. The equivalent temperature is in the left column, if converting to Celsius, and in the right column, if converting to Fahrenheit.

Fahrenheit and Celsius (Centigrade) continued

C	*	F	C	*	F	C	*	F	C	*	F
543	1010	1850	821	1510	2750	1099	2010	3650	1377	2510	4550
549	1020	1868	827	1520	2768	1104	2020	3668	1382	2520	4568
554	1030	1886	832	1530	2786	1110	2030	3686	1388	2530	4586
560	1040	1904	838	1540	2804	1116	2040	3704	1393	2540	4604
566	1050	1922	843	1550	2822	1121	2050	3722	1399	2550	4622
571	1060	1940	849	1560	2840	1127	2060	3740	1404	2560	4640
577	1070	1958	854	1570	2858	1132	2070	3758	1410	2570	4658
582	1080	1976	860	1580	2876	1138	2080	3776	1416	2580	4676
588	1090	1994	866	1590	2894	1143	2090	3794	1421	2590	4694
593	1100	2012	871	1600	2912	1149	2100	3812	1427	2600	4712
599	1110	2030	877	1610	2930	1154	2110	3830	1432	2610	4730
604	1120	2048	882	1620	2948	1160	2120	3848	1438	2620	4748
610	1130	2066	888	1630	2966	1166	2130	3866	1443	2630	4766
616	1140	2084	893	1640	2984	1171	2140	3884	1449	2640	4784
621	1150	2102	899	1650	3002	1177	2150	3902	1454	2650	4802
627	1160	2120	904	1660	3020	1182	2160	3920	1460	2660	4820
632	1170	2138	910	1670	3038	1188	2170	3938	1466	2670	4838
638	1180	2156	916	1680	3056	1193	2180	3956	1471	2680	4856
643	1190	2174	921	1690	3074	1199	2190	3974	1477	2690	4874
649	1200	2192	927	1700	3092	1204	2200	3992	1482	2700	4892
654	1210	2210	932	1710	3110	1210	2210	4010	1488	2710	4910
660	1220	2228	938	1720	3128	1216	2220	4028	1493	2720	4928
666	1230	2246	943	1730	3146	1221	2230	4046	1499	2730	4946
671	1240	2264	949	1740	3164	1227	2240	4064	1504	2740	4964
677	1250	2282	954	1750	3182	1232	2250	4082	1510	2750	4982
682	1260	2300	960	1760	3200	1238	2260	4100	1516	2760	5000
688	1270	2318	966	1770	3218	1243	2270	4118	1521	2770	5018
693	1280	2336	971	1780	3236	1249	2280	4136	1527	2780	5036
699	1290	2354	977	1790	3254	1254	2290	4154	1532	2790	5054
704	1300	2372	982	1800	3272	1260	2300	4172	1538	2800	5072

C		F	C		F	C		F	C		F
710	1310	2390	988	1810	3290	1266	2310	4190	1543	2810	5090
716	1320	2408	993	1820	3308	1271	2320	4208	1549	2820	5108
721	1330	2426	999	1830	3326	1277	2330	4226	1554	2830	5126
727	1340	2444	1004	1840	3344	1282	2340	4244	1560	2840	5144
732	1350	2462	1010	1850	3362	1288	2350	4262	1566	2850	5162
738	1360	2480	1016	1860	3380	1293	2360	4280	1571	2860	5180
743	1370	2498	1021	1870	3398	1299	2370	4298	1577	2870	5198
749	1380	2516	1027	1880	3416	1304	2380	4316	1582	2880	5216
754	1390	2534	1032	1890	3434	1310	2390	4334	1588	2890	5234
760	1400	2552	1038	1900	3452	1316	2400	4352	1593	2900	5252
766	1410	2570	1043	1910	3470	1321	2410	4370	1599	2910	5270
771	1420	2588	1049	1920	3488	1327	2420	4388	1604	2920	5288
777	1430	2606	1054	1930	3506	1332	2430	4406	1610	2930	5306
782	1440	2624	1060	1940	3524	1338	2440	4424	1616	2940	5324
788	1450	2642	1066	1950	3542	1343	2450	4442	1621	2950	5342
793	1460	2660	1071	1960	3560	1349	2460	4460	1627	2960	5360
799	1470	2678	1077	1970	3578	1354	2470	4478	1632	2970	5378
804	1480	2696	1082	1980	3596	1360	2480	4496	1638	2980	5396
810	1490	2714	1088	1990	3614	1366	2490	4514	1643	2990	5414
816	1500	2732	1093	2000	3632	1371	2500	4532	1649	3000	5432

Temperature Conversion Formulas

Degrees Celsius (formerly Centigrade) C

$C + 273.15 = K$ Kelvin

$(C \times {}^{9}/_{5}) + 32 = F$ Fahrenheit

$C \times {}^{4}/_{5} = R$ Réaumur

Degrees Fahrenheit — F

$F + 459.67 =$ Rankine

$(F - 32) \times {}^{5}/_{9} = C$ Celsius

$(F - 32) \times {}^{4}/_{9} = R$ Réaumur

Degrees Réaumur — R

$R \times {}^{5}/_{4} = C$ Celsius

$(R \times {}^{9}/_{4}) + 32 = F$ Fahrenheit

PROPERTIES OF SATURATED STEAM AND SATURATED WATER

Press.	Temp.	Volume, ft³/lbm			Enthalpy, Btu/lbm			Entropy, Btu/lbm x F			Energy, Btu/lbm	
psia	F	Water	Evap.	Steam	Water	Evap.	Steam	Water	Evap.	Steam	Water	Steam
		v_f	v_{fg}	v_g	h_f	h_{fg}	h_g	s_f	s_{fg}	s_g	u_f	u_g
3208.2	705.47	0.05078	0.00000	0.05078	906.0	0.0	906.0	1.0612	0.0000	1.0612	875.9	875.9
3094.3	700.0	0.03662	0.03857	0.07519	822.4	172.7	995.2	0.9901	0.1490	1.1390	801.5	952.2
3000.0	695.33	0.03428	0.05073	0.08500	801.8	218.4	1020.3	0.9728	0.1891	1.1619	782.8	973.1
2708.6	680.0	0.03037	0.08080	0.11117	758.5	310.1	1068.5	0.9365	0.2720	1.2086	743.2	1012.8
2500.0	668.11	0.02859	0.10209	0.13068	731.7	361.6	1093.3	0.9139	0.3206	1.2345	718.5	1032.9
2365.7	660.0	0.02768	0.11663	0.14431	714.9	392.1	1107.0	0.8995	0.3502	1.2498	702.8	1043.9
2059.9	640.0	0.02595	0.15427	0.18021	679.1	454.6	1133.7	0.8686	0.4134	1.2821	669.2	1065.0
2000.0	635.80	0.02565	0.16266	0.18831	672.1	466.2	1138.3	0.8625	0.4256	1.2881	662.6	1068.6
1786.9	620.0	0.02466	0.19615	0.22081	646.9	506.3	1153.2	0.8403	0.4689	1.3092	638.8	1080.2
1543.2	600.0	0.02364	0.24384	0.26748	617.1	550.6	1167.7	0.8134	0.5196	1.3330	610.4	1091.4
1500.0	596.20	0.02346	0.25372	0.27719	611.7	558.4	1170.1	0.8085	0.5288	1.3373	605.2	1093.1
1326.17	580.0	0.02279	0.29937	0.32216	589.1	589.9	1179.0	0.7876	0.5673	1.3550	583.5	1099.9
1200.0	567.19	0.02232	0.34013	0.36245	571.9	613.0	1184.8	0.7714	0.5969	1.3683	566.9	1104.3
1133.38	560.0	0.02207	0.36507	0.38714	562.4	625.3	1187.7	0.7625	0.6132	1.3757	557.8	1106.5
1000.0	544.58	0.02159	0.42236	0.44596	542.6	650.4	1192.9	0.7434	0.6476	1.3910	538.6	1110.4
962.79	540.0	0.02146	0.44367	0.46513	536.8	657.5	1194.3	0.7378	0.6577	1.3954	532.9	1111.4
812.53	520.0	0.02091	0.53864	0.55956	512.0	687.0	1199.0	0.7133	0.7013	1.4146	508.8	1115.0
800.0	518.21	0.02087	0.54809	0.56896	509.8	689.6	1199.4	0.7111	0.7051	1.4163	506.7	1115.2
680.86	500.0	0.02043	0.65448	0.67492	487.9	714.3	1202.2	0.6890	0.7443	1.4333	485.4	1117.2
600.0	486.20	0.02013	0.74962	0.76975	471.7	732.0	1203.7	0.6723	0.7738	1.4461	469.5	1118.2
566.15	480.0	0.02000	0.79716	0.81717	464.5	739.6	1204.1	0.6648	0.7871	1.4518	462.4	1118.5
500.0	467.01	0.01975	0.90787	0.92762	449.5	755.1	1204.7	0.6490	0.8148	1.4639	447.7	1118.8
466.87	460.0	0.01961	0.97463	0.99424	441.5	763.2	1204.8	0.6405	0.8299	1.4704	439.8	1118.9
400.0	444.60	0.01934	1.1416	1.1610	424.2	780.4	1204.6	0.6217	0.8630	1.4847	422.7	1118.7
381.54	440.0	0.01926	1.1976	1.2169	419.0	785.4	1204.4	0.6161	0.8729	1.4890	417.6	1118.5
308.780	420.0	0.01894	1.4808	1.4997	396.9	806.2	1203.1	0.5915	0.9165	1.5080	395.8	1117.5
300.0	417.35	0.01889	1.5238	1.5427	394.0	808.9	1202.9	0.5882	0.9223	1.5105	392.9	1117.2
250.0	400.97	0.01865	1.8245	1.8432	376.1	825.0	1201.1	0.5679	0.9585	1.5264	375.3	1115.8
247.259	400.0	0.01864	1.8444	1.8630	375.1	825.9	1201.0	0.5667	0.9607	1.5274	374.3	1115.7
200.0	381.80	0.01839	2.2689	2.2873	355.5	842.8	1198.3	0.5438	1.0016	1.5454	354.8	1113.7
195.729	380.0	0.01836	2.3170	2.3353	353.6	844.5	1198.0	0.5416	1.0057	1.5473	352.9	1113.5
153.010	360.0	0.01811	2.9392	2.9573	332.3	862.1	1194.4	0.5161	1.0517	1.5678	331.8	1110.6
150.0	358.43	0.01809	2.9958	3.0139	330.6	863.4	1194.1	0.5141	1.0554	1.5695	330.1	1110.4
120.0	341.27	0.01789	3.7097	3.7275	312.6	877.8	1190.4	0.4919	1.0960	1.5879	312.2	1107.6
117.992	340.0	0.01787	3.7699	3.7878	311.3	878.8	1190.1	0.4902	1.0990	1.5892	310.9	1107.4

100.0	327.82	0.01774	4.4133	4.4310	298.5	888.6	1187.2	0.4743	1.1284	1.6027	298.2	1105.2
89.643	320.0	0.017766	4.8961	4.9138	290.4	894.8	1185.2	0.4640	1.1477	1.6116	290.1	1103.7
80.0	312.04	0.01757	5.4536	5.4711	282.1	900.9	1183.1	0.4534	1.1675	1.6208	281.9	1102.1
70.0	302.93	0.01748	6.1875	6.2050	272.7	907.8	1180.6	0.4411	1.1905	1.6316	272.5	1100.2
67.005	300.0	0.01745	6.4483	6.4658	269.7	910.0	1179.7	0.4372	1.1979	1.6351	269.5	1099.6
60.0	292.71	0.017383	7.1562	7.1736	262.2	915.4	1177.6	0.4273	1.2167	1.6440	262.0	1098.0
50.0	281.02	0.017274	8.4967	8.5140	250.2	923.9	1174.1	0.4112	1.2474	1.6586	250.1	1095.3
49.200	280.0	0.017264	8.6267	8.6439	249.2	924.6	1173.8	0.4098	1.2501	1.6599	249.1	1095.1
40.0	267.25	0.017151	10.479	10.496	236.1	933.6	1169.8	0.3921	1.2844	1.6765	236.0	1092.1
35.427	260.0	0.017089	11.745	11.762	228.8	938.6	1167.4	0.3819	1.3043	1.6862	228.6	1090.3
30.0	250.34	0.017009	13.727	13.744	218.9	945.2	1164.1	0.3682	1.3313	1.6995	218.8	1087.9
25.0	240.07	0.016927	16.284	16.301	208.52	952.1	1160.6	0.3535	1.3607	1.7141	208.4	1085.2
24.968	240.0	0.016926	16.304	16.321	208.45	952.1	1160.6	0.3533	1.3609	1.7142	208.3	1085.2
20.0	227.96	0.016834	20.070	20.087	196.27	960.1	1156.3	0.3358	1.3962	1.7320	196.21	1082.0
17.186	220.0	0.016775	23.131	23.148	188.23	965.2	1153.4	0.3241	1.4201	1.7442	188.18	1079.8
15.0	213.03	0.016726	26.274	26.290	181.21	969.7	1150.9	0.3137	1.4415	1.7552	181.16	1077.9
14.696	212.00	0.016719	26.782	26.799	180.17	970.3	1150.5	0.3121	1.4447	1.7568	180.12	1077.6
11.526	200.0	0.016637	33.622	33.639	168.09	977.9	1146.0	0.2940	1.4824	1.7764	168.05	1074.2
10.0	193.21	0.016592	38.404	38.420	161.26	982.1	1143.3	0.2836	1.5043	1.7879	161.23	1072.3
8.0	182.86	0.016527	47.328	47.345	150.87	988.5	1139.3	0.2676	1.5384	1.8060	150.84	1069.2
7.5110	180.0	0.016510	50.208	50.225	148.00	990.2	1138.2	0.2631	1.5480	1.8111	147.98	1068.4
6.0	170.05	0.016451	61.967	61.984	138.03	996.2	1134.2	0.2474	1.5820	1.8294	138.01	1065.4
5.0	162.24	0.016407	73.515	73.532	130.20	1000.9	1131.1	0.2313	1.6094	1.8443	130.18	1063.1
4.7414	160.0	0.016395	77.27	77.29	127.96	1002.2	1130.2	0.2313	1.6174	1.8487	127.94	1062.4
4.0	152.96	0.016358	90.63	90.64	120.92	1006.4	1127.3	0.2199	1.6428	1.8626	120.90	1060.2
3.0	141.47	0.016300	118.71	118.73	109.42	1013.2	1122.6	0.2009	1.6854	1.8864	109.41	1056.7
2.8892	140.0	0.016293	122.98	123.00	107.95	1014.0	1122.0	0.1985	1.6910	1.8895	107.94	1056.2
2.0	126.07	0.016230	173.74	173.76	94.03	1022.1	1116.2	0.1750	1.7450	1.9200	94.03	1051.8
1.6927	120.0	0.016204	203.25	203.26	87.97	1025.6	1113.6	0.1646	1.7693	1.9339	87.96	1049.9
1.0	101.74	0.016136	333.59	333.60	69.732	1036.1	1105.8	0.1326	1.8455	1.9781	69.73	1044.1
0.94924	100.0	0.016130	350.4	350.4	67.999	1037.1	1105.1	0.1295	1.8530	1.9825	68.00	1043.5
0.50683	80.0	0.016072	633.3	633.3	48.037	1048.4	1096.4	0.0932	1.9426	2.0359	48.036	1037.0
0.25611	60.0	0.016033	1207.6	1207.6	28.060	1059.7	1087.7	0.0555	2.0391	2.0946	28.060	1030.5
0.12163	40.0	0.016019	2445.8	2445.8	8.027	1071.0	1079.0	0.0162	2.1432	2.1594	8.027	1024.0
0.08865	32.018	0.016022	3302.4	3302.4	0.0003	1075.5	1075.5	0.0000	2.1872	2.1872	0.000	1021.3

Derived and Abridged from the 1967 ASME Steam Tables. Copyright 1967 by the American Society of Mechanical Engineers.

LIQUID GRAVITY TABLES AND WEIGHT FACTORS

Liquid Lighter than Water

Sp Gr 60 F/60 F	°Be	°API	Lb per gal at 60 F in vacuo	Lb per cu ft at 60 F in vacuo
.500	150.00	151.50	4.169	31.18
.505	147.23	148.70	4.210	31.50
.510	144.51	145.95	4.252	31.81
.515	141.84	143.26	4.294	32.12
.520	139.23	140.62	4.335	32.43
.525	136.67	138.02	4.377	32.74
.530	134.15	135.48	4.419	33.05
.535	131.68	132.99	4.460	33.37
.540	129.26	130.34	4.502	33.68
.545	126.88	128.13	4.544	33.99
.550	124.55	125.77	4.585	34.30
.555	122.25	123.45	4.627	34.61
.560	120.00	121.18	4.669	34.93
.565	117.79	118.94	4.711	35.24
.570	115.61	116.75	4.752	35.55
.575	113.48	114.59	4.794	35.86
.580	111.38	112.47	4.836	36.17
.585	109.32	110.38	4.877	36.48
.590	107.29	108.33	4.919	36.80
.595	105.29	106.32	4.961	37.11
.600	103.33	104.33	5.002	37.42
.605	101.40	102.38	5.044	37.73
.610	99.51	100.47	5.086	38.04
.615	97.64	98.58	5.127	38.36
.620	95.81	96.73	5.169	38.67
.625	94.00	94.90	5.211	38.98
.630	92.22	93.10	5.252	39.29
.635	90.47	91.33	5.294	39.60
.640	88.75	89.59	5.336	39.91
.645	87.05	87.88	5.377	40.23
.650	85.38	86.19	5.419	40.54
.655	83.74	84.53	5.461	40.85
.660	82.12	82.89	5.503	41.16
.665	80.53	81.28	5.544	41.47
.670	78.96	79.69	5.586	41.79
.675	77.41	78.13	5.628	42.10
.680	75.88	76.59	5.669	42.41
.685	74.38	75.07	5.711	42.72
.690	72.90	73.57	5.753	43.03
.695	71.44	72.10	5.794	43.34
.700	70.00	70.64	5.836	43.66
.705	68.58	69.21	5.878	43.97
.710	67.18	67.80	5.919	44.28

Liquid Heavier than Water

Sp Gr 60 F/60 F	°Be	°Tw	Lb per gal at 60 F in vacuo	Lb per cu ft at 60 F in vacuo
1.000	0.00	0	8.337	62.37
1.005	.72	1	8.379	62.68
1.010	1.44	2	8.421	62.99
1.015	2.14	3	8.462	63.30
1.020	2.84	4	8.504	63.61
1.025	3.54	5	8.546	63.93
1.030	4.22	6	8.587	64.24
1.035	4.90	7	8.629	64.55
1.040	5.58	8	8.671	64.86
1.045	6.24	9	8.712	65.17
1.050	6.90	10	8.754	65.48
1.055	7.56	11	8.796	65.80
1.060	8.21	12	8.837	66.11
1.065	8.85	13	8.879	66.42
1.070	9.49	14	8.921	66.73
1.075	10.12	15	8.962	67.04
1.080	10.74	16	9.004	67.36
1.085	11.36	17	9.046	67.67
1.090	11.97	18	9.088	67.98
1.095	12.58	19	9.129	68.29
1.100	13.18	20	9.171	68.60
1.105	13.78	21	9.213	68.91
1.110	14.37	22	9.254	69.23
1.115	14.96	23	9.296	69.54
1.120	15.54	24	9.338	69.85
1.125	16.11	25	9.379	70.16
1.130	16.68	26	9.421	70.47
1.135	17.25	27	9.463	70.79
1.140	17.81	28	9.504	71.10
1.145	18.36	29	9.546	71.41
1.150	18.91	30	9.588	71.72
1.155	19.46	31	9.629	72.03
1.160	20.00	32	9.671	72.35
1.165	20.54	33	9.713	72.66
1.170	21.07	34	9.755	72.97
1.175	21.60	35	9.796	73.28
1.180	22.12	36	9.838	73.59
1.185	22.64	37	9.880	73.90
1.190	23.15	38	9.921	74.22
1.195	23.66	39	9.963	74.53
1.200	24.17	40	10.005	74.84
1.205	24.67	41	10.046	75.15
1.210	25.17	42	10.088	75.46

Sp Gr 60 F/60 F	°Be	°Tw	Lb per gal at 60 F in vacuo	Lb per cu ft at 60 F in vacuo
1.500	48.33	100	12.506	93.55
1.505	48.65	101	12.547	93.86
1.510	48.97	102	12.589	94.17
1.515	49.29	103	12.631	94.49
1.520	49.61	104	12.673	94.80
1.525	49.92	105	12.714	95.11
1.530	50.23	106	12.756	95.42
1.535	50.54	107	12.798	95.73
1.540	50.84	108	12.839	96.04
1.545	51.15	109	12.881	96.36
1.550	51.45	110	12.923	96.67
1.555	51.75	111	12.964	96.98
1.560	52.05	112	13.006	97.29
1.565	52.35	113	13.048	97.60
1.570	52.64	114	13.089	97.92
1.575	52.94	115	13.131	98.23
1.580	53.23	116	13.173	98.54
1.585	53.52	117	13.214	98.85
1.590	53.81	118	13.256	99.16
1.595	54.09	119	13.298	99.47
1.600	54.37	120	13.339	99.79
1.605	54.66	121	13.381	100.10
1.610	54.94	122	13.423	100.41
1.615	55.22	123	13.465	100.72
1.620	55.49	124	13.506	101.03
1.625	55.77	125	13.548	101.35
1.630	56.04	126	13.590	101.66
1.635	56.31	127	13.631	101.97
1.640	56.59	128	13.673	102.28
1.645	56.85	129	13.715	102.59
1.650	57.12	130	13.756	102.90
1.655	57.39	131	13.798	103.22
1.660	57.65	132	13.840	103.53
1.665	57.91	133	13.881	103.84
1.670	58.17	134	13.923	104.15
1.675	58.43	135	13.965	104.46
1.680	58.69	136	14.006	104.78
1.685	58.95	137	14.048	105.09
1.690	59.20	138	14.090	105.40
1.695	59.45	139	14.132	105.71
1.700	59.71	140	14.173	106.02
1.705	59.96	141	14.215	106.33

Specific gravity conversion table — lower range (specific gravity .750 to 1.000):

.750	56.67	57.17	6.253	46.77
.755	55.43	55.92	6.295	47.09
.760	54.21	54.68	6.336	47.40
.765	53.01	53.47	6.378	47.71
.770	51.82	52.27	6.420	48.02
.775	50.65	51.08	6.461	48.33
.780	49.49	49.91	6.503	48.65
.785	48.34	48.75	6.545	48.96
.790	47.22	47.61	6.586	49.27
.795	46.10	46.49	6.628	49.58
.800	45.00	45.38	6.670	49.89
.805	43.91	44.28	6.711	50.21
.810	42.84	43.19	6.753	50.52
.815	41.78	42.12	6.795	50.83
.820	40.73	41.06	6.836	51.14
.825	39.70	40.02	6.878	51.45
.830	38.67	38.98	6.920	51.76
.835	37.66	37.96	6.962	52.08
.840	36.67	36.95	7.003	52.39
.845	35.68	35.96	7.045	52.70
.850	34.71	34.97	7.087	53.01
.855	33.74	34.00	7.128	53.32
.860	32.79	33.03	7.170	53.64
.865	31.85	32.08	7.212	53.95
.870	30.92	31.12	7.253	54.26
.875	30.00	30.21	7.295	54.57
.880	29.09	29.30	7.337	54.88
.885	28.19	28.39	7.378	55.19
.890	27.30	27.49	7.420	55.51
.895	26.42	26.60	7.462	55.82
.900	25.56	25.72	7.503	56.13
.905	24.70	24.85	7.545	56.44
.910	23.85	23.99	7.587	56.75
.915	23.01	23.14	7.629	57.07
.920	22.17	22.30	7.670	57.38
.925	21.35	21.47	7.712	57.69
.930	20.54	20.65	7.754	58.00
.935	19.73	19.84	7.795	58.31
.940	18.94	19.03	7.837	58.62
.945	18.15	18.24	7.879	58.94
.950	17.37	17.45	7.920	59.25
.955	16.60	16.67	7.962	59.56
.960	15.83	15.90	8.004	59.87
.965	15.08	15.13	8.045	60.18
.970	14.33	14.38	8.087	60.50
.975	13.59	13.63	8.129	60.81
.980	12.86	12.89	8.170	61.12
.985	12.13	12.15	8.212	61.43
.990	11.41	11.43	8.254	61.74
.995	10.70	10.71	8.295	62.05
1.000	10.00	10.00	8.337	62.37

Specific gravity conversion table — upper range (specific gravity 1.250 to 1.500):

1.250	29.00	50	10.421	77.96	1.750	62.14	150	14.590	109.14
1.255	29.46	51	10.463	78.27	1.755	62.38	151	14.632	109.45
1.260	29.92	52	10.505	78.58	1.760	62.61	152	14.673	109.76
1.265	30.38	53	10.547	78.89	1.765	62.85	153	14.715	110.08
1.270	30.83	54	10.588	79.21	1.770	63.08	154	14.757	110.39
1.275	31.27	55	10.630	79.52	1.775	63.31	155	14.799	110.70
1.280	31.72	56	10.672	79.83	1.780	63.54	156	14.840	111.01
1.285	32.16	57	10.713	80.14	1.785	63.77	157	14.882	111.32
1.290	32.60	58	10.755	80.45	1.790	63.99	158	14.924	111.64
1.295	33.03	59	10.797	80.76	1.795	64.22	159	14.965	111.95
1.300	33.46	60	10.838	81.08	1.800	64.44	160	15.007	112.26
1.305	33.89	61	10.880	81.39	1.805	64.67	161	15.049	112.57
1.310	34.31	62	10.922	81.70	1.810	64.89	162	15.090	112.88
1.315	34.73	63	10.963	82.01	1.815	65.11	163	15.132	113.20
1.320	35.15	64	11.005	82.32	1.820	65.33	164	15.174	113.51
1.325	35.57	65	11.047	82.64	1.825	65.55	165	15.215	113.82
1.330	35.98	66	11.088	82.95	1.830	65.77	166	15.257	114.13
1.335	36.39	67	11.130	83.26	1.835	65.98	167	15.299	114.44
1.340	36.79	68	11.172	83.57	1.840	66.20	168	15.340	114.75
1.345	37.19	69	11.214	83.88	1.845	66.41	169	15.382	115.07
1.350	37.59	70	11.255	84.19	1.850	66.62	170	15.424	115.38
1.355	37.99	71	11.297	84.51	1.855	66.83	171	15.465	115.69
1.360	38.38	72	11.339	84.82	1.860	67.04	172	15.507	116.00
1.365	38.77	73	11.380	85.13	1.865	67.25	173	15.549	116.31
1.370	39.16	74	11.422	85.44	1.870	67.46	174	15.591	116.63
1.375	39.55	75	11.464	85.75	1.875	67.67	175	15.632	116.94
1.380	39.93	76	11.505	86.07	1.880	67.87	176	15.674	117.25
1.385	40.31	77	11.547	86.38	1.885	68.08	177	15.716	117.56
1.390	40.68	78	11.589	86.69	1.890	68.28	178	15.757	117.87
1.395	41.06	79	11.630	87.00	1.895	68.48	179	15.799	118.18
1.400	41.43	80	11.672	87.31	1.900	68.68	180	15.841	118.50
1.405	41.80	81	11.714	87.64	1.905	68.88	181	15.882	118.81
1.410	42.16	82	11.755	87.94	1.910	69.08	182	15.924	119.12
1.415	42.53	83	11.797	88.25	1.915	69.28	183	15.966	119.43
1.420	42.89	84	11.839	88.56	1.920	69.48	184	16.007	119.74
1.425	43.25	85	11.880	88.87	1.925	69.68	185	16.049	120.06
1.430	43.60	86	11.922	89.18	1.930	69.87	186	16.091	120.37
1.435	43.95	87	11.964	89.50	1.935	70.06	187	16.132	120.68
1.440	44.31	88	12.006	89.81	1.940	70.26	188	16.174	121.00
1.445	44.65	89	12.047	90.12	1.945	70.45	189	16.216	121.30
1.450	45.00	90	12.089	90.43	1.950	70.64	190	16.258	121.61
1.455	45.34	91	12.131	90.74	1.955	70.83	191	16.299	121.93
1.460	45.68	92	12.172	91.06	1.960	71.02	192	16.341	122.24
1.465	46.02	93	12.214	91.37	1.965	71.21	193	16.383	122.55
1.470	46.36	94	12.256	91.68	1.970	71.40	194	16.424	122.86
1.475	46.69	95	12.297	91.99	1.975	71.58	195	16.466	123.17
1.480	47.03	96	12.339	92.30	1.980	71.77	196	16.508	123.49
1.485	47.36	97	12.381	92.61	1.985	71.95	197	16.549	123.80
1.490	47.68	98	12.422	92.93	1.990	72.14	198	16.591	124.11
1.495	48.01	99	12.464	93.24	1.995	72.32	199	16.633	124.42
1.500	48.33	100	12.506	93.55	2.000	72.50	200	16.674	124.73

NOTE: To compute weight of water in air at 60 F, subtract weight of air at 60 F, 0.010 pounds per gallon or 0.076 pounds per cubic foot.
When weighing water on an equal arm balance using brass weights having a specific gravity of 8.4, add 145 parts per million by weight to compensate for the volume of air displaced by the brass weights.

Units and Conversion Tables 481

VISCOMETER COMPARISON CHART

FOR NEWTONIAN LIQUIDS

Centistokes Reference	0 100 200 300 400 500 600 700 800 900 1000 1100 1200 1300 1400 1500
Mobilometer sec. 100g 10cm	10 20 30 40 50 60 70
Engler degrees	25 50 75 100 125 150 175
Ford 4 sec.	25 50 75 100 150 175 200 225 250 275 300 325 350 375
Ford 3 sec.	25 50 75 125 150 175 200 225 250 275 300 325 350 375 400 425 450 475 500 550 600 650
Saybolt Universal sec.	500 1000 1500 2000 2500 3000 3500 4000 4500 5000 5500 6000
Saybolt Furol sec.	50 100 150 200 250 300 350 400 450 500 550 600 650
Redwood Standard 1 sec.	250 500 750 1000 1250 1500 2000 2500 3000 3500 4000 4500 5000 5500 6000 6500
Ubbelohde cks.	200 300 400 500 600 700 800 900 1000
Gardner Holts cks.	A2 A4 ABCDEF G HIJKLMN O P Q R S T U V W X A3 A1
Zahn 5 sec.	13 20 30 40 50 60
Zahn 3 sec.	23 30 40 50 60

SCALES ABOVE COMPARE TO CENTISTOKES REFERENCE (TO CONVERT INTO CENTIPOISE MULTIPLY BY LIQUID SPECIFIC GRAVITY.)

SCALES BELOW COMPARE DIRECTLY TO CENTIPOISE REFERENCE

Kreb Stormer 200G K.U.	50 60 70 80 90
Stormer Cyl. 150G sec.	27 50 115 223
Brookfield Cps.	0 100 200 300 400 500 600 700 800 900 1000 1100 1200 1300 1400 1500
Centipoise Reference	0 100 200 300 400 500 600 700 800 900 1000 1100 1200 1300 1400 1500

BROOKFIELD ENGINEERING LABORATORIES, INC.
STOUGHTON, MASSACHUSETTS, U.S.A.

NOTE: This chart is intended to be an aid in comparing viscometer measurements of Newtonian liquids by referencing to absolute

AR-15
67-1024

INDEX